Linear Optimal Control

$\mathcal{H}_2$ and $\mathcal{H}_\infty$ Methods

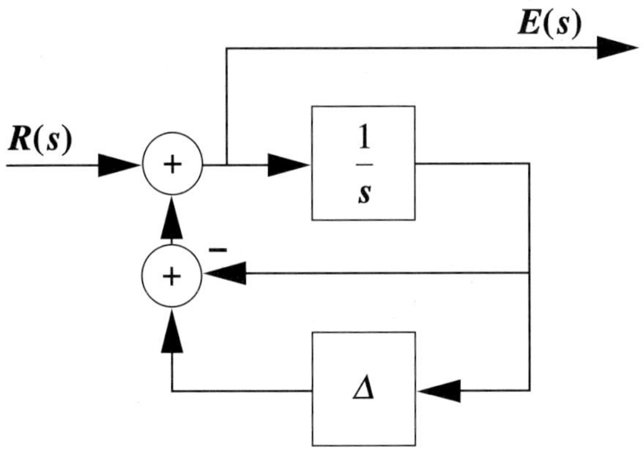

Jeffrey B. Burl
Michigan Technological University

ADDISON-WESLEY
An imprint of Addison Wesley Longman, Inc.

Menlo Park, California • Reading, Massachusetts • Harlow, England
Berkeley, California • Don Mills, Ontario • Sydney • Bonn • Amsterdam • Tokyo • Mexico City

Acquisitions Editor: Paul Becker
Assistant Editor: Anna Eberhard Friedlander
Production Manager: Pattie Myers
Associate Production Editor: Kamila Storr
Art and Design Supervisor: Kevin Berry
Composition: G & S Typesetters
Cover Design: Karl Miyajima
Cover Art: Jeff Burl
Text Printer and Binder: MapleVail
Cover Printer: Phoenix Color

Library of Congress Cataloging-in-Publication Data
Burl, Jeff B.
Linear optimal control : H_2 and H[infinity] methods / Jeff B. Burl.
p. cm.
On t.p. "[infinity]" appears as the infinity symbol.
Includes index.
ISBN 0-201-80868-4
1. Linear control systems. 2. H[infinity symbol] control. 3. H_2 control.
I. Title.

TJ220.B87 1998
629.8′32–dc21 98-36516
CIP

Instructional Material Disclaimer
The programs presented with this book have been included for their instructional value. They have been tested with care but are not guaranteed for any particular purpose. Neither the publisher nor the author offers any warranties or representations, nor do they accept any liabilities with respect to the programs.

The full complement of supplemental teaching materials is available to qualified instructors.

ISBN 0-201-80868-4

1 2 3 4 5 6 7 8 9 10—MA—02 01 00 99 98

Addison Wesley Longman, Inc.
2725 Sand Hill Road
Menlo Park, California 94025

To my girls: Alyse, Lily, and Sue

PREFACE

The last ten to fifteen years have seen a resurgence of interest in control systems designed to meet specifications for robustness and disturbance rejection. Robust stability and robust performance tests, based on the structured singular value, have been developed. The robustness of the linear quadratic regulator and the linear quadratic Gaussian controllers have been analyzed. Loop transfer recovery was developed as a means to improve the robustness properties of the linear quadratic Gaussian controller. $\mathcal{H}_\infty$ control was developed to provide a means of incorporating frequency-domain specifications in control system designs. $\mathcal{H}_\infty$ control also provides an *ad hoc* means of incorporating robustness specifications into control system designs. Finally, μ-synthesis was developed as a powerful procedure for designing control systems that satisfy both robustness and performance specifications.

This book is being written to accomplish a number of tasks. First and foremost, I wish to make the new material on robustness, $\mathcal{H}_\infty$ control, and μ-synthesis accessible to both students and professionals. Second, I wish to combine these new results with previous work on optimal control to form a more complete picture of control system design and analysis. Third, I wish to incorporate recent results on robust stability and robust performance analysis into the presentation of linear quadratic Gaussian optimal control. Lastly, I wish to acquaint the reader with the CAD tools available for robust optimal controller design.

Special Features

This book has been written to provide students and professionals with access to relatively recent research results on robustness analysis, $\mathcal{H}_\infty$ optimal control, and μ-synthesis. In addition, this material is integrated with linear quadratic Gaussian ($\mathcal{H}_2$) optimal control results. The overall treatment is organized in a logical manner rather than along the lines of historical development. A number of more specific features enhance the value of this book as a teaching text.

The results and derivations are simplified by treating special cases whenever this can be done without compromising the clarity of results or methods. In addition, mathematical developments that provide little insight into key derivations, results, and/or applications are relegated to the appendix. This approach allows the reader to develop a solid grounding in the basics before tackling the mathematical subtleties required to derive the most general results. Practicing engineers can then augment their understanding using more advanced books and research papers, or use computer-aided design software to handle the more general cases. While this approach does simplify the derivations, the mathematical level of this text is still quite challenging.

The solutions of both the $\mathcal{H}_2$ (linear quadratic Guassian) control problem and the $\mathcal{H}_\infty$ control problem are based on a common variational approach. The variational approach adds more insight into the optimization process than completing the square. Using a common variational approach in both of these problems also tends to demystify the $\mathcal{H}_\infty$ theory.

The use of computer-aided design (CAD) tools is integrated into the presentation and problems. The CAD software employed is MATLAB® with the Control System Toolbox and the μ-Synthesis and Analysis Toolbox. Software is available via ftp for almost every example in the text. For examples that are done analytically, software is included for numerically checking the result. These software programs and their documentation provide a significant learning resource (and also a significant reference source), since virtually all optimal controller design and analysis is performed with the aid of CAD software.

A general treatment of performance, including transient performance, tracking performance, and disturbance rejection, is given up front along with a treatment of robustness. This organization provides a solid foundation in control system analysis. A thorough performance and robustness analysis can then be performed on the controllers developed subsequently. This approach develops an understanding of what each design does well and what each design does poorly, as opposed to simply showing that a design is optimal.

Tracking and disturbance rejection are presented in the linear quadratic Gaussian setting, as opposed to the linear quadratic regulator setting. Presenting this material in the linear quadratic regulator setting leaves many students unsure of how to incorporate estimation within a tracking system design or within a design tailored for disturbance rejection. The organization of this book solves this problem by treating the idiosyncrasies involving estimation in these systems.

The book concludes with a case study that compares a design obtained via linear quadratic Gaussian-loop transfer recovery with a design obtained via μ-synthesis. The insight gained through this comparison yields a better understanding of both design methodologies, and provides guidelines on when to apply which design method.

Computer exercises are included for each chapter. These computer exercises develop familiarity with current CAD software and allow the exploration of design options and "what if" questions concerning the results. My students have frequently told me that a significant part of their learning comes as a result of performing the computer exercises.

A symbol list is included to help the reader with the notation. When writing this text, I became painfully aware that the English language has only twenty-six letters, and the Greek language has even fewer. While much effort has been made to keep the notation simple and consistent, of necessity some symbols appear in multiple roles. In general, I have rendered time-domain functions in lowercase. Laplace-domain and Fourier-domain functions are rendered in uppercase. These practices are consistent with most introductory control texts and should be familiar to the reader. Matrices are given in bold type, while both scalars and vectors are not bolded. Most introductory books bold both vectors and matrices, but almost everything in this text is a vector, so I have instead used bolding to highlight matrices, a practice I feel will better serve the reader. There are some exceptions to these basic formatting rules in order to make the results in this book match those appearing widely in the open literature. In these cases, the notation used should be clearly delineated and not lead to confusion.

Supplemental material consists of a solutions manual available to instructors and software used for the examples, which is available via ftp at ftp://ftp.aw.com/cseng/authors/burl/loc/mfiles.

Prerequisites

Prerequisites include an introduction to control systems (classical control), probability, state-space linear systems, and a working knowledge of linear algebra. In addition, an introduction to random processes is desirable.

Classical control, probability, and some linear algebra are part of the undergraduate education of most incoming engineering graduate students. An introduction to state-space linear systems is typically accomplished (along with converting a linear algebra background into working knowledge) by an introductory graduate course from a book such as Chen or Kailath. This book begins with a review of the relevant state-space linear systems material in a multivariable setting (Chapter 2). I usually only present the sections on singular value decomposition, principle gains, and internal stability, since this material is new to most of my students. But I recommend that my students skim the remainder of Chapter 2 as a review. I then ask them to inform me of any topics with which they are not familiar, and I provide references, when necessary, to bring students up to speed.

The book contains a terse but self-contained introduction to random processes (Chapter 3). This chapter also contains many state-space random process results that are not typically included in introductory random process courses. Therefore, I usually cover this chapter thoroughly.

Organization

This book consists of three parts. The first part covers the analysis of control systems. It contains a review of multivariable linear systems (Chapter 2), an introduction to vector random processes (Chapter 3), control system performance analysis (Chapter 4), and robustness analysis (Chapter 5). The performance analysis chapter includes transient performance analysis, tracking system analysis, and disturbance rejection. Cost functions are also presented as a means of quantifying performance analysis.

The robustness analysis chapter begins with a review of the Nyquist stability criterion. The Nyquist plot is used to develop the gain margin, the phase margin, and the downside gain margin. The stability robustness interpretation of these classical control stability margins is clearly illustrated. The small-gain theorem is presented as a means of determining stability robustness to unstructured perturbations. The structured singular value is then presented as a means of determining both stability and performance robustness to more general structured perturbations.

The second part of this book is devoted to $\mathcal{H}_2$, (i.e., linear quadratic Gaussian) optimal control. This part is divided into the linear quadratic regulator (Chapter 6), Kalman filtering (Chapter 7), and linear quadratic Gaussian control (Chapter 8).

The chapter on the linear quadratic regulator (LQR) begins with a brief introduction to optimization using variational theory. The results are then used to derive the LQR, both time-varying and steady-state. Application of the LQR is discussed along with cost function selection. Performance and robustness of the steady-state LQR are evaluated in some detail.

The chapter on Kalman filtering begins with an introduction to minimum mean square estimation theory and the orthogonality principle. The Kalman filter, both time-varying and steady-state, is then developed. Application of the Kalman filter is discussed in some detail. Kalman filter performance and robustness are also discussed.

The chapter on linear quadratic Gaussian (LQG) control begins with the development of the stochastic separation principle leading to the structure of the LQG controller. Performance and robustness of the LQG control system are discussed. Loop transfer recovery is presented as a means of increasing LQG robustness when needed. Tailoring the LQG control system for tracking and disturbance rejection is also discussed in this chapter.

The last part of this book is devoted to $\mathcal{H}_\infty$ control. Chapter 9 begins with an introduction to differential games. Differential game theory is used to derive the solution of the suboptimal $\mathcal{H}_\infty$ full information controller. The $\mathcal{H}_\infty$ output estimator is then derived using duality.

The $\mathcal{H}_\infty$ output feedback controller is presented in Chapter 10. The application of this controller to tracking systems, disturbance rejection, and robustness optimization is discussed in detail. The $\mathcal{D}$-$\mathcal{K}$ iteration algorithm for μ-synthesis is then presented. A significant case study is presented as a means of contrasting the μ-synthesis and the LQG loop transfer recovery design methodologies.

The generation of reduced-order controllers is presented in the final chapter. This chapter begins by showing that reduced-order controller approximation can often be evaluated using a frequency-weighted ∞-norm. The general properties of a desirable reduced-order approximation are then gleaned from a few examples. Pole-zero truncation and balance truncation are both presented as methods of generating reduced-order controllers.

Usage

This book is recommended for use as a text in a two-semester sequence covering linear optimal control. The entire book can be covered thoroughly in two semesters. A two-quarter course can also be formed from this material if the students are well prepared. The two-quarter course would also necessitate that the material be covered in less depth.

A one-semester course that covers robust optimal control can also be based on the material in this book. Such a course would be composed of a review of Chapter 2 followed by a thorough treatment of Chapters 4, 5, 9, 10, and 11.

An additional one-semester course on linear quadratic Gaussian control can be taught using this material. This course would be composed of a review of Chapter 2, a thorough treatment of Chapter 3, the cost function presentation in Chapter 4, the material on unstructured perturbations and the small-gain theorem in Chapter 5, and a thorough treatment of Chapters 6, 7, and 8.

Acknowledgments

I would like to express my appreciation to several people who contributed to this project. I am greatly indebted to Roberto Cristi for providing me with many useful suggestions associated with his use of a preliminary version of this text in classes at the Naval Postgraduate School. I would also like to thank all my students at Michigan Technological University for finding errors (big and small) and for letting me know when an explanation was unsatisfactory. I am especially indebted to two students, Willem Van Marian and John Pakkala, who used special diligence in finding errors in earlier versions of this manuscript. My appreciation is also extended to the reviewers: Roberto Cristi, Naval Postgraduate School; Faryar Jabarri, University of California, Irvine; Charles P. Neuman, Carnegie Mellon University; and Ronald A. Perez, University of Wisconsin-Milwaukee for their valuable suggestions. Last, but not least, I wish to thank the editorial and production staff at Addison-Wesley, especially Royden Tonomura, Paul Becker, and Kamila Storr for their help in improving the manuscript and in bringing it to production.

CONTENTS

CHAPTER 3 Vector Random Processes 54

CHAPTER 4 Performance 88

CHAPTER 5 Robustness 125

PART 2 $\mathcal{H}_2$ CONTROL

CHAPTER 6 The Linear Quadratic Regulator 179

CHAPTER 7 The Kalman Filter 231

CHAPTER 8 Linear Quadratic Gaussian Control 280

APPENDIX: Mathematical Notes 433

LIST OF SYMBOLS

Symbol	Meaning
$1(t)$	Unit step function
α	Real part of pole, coefficient
$\alpha(s)$	Characteristic equation
$\mathbf{A}(t)$, $\mathbf{A}$	State matrix
a, b	First-order shaping filter coefficients
$\mathbf{A}_c$	Controller state matrix
$\mathbf{A}_{cl}$	Closed-loop system state matrix
adj(•)	Adjugate of a matrix
$\mathbf{A}_f$	State matrix for shaping filter
$\mathcal{B}$	Linear space on which a norm is defined (Banach space)
$\mathbf{B}(t)$, $\mathbf{B}$	Input matrix
$\mathbf{B}_{cl}$	Closed-loop system input matrix
$\mathbf{B}_{cr}$	Controller input matrix for the input r
$\mathbf{B}_f$	Input matrix for shaping filter
$\mathcal{BW}$	Bandwidth
$\mathbf{B}_w$	Input matrix for the input w
$\mathbf{B}_{w_0}$	Input matrix associated with a constant input
$c(t)$	Correction term in unbiased estimator
$\mathbf{C}(t)$, $\mathbf{C}$	Output matrix
$c(\bullet, \bullet, \ldots)$	Constraint equation
$\mathbf{C}_c$	Controller output matrix
$\mathbf{C}_{cl}$	Closed-loop system output matrix
$\mathbf{C}_e$	Output matrix associated with the output error
$\mathbf{C}_f$	Output matrix for shaping filter
$\mathbf{C}_x(\tau)$	Covariance function of the stationary random process $x(t)$
$\mathbf{C}_x(t_1, t_2)$	Covariance function of the random process $x(t)$
$\mathbf{C}_y$	Output matrix for the output y
$\mathbf{C}_z$	Output matrix for constructed measurement
$\mathcal{C}^{m \times n}$	Set of complex $m \times n$ matrices
$\mathcal{C}^n$	Set of complex n-dimensional vectors
$d(t)$	Output disturbance
$\mathbf{D}(t)$, $\mathbf{D}$	Input-to-output coupling matrix
$\mathbf{D}_{cl}$	Closed-loop system input-to-output coupling matrix
$\mathbf{D}_{cr}$	Controller input-to-output coupling matrix from r to u
den(s)	Transfer function denominator

Symbol	Meaning
$\det(\bullet)$	Determinant of a matrix
$\mathbf{D}_{yw}$	Input-to-output coupling matrix from w to y
$\mathfrak{D}_L(s)$	Diagonal scaling matrix (left $\mathfrak{D}$-scaling matrix)
$\mathfrak{D}_R(s)$	Diagonal scaling matrix (right $\mathfrak{D}$-scaling matrix)
$\boldsymbol{\Delta}(s)$	Normalized general perturbation
$\boldsymbol{\Delta}'(s)$	Unnormalized general perturbation
$\boldsymbol{\Delta}_a(s)$	Additive perturbation
$\boldsymbol{\Delta}_{fi}(s)$	Input feedback perturbation
$\boldsymbol{\Delta}_{fo}(s)$	Output feedback perturbation
$\boldsymbol{\Delta}_{G_{yr}}$	Change in the closed-loop transfer function due to a perturbation
$\boldsymbol{\Delta}_i(s)$	Input-multiplicative perturbation
$\boldsymbol{\Delta}_{\max}(j\omega)$	Frequency-dependent perturbation bound
$\boldsymbol{\Delta}_o(s)$	Output multiplicative perturbation
$\boldsymbol{\Delta}_p$	Normalized perturbation augmented with performance block
$\tilde{\boldsymbol{\Delta}}$	Specific perturbation
$\bar{\boldsymbol{\Delta}}$	Set of perturbations with a given block diagonal structure
$\bar{\boldsymbol{\Delta}}_s$	Set of block diagonal perturbations (square blocks)
$\Delta J(\bullet, \delta\bullet)$	Increment of J
d_i	$\mathfrak{D}$-scale
$\delta J(\bullet, \delta\bullet)$	Variation of J
δx	Variation of x, differential of x
δ_{ij}	Kronecker delta function
$\delta(t)$	Dirac delta (impulse) function
ε	Small, positive constant
$e(t)$	Tracking error, estimation error
e_I	Integral of the tracking error
$E[\bullet]$	Expectation operator
ϕ	Phase perturbation
$\boldsymbol{\Phi}(t, t_0)$, $\boldsymbol{\Phi}(t)$	State-transition matrix of a time-varying, time-invariant system
$\boldsymbol{\Phi}$, $\boldsymbol{\Phi}(T)$	Discrete-time state matrix
$\mathbf{F}$, $\mathbf{F}(t, \tau)$	Linear estimator weights
$f_{x,y}(x, y)$	Joint density function
$f_{x\|y}(x \| y)$	Conditional density function
$\phi_{\max}$	Maximum-phase perturbation
$\phi_{\min}$	Minimum-phase perturbation
$f_x(x; t)$	Density function of the random process $x(t)$
g	Uncertain gain
γ	∞-norm performance bound
$\mathbf{G}(s)$	Laplace transfer function of a generic system or a plant
$\mathbf{G}(t)$	Kalman gain
$\mathbf{g}(t)$	Impulse response matrix (generic system)
$\tilde{\mathbf{G}}_c(s)$	Reduced-order controller transfer function
$\mathcal{G}(\bullet)$	Linear system (possibly time-varying)
$\boldsymbol{\Gamma}$, $\boldsymbol{\Gamma}(T)$	Discrete-time input matrix
$\mathbf{G}_0(s)$	Nominal plant transfer function
$\mathbf{G}_1(t)$	State feedback gain for normalized measurement
$\mathbf{G}_{ab}(s)$	Transfer function from input b to output a
$\mathbf{G}_{cl}(s)$	Closed-loop system transfer function
$\mathbf{g}_{cl}(s)$	Closed-loop system impulse response
G_{Δ}	System with input $w - \gamma^{-2}\mathbf{B}_w^T\mathbf{P}x$ and output $u + \mathbf{B}_w^T\mathbf{P}x$
$\mathbf{G}_l(s)$	Loop transfer function
GM^+	Gain margin
Gm^-	Downside gain margin

Symbol	Meaning
$g_{\max}$	Maximum uncertain gain
$g_{\min}$	Minimum uncertain gain
$\mathbf{G}_s(s)$	Stable part of a transfer function
$\mathbf{G}_u(s)$	Unstable part of a transfer function
$\mathcal{H}$	Hardy space
$\mathbf{H}(s)$	Perturbed closed-loop system
$\mathbf{I}$	Identity matrix
$\inf(\bullet)$	Infimum
j	Square root of -1, discrete index
$J(\bullet)$	Cost function, objective function
J_2	2-norm cost function
$J_a(\bullet, p)$	Augmented cost function, augmented objective function
J_γ	Objective function for suboptimal control
$J_{\gamma a}$	Augmented objective function for suboptimal control
J_{LQR}	LQR cost function
J_{SR}	Stochastic regulator cost function
J_{SS}	Cost for suboptimal steady-state control
J_{TV}	Cost for time-varying optimal control
k	Discrete time index
$\mathbf{K}(s)$, $\mathbf{K}$	Controller transfer function, controller gain matrix
$\mathbf{K}_r$	Feedforward control gain (tracking input)
$\mathbf{K}_w$	Feedforward control gain (disturbance feedforward)
$\mathbf{\Lambda}$	Diagonal matrix containing eigenvalues
$\mathcal{L}$	Laplace transform
$\mathcal{L}_2$	Space of signals with finite 2-norms
λ	Eigenvalue
$l_1 \times n_i$	Matrix dimensions
$\mathbf{L}_c$	Controllability grammian
$\lambda_i(\bullet)$	ith eigenvalue of a matrix
lim	Limit
$\mathbf{L}_o$	Observability grammian
$m(t)$	Measured output
$\max\{\bullet\}$	Maximum operator
$\mathbf{M}_{ij}$	Element of the matrix **M** (row i, column j)
$\min\{\bullet\}$	Minimum operator
$m_x(t)$	Mean of the random process $x(t)$
$m_{x\|y}$	Conditional mean
$\tilde{m}(t)$	Kalman innovations process
$\mu_{\bar{\Delta}}(\bullet)$	Structured singular value
$\mathcal{N}$	Observability matrix
n	Discrete-time index, matrix dimension
N_c	Number of encirclements of the point minus one by the Nyquist locus
N_p	Number of right half-plane poles of $\{1 + G(s)K(s)\}$
$\mathbf{N}_s$	Square version of a matrix
$\mathbf{N}(s)$	Nominal closed-loop transfer function (standard form)
num(s)	Transfer function numerator
n_v	Dimension of the vector v
N_z	Number of right half-plane zeros of $\{1 + G(s)K(s)\}$
$\mathbf{P}(s)$	Plant transfer function in standard form
$\mathbf{P}(t)$, $\mathbf{P}$	Riccati solution (LQR, $\mathcal{H}_\infty$ full information control)
p, $p(t)$	Lagrange multiplier
$p_1(t)$	Pulse function
p_i	Poles of a system

Symbol	Meaning
PM	Phase margin
$\mathfrak{Q}$	Controllability matrix
θ	Angle, phase
$\mathbf{Q}(t)$	State weighting function
$\mathbf{Q}_R$	Combined state and control weighting matrix
$\mathbf{R}$	Factor in Cholsky decomposition of controllability grammian
R	Rank
ρ	Measurement weighting coefficient (output LTR)
$\mathbf{R}(t)$	Control weighting function
$r(t)$	Reference input
$\rho(\bullet)$	Spectral radius
$\text{rank}(\bullet)$	Rank of a matrix
$\text{Ric}(\bullet)$	Ricci operator
$\mathbf{R}_x(\tau)$	Correlation function of the stationary random process $x(t)$
$\mathbf{R}_x(t_1, t_2)$	Correlation function of the random process $x(t)$
$\mathbf{R}_w^{(d)}(n)$	Discrete-time correlation function of the random sequence $w(k)$
$\Re^{m\times n}$	Set of real $m \times n$ matrices
$\Re^n$	Set of real n-dimensional vectors
$\mathbf{S}$	Matrix of singular values
s	Laplace variable
σ_i	Singular value
$\bar{\sigma}$	Maximum singular value
$\underline{\sigma}$	Minimum singular value
$\sup(\bullet)$	Supremum
$\mathbf{S}_w(\omega)$	Spectral density of the random process $w(t)$
$\mathbf{\Sigma}_x(t)$	Correlation matrix of the random process $x(t)$
$\mathbf{\Sigma}_{xy}$	Cross-correlation matrix
$\mathbf{\Sigma}_w^{(d)}(n)$	Discrete-time correlation matrix of the random sequence $w(k)$
T	Sampling time (for discrete-time systems)
t	Time
τ	Time variable; time difference in correlation function
$\mathbf{T}$, $\mathbf{T}(t)$	Transformation matrix for change of basis
t_0	Initial time
T_1, T_2	Components of matrix fraction decomposition
τ_c	Correlation time
t_f	Final time
$\text{tr}(\bullet)$	Trace operator
T_s	Settling time
$\mathbf{U}$	Matrix of left singular vectors
$u(t)$	Control input; generic system input
u_1	Normalized control input
$\bar{u}_i$	Upper bound on the ith element of the control
$\mathbf{V}$	Matrix of right singular vectors
v	Generic vector, measurement error, measurement noise
$v(t)$	Measurement noise
v_1	White shaping filter input, normalized measurement input
v_2	Noise on constructed measurement
v_k	kth element of the vector v
v^T	Transpose of the vector v
$v^\dagger$	Conjugate transpose of v
$\mathbf{\Omega}$	$(s\mathbf{I} - \mathbf{A}_{11})^{-1}$
$\mathbf{W}$	Weighted matrix
ω	Frequency

Symbol	Meaning
$\mathbf{W}(t)$	Weighted function
$w(t)$	Disturbance input
w_0	Constant disturbance input
$\mathbf{W}_o(t)$	Output weighted function
w_1	White shaping filter input, normalized plant input
w_f	Fictitious noise for LTR
$\mathbf{W}_i(t)$	Input-weighted function
$\hat{x}$	Estimate of x
$\hat{x}_\varepsilon$	Estimate of x given data through $x - \varepsilon$
$X(s)$	Laplace transform of $x(t)$
$x(t)$	State
x^*	Extremal of x
x_0	Initial state
x_d	Desired state
x_d	Desired state (tracking systems)
x_f	Final state
$\tilde{x}(t)$	State after change of basis or coordinate translation, adjoint state
$\boldsymbol{\Xi}_x(t)$	Covariance matrix of the random process $x(t)$
$\boldsymbol{\Xi}_{xy}$	Cross-covariance matrix
$\boldsymbol{\Psi}$	Eigenvector matrix
$\mathbf{Y}(t)$	Reference output weighting function
$y(t)$	Reference output; generic system output
y_1	Normalized reference output
y_w	Reference output due to w
$\mathfrak{Y}$	Hamiltonian matrix (Kalman filter)
$\mathfrak{Y}_\infty$	Hamiltonian matrix ($\mathcal{H}_\infty$ estimation)
ζ	Damping ratio
$\mathbf{Z}(t)$	Output error weighting function
$z(t)$	Constructed measurement (nonwhite measurement noise)
$z(t)$	Coordinate translation of the state
$z(t)$	Transformed state
z_i	Zeros of a system
$\mathfrak{X}$	Hamiltonian matrix (LQR)
$\mathfrak{X}_\infty$	Hamiltonian matrix (full information control)
$[a, b)$	Real interval, closed at a and open at b
∞	Infinity
$\lVert\bullet\rVert_2$	2-norm (vector, signal, or system)
$\lVert\bullet\rVert$	Euclidean vector norm, generic norm
$\lVert\bullet\rVert_{2,[t_0,t_f]}$	Finite-time signal 2-norm
$\lVert\bullet\rVert_{\infty,[t_0,t_f]}$	Finite-time signal ∞-norm
$\lVert\bullet\rVert_E$	Vector Euclidian norm
$\lVert\bullet\rVert_{W(t)}$	Weighted signal 2-norm
$\lVert\bullet\rVert_W$	Weighted vector 2-norm
$\lVert\bullet\rVert_\infty$	∞-norm (vector, signal, or system)
$\angle\{\bullet\}$	Argument of a complex number
$\otimes$	Convolution operator
$\in$	Element of a set
$\lvert\bullet\rvert$	Magnitude of a complex number

1 Introduction

Control systems are found throughout the natural and the technological worlds. Man-made control systems range from feedback controllers that regulate the water level in the tank of most flush toilets to autopilots capable of landing the Space Shuttle. For example, control systems regulate the temperature of our houses, the color of our toast, the dryness of our clothes, the metering of fuel into our automobile engines, the speed of our automobiles, the read head position in our hard disc drives, and the trajectories of the planes on which we fly. Control systems allow a large robot to paint an automobile, allow a tiny robot to move across the surface of Mars, and also allow the Hubble space telescope to get remarkably sharp pictures of the heavens. As with all technologies, there is a continual push for ever more sophisticated control systems, that is, control systems that provide better accuracy and allow more sophisticated functions.

Control systems are also found throughout nature. For example, natural feedback control systems regulate thyroid levels in the human body, allow the eye to track a baseball, and allow a human hand to pick up an egg without shattering the shell. Natural control systems have evolved over millions, or even billions, of years and some have obtained an unequaled degree of sophistication. Unfortunately, the time required for their development is inconsistent with the demands of the modern technological world. This book presents methodologies that can be used to systematically design sophisticated controllers in a reasonable time for a wide range of applications.

1.1 The Fundamental Objectives of Feedback Control

Feedback control is the generation of plant (the physical object being controlled) inputs based, in part, on measured plant outputs, and is used to force the plant to exhibit some desirable behavior. A block diagram of a generic feedback control system is given in Figure 1.1. The control input $u(t)$ is generated by the controller and applied to the plant to create the desired behavior. The measured plant output $m(t)$ is fed back into the controller for use in generating the control input. The reference output $y(t)$ is the plant output to be controlled, while the reference input $r(t)$ specifies the desired behavior of the reference output. The disturbance input $w(t)$ consists of those inputs to the plant that are generated by the environment and typically force the plant to exhibit undesirable behavior. This generic model is presented to enable a discussion of feedback objectives. A more general feedback model is given in Chapter 4 along with additional discussion of the model variables and feedback control objectives.

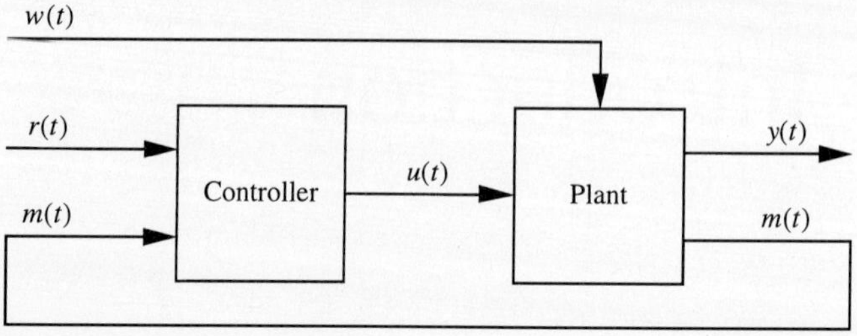

FIGURE 1.1 A generic feedback control system

The fundamental objective of the control system is to keep the reference output close to the desired value, that is, the reference input. Tracking performance, or steady-state performance, is used to describe the control system's ability to meet this objective during normal operation when the reference input is slowly varying. Abrupt changes in the reference input generate transients in the system, since physical systems cannot react instantaneously. The controller should force these transients to decay out in a reasonable period of time. *Transient performance* is the term used to describe the decay of transients in the control system. Note that the feedback system is required to be stable in order to guarantee this decay. The disturbance inputs tend to force the plant to exhibit undesirable behavior. The ability of the control system to mitigate the effects of disturbance inputs is called disturbance rejection.

The control input is supplied to an actuator, that is, a physical system that generates forces, torques, and so forth on the plant. Actuators typically have an operating range. For example, the maximum voltage applied to a motor that controls a fluid valve may be 15 volts. Control inputs outside of the normal operating range may damage the equipment or lead to saturation where the additional input does not generate additional force, torque, and so forth. Therefore, the controller should only generate inputs to the actuators that are of reasonable magnitude.

The controller is designed with reference to a mathematical model of the plant. In general, this model only approximates the behavior of the actual physical plant. In addition, the physical controller constructed is never exactly equal to the desired design. The control system should, therefore, be insensitive to errors in the mathematical models of the plant and the controller, a property known as robustness.

In summary, a well-designed control system should be stable and exhibit good tracking performance, transient performance, disturbance rejection, and robustness, all while using a reasonable amount of control.

1.2 A Brief History of Modern Controller Design

Classical control system design was mostly developed during the first sixty years of the twentieth century, and is characterized by the use of root locus, Bode plots, Nichols charts, and Nyquist plots for design and analysis. Desired transient performance is obtained using the root locus and/or frequency response methods. Desired steady-state performance and disturbance rejection is obtained by increasing system type and/or modifying the frequency response of the loop transfer function. Robustness is analyzed using the phase and gain margins, where acceptable robustness is characterized by minimum gain and phase margins. Phase and gain margins are primarily manipulated using frequency response methods, resulting in the addition of lead and lag compensators. The resulting system performance is, typically, a compromise between performance, robustness, and the limitations on control input magnitude.

The classical controller design methodology is iterative. The designer starts with a reasonable design and reasonable inputs, and then analyzes performance, disturbance rejection, and robustness, and also determines the resulting control magnitude. The design is then modified by adjusting parameters or adding compensation until acceptable performance, robustness, and so forth are obtained. Root locus, Bode plots, Nichols charts, and Nyquist plots are used as tools for understanding the cause-and-effect relationships between controller modifications and final system performance, and also for selecting parameters, compensators, and so on. The strength of classical controller design is the intuitive insight obtained from this process. The disadvantage of classical design is that this intuitive insight often fails for high-order systems and for multi-input, multi-output (MIMO) systems.[1] In addition, many design engineers have asked the question, Is there a way to directly find the single best, or optimal, control system?

Linear quadratic Gaussian (LQG) optimal control, also known as $\mathcal{H}_2$ optimal control, was developed during the 1960s largely for aerospace applications, with refinements continuing into the 1970s.[2] This design method has been successfully used in a wide range of applications, both in aerospace and in many other areas.

Linear quadratic Gaussian optimal control utilizes the state model of the plant as opposed to the transfer function model employed in classical control design. The use of a state model allows the results to be more easily extended to high-order systems and to MIMO systems. Optimal control is characterized by the specification of a cost function that combines all of the available performance specifications. This cost function is a real, scalar-valued function of the system that can be minimized to obtain the single best controller. This approach appears to eliminate the iteration required in classical control system design and is, therefore, aesthetically more satisfying. In practice, iteration is still a part of the design process since, in the vast majority of applications, the cost function is not firmly constrained. Iteration in optimal control consists of modifying the cost function until a reasonable design is obtained. The number of parameters in the cost function that are modified is typically small, even for high-order MIMO systems, since many firm constraints on these parameters can be specified. LQG optimal control, therefore, provides a very systematic controller design methodology for high-order plants and for MIMO plants.

The disadvantage of LQG design is that much of the intuitive insight about the effects of controller parameter modification is lost. The cost functions used in LQG design do not include robustness specifications, although a procedure known as loop transfer recovery can be applied to improve the robustness of the resulting design. LQG cost

[1] A great number of papers have been written on modifying classical control techniques for use with MIMO systems. Many of these modifications are useful and can be used for some MIMO designs, but the fact remains that the classical design approach becomes cumbersome for high-order plants and for MIMO plants.

[2] Some of the fundamental papers on optimal control were published in the late 1950s.

functions include only very limited disturbance rejection specifications, and applying LQG design to tracking systems requires the use of classical control concepts like integral control. An additional disadvantage is that the resulting controller has at least as high an order as the plant, which often makes implementation difficult. Implementation can be simplified by approximating the resulting controller with a lower order controller, that is, by controller order reduction.

$\mathcal{H}_\infty$ control and μ-synthesis were developed in the 1980s, and continue to be refined in the 1990s in response to requirements for more robust control designs. These methods have been successfully used in a diverse set of difficult applications, including a tokamak plasma control system for use in nuclear fusion power generation; a flight control system for the space shuttle; an ultrahigh performance missile control system; and a controller for a high-purity distillation system.

The cost functions used in $\mathcal{H}_\infty$ control and μ-synthesis are very general and can directly include performance specifications, disturbance rejection specifications, control input magnitude limitations, and robustness requirements. The strength of these design methodologies is the generality of the cost function. The disadvantage is that the $\mathcal{H}_\infty$ cost function must be optimized iteratively, which is much less efficient than LQG optimization. In addition, no general, globally optimal method of minimizing the μ-synthesis cost function is currently available, even though this cost function provides the most general description of robust performance. The controllers obtained with $\mathcal{H}_\infty$ control and μ-synthesis also tend to be of very high order even for simple problems. Controller order reduction is, therefore, typically required when applying these design methods.

1.3 Scope and Objectives

The advantages and disadvantages of classical control and several optimal controller design methods were very briefly discussed in the preceding section. The best method depends on the particular plant, operating environment, required reliability, and so on. The skilled control system designer, expecting to work on a variety of design projects, should, therefore, have a working understanding of classical control, LQG optimal control, $\mathcal{H}_\infty$ control, μ-synthesis, and also nonlinear controller design and analysis (a subject beyond the scope of this text).[3] The classical control design methodology is covered in a number of excellent texts, for example [4]–[8]. A number of texts written over the last 30 years cover LQG optimal control, for example [9]–[11]. Some recent texts cover $\mathcal{H}_\infty$ optimal control and μ-synthesis, for example [12]–[14]. The question that immediately comes to mind is, Why write another book?

This book is offered to accomplish a number of tasks. First, the author wishes to make new material on robustness analysis, $\mathcal{H}_\infty$ control, and μ-synthesis accessible to first-year graduate students. The current books covering this material are geared toward research-oriented students. This text is geared toward providing a sufficient background so that the practicing engineer can use these design and analysis methods

[3] Nonlinear controller design and analysis is addressed in a number of textbooks, for example [1] and [2]. Nonlinear optimal controller design is also addressed in [3].

in real-world applications. Derivations of all key results are included, since an understanding of these derivations provides insight into the design process and allows the engineer to modify the results when necessary. However, the presentation of this material is accomplished with a minimum of mathematics and an emphasis on the intuitive development of concepts. Second, the author wishes to combine the newer results on robustness analysis, $\mathcal{H}_\infty$ control, and μ-synthesis with previous work on LQG control to form a more complete picture of linear optimal control system design and analysis. Research in $\mathcal{H}_\infty$ control and robustness has provided new insight into LQG optimal control and also yielded modifications to this design methodology. The ties between these two design methods are explored and exploited. Finally, the author wishes to acquaint the reader with the CAD tools available for linear optimal controller design. This is accomplished by utilizing CAD software in the examples and providing the software as a resource for the reader.

1.4 Organization

The book is divided into three parts: Analysis of control systems, $\mathcal{H}_2$ (LQG) control, and $\mathcal{H}_\infty$ control. Part 1 presents material on multivariable linear system analysis, vector random processes, control system performance analysis, and robustness analysis. Linear quadratic regulator design, Kalman filtering, and linear quadratic Gaussian controller design are developed in Part 2. $\mathcal{H}_\infty$ control and μ-synthesis are presented in Part 3. Part 3 also includes material on reduced-order controller synthesis.

Chapter 2 provides a summary of the fundamentals of analyzing multi-input/multi-output (MIMO) linear systems using state space models. Continuous-time and discrete-time linear state models are presented along with their general solutions in a MIMO setting. The transfer function and frequency response are defined and discussed. A more detailed presentation is given of the frequency response of MIMO systems, including singular values and principal gains, since this may be new material to many readers. Poles, zeros, and modes are also defined, and stability is reviewed. Internal stability of linear feedback systems is examined in depth. Similarity transformations resulting from a change of basis of the state vector are defined. Controllability and observability are defined, and tests for these properties are given. State feedback design and observer design are presented. This chapter concludes with a discussion of observer feedback and the deterministic separation principle.

Random signals appear in control system analysis as models of reference inputs, disturbance inputs, and measurement noise. Chapter 3 presents a brief overview of random-signal analysis. This chapter begins with a review of random signal basics, which includes the definitions of moments, correlation, orthogonality, conditional expectation, stationarity, and the spectral density. These definitions are all provided for the case of vector random signals. A fairly complete treatment of the response of linear systems to white noise inputs is included. These results can also be applied in evaluating the response of a linear system to a nonwhite random input modeled as the output of a shaping filter driven by white noise. This chapter concludes with material on the generation and application of shaping filters.

An overview of control system performance analysis is given in Chapter 4. This chapter begins by defining a very general model of a feedback control system that explicitly includes desired outputs and disturbance inputs. Transient performance, tracking performance, and disturbance rejection are then discussed. A number of cost functions that are useful in control system design are also presented. This chapter concludes with a discussion of methods for computing the various performance criteria and costs.

Control system robustness analysis is presented in Chapter 5. The stability robustness of a single-input, single-output system subject to gain and phase perturbations is addressed. This analysis leads to the gain and phase margins of classical control. Two more general types of perturbations are then presented: transfer functions (unstructured uncertainty) and multiple transfer functions and/or parameters (structured uncertainty). The stability and performance robustness of general systems subject to these perturbations is then addressed.

Chapter 6 addresses the problem of finding an optimal control law for plants governed by linear state equations and quadratic cost functions. Chapter 6 begins with a brief summary of the mathematics of optimization and the calculus of variations. The linear quadratic regulator is then developed as the optimal controller for the specified problem. The performance and robustness of the linear quadratic regulator are discussed, along with some interesting limiting properties of this control system.

The linear quadratic regulator requires that the entire state be available for feedback. This is impractical in many applications, but it is reasonable to estimate the state from available measurements and use the estimated state for feedback. The Kalman filter is an optimal estimator of the plant state, and is developed in Chapter 7.

Linear quadratic Gaussian optimal control is presented in Chapter 8. The performance and robustness of LQG optimal control is addressed in detail. The loop transfer recovery procedure for increasing robustness is presented. The addition of integral control, compensators, and general methods of shaping the frequency response of LQG systems are also presented. This chapter concludes by developing the correspondence between LQG optimal control and an $\mathcal{H}_2$ optimal control.

$\mathcal{H}_\infty$ full information control is developed along with $\mathcal{H}_\infty$ estimation in Chapter 9. The chapter begins with a brief summary of the mathematics of differential games. These results are then used to develop suboptimal solutions to the $\mathcal{H}_\infty$ full information control problem. The $\mathcal{H}_\infty$ full information control problem is defined: Find a feedback controller that yields a given bound on the closed-loop system ∞-norm from the disturbance input to the reference output. The measurements available to the controller are assumed to be the entire plant state and the disturbance input (hence the term *full information*).

The suboptimal $\mathcal{H}_\infty$ estimation problem is defined: Find an estimator that yields a given bound on the gain from the plant disturbance inputs (including measurement error) to the estimation error. The $\mathcal{H}_\infty$ estimation results are developed using the full information results and duality.

The $\mathcal{H}_\infty$ output feedback controller is developed in Chapter 10. The application of $\mathcal{H}_\infty$ control is also discussed in detail. Tracking system design, integral control, and designing for robustness are all addressed. Performance limitations are also presented, since these limitations frequently affect the specification of $\mathcal{H}_\infty$ control problems. A very general procedure for the development of robust controllers, known as μ-synthesis, is also presented in Chapter 10. This chapter concludes with a design case study that compares the LQG and μ-synthesis design methodologies.

State-space design methodologies, especially μ-synthesis, tend to result in unreasonably high-order controllers. Methods of generating reduced-order controllers are presented in Chapter 11.

REFERENCES

[1] J.-J. E. Slotine and W. Li, *Applied Nonlinear Control,* Prentice-Hall, Englewood Cliffs, NJ, 1991.

[2] M. Vidyasagar, *Nonlinear Systems Analysis,* 2d ed., Prentice-Hall, Englewood Cliffs, NJ, 1993.

[3] D. E. Kirk, *Optimal Control Theory: An Introduction,* Prentice-Hall, Englewood Cliffs, NJ, 1970.

[4] N. S. Nise, *Control Systems Engineering,* 2d ed., Addison-Wesley, Menlo Park, CA, 1995.

[5] R. C. Dorf, *Modern Control Systems,* 5th ed., Addison-Wesley, Reading, MA, 1989.

[6] R. T. Steffani, C. J. Savant, Jr., B. Shahian, and G. H. Hostetter, *Design of Feedback Control Systems,* 3d ed., Saunders College Publishing, New York, 1994.

[7] B. C. Kuo, *Automatic Control Systems,* 4th ed., Prentice-Hall, Englewood Cliffs, NJ, 1982.

[8] C. L. Phillips and R. D. Harbor, *Feedback Control Systems,* 2d ed., Prentice-Hall, Englewood Cliffs, NJ, 1991.

[9] H. Kwakernaak and R. Sivan, *Linear Optimal Control Systems,* Wiley-Interscience, New York, 1972.

[10] B. Friedland, *Control System Design: An Introduction to State-Space Methods,* McGraw-Hill, New York, 1986.

[11] B. D. O. Anderson and J. B. Moore, *Optimal Control: Linear Quadratic Methods,* Prentice-Hall, Englewood Cliffs, NJ, 1990.

[12] K. Zhou with J. C. Doyle and K. Glover, *Robust and Optimal Control,* Prentice-Hall, Upper Saddle River, NJ, 1996.

[13] M. Green and D. J. N. Limebeer, *Linear Robust Control,* Prentice-Hall, Englewood Cliffs, NJ, 1995.

[14] J. M. Maciejowski, *Multivariable Feedback Design,* Addison-Wesley, Wokingham, England, 1989.

PART 1
Analysis of Control Systems

2 Multivariable Linear Systems

Control system design requires a mathematical model of the plant. Mathematical models of plants come in many forms, including differential equation models, transfer function models, block diagrams (a graphical form of a mathematical model), and state models. The state model is a collection of first-order, linear differential equations placed in vector-matrix form. The state model forms an ideal setting when using computer-aided design software since computers typically work well with matrices and vectors. In addition, many powerful results from linear algebra can be applied when using state models. The state model is particularly convenient for control system design and analysis because of the powerful mathematical results available and the available computer-aided design software, and because the same basic equations can be used to describe low-order systems, high-order systems, single-input/single-output (SISO) systems, and multi-input/multi-output (MIMO)[1] systems.

This chapter provides a summary of the fundamentals of analyzing MIMO linear systems using state models. The theory of state-space linear systems is an elegant area of mathematics from which comes a very useful set of tools for doing engineering design and analysis. This summary is intended to present only a subset of this material. This subset is defined by what is required to perform the design and analysis tasks presented in the remainder of this book.

The presentations in this chapter are quite terse and meant as a review of the material. Notable exceptions to this format are the developments of the singular value decomposition, the principal gains, internal stability, and observer feedback control. This material will probably be new to many readers and is discussed in detail. More thorough treatments of other portions of this chapter, particularly those that are unfamiliar to the reader, can be found in most good, graduate-level, linear systems textbooks, a number of which are listed in the references.

Continuous-time and discrete-time linear state equations are presented along with their general solutions in a MIMO setting. The transfer function and frequency response are defined and discussed. A more detailed presentation of the frequency response of MIMO systems, including singular values and principal gains, is given since this is probably new material to many readers. Poles, zeros, and modes are also defined, and stability is reviewed. Internal stability of linear feedback systems is examined in depth

[1] MIMO will also be used to designate single-input/multiple-output and multiple-input/single-output systems unless otherwise noted.

since this material is probably new to some students. Similarity transformations resulting from a change of basis of the state vector are defined. The definitions of controllability and observability are presented along with a practical means of testing a state model for controllability and observability. State feedback and observer design are discussed. The chapter concludes with a discussion of observer feedback and the deterministic separation principle.

2.1 The Continuous-Time State Model

A continuous-time linear system is a transformation between the input (and the initial conditions) and the output, both being vector functions of time in general. A mathematical model of a continuous-time linear system is typically in the form of one or more ordinary differential equations involving the input, the output, and possibly some additional variables that are intermediary between the input and output. The mathematical model can be put into a standard form known as the state model:

$$\dot{x}(t) = \mathbf{A}(t)x(t) + \mathbf{B}(t)u(t); \tag{2.1a}$$

$$y(t) = \mathbf{C}(t)x(t) + \mathbf{D}(t)u(t). \tag{2.1b}$$

The variable $x(t) \in \Re^{n_x}$ is called the state of the system, $u(t) \in \Re^{n_u}$ is the input to the system, and $y(t) \in \Re^{n_y}$ is the output of the system. The matrix $\mathbf{A}(t) \in \Re^{n_x \times n_x}$ is the state matrix, $\mathbf{B}(t) \in \Re^{n_x \times n_u}$ is the input matrix, $\mathbf{C}(t) \in \Re^{n_y \times n_x}$ is the output matrix, and $\mathbf{D}(t) \in \Re^{n_y \times n_u}$ is the input-to-output coupling matrix. Equation (2.1a) is called the state equation, and (2.1b) is the measurement or output equation.

The state, after the initial time t_0, can be found from the input and the initial conditions:

$$x(t) = \mathbf{\Phi}(t, t_0)x(t_0) + \int_{t_0}^{t} \mathbf{\Phi}(t, \tau)\mathbf{B}(\tau)u(\tau)d\tau. \tag{2.2}$$

The state-transition matrix of the system $\mathbf{\Phi}(t, t_0)$ is the solution of the homogeneous initial value problem:

$$\dot{\mathbf{\Phi}}(t, t_0) = \mathbf{A}(t)\mathbf{\Phi}(t, t_0), \text{ where } \mathbf{\Phi}(t_0, t_0) = \mathbf{I}.$$

Substituting (2.2) into the measurement equation yields the output as a function of the initial conditions and the input:

$$y(t) = \mathbf{C}(t)\mathbf{\Phi}(t, t_0)x(t_0) + \int_{t_0}^{t} \mathbf{C}(t)\mathbf{\Phi}(t, \tau)\mathbf{B}(\tau)u(\tau)d\tau + \mathbf{D}(t)u(t). \tag{2.3}$$

The first term in this equation depends on the initial conditions and is called the initial condition response. The last two terms in this equation depend on the input and collectively are termed the forced response.

The state model and its solution can be simplified when the system is time-invariant (i.e., the system matrices in the state model do not depend on time):

$$\begin{aligned} \dot{x}(t) &= \mathbf{A}x(t) + \mathbf{B}u(t); \qquad (2.4a) \\ y(t) &= \mathbf{C}x(t) + \mathbf{D}u(t). \qquad (2.4b) \end{aligned}$$

The state at all positive times can be found from the input and the initial conditions:

$$x(t) = e^{\mathbf{A}t}x(0) + \int_0^t e^{\mathbf{A}(t-\tau)}\mathbf{B}u(\tau)d\tau. \qquad (2.5)$$

Note that the initial time is typically set equal to zero when dealing with time-invariant systems.[2] The state-transition matrix for the time-invariant system is the matrix exponential, which can be defined by the power series:

$$\mathbf{\Phi}(t) = e^{\mathbf{A}t} = \mathbf{I} + \mathbf{A}t + \frac{1}{2!}\mathbf{A}^2t^2 + \frac{1}{3!}\mathbf{A}^3t^3 + \cdots.$$

Note that the second argument (the initial time) for the state-transition matrix is dropped because this time is set to zero. The output is generated as a function of the initial conditions and the input:

$$y(t) = \mathbf{C}e^{\mathbf{A}t}x(0) + \int_0^t \mathbf{C}e^{\mathbf{A}(t-\tau)}\mathbf{B}u(\tau)d\tau + \mathbf{D}u(t). \qquad (2.6)$$

The impulse response is a matrix,

$$\mathbf{g}(t) = \begin{bmatrix} g_{11}(t) & \cdots & g_{1n_u}(t) \\ \vdots & \ddots & \vdots \\ g_{n_y1}(t) & \cdots & g_{n_yn_u}(t) \end{bmatrix},$$

whose elements are the single-input/single-output (SISO) impulse responses from each input to each output:

$$g_{ij}(t) = y_i(t), \text{ where } u_k(t) = \begin{cases} \delta(t) & k = j \\ 0 & k \neq j \end{cases}; \; x(0) = 0,$$

and $\delta(t)$ is the Dirac delta (impulse) function. An analytic expression for the impulse response is obtained by applying impulses to each system input (one at a time) and using (2.6) to compute the outputs. Combining these outputs yields the impulse response matrix:

$$\mathbf{g}(t) = \begin{cases} \mathbf{C}e^{\mathbf{A}t}\mathbf{B} + \mathbf{D}\delta(t) & t \geq 0 \\ \mathbf{0} & t < 0 \end{cases}. \qquad (2.7)$$

[2] The response of a time-invariant system only depends on the length of time since applying the input, and not on the initial time. Therefore, the initial time can be set equal to zero without loss of generality.

This expression for the impulse response matrix is useful theoretically, but typically too cumbersome for use in computation. When numerical results are desired, the impulse response can be found either by performing the real or simulated experiment of putting impulses into each input and observing each output.

The forced response, consisting of the last two terms in (2.6), can be written as the impulse response convolved with the input:

$$y(t) = \mathbf{C}e^{\mathbf{A}t}x(0) + \int_0^t \mathbf{g}(t-\tau)u(\tau)d\tau = \mathbf{C}e^{\mathbf{A}t}x(0) + \mathbf{g}(t) \otimes u(t). \tag{2.8}$$

The symbol $\otimes$ denotes convolution, which is defined by the integral in (2.8). Note that the input must be assumed equal to zero at all negative times for convolution to yield the correct forced response. This is reasonable, since the initial conditions are defined at time zero, and any nonzero inputs before this time are incorporated into the initial conditions and, therefore, into the initial condition response.

EXAMPLE 2.1 We are given the system described by the state model:

$$\begin{bmatrix} \dot{x}_1(t) \\ \dot{x}_2(t) \end{bmatrix} = \begin{bmatrix} -1 & 2 \\ 0 & -3 \end{bmatrix}\begin{bmatrix} x_1(t) \\ x_2(t) \end{bmatrix} + \begin{bmatrix} 1 & 0 \\ 0 & 4 \end{bmatrix}\begin{bmatrix} u_1(t) \\ u_2(t) \end{bmatrix};$$

$$\begin{bmatrix} y_1(t) \\ y_2(t) \end{bmatrix} = \begin{bmatrix} 1 & 1 \\ 1 & -1 \end{bmatrix}\begin{bmatrix} x_1(t) \\ x_2(t) \end{bmatrix};$$

with the input

$$\begin{bmatrix} u_1(t) \\ u_2(t) \end{bmatrix} = \begin{bmatrix} 5 & \delta(t) \\ 2 & 1(t) \end{bmatrix},$$

where $1(t)$ is the unit step function. The state-transition matrix is

$$e^{\mathbf{A}t} = \begin{bmatrix} 1 + (-t) + \frac{(-t)^2}{2!} + \cdots & 1 - 1 + (-t) - (-3t) + \frac{(-t)^2}{2!} - \frac{(-3t)^2}{2!} + \cdots \\ 0 & 1 + (-3t) + \frac{(-3t)^2}{2!} + \cdots \end{bmatrix}$$

$$= \begin{bmatrix} e^{-t} & e^{-t} - e^{-3t} \\ 0 & e^{-3t} \end{bmatrix},$$

where the power series for the exponential function is used in generating the above result. The impulse response is

$$\mathbf{g}(t) = \begin{bmatrix} 1 & 1 \\ 1 & -1 \end{bmatrix}\begin{bmatrix} e^{-t} & e^{-t} - e^{-3t} \\ 0 & e^{-3t} \end{bmatrix}\begin{bmatrix} 1 & 0 \\ 0 & 4 \end{bmatrix} = \begin{bmatrix} e^{-t} & 4e^{-t} \\ e^{-t} & 4e^{-t} - 8e^{-3t} \end{bmatrix},$$

which is plotted in Figure 2.1. The forced output is

$$\begin{bmatrix} y_1(t) \\ y_2(t) \end{bmatrix} = \int_0^t \begin{bmatrix} e^{-(t-\tau)} & 4e^{-(t-\tau)} \\ e^{-(t-\tau)} & 4e^{-(t-\tau)} - 8e^{-3(t-\tau)} \end{bmatrix}\begin{bmatrix} 5 & \delta(\tau) \\ 2 & 1(\tau) \end{bmatrix} d\tau$$

$$= \begin{bmatrix} 8 - 3e^{-t} \\ \frac{8}{3} - 3e^{-t} + \frac{16}{3}e^{-3t} \end{bmatrix}.$$

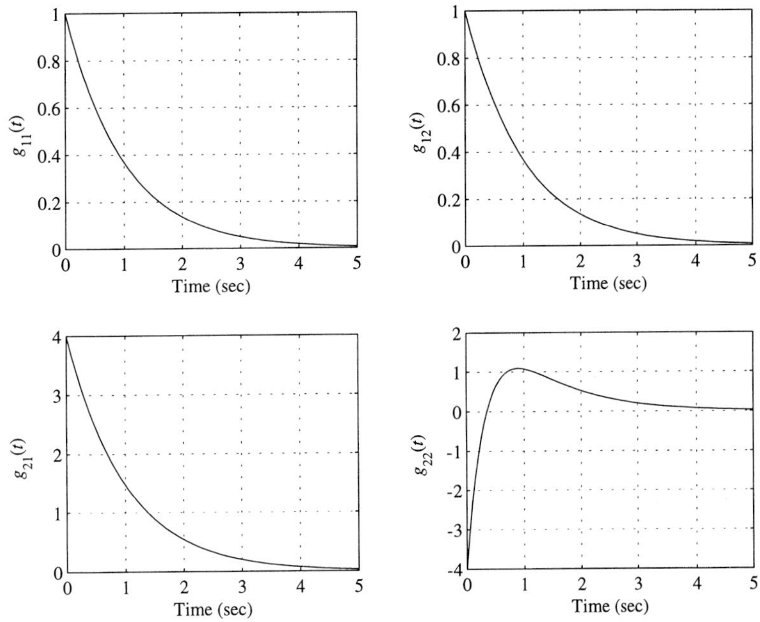

FIGURE 2.1 Impulse response for Example 2.1

2.2 The Discrete-Time State Model and Simulation

Computer simulation is frequently used to evaluate control system performance. Computer simulation can be used to solve for outputs when the system is of high order or is subject to complicated inputs that are not easily amenable to analytic solutions. In addition, the effects of time-variations, delays, and nonlinearities can be easily evaluated using simulation. Computer simulation is also typically used to verify analytic results prior to hardware implementation.

Computer simulation requires that the continuous-time plant and controller models be approximated by discrete-time systems, that is, by difference equations. Digital controllers also utilize difference equation models when operating. One method of generating a digital controller is to design a continuous-time controller and then approximate this controller in discrete time. The generation of discrete-time approximations of continuous-time systems is, therefore, of fundamental importance in simulation and control system design.

A number of methods can be used to form a discrete-time approximation of a continuous-time system: Euler's method, the zero-order hold approximation, the bilinear transformation, the impulse invariant approximation, and so on. Two of these approximation methods are considered here: Euler's method and the zero-order hold approximation. Euler's method provides a means of generating discrete-time state equations for a wide range of applications, including systems that are time-varying and nonlinear. The zero-order hold approximation is also presented because this method tends to provide better results for linear, time-invariant systems.

A discrete-time state equation can be obtained from the state equation (2.4a) using Euler's method of approximating the derivative:

$$\frac{x(kT + T) - x(kT)}{T} \approx \dot{x}(kT) = \mathbf{A}x(kT) + \mathbf{B}u(kT).$$

A difference equation for the state can then be obtained:[3]

$$x(kT + T) = x(kT) + T\,\{\mathbf{A}x(kT) + \mathbf{B}u(kT)\}. \tag{2.9}$$

A discrete-time state equation can also be obtained from (2.4a) using the zero-order hold approximation. In this case, the state difference equation is obtained by solving the state equation (2.4a) while assuming the input is constant over the sampling time:

$$u(t) = u(0) \quad \text{for all } t \in [0, T).$$

The state at the sampling time T is then

$$x(T) = e^{\mathbf{A}T}x(0) + \int_0^T e^{\mathbf{A}(T-\tau)}\mathbf{B}u(0)d\tau.$$

The input can be taken out of the integral:

$$x(T) = [e^{\mathbf{A}T}]x(0) + \left[\int_0^T e^{\mathbf{A}(T-\tau)}\mathbf{B}d\tau\right]u(0) = [e^{\mathbf{A}T}]x(0) + \left[\int_0^T e^{\mathbf{A}\tau}\mathbf{B}d\tau\right]u(0),$$

where the second integral in this equation is generated from the first by performing a change of variables. This equation can be written as follows:

$$x(T) = \mathbf{\Phi}(T)x(0) + \mathbf{\Gamma}(T)u(0),$$

where

$$\mathbf{\Phi}(T) = e^{\mathbf{A}T}; \qquad \mathbf{\Gamma}(T) = \int_0^T e^{\mathbf{A}\tau}\mathbf{B}d\tau.$$

Note that $\mathbf{\Phi}(T)$ is the state-transition matrix evaluated at one sampling time. The matrix $\mathbf{\Gamma}(T)$ is the discrete-time input matrix, which is often referred to as simply the input matrix when the system can be implicitly assumed to be discrete-time. The state at any time $(k + 1)T$ can be found given the initial conditions on the state at time kT because the system (2.4) is time-invariant:

$$x(kT + T) = \mathbf{\Phi}x(kT) + \mathbf{\Gamma}u(kT). \tag{2.10}$$

The output equation in the continuous-time state model is an algebraic equation. A discrete-time version of this equation is obtained by simply evaluating (2.4b) at the sample times:

$$y(kT) = \mathbf{C}x(kT) + \mathbf{D}u(kT). \tag{2.11}$$

[3] This method can also be applied in a straightforward manner to time-varying and nonlinear state equations.

The response of a continuous-time system can be approximated by implementing the discrete-time system on a digital computer, a process known as simulation. The system is simulated by recursively implementing either (2.9) or (2.10) to generate a time history, or trajectory, of the state. The output trajectory is then generated from the state trajectory by applying (2.11).

The state and output trajectories obtained using simulation are good approximations of the continuous-time trajectories if the sampling time is sufficiently short. When using Euler's method, the sampling time must be short enough that the finite-time derivative is an accurate approximation of the true derivative. For the zero-order hold approximation, the sampling time must be short enough that the input is approximately piecewise constant; that is, the input does not change appreciably over the sampling time.[4] Generally, the zero-order hold approximation yields more accurate results for linear, time-invariant systems, but becomes cumbersome for time-varying systems and is not applicable (as presented) for nonlinear systems. Therefore, the zero-order hold approximation is typically used for linear, time-invariant systems, and Euler's method is usually applied to time-varying and nonlinear systems.

The continuous-time state model and the discrete-time state model form two models of the same physical system. Both of these models have a similar form. Due to this parity, similar and sometimes identical expressions result when performing analysis and design with reference to each of these models. The remainder of this book will only derive and discuss results obtained for the continuous-time model. Since the design may be realized with either digital or analog hardware, equivalent discrete-time expressions for key equations are given in the summary sections of each chapter. These results are given with reference to the zero-order hold approximation unless otherwise stated.

2.3 Transfer Functions

The Laplace transform is a useful tool for the analysis and solution of time-invariant differential equations. Taking the Laplace transform of the state and output equations (2.4) yields

$$sX(s) - x(0) = \mathbf{A}X(s) + \mathbf{B}U(s);$$
$$Y(s) = \mathbf{C}X(s) + \mathbf{D}U(s),$$

where $X(s)$, $U(s)$, and $Y(s)$ are the Laplace transforms of $x(t)$, $u(t)$, and $y(t)$, respectively. Note that $x(0)$ is the initial condition on the state. Solving for $Y(s)$ as a function of $U(s)$ and $x(0)$ yields

$$Y(s) = \mathbf{C}(s\mathbf{I} - \mathbf{A})^{-1}x(0) + \{\mathbf{C}(s\mathbf{I} - \mathbf{A})^{-1}\mathbf{B} + \mathbf{D}\}U(s). \tag{2.12}$$

The Laplace transfer function (or simply the transfer function) of the linear system is the matrix of gains between the Laplace transform of the input vector and the Laplace transform of the output vector:

$$\boxed{\mathbf{G}(s) = \mathbf{C}(s\mathbf{I} - \mathbf{A})^{-1}\mathbf{B} + \mathbf{D}.} \tag{2.13}$$

[4] A more detailed discussion of the selection of sampling times can be found in [1], page 180.

Assuming the initial conditions are zero,

$$\boxed{Y(s) = \mathbf{G}(s)U(s).} \tag{2.14}$$

Note that this equation is often used to define the transfer function.

The transfer function can be written as a matrix of ratios of polynomials in s:

$$\mathbf{G}(s) = \begin{bmatrix} G_{11}(s) & \cdots & G_{1n_u}(s) \\ \vdots & \ddots & \vdots \\ G_{n_y1}(s) & \cdots & G_{n_yn_u}(s) \end{bmatrix} = \begin{bmatrix} \dfrac{\text{num}_{11}(s)}{\text{den}_{11}(s)} & \cdots & \dfrac{\text{num}_{1n_u}(s)}{\text{den}_{1n_u}(s)} \\ \vdots & \ddots & \vdots \\ \dfrac{\text{num}_{n_y1}(s)}{\text{den}_{n_y1}(s)} & \cdots & \dfrac{\text{num}_{n_yn_u}(s)}{\text{den}_{n_yn_u}(s)} \end{bmatrix}.$$

Note that each term in $\mathbf{G}(s)$ is a proper ratio of polynomials; that is, the order of the numerator is less than or equal to the order of the denominator. When there is no input-to-output coupling ($\mathbf{D} = \mathbf{0}$), each term in $\mathbf{G}(s)$ is a strictly proper ratio of polynomials; that is, the order of the numerator is less than the order of the denominator.

The system output can be found by taking the inverse Laplace transform of (2.12):

$$\begin{aligned} y(t) &= \mathscr{L}^{-1}\{Y(s)\} \\ &= \mathscr{L}^{-1}\{\mathbf{C}(s\mathbf{I} - \mathbf{A})^{-1}\}x(0) + \mathscr{L}^{-1}\{\mathbf{C}(s\mathbf{I} - \mathbf{A})^{-1}\mathbf{B} + \mathbf{D}\} \otimes \mathscr{L}^{-1}\{U(s)\}, \end{aligned}$$

where the linearity of the Laplace transform and the convolution property of the Laplace transform are used in generating this expression. Equating this expression for $y(t)$ with that in (2.8) yields:

$$\begin{aligned} &\mathbf{C}e^{\mathbf{A}t}x(0) + \mathbf{g}(t) \otimes u(t) \\ &\quad = \mathbf{C}\mathscr{L}^{-1}\{(s\mathbf{I} - \mathbf{A})^{-1}\}x(0) + \mathscr{L}^{-1}\{\mathbf{C}(s\mathbf{I} - \mathbf{A})^{-1}\mathbf{B} + \mathbf{D}\} \otimes u(t). \end{aligned}$$

The following correspondences are apparent:

$$e^{\mathbf{A}t} = \mathscr{L}^{-1}\{(s\mathbf{I} - \mathbf{A})^{-1}\}; \tag{2.15}$$

$$\mathbf{g}(t) = \mathscr{L}^{-1}\{\mathbf{C}(s\mathbf{I} - \mathbf{A})^{-1}\mathbf{B} + \mathbf{D}\} = \mathscr{L}^{-1}\{\mathbf{G}(s)\}. \tag{2.16}$$

Equation (2.15) provides a means of determining the state-transition matrix, while (2.16) provides a means of determining the impulse response. The transfer function can also be found by taking the Laplace transform of the impulse response matrix:

$$\mathbf{G}(s) = \mathscr{L}\{\mathbf{g}(t)\}.$$

EXAMPLE 2.2 Given the system described by the state model,

$$\begin{bmatrix} \dot{x}_1(t) \\ \dot{x}_2(t) \end{bmatrix} = \begin{bmatrix} -1 & 2 \\ 0 & -3 \end{bmatrix}\begin{bmatrix} x_1(t) \\ x_2(t) \end{bmatrix} + \begin{bmatrix} 1 & 0 \\ 0 & 4 \end{bmatrix}\begin{bmatrix} u_1(t) \\ u_2(t) \end{bmatrix};$$

$$\begin{bmatrix} y_1(t) \\ y_2(t) \end{bmatrix} = \begin{bmatrix} 1 & 1 \\ 1 & -1 \end{bmatrix}\begin{bmatrix} x_1(t) \\ x_2(t) \end{bmatrix},$$

the transfer function matrix is

$$\mathbf{G}(s) = \begin{bmatrix} G_{11}(s) & G_{12}(s) \\ G_{21}(s) & G_{22}(s) \end{bmatrix} = \begin{bmatrix} 1 & 1 \\ 1 & -1 \end{bmatrix} \begin{bmatrix} s+1 & -2 \\ 0 & s+3 \end{bmatrix}^{-1} \begin{bmatrix} 1 & 0 \\ 0 & 4 \end{bmatrix}$$

$$= \begin{bmatrix} \dfrac{1}{s+1} & \dfrac{4}{s+1} \\ \dfrac{1}{s+1} & \dfrac{-4(s-1)}{(s+1)(s+3)} \end{bmatrix}.$$

Note that the inverse Laplace transform of this matrix equals the impulse response matrix:

$$\mathbf{g}(t) = \mathscr{L}^{-1}\{\mathbf{G}(s)\} = \begin{bmatrix} e^{-t} & 4e^{-t} \\ e^{-t} & 4e^{-t} - 8e^{-3t} \end{bmatrix}.$$

◆

The transfer function is used in this example as a means of finding the impulse response. The transfer function is also useful in finding the system response to sinusoidal inputs (i.e., the frequency response).

2.4 Frequency Response

The forced output of a linear system with a sinusoidal input can be computed very simply from the transfer function. The linear system changes the magnitude and phase of the sinusoid while leaving the frequency unchanged. For SISO systems, the change in magnitude (the gain) and the change of phase (the phase shift) of the linear system are termed the frequency response. For MIMO systems, the gain and phase shift depend on which input is used and which output is observed. A more general concept of frequency response is therefore required. The singular value decomposition and the principal gains provide this generalization. Before presenting these results, we must develop some elementary properties of linear systems subject to complex exponential inputs and sinusoidal inputs.

The steady-state output of a linear system with a complex exponential input is the product of the input and the transfer function evaluated at the corresponding complex frequency. To demonstrate this, consider the following complex exponential input:

$$u(t) = u_0 e^{st},$$

where $u_0 \in \mathscr{C}^{n_u}$. The steady-state output is given by (2.8) under the assumption that the initial conditions occurred infinitely far in the past:

$$y(t) = \int_{-\infty}^{t} \mathbf{g}(t-\tau) u_0 e^{s\tau} d\tau = \int_{-\infty}^{t} \mathbf{g}(t-\tau) e^{s\tau} d\tau u_0.$$

Performing the change of variables $t_1 = t - \tau$ yields

$$y(t) = \int_{0}^{\infty} \mathbf{g}(t_1) e^{s(t-t_1)} dt_1 u_0 = \int_{0}^{\infty} \mathbf{g}(t_1) e^{-st_1} dt_1 u_0 e^{st}.$$

Recognizing that the last integral in this equation is the Laplace transform of the impulse response, the output is the product of the input and the transfer function:

$$y(t) = \mathbf{G}(s)u_0 e^{st}. \tag{2.17}$$

The result (2.17) can be directly applied to the case when the exponent of the input is purely imaginary; that is, the input is a pure tone:

$$u(t) = u_0 e^{j\omega t}.$$

The steady-state forced response is then

$$y(t) = \mathbf{G}(j\omega)u_0 e^{j\omega t}. \tag{2.18}$$

The function $\mathbf{G}(j\omega)$ is called the Fourier transfer function of the system and is simply the Laplace transfer function evaluated on the imaginary axis.

2.4.1 Frequency Response for SISO Systems

Equation (2.18) can be expanded for SISO systems:

$$y(t) = G(j\omega)u_0 e^{j\omega t} = |G(j\omega)|u_0 e^{j\omega t + j\angle\{G(j\omega)\}},$$

where $|\bullet|$ and $\angle\{\bullet\}$ denote the magnitude and the argument of a complex number, respectively. The magnitude and argument of the Fourier transfer function are termed the system frequency response. The system frequency response can be used to compute the steady-state output of the system when the input is a general sinusoid:

$$u(t) = \alpha \cos(\omega t + \theta).$$

The steady-state output is then a sinusoid of the same frequency with a different amplitude and phase:

$$y(t) = |G(j\omega)|\alpha \cos[\omega t + \theta + \angle\{G(j\omega)\}].$$

The frequency response may be determined by applying sinusoidal inputs to the system over a range of frequencies and computing the resulting gains and phase shifts.

2.4.2 Frequency Response for MIMO Systems

The definition of gain and phase shift becomes more involved for MIMO systems. The gain of a MIMO system can be defined as the length of the output vector divided by the length of the input vector. When the input is a pure tone,

$$\text{Gain} = \frac{\|y(t)\|}{\|u(t)\|} = \frac{\|\mathbf{G}(j\omega)u_0 e^{j\omega t}\|}{\|u_0 e^{j\omega t}\|} = \frac{\|\mathbf{G}(j\omega)u_0\|}{\|u_0\|},$$

where $\|\bullet\|$ denotes the Euclidean vector norm:[5]

$$\|v\| = \sqrt{v_1^2 + v_2^2 + \cdots + v_{n_v}^2}.$$

This gain depends on both u_0 and ω, so a unique gain as a function of frequency cannot be defined. Instead, a range of gains can be defined:

[5] A general discussion of norms is presented in Chapter 4.

$$\min_{u_0} \frac{\|\mathbf{G}(j\omega)u_0\|}{\|u_0\|} \le \text{Gain} \le \max_{u_0} \frac{\|\mathbf{G}(j\omega)u_0\|}{\|u_0\|}; \tag{2.19a}$$

$$\min_{\|u_0\|=1} \|\mathbf{G}(j\omega)u_0\| \le \text{Gain} \le \max_{\|u_0\|=1} \|\mathbf{G}(j\omega)u_0\|. \tag{2.19b}$$

Note that multiplying the input by a scalar does not change the gain of a linear system. Therefore, finding the minimum (or maximum) over all u_0 with unity magnitude is equivalent to finding the minimum (or maximum) over all u_0.

This range of gains is also valid when the input vector consists of sinusoids at a single frequency:

$$u(t) = \begin{bmatrix} \alpha_1 \cos(\omega t + \theta_1) \\ \vdots \\ \alpha_{n_u} \cos(\omega t + \theta_{n_u}) \end{bmatrix}.$$

The output vector then consists of sinusoids at this same frequency:

$$y(t) = \begin{bmatrix} \beta_1 \cos(\omega t + \varphi_1) \\ \vdots \\ \beta_{n_y} \cos(\omega t + \varphi_{n_y}) \end{bmatrix}.$$

A gain can be defined based on the amplitudes of the input and output sinusoids:

$$\text{Gain} = \frac{\|\beta\|}{\|\alpha\|}.$$

This gain can be shown to lie within the bounds given by (2.19).

A reasonable definition for the phase shift of a MIMO system is not available. In general, the phase shift from each input to each output is different. Specifying a range for the phase shifts, as was done with gain, is not useful since the individual phase shifts typically come close to spanning the entire range of possible values.[6] Therefore, frequency response information for a MIMO system is displayed by plotting only the maximum and minimum gains of the system. The direct computation of this range using (2.19) proves very tedious. A simpler method of computing the range of gains is based on the singular value decomposition.

The Singular Value Decomposition The singular value decomposition (SVD) is a matrix factorization that has found a number of applications to engineering problems. The SVD of a matrix $\mathbf{M} \in \mathscr{C}^{n_y \times n_u}$ is defined as follows:

$$\boxed{\mathbf{M} = \mathbf{USV}^\dagger = \sum_{i=1}^{p} \sigma_i U_i V_i^\dagger,} \tag{2.20}$$

[6] The phase shift is ambiguous with respect to changes of 2π. A range of phase shifts from any angle ϕ to $\phi + 2\pi$, therefore, spans the entire range of possible phase shifts.

where $\mathbf{U} \in \mathscr{C}^{n_y \times n_y}$ and $\mathbf{V} \in \mathscr{C}^{n_u \times n_u}$ are unitary matrices, p equals the minimum of n_y and n_u, superscript $\dagger$ denotes the conjugate transpose, and U_i and V_i are the ith columns of $\mathbf{U}$ and $\mathbf{V}$, respectively. A unitary matrix $\mathbf{V}$ is a matrix which has the property

$$\mathbf{V}^\dagger \mathbf{V} = \mathbf{V}\mathbf{V}^\dagger = \mathbf{I}, \tag{2.21a}$$

or equivalently, a matrix whose columns are orthonormal:

$$V_i^\dagger V_j = \delta_{ij} = \begin{cases} 1 & i = j \\ 0 & i \neq j \end{cases}. \tag{2.21b}$$

The expression δ_{ij} denotes the Kronecker delta function, which is defined by (2.21b). The vectors $\{U_i\}$ and $\{V_i\}$ are called the left and right singular vectors of $\mathbf{M}$, respectively.

The matrix $\mathbf{S} \in \Re^{n_y \times n_u}$ is diagonal:

$$\mathbf{S} = \left[\begin{array}{ccc|c} \sigma_1 & & 0 & \\ & \ddots & & \mathbf{0} \\ 0 & & \sigma_p & \end{array}\right] \text{ or } \mathbf{S} = \begin{bmatrix} \sigma_1 & & 0 \\ & \ddots & \\ 0 & & \sigma_p \end{bmatrix} \text{ or } \mathbf{S} = \left[\begin{array}{ccc} \sigma_1 & & 0 \\ & \ddots & \\ 0 & & \sigma_p \\ \hline & \mathbf{0} & \end{array}\right]$$

when $n_y < n_u$, $n_y = n_u$, and $n_y > n_u$, respectively. The parameters $\{\sigma_i\}$ are called the singular values of $\mathbf{M}$. These singular values are ordered (by convention):

$$\sigma_1 \geq \sigma_2 \geq \cdots \geq \sigma_{n_u} \geq 0.$$

The singular value decomposition provides a detailed picture of how the matrix operates on a vector (termed the input):

$$\mathbf{M}x = \sum_{i=1}^{p} \sigma_i U_i V_i^\dagger x.$$

The term $V_i^\dagger x$ gives the length of the input in the direction defined by the given right singular vector. This length is then multiplied by the associated singular value. This product is the length of the output vector in the direction defined by the left singular vector. The matrix-vector product is then the sum of these terms over the various input directions specified by the right singular vectors. Note that the gain for an input in the direction of a right singular vector is given by the associated singular value.

The gains for a complete orthonormal basis for the input are useful in generating the range of matrix gains. The p singular values (given above) provide this information when the number of inputs is less than or equal to the number of outputs. When the number of inputs is greater than the number of outputs, there are input basis vectors that do not appear in the summation in (2.20). In this case, the gain for these additional input basis vectors is zero. To simplify future expressions, additional zero singular values are defined for these input directions. The complete set of singular values (numbering n_u) is then defined as the positive square roots of the diagonal elements of $\mathbf{S}^T\mathbf{S}$, where the superscript T denotes the transpose.

The singular vectors and singular values can be computed by solving the following pair of eigenvalue problems:

$$\mathbf{M}\mathbf{M}^\dagger U_i = \sigma_i^2 U_i; \tag{2.22a}$$

$$\mathbf{M}^\dagger \mathbf{M} V_i = \sigma_i^2 V_i. \tag{2.22b}$$

The nonzero singular values are the non-negative square roots of these eigenvalues, and they can be found from either (2.22a) or (2.22b). The complete set of singular values, including all zero singular values, consists of the non-negative square roots of the eigenvalues in (2.22b). The eigenvectors in (2.22) are the singular vectors, provided they are normalized as given in (2.21). Note that a sign ambiguity exists for the normalized eigenvectors since multiplication by -1 yields a distinct normalized eigenvector. For the SVD, the sign of the product $U_i V_i^\dagger$ is constrained by (2.20) for singular vectors associated with nonzero singular values. Therefore, the sign of one of these vectors can be chosen arbitrarily, while the sign of the other is specified by the following constraint equation:

$$\mathbf{M}V_i = \sigma_i U_i.$$

This constraint equation can also be used to simplify the computation of the SVD. For example, the right singular vectors can be computed using (2.22b), and then the left singular vectors (associated with nonzero singular values) can be found:[7]

$$U_i = \frac{1}{\sigma_i}\mathbf{M}V_i.$$

There is no sign constraint on singular vectors associated with zero singular values, and any of the associated eigenvectors can be used. Other numerically reliable and computationally efficient algorithms exist for finding the SVD,[8] but a discussion of these algorithms is beyond the scope of this text.

The singular value decomposition has the property that the matrix gain for all inputs is less than the largest singular value (denoted $\bar{\sigma}$):

$$\frac{\|\mathbf{M}x\|}{\|x\|} \le \sigma_1 = \bar{\sigma}. \tag{2.23}$$

A formal derivation of this result is given in the Appendix, see Appendix equation (A3.3). The gain bound (2.23) can be intuitively understood by noting that σ_1 is the largest gain over the orthogonal set of input directions defined by the right singular vectors. As such, the maximum gain of the matrix is achieved when the input x is proportional to V_1. The direct calculation of matrix gain with the input (αV_1) yields

$$\frac{\|\mathbf{M}(\alpha V_1)\|}{\|\alpha V_1\|} = \frac{\left\|\sum_{i=1}^{p} \sigma_i U_i V_i^\dagger (\alpha V_1)\right\|}{\|\alpha V_1\|} = \frac{\|\alpha\sigma_1 U_1\|}{\|\alpha V_1\|} = \frac{|\alpha|\sigma_1}{|\alpha|} = \sigma_1,$$

where the orthonormality of the singular vectors (2.21b) is used in generating this result.

The minimum matrix gain can also be determined from the singular value decomposition:

$$\frac{\|\mathbf{M}x\|}{\|x\|} \ge \sigma_{n_u} = \underline{\sigma} = \begin{cases} \sigma_p & \text{if } n_y \ge n_u \\ 0 & \text{if } n_y < n_u \end{cases}. \tag{2.24}$$

---

[7] The roles of V_i and U_i can be reversed by replacing $\mathbf{M}$ with $\mathbf{M}^\dagger$.

[8] See for example [2], page 427.

A formal derivation of this result is given in the Appendix, see Appendix equation (A3.6). The minimum gain equals zero whenever the rank of **M** is less than the number of inputs. A gain of zero indicates that there is an input that yields a zero output; that is, the null space of **M** is not empty. The dimension of this null space is greater than or equal to $(n_u - n_y)$, by the fundamental theorem of linear algebra (see [3], page 69). Therefore, the minimum gain is always zero whenever $n_y < n_u$. The minimum matrix gain is achieved when the input x is proportional to V_{n_u}. Further, the output for this input is proportional to U_{n_u} whenever $n_y \geq n_u$, or the output is zero (as discussed previously) when $n_y < n_u$.

An additional property of the singular value decomposition will prove useful in deriving subsequent results. The maximum singular value of a product of matrices is bounded by the maximum singular values of the individual matrices:

$$\bar{\sigma}(\mathbf{MA}) \leq \bar{\sigma}(\mathbf{M})\bar{\sigma}(\mathbf{A}) \tag{2.25}$$

This fact is derived by noting that the maximum gain of (**MA**) occurs when the input vector is the first right singular vector of (**MA**), which is the singular vector associated with the largest singular value:

$$\bar{\sigma}(\mathbf{MA}) = \frac{\|\mathbf{MA}V_1\|}{\|V_1\|} = \frac{\|\mathbf{M}(\mathbf{A}V_1)\|}{\|\mathbf{A}V_1\|}\frac{\|\mathbf{A}V_1\|}{\|V_1\|} \leq \bar{\sigma}(\mathbf{M})\bar{\sigma}(\mathbf{A}).$$

Intuitively, this inequality results from the fact that the vector V_1 maximizes the gain of the product (**MA**) but may not maximize the gain of the individual matrices **A** and **M**.

The Principal Gains The steady-state output resulting from a pure tone input $u(t) = u_0 e^{j\omega t}$ is given in terms of the transfer function matrix:

$$y(t) = \mathbf{G}(j\omega)u_0 e^{j\omega t}.$$

The matrix-vector product $\mathbf{G}(j\omega)u_0$ defines the output amplitude and, therefore, the system gain at a particular frequency and input direction. For a given frequency, the range of gains over input direction can be found from the singular value decomposition of the transfer function matrix.

The SVD of the transfer function matrix is

$$\mathbf{G}(j\omega) = \sum_{i=1}^{p} \sigma_i(\omega)U_i(\omega)V_i^{\dagger}(\omega), \tag{2.26}$$

where p equals the minimum of n_y and n_u. The frequency-dependent singular values of the transfer function matrix are called the principal gains of the system. A plot of the principal gains (which are continuous functions of frequency) provides frequency-dependent information on the maximum and minimum system gain. The principal gains also provide an indication of the likelihood of observing a particular gain for a system with many inputs and many outputs. When many of the principal gains are clustered together, this indicates that many inputs yield a similar gain and that this gain is more likely to occur.

The right singular vectors of the transfer function matrix define which inputs yield the maximum and minimum gains. The left singular vectors of the transfer function matrix define what outputs result when the maximum and minimum gains are achieved.

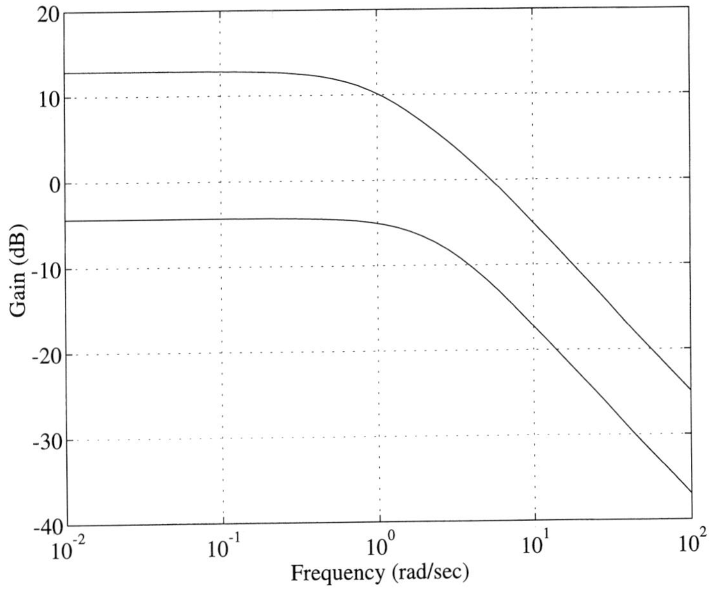

FIGURE 2.2 Principal gains for Example 2.3

EXAMPLE 2.3 Given the system described by the following state model,

$$\begin{bmatrix} \dot{x}_1(t) \\ \dot{x}_2(t) \end{bmatrix} = \begin{bmatrix} -1 & 2 \\ 0 & -3 \end{bmatrix} \begin{bmatrix} x_1(t) \\ x_2(t) \end{bmatrix} + \begin{bmatrix} 1 & 0 \\ 0 & 4 \end{bmatrix} \begin{bmatrix} u_1(t) \\ u_2(t) \end{bmatrix};$$

$$\begin{bmatrix} y_1(t) \\ y_2(t) \end{bmatrix} = \begin{bmatrix} 1 & 1 \\ 1 & -1 \end{bmatrix} \begin{bmatrix} x_1(t) \\ x_2(t) \end{bmatrix},$$

the frequency response is

$$\mathbf{G}(j\omega) = \begin{bmatrix} \dfrac{1}{j\omega + 1} & \dfrac{4}{j\omega + 1} \\ \dfrac{1}{j\omega + 1} & \dfrac{-4(j\omega - 1)}{(j\omega + 1)(j\omega + 3)} \end{bmatrix},$$

and the principal gains are plotted in Figure 2.2.

2.5 Poles, Zeros, and Modes

A number of definitions are useful when discussing frequency response and stability of linear, time-invariant systems.

2.5.1 Poles and Zeros for SISO Systems

The transfer function of a SISO system is a ratio of polynomials in the Laplace variable:

$$G(s) = \mathbf{C}(s\mathbf{I} - \mathbf{A})^{-1}\mathbf{B} + \mathbf{D} = \frac{\mathbf{C}\text{adj}(s\mathbf{I} - \mathbf{A})\mathbf{B} + \mathbf{D}\det(s\mathbf{I} - \mathbf{A})}{\det(s\mathbf{I} - \mathbf{A})} = \frac{\text{num}(s)}{\text{den}(s)}$$

where adj(•) denotes the adjugate (as defined in [3], page 163) and det(•) denotes the determinant.[9] The transfer function is termed minimal if there are no common factors in both the numerator and denominator polynomials. The state-space realization {**A**, **B**, **C**, **D**} is termed minimal if $\det(s\mathbf{I} - \mathbf{A})$ equals the denominator of the minimal transfer function. Transfer functions and state-space realizations are subsequently assumed to be minimal unless it is specifically stated to the contrary. The roots of the numerator—solutions of the equation $\text{num}(s) = 0$—are called the system zeros. The zeros can also be defined as the values of $\{z_i\}$ such that

$$G(z_i) = 0. \tag{2.27}$$

The roots of the denominator—solutions of the equation $\text{den}(s) = 0$—are called the system poles. The poles can also be defined as the values $\{p_i\}$ such that

$$|G(p_i)| = \infty$$

For minimal realizations, the poles are the solutions of the *characteristic equation:*

$$\det(s\mathbf{I} - \mathbf{A}) = 0. \tag{2.28}$$

Equation (2.28) also defines the eigenvalues of **A**. The poles are, therefore, the eigenvalues of the system state matrix.

The poles of a real system, in general, consist of real poles and complex poles. The complex poles always exist in complex conjugate pairs. The same is true for the complex zeros of a real system.

2.5.2 Poles and Zeros for MIMO Systems

The transfer function of a MIMO system is a matrix of SISO transfer functions. The system poles are defined as the union of the poles of each of these SISO transfer functions. This definition is sufficient for our purposes but does not include information on the multiplicities of the poles.[10] For a state-space realization of the system, the poles are the eigenvalues of the state matrix (2.28). For a minimal realization, the multiplicities of the poles can then be defined as equal to the multiplicities of the eigenvalues.

The zeros of a MIMO system are the values of s such that the transfer function matrix has less than full rank:

$$\text{rank}[\mathbf{G}(s)] < \min\{n_y, n_u\}. \tag{2.29}$$

This definition is a generalization of (2.27).[11]

For the special case of square transfer functions, the zeros are the values of s such that

$$\det[\mathbf{G}(s)] = 0.$$

---

[9] The adjugate matrix appearing in the inverse is called the adjoint matrix by many authors. I prefer the designation *adjugate* because adjoint, as defined in functional analysis, leads to an alternative definition of the matrix adjoint.

[10] A more precise definition of the poles of MIMO systems appears in [4], page 446.

[11] The zeros of a MIMO system are defined in a number of ways in the literature. A detailed discussion of zeros of MIMO systems appears in [4], page 446.

This condition implies that the output is zero for some nonzero input (the initial conditions are assumed to equal zero):

$$Y(s) = 0 = \mathbf{G}(s)U(s).$$

The Laplace transform of the state model (2.4) is

$$sX(s) = \mathbf{A}X(s) + \mathbf{B}U(s);$$

$$Y(s) = 0 = \mathbf{C}X(s) + \mathbf{D}U(s).$$

Rewriting these equations as a matrix equation, the system has a zero at s provided there is a nontrivial solution to the following equation:

$$\begin{bmatrix} s\mathbf{I} - \mathbf{A} & -\mathbf{B} \\ \mathbf{C} & \mathbf{D} \end{bmatrix} \begin{bmatrix} X(s) \\ U(s) \end{bmatrix} = \begin{bmatrix} 0 \\ 0 \end{bmatrix},$$

or equivalently,

$$\det \begin{bmatrix} s\mathbf{I} - \mathbf{A} & -\mathbf{B} \\ \mathbf{C} & \mathbf{D} \end{bmatrix} = 0. \quad (2.30)$$

This equation provides an alternate test for the zeros of a square system. Note that this test is only valid when the plant is both observable and controllable. Otherwise, this determinant is zero on the entire complex plane.

Certain pathological systems, termed degenerate systems, have zeros everywhere in the complex plane even when they are both observable and controllable. Degenerate systems have rank-deficient transfer functions that result from poor system modeling. These systems rarely appear in applications, and it will be assumed throughout this book that all systems are nondegenerate. The number of zeros of a nondegenerate system is less than or equal to the system order. In practice, transfer functions of systems that are square often have zeros, while transfer functions of systems that are not square rarely have any zeros.

EXAMPLE 2.4 Given the system described by the state model

$$\begin{bmatrix} \dot{x}_1(t) \\ \dot{x}_2(t) \end{bmatrix} = \begin{bmatrix} -1 & 2 \\ 0 & -3 \end{bmatrix} \begin{bmatrix} x_1(t) \\ x_2(t) \end{bmatrix} + \begin{bmatrix} 1 & 0 \\ 0 & 4 \end{bmatrix} \begin{bmatrix} u_1(t) \\ u_2(t) \end{bmatrix};$$

$$\begin{bmatrix} y_1(t) \\ y_2(t) \end{bmatrix} = \begin{bmatrix} 1 & 1 \\ 1 & -1 \end{bmatrix} \begin{bmatrix} x_1(t) \\ x_2(t) \end{bmatrix} + \begin{bmatrix} -1 & 0 \\ 0 & -1 \end{bmatrix} \begin{bmatrix} u_1(t) \\ u_2(t) \end{bmatrix},$$

the transfer function is

$$\mathbf{G}(s) = \begin{bmatrix} \dfrac{-s}{s+1} & \dfrac{4}{s+1} \\ \dfrac{1}{s+1} & \dfrac{-s^2 - 8s + 1}{(s+1)(s+3)} \end{bmatrix}.$$

The poles are the solutions of (2.28):

$$\det \begin{bmatrix} s+1 & -2 \\ 0 & s+3 \end{bmatrix} = (s+1)(s+3) = 0,$$

which are $p_1 = -1$ and $p_2 = -3$. The zeros are the solutions of

$$\det\begin{bmatrix} \dfrac{-s}{s+1} & \dfrac{4}{s+1} \\ \dfrac{1}{s+1} & \dfrac{-s^2-8s+1}{(s+1)(s+3)} \end{bmatrix} = \frac{s^2+7s-12}{(s+1)(s+3)} = 0,$$

which are $z_1 = -8.42$ and $z_2 = 1.42$. The direct computation of this determinant yields a numerator and denominator of order 3. A cancellation between these terms results in the determinant being a ratio of second-order polynomials. Cancellations always result in the determinant of the transfer function being a ratio of polynomials with orders less than or equal to the order of the state model; that is, the number of zeros is less than or equal to the order of the state model.

The zeros of this system can also be found using (2.30):

$$\det\left[\begin{array}{cc|cc} s+1 & -2 & -1 & 0 \\ 0 & s+3 & 0 & -4 \\ \hline 1 & 1 & -1 & 0 \\ 1 & -1 & 0 & -1 \end{array}\right] = s^2+7s-12 = 0.$$

Solving this equation yields the same zeros as found above. ◆

2.5.3 Modes

The initial condition response of each state and each output consists of linear combinations of terms called the natural modes, or just modes, of the system. The modes are determined by the system poles p_i:

$$\{e^{p_i t} \mid i = 1, \ldots, n_x\}, \tag{2.31}$$

where all poles are assumed to be simple poles; that is, they have a multiplicity of 1.[12] The impulse response between each input/output pair also consists of linear combinations of the modes of the system and the impulse function. For a complex conjugate pair of poles, the associated pair of complex modes, defined by (2.31), can be written in an alternative form:

$$\{e^{\mathrm{Re}(p_i)t}\cos[\mathrm{Im}(p_i)t],\ e^{\mathrm{Re}(p_i)t}\sin[\mathrm{Im}(p_i)t]\}, \tag{2.32}$$

where Re(•) and Im(•) are the real and imaginary parts of •, respectively. The initial condition response and the impulse response can be written as linear combinations of the natural modes as given by either (2.31) or (2.32) when the poles are complex. The impulse function may also appear in the impulse response.

EXAMPLE 2.5 The impulse response of the system

$$G(s) = \frac{s+1}{(s+2)(s^2+2s+5)}$$

[12] The natural modes must be modified when poles are repeated. For a pole with multiplicity k, there are k associated natural modes:

$$\{e^{p_i t}, te^{p_i t}, \ldots, t^{k-1}e^{p_i t}\}.$$

can be expressed as a linear combination of the modes:

$$\begin{aligned} g(t) &= c_1 e^{-2t} + c_2 e^{-(1-2j)t} + c_3 e^{-(1+2j)t} + c_4 \delta(t) \\ &= c_1 e^{-2t} + c_5 e^{-t} \cos(2t) + c_6 e^{-t} \sin(2t) + c_4 \delta(t). \end{aligned}$$

This expression, even without solving for the weights, contains most of the information about the impulse response that is required for control system design. If the weights are needed, they can be found by performing the inverse Laplace transform of the transfer function. Note that the weight on the impulse function equals zero whenever the transfer function is strictly proper, as in this example. ◆

2.6 Stability

An important property of feedback control systems (indeed, one that is absolutely essential) is stability. Stability guarantees that the system output remains finite if the input is finite.

DEFINITION: A system is bounded input/bounded output–stable (or simply stable) if for every bounded input,

$$|u_i(t)| < M_1 \quad \text{for all } t \text{ and all } i,$$

the output is bounded:

$$|y_j(t)| < M_2 \quad \text{for all } t \text{ and all } j,$$

provided that the initial conditions are zero.[13]

The stability of a causal, linear, time-invariant system depends on the system impulse response. The jth element of the output is a function of the input and the impulse response:

$$|y_j(t)| = \left| \sum_{i=1}^{n_u} \left\{ \int_0^t g_{ji}(\tau) u_i(t-\tau) d\tau \right\} \right|.$$

Each element of the output is bounded by the expression

$$|y_j(t)| \leq M_1 \sum_{i=1}^{n_u} \left\{ \int_0^t |g_{ji}(\tau)| d\tau \right\},$$

if and only if each element of the impulse response matrix is absolutely integrable:

$$\int_0^\infty |g_{ji}(\tau)| d\tau < \infty. \tag{2.33}$$

This equation provides a test for stability and demonstrates that the impulse response of a stable system must approach zero as time approaches infinity; that is, the output of a stable system returns to zero after being subjected to a temporary disturbance.

---

[13] An alternative definition of input-output stability, known as $\mathcal{L}_2$ stability, is given in Appendix A5. This alternative stability definition will prove useful when discussing robustness.

The elements of the impulse response matrix can be expanded as a linear combination of the natural modes and the impulse function:

$$g_{ji}(t) = \alpha_1 e^{\mathrm{Re}[p_1]t} e^{j\mathrm{Im}[p_1]t} + \cdots + \alpha_{n_x} e^{\mathrm{Re}[p_{n_x}]t} e^{j\mathrm{Im}[p_{n_x}]t} + D_{ji}\delta(t) \quad \text{for } t \geq 0.$$

The term involving the delta function is absolutely integrable. The α_i terms are constants, and the exponential of an imaginary number has a magnitude of 1 for all time. Therefore, the terms involving each pole are absolutely integrable if and only if the real part of the pole is negative, and

> A causal, linear, time-invariant system is stable if and only if all of its poles have negative real parts.[14]

This test for stability is used extensively for linear, time-invariant systems.

2.6.1 Internal Stability

The above definitions of stability are based on the input/output behavior of the system. A system may be input/output stable and still have internal signals that are unbounded. This situation is typically catastrophic to a real-world system, since these unbounded signals cause loss of linearity, damage to the system, or both. Unbounded internal signals in stable, linear, time-invariant systems are the result of internal, unstable pole-zero cancellations.

EXAMPLE 2.6 Consider the system given in Figure 2.3, where $v(t)$ is the measurement error. The transfer function between the reference input and the reference output is

$$\frac{Y(s)}{R(s)} = \frac{\dfrac{s}{s-1}}{1 + \dfrac{2}{s}\dfrac{s}{s-1}} = \frac{s}{s-1+2} = \frac{s}{s+1},$$

which represents a stable system (this system is a high-pass filter and provides good tracking of high-frequency reference inputs). The transfer function between the measurement error and the control input is

$$\frac{U(s)}{V(s)} = \frac{\dfrac{-2}{s}}{1 + \dfrac{2}{s}\dfrac{s}{s-1}} = \frac{\dfrac{-2(s-1)}{s}}{s-1+2} = \frac{-2(s-1)}{s(s+1)},$$

which is unstable. In particular, the control input is unbounded if the measurement error contains a constant bias.

[14] This result is only strictly correct for linear, time-invariant systems described by a state model. An additional requirement that the transfer function be proper (the order of the numerator be less than or equal to the order of the denominator) must be added to this test for general linear, time-invariant systems described by transfer functions. Note that state models always result in transfer function models that are proper.

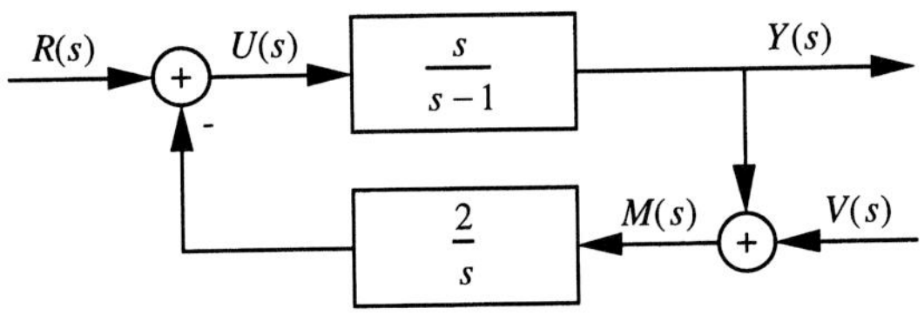

FIGURE 2.3 A system with pole-zero cancellation

A linear feedback system is termed internally stable if all internal signals and all possible outputs remain bounded given that all possible inputs are bounded. Internal stability is evaluated by considering all of the possible transfer functions associated with the feedback system. A general feedback system is shown in Figure 2.4. The inputs u_1 and u_2 are applied at each of the two possible locations between blocks. The four possible outputs y_1, y_2, e_1, and e_2 are the output of each block and the output of each summing junction. The inputs can represent input disturbances, output disturbances, reference inputs, and measurement noise. The outputs represent the plant output, the controller output, the plant input, and the controller input, respectively. The eight possible transfer functions are

$$\begin{bmatrix} y_1 \\ \hline y_2 \end{bmatrix} = \left[\begin{array}{c|c} (\mathbf{I} + \mathbf{GK})^{-1}\mathbf{G} & -(\mathbf{I} + \mathbf{GK})^{-1}\mathbf{GK} \\ \hline (\mathbf{I} + \mathbf{KG})^{-1}\mathbf{KG} & (\mathbf{I} + \mathbf{KG})^{-1}\mathbf{K} \end{array}\right] \begin{bmatrix} u_1 \\ \hline u_2 \end{bmatrix} \tag{2.34a}$$

$$= \left[\begin{array}{c|c} \mathbf{G}_{y_1u_1} & \mathbf{G}_{y_1u_2} \\ \hline \mathbf{G}_{y_2u_1} & \mathbf{G}_{y_2u_2} \end{array}\right] \begin{bmatrix} u_1 \\ \hline u_2 \end{bmatrix};$$

$$\begin{bmatrix} e_1 \\ \hline e_2 \end{bmatrix} = \left[\begin{array}{c|c} (\mathbf{I} + \mathbf{KG})^{-1} & -(\mathbf{I} + \mathbf{KG})^{-1}\mathbf{K} \\ \hline (\mathbf{I} + \mathbf{GK})^{-1}\mathbf{G} & (\mathbf{I} + \mathbf{GK})^{-1} \end{array}\right] \begin{bmatrix} u_1 \\ \hline u_2 \end{bmatrix} \tag{2.34b}$$

$$= \left[\begin{array}{c|c} \mathbf{G}_{e_1u_1} & \mathbf{G}_{e_1u_2} \\ \hline \mathbf{G}_{e_2u_1} & \mathbf{G}_{e_2u_2} \end{array}\right] \begin{bmatrix} u_1 \\ \hline u_2 \end{bmatrix}.$$

The concept of internal stability is formally defined as follows.

DEFINITION: The feedback system consisting of the plant $\mathbf{G}(s)$ and the controller $\mathbf{K}(s)$ (either in the feedback path or the forward path) is internally stable if each of the eight transfer functions in (2.34) are stable.

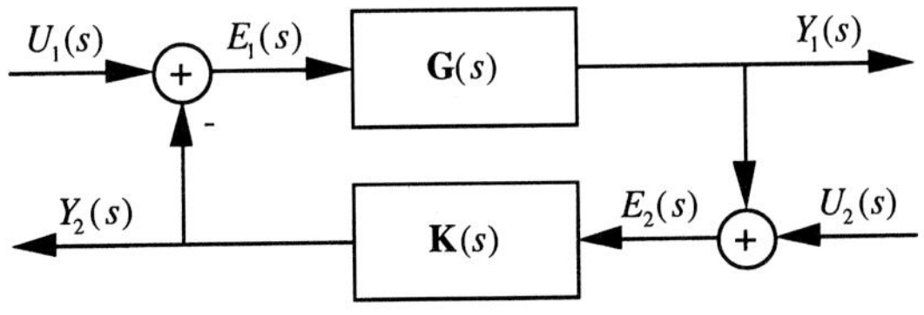

FIGURE 2.4 Block diagram for evaluating internal stability of a feedback system

Internal stability is a stronger condition than stability and will be required when designing feedback systems.

The determination of internal stability is simplified by considering:

$$\begin{bmatrix} e_1 \\ \hline e_2 \end{bmatrix} = \begin{bmatrix} u_1 \\ \hline u_2 \end{bmatrix} + \begin{bmatrix} -y_2 \\ \hline y_1 \end{bmatrix} = \left[\begin{array}{c|c} \mathbf{I} & \mathbf{0} \\ \hline \mathbf{0} & \mathbf{I} \end{array}\right] \begin{bmatrix} u_1 \\ \hline u_2 \end{bmatrix} + \left[\begin{array}{c|c} \mathbf{0} & -\mathbf{I} \\ \hline \mathbf{I} & \mathbf{0} \end{array}\right] \begin{bmatrix} y_1 \\ \hline y_2 \end{bmatrix}.$$

Solving for the vector y yields:

$$\begin{bmatrix} y_1 \\ \hline y_2 \end{bmatrix} = \left[\begin{array}{c|c} \mathbf{0} & \mathbf{I} \\ \hline -\mathbf{I} & \mathbf{0} \end{array}\right] \left(\begin{bmatrix} e_1 \\ \hline e_2 \end{bmatrix} - \left[\begin{array}{c|c} \mathbf{I} & \mathbf{0} \\ \hline \mathbf{0} & \mathbf{I} \end{array}\right] \begin{bmatrix} u_1 \\ \hline u_2 \end{bmatrix} \right).$$

The transfer functions from u to y are then related to the transfer functions from u to e:

$$\left[\begin{array}{c|c} \mathbf{G}_{y_1u_1} & \mathbf{G}_{y_1u_2} \\ \hline \mathbf{G}_{y_2u_1} & \mathbf{G}_{y_2u_2} \end{array}\right] = \left[\begin{array}{c|c} \mathbf{0} & \mathbf{I} \\ \hline -\mathbf{I} & \mathbf{0} \end{array}\right] \left(\left[\begin{array}{c|c} \mathbf{G}_{e_1u_1} & \mathbf{G}_{e_1u_2} \\ \hline \mathbf{G}_{e_2u_1} & \mathbf{G}_{e_2u_2} \end{array}\right] - \left[\begin{array}{c|c} \mathbf{I} & \mathbf{0} \\ \hline \mathbf{0} & \mathbf{I} \end{array}\right] \right).$$

Since all of the transformation matrices relating $\mathbf{G}_{e_iu_j}$ to $\mathbf{G}_{y_iu_j}$ are constants and therefore stable, the following result is obtained:

> The feedback system consisting of the plant $\mathbf{G}(s)$ and the controller $\mathbf{K}(s)$ (either in the feedback path or in the forward path) is internally stable if and only if each of the four transfer functions, $\mathbf{G}_{e_1u_1}$, $\mathbf{G}_{e_1u_2}$, $\mathbf{G}_{e_2u_1}$, and $\mathbf{G}_{e_2u_2}$ are stable.

The feedback system is also internally stable if and only if the four transfer functions from the inputs u_1 and u_2 to the outputs y_1 and y_2 are stable. As a matter of convention, however, the transfer functions from the inputs to the outputs e_1 and e_2 are typically used to evaluate internal stability.

2.7 Change of Basis: Similarity Transformations

The state equations of a system are not unique. In fact, there exists an infinite number of state representations for a given physical system. A new state model can be obtained by defining a new state vector consisting of linear combinations of the original state vector. This process is known as performing a change of basis on the state vector and results in a new state model, which is said to be related to the original model by a similarity transformation. The definitions of the system input and the system output remains unchanged, so all of these state models represent the same physical system in terms of input/output behavior. Similarity transformation can be used to generate special state models that have nice algebraic and numerical properties that aid in design and analysis tasks.

The generic state model of a system is given in Section 2.1 and is also reproduced here:

$$\dot{x}(t) = \mathbf{A}x(t) + \mathbf{B}u(t); \tag{2.4a}$$

$$y(t) = \mathbf{C}x(t) + \mathbf{D}u(t). \tag{2.4b}$$

A new state vector can be defined:

$$\tilde{x}(t) = \mathbf{T}^{-1}x(t) \tag{2.35}$$

where $\mathbf{T}$ is a constant, invertible transformation matrix.[15] The fact that $\mathbf{T}$ is invertible means that $x(t)$ can be computed from $\tilde{x}(t)$, and also $\tilde{x}(t)$ can be computed from $x(t)$. Therefore, the new state vector contains all the information contained in $x(t)$. From (2.35), we see that

$$x(t) = \mathbf{T}\tilde{x}(t);$$
$$\dot{x}(t) = \mathbf{T}\dot{\tilde{x}}(t).$$

Making these substitutions in the state model (2.4) yields

$$\mathbf{T}\dot{\tilde{x}}(t) = \mathbf{AT}\tilde{x}(t) + \mathbf{B}u(t); \tag{2.36}$$
$$y(t) = \mathbf{CT}\tilde{x}(t) + \mathbf{D}u(t).$$

Multiplying (2.36) by $\mathbf{T}^{-1}$ yields a state model in terms of the new state $\tilde{x}(t)$:

$$\dot{\tilde{x}}(t) = (\mathbf{T}^{-1}\mathbf{AT})\tilde{x}(t) + (\mathbf{T}^{-1}\mathbf{B})u(t); \tag{2.37a}$$
$$y(t) = (\mathbf{CT})\tilde{x}(t) + \mathbf{D}u(t). \tag{2.37b}$$

In summary, the new state model is generated by making the following substitutions:

$$\mathbf{A} \Rightarrow \mathbf{T}^{-1}\mathbf{AT};\ \mathbf{B} \Rightarrow \mathbf{T}^{-1}\mathbf{B};\ \mathbf{C} \Rightarrow \mathbf{CT};\ \mathbf{D} \Rightarrow \mathbf{D}. \tag{2.38}$$

These transformations are collectively known as a similarity transformation.[16]

The input and output in (2.37) are identical to the input and output in (2.4); that is, the input and output are invariant under a similarity transformation. The transfer function must be invariant under a similarity transformation since the transfer function describes how the output is related to the input. The poles and zeros are therefore also invariant under a similarity transformation.

EXAMPLE 2.7 We are given

$$\begin{bmatrix}\dot{x}_1(t)\\ \dot{x}_2(t)\end{bmatrix} = \begin{bmatrix}-3 & 1\\ 1 & -3\end{bmatrix}\begin{bmatrix}x_1(t)\\ x_2(t)\end{bmatrix} + \begin{bmatrix}2\\ 0\end{bmatrix}u(t);$$
$$y(t) = [4 \quad -2]\begin{bmatrix}x_1(t)\\ x_2(t)\end{bmatrix} + 3u(t).$$

Generate a new model with states that are the sum and difference of the original states:

$$\begin{bmatrix}\tilde{x}_1(t)\\ \tilde{x}_2(t)\end{bmatrix} = \begin{bmatrix}1 & 1\\ 1 & -1\end{bmatrix}\begin{bmatrix}x_1(t)\\ x_2(t)\end{bmatrix}.$$

The transformation matrices are then

$$\mathbf{T}^{-1} = \begin{bmatrix}1 & 1\\ 1 & -1\end{bmatrix};\ \mathbf{T} = \begin{bmatrix}\frac{1}{2} & \frac{1}{2}\\ \frac{1}{2} & -\frac{1}{2}\end{bmatrix}.$$

---

[15] More complicated equations result when using a time-varying transformation matrix, as discussed in the Appendix (see A6).

[16] The transformation of a square matrix, $\mathbf{A} \Rightarrow \mathbf{T}^{-1}\mathbf{AT}$, is also referred to as a similarity transformation.

Performing the similarity transformation yields the new model:

$$\begin{bmatrix} \dot{\tilde{x}}_1(t) \\ \dot{\tilde{x}}_2(t) \end{bmatrix} = \begin{bmatrix} -2 & 0 \\ 0 & -4 \end{bmatrix} \begin{bmatrix} \tilde{x}_1(t) \\ \tilde{x}_2(t) \end{bmatrix} + \begin{bmatrix} 2 \\ 2 \end{bmatrix} u(t);$$

$$y(t) = [1 \quad 3] \begin{bmatrix} \tilde{x}_1(t) \\ \tilde{x}_2(t) \end{bmatrix} + 3u(t).$$

Note that the new state model has no coupling between states; that is, the state matrix is diagonal. The eigenvalues (poles) of the system are then simply the diagonal elements of this state matrix. ◆

This example demonstrates that certain tasks are simpler after performing a similarity transformation. The informed reader will know that the determination of the transformation matrix that yields a diagonal state matrix typically requires more work than finding the poles. Still, similarity transformations prove useful in simplifying many calculations, including finding the poles when the state matrix is defined as a block matrix.

2.8 Controllability and Observability

The problems of control and estimation were presented in Chapter 1. The present section addresses the following questions: Under what conditions can a system be controlled from the input, and under what conditions can the state of a system be estimated from the knowledge of the input and output? The answers to these questions depend on the system being controlled/estimated and the actuators/sensors that are employed. The theory developed in answering these questions provides a guide to the selection of actuators and sensors, and also proves useful in generating controllers and estimators.

2.8.1 Controllability

The question of controllability arises in a number of applications. For example, is it possible to maintain roll control of an aircraft using throttles and rudder if the ailerons jam? Questions of this type can be addressed only after defining precisely what is meant when saying it is possible to maintain roll control. Controllability is formally defined as follows:

DEFINITION: A system is said to be controllable if and only if it is possible, by means of the input, to transfer the system from *any* initial state $x(t_0) = x_0$ to *any* other state $x(t_f) = x_f$ in a finite time $0 \leq t_f - t_0 < \infty$.

Controllability is defined in terms of the ability to drive the state to a given value, but the implications of controllability extend far beyond this simple definition. The concept of controllability is frequently encountered in linear system theory and controller design. An example of particular importance to controller design is the fact that the closed-loop poles can be placed at any desired location using state feedback if and only if the plant is controllable. Other applications of controllability will be encountered later in this text.

A simple test for controllability exists when the system is linear and time-invariant. A linear, time-invariant system is controllable if and only if the controllability matrix, which is defined as

$$\mathfrak{Q} = [\mathbf{B} \,\vdots\, \mathbf{AB} \,\vdots\, \cdots \,\vdots\, \mathbf{A}^{(n_x-1)}\mathbf{B}], \tag{2.39}$$

has full rank, that is, a rank equal to the system order n_x. This test for controllability is derived in most linear systems books (see for example [4], page 355). The controllability matrix $\mathfrak{Q} \in \Re^{n_x \times n_x n_u}$ is square when the system has a single input (i.e., $n_u = 1$). In this case, the system is controllable if and only if $\det(\mathfrak{Q})$ is not equal to zero.

2.8.2 Observability

The question of observability arises in a number of applications. For example, is it possible to determine the gasoline consumption of an automobile from a measurement of the revolutions per minute of the engine? Observability is formally defined as follows:

DEFINITION: A system is said to be observable if and only if its state $x(t_0)$, at any time t_0, can be determined from knowledge of the input and output over a finite period of time, that is, $u(t)$ and $y(t)$, where $t_0 \leq t \leq t_f$.

Observability is defined in terms of the ability to estimate the state, but the implications of observability extend far beyond this simple definition. The concept of observability is frequently encountered in linear system theory, estimator design, and controller design. An example of particular importance occurs in the design of Luenberger observers. The observer poles, which control the rate of convergence of the estimates, can be placed at any location if and only if the plant is observable. Other applications of observability will be encountered later in this text.

A simple test for observability exists for the case where the system is linear and time-invariant. A linear, time-invariant system is observable if and only if the observability matrix, which is defined as

$$\mathcal{N} = \begin{bmatrix} \mathbf{C} \\ \hdashline \mathbf{CA} \\ \hdashline \vdots \\ \hdashline \mathbf{CA}^{(n_x-1)} \end{bmatrix}, \tag{2.40}$$

has full rank. This test for observability is derived in most linear systems books (see for example [4], page 353). The observability matrix $\mathcal{N} \in \Re^{n_x n_y \times n_x}$ is square when the system has a single output (i.e., $n_y = 1$). In this case, the system is observable if and only if $\det(\mathcal{N})$ is not equal to zero.

EXAMPLE 2.8 We are given the system described by the following state model:

$$\begin{bmatrix} \dot{x}_1(t) \\ \dot{x}_2(t) \end{bmatrix} = \begin{bmatrix} -1 & 2 \\ 0 & -3 \end{bmatrix} \begin{bmatrix} x_1(t) \\ x_2(t) \end{bmatrix} + \begin{bmatrix} 1 & 0 \\ 0 & 4 \end{bmatrix} \begin{bmatrix} u_1(t) \\ u_2(t) \end{bmatrix};$$

$$\begin{bmatrix} y_1(t) \\ y_2(t) \end{bmatrix} = \begin{bmatrix} 1 & 1 \\ 1 & -1 \end{bmatrix} \begin{bmatrix} x_1(t) \\ x_2(t) \end{bmatrix}.$$

The controllability test matrix is

$$\mathfrak{Q} = \begin{bmatrix} 1 & 0 & -1 & 8 \\ 0 & 4 & 0 & -12 \end{bmatrix},$$

which has full rank (a rank of 2). The system is therefore controllable. The observability test matrix is

$$\mathcal{N} = \begin{bmatrix} 1 & 1 \\ 1 & -1 \\ -1 & -1 \\ -1 & 5 \end{bmatrix},$$

which has full rank (a rank of 2). The system is therefore observable. ◆

2.9 Observer Feedback

A simple controller design philosophy is to place the closed-loop poles at desired locations. The poles are placed at locations selected to meet specific transient performance specifications like settling time and damping ratio. Root locus is a familiar example of a method of placing the closed-loop poles. The poles can also be placed using equations involving the state model presented in this chapter. A very systematic method of pole placement within a state-space setting is observer feedback.

Observer feedback is a powerful control system design methodology that has been employed in a wide range of applications. As such, observer feedback is certainly worthy of study for its own sake. But this book is focused on optimal control system design methods. Why then present observer feedback, which is a non-optimal design method? The answer is that many of the optimal control systems developed later in this book have the same structure as the observer feedback system. While these designs are not based on pole placement, many aspects of the resulting controllers are related to the observer feedback system structure.

The observer feedback is designed with reference to a state model of the plant:

$$\dot{x}(t) = \mathbf{A}x(t) + \mathbf{B}u(t); \tag{2.41a}$$

$$m(t) = \mathbf{C}x(t) + \mathbf{D}u(t), \tag{2.41b}$$

where $u(t)$ is the control input and $m(t)$ is the measurement. The observer feedback controller is synthesized in two stages: state feedback design and observer design. The state feedback controller is designed assuming that all of the states are measured exactly. The observer is then used to estimate the state from the actual measurements. Using the estimates provided by the observer in place of the state in the state feedback controller completes the design.

2.9.1 State Feedback

Pole placement using full state feedback is one method of synthesizing a feedback controller that is tasked with driving the plant state to zero.[17] As mentioned above, state

[17] State feedback can be generalized to yield outputs that track reference inputs. A detailed discussion of modifying a state feedback regulator to yield a tracking system is given in Section 8.4.

feedback provides a component in the more general observer feedback system. State feedback is also a viable stand-alone control system design method for those plants where the entire state is measured.

The state feedback controller is designed with reference to the plant state equation (2.41a). In addition, the entire state is assumed to be available for feedback. A simple controller is then generated as a generalization of proportional control:

$$\boxed{u(t) = -\mathbf{K}x(t).} \tag{2.42}$$

The control law (2.42) is called state feedback since the entire state is used for feedback. This control law is totally specified by the gain matrix **K**.

The closed-loop state equation is obtained by substituting (2.42) into (2.41a):

$$\dot{x}(t) = (\mathbf{A} - \mathbf{BK})x(t). \tag{2.43}$$

Note that this state equation describes the system formed by combining the plant and the controller. Equation (2.43) is a homogeneous state equation; that is, it has no input. The solution of this state equation is given by (2.5):

$$x(t) = e^{(\mathbf{A}-\mathbf{BK})t}x(0).$$

The state feedback controller (2.42) drives the state to zero for arbitrary initial conditions, provided that the closed-loop poles—the eigenvalues of $(\mathbf{A} - \mathbf{BK})$—all have negative real parts.

Closed-loop control systems are required to be stable, but stability alone is not sufficient to guarantee controller acceptability. It is also desirable that the state go to zero in a reasonable amount of time and exhibit other nice properties like having a small overshoot, a given damping ratio, and so on. One method of achieving this end is to set the closed-loop poles at locations that result in these desired properties.[18] Note that pole locations that satisfy a given set of transient specifications are typically not unique. Pole locations should be selected to meet but not greatly exceed the specifications. Greatly exceeding the transient performance specifications usually results in a controller that uses unnecessarily large actuator inputs.

A state feedback gain matrix that yields the closed-loop poles $\{p_1, p_2, \ldots, p_{n_x}\}$ is obtained by solving the equation

$$\boxed{\det(s\mathbf{I} - \mathbf{A} + \mathbf{BK}) = (s - p_1)(s - p_2)\cdots(s - p_{n_x})} \tag{2.44}$$

for **K**.[19] In general, there exists a feedback gain matrix that satisfies this equation for arbitrary pole locations if the system is controllable.

EXAMPLE 2.9 An ac motor is described by the state equation

$$\dot{x}(t) = \begin{bmatrix} 0 & 1 \\ 0 & -1 \end{bmatrix} x(t) + \begin{bmatrix} 0 \\ 1 \end{bmatrix} u(t),$$

[18] The relationship between pole locations and transient performance specifications like settling time and damping ratio is discussed in Section 4.2.

[19] A number of computational algorithms have been developed for solving this equation. See, for example [1], page 309, and [5].

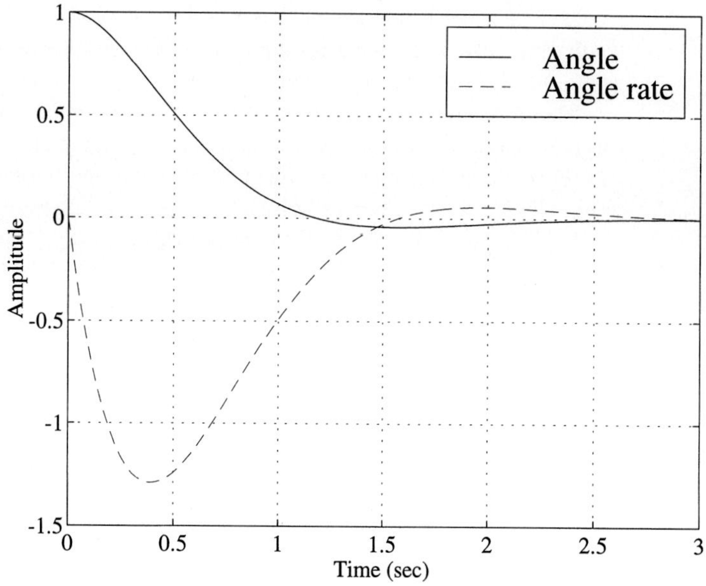

FIGURE 2.5 State feedback control of an ac motor

where the two states are motor shaft angle and angle rate, and the control input is the applied voltage. A state feedback controller is desired that yields the following closed-loop poles:[20]

$$p_{1,2} = -2 \pm 2j.$$

The feedback gain matrix is found by solving (2.44):

$$\det\left(s\begin{bmatrix} 1 & 0 \\ 0 & 1 \end{bmatrix} - \begin{bmatrix} 0 & 1 \\ 0 & -1 \end{bmatrix} + \begin{bmatrix} 0 \\ 1 \end{bmatrix}[k_1 \quad k_2]\right) = (s + 2 - 2j)(s + 2 + 2j)$$
$$= s^2 + 4s + 8.$$

Reducing, we have

$$\det\begin{bmatrix} s & -1 \\ k_1 & s + 1 + k_2 \end{bmatrix} = s^2 + (1 + k_2)s + k_1 = s^2 + 4s + 8.$$

Solving for the feedback gains, $k_1 = 8$ and $k_2 = 3$, the controller is

$$u(t) = -[8 \quad 3]x(t).$$

The closed-loop system is simulated with an initial state of $x(0) = [1 \quad 0]^T$, and the results are plotted in Figure 2.5. Note that states converge to zero, the settling time is roughly 2 seconds, and there is only a small overshoot in the motor shaft angle. ◆

[20] These poles are selected to yield a damping ratio of 0.707 and an approximate 2% settling time of 2 seconds.

A unique state feedback gain matrix resulted when employing (2.44) in this example. In general, the feedback gain matrix is uniquely determined for single-input plants. For multiple-input plants, (2.44) does not uniquely specify the controller gains. In this case, some of the gains can be selected at will by the designer, and the rest of the gains are determined. Alternatively, constraint equations involving the gains can be imposed to yield a unique solution to (2.44). In general, having additional control inputs makes the job of control easier but makes the job of designing the control system more difficult.

EXAMPLE 2.10 Consider the following generic, two-input plant:

$$\dot{x}(t) = \begin{bmatrix} 0 & 1 & 0 \\ 0 & -2 & 1 \\ -3 & 0 & -1 \end{bmatrix} x(t) + \begin{bmatrix} 0 & 0 \\ 1 & 0 \\ 0 & 1 \end{bmatrix} u(t).$$

A state feedback controller

$$u(t) = -\mathbf{K}x(t) = \begin{bmatrix} u_1(t) \\ u_2(t) \end{bmatrix} = -\begin{bmatrix} k_{11} & k_{12} & k_{13} \\ k_{21} & k_{22} & k_{23} \end{bmatrix} x(t)$$

is desired that yields the following closed-loop poles:

$$p_{1,2} = -2 \pm 3j \quad \text{and} \quad p_3 = -4.$$

The controller design consists of finding the feedback gain matrix such that (2.44) is satisfied:

$$\det(s\mathbf{I} - \mathbf{A} + \mathbf{BK}) = (s + 2 - 3j)(s + 2 + 3j)(s + 4) = s^3 + 8s^2 + 29s + 52.$$

Performing the determinant on the left side of this equation gives

$$\begin{aligned} &s^3 + (k_{12} + k_{23} + 3)s^2 + (k_{12}k_{23} + 2k_{23} + k_{12} + k_{11} + 2)s \\ &\quad + (k_{11} + k_{11}k_{23} + k_{22} - k_{22}k_{13} - 3k_{13} - k_{21}k_{13} + k_{21} + 3) \\ &= s^3 + 8s^2 + 29s + 52. \end{aligned}$$

Equating coefficients yields three equations containing the six unknown feedback gains:

$$k_{12} + k_{23} + 3 = 8;$$

$$k_{12}k_{23} + 2k_{23} + k_{12} + k_{11} + 2 = 29;$$

$$k_{11} + k_{11}k_{23} + k_{22} - k_{22}k_{13} - 3k_{13} - k_{21}k_{13} + k_{21} + 3 = 52.$$

These equations yield an infinite number of solutions for the feedback gains, all of which result in the desired closed-loop poles.

In general, an infinite number of gain matrices exist that yield the same closed-loop poles whenever the plant has multiple inputs and is controllable.

The problem of state feedback for multiple-input systems is to select one of the infinite number of gain matrices that meets the design objectives. The extra freedom in the gain selection for multiple-input systems can be used to achieve design criteria other than pole placement. For example, the extra gains (or the gain constraints) can be selected to yield reasonable scaling of the control inputs, acceptable robustness, fault tolerance to actuator failures, and so on. Unfortunately, it is often computationally quite difficult to achieve these additional design objectives within the pole placement framework. The alternative design approach of optimal control allows the selection of feedback gains that are optimized to achieve one or more of these more general design objectives.

In summary, state feedback via pole placement is a powerful methodology for controller design. The primary limitation of this method is that using state feedback requires all of the plant states to be measured. This is expensive in many applications, and even impossible in some applications. A more general design is obtained by using an observer to estimate the state and using the estimated state in the state feedback controller. Before presenting this complete design, we examine observer design.

2.9.2 Observers

The primary purpose of using an observer is to estimate the plant states that are not measured directly. The observer has the secondary benefit that it can also be used to reduce noise on states that are measured. An observer is designed to estimate the plant state in (2.41a) given the control input and the measured output in (2.41b).

The observer is a system whose state is an approximation of the plant state. A state equation for the observer is

$$\boxed{\dot{\hat{x}}(t) = \mathbf{F}\hat{x}(t) + \mathbf{G}m(t) + \mathbf{H}u(t),} \tag{2.45}$$

where $\hat{x}(t)$ is called the state estimate. The designer is free to select the matrices $\mathbf{F}$, $\mathbf{G}$, and $\mathbf{H}$. These matrices are chosen in order to drive the error between the plant state and the estimated state to zero. This error is defined mathematically:

$$e(t) = x(t) - \hat{x}(t).$$

Differentiating this error and substituting in the state equations (2.41a) and (2.45) yields the following differential equation:

$$\dot{e}(t) = \dot{x}(t) - \dot{\hat{x}}(t) = \mathbf{A}x(t) + \mathbf{B}u(t) - \mathbf{F}\hat{x}(t) - \mathbf{G}m(t) - \mathbf{H}u(t).$$

Substituting (2.41b) into this equation yields

$$\dot{e}(t) = \mathbf{A}x(t) + \mathbf{B}u(t) - \mathbf{F}\hat{x}(t) - \mathbf{GC}x(t) - \mathbf{GD}u(t) - \mathbf{H}u(t);$$

$$\dot{e}(t) = (\mathbf{A} - \mathbf{GC})x(t) - \mathbf{F}\hat{x}(t) + (\mathbf{B} - \mathbf{GD} - \mathbf{H})u(t). \tag{2.46}$$

Selecting

$$\mathbf{F} = \mathbf{A} - \mathbf{GC}, \tag{2.47}$$

(2.46) becomes a state equation for the error:

$$\dot{e}(t) = (\mathbf{A} - \mathbf{GC})e(t) + (\mathbf{B} - \mathbf{GD} - \mathbf{H})u(t). \tag{2.48}$$

A stable linear system always approaches zero as time approaches infinity when no input is present. The input can be removed from (2.48) by selecting

$$\mathbf{H} = \mathbf{B} - \mathbf{GD}. \tag{2.49}$$

Equation (2.48) then becomes the homogeneous state equation:

$$\dot{e}(t) = (\mathbf{A} - \mathbf{GC})e(t). \tag{2.50}$$

To complete the observer design, the observer gain **G** must be selected to make the system (2.50) stable. When the original system (2.41) is observable, the observer gain can be selected not only to make (2.50) stable, but also to place the poles at arbitrary locations. An observer can then be designed such that the error approaches zero at a rate that is convenient, with desired overshoot characteristics, and so forth.

The poles of (2.50) are termed the observer poles since they are also the poles of the observer (2.45). Given the desired observer poles $\{p_{o1}, p_{o2}, \ldots, p_{on}\}$, the observer gain **G** is found by solving the following equation:

$$\det(s\mathbf{I} - \mathbf{A} + \mathbf{GC}) = (s - p_{o1})(s - p_{o2}) \cdots (s - p_{on_x}). \tag{2.51}$$

Equation (2.51) yields a unique solution for the observer gain provided that the plant has a single output and is observable. Multiple gain matrices satisfy (2.51) when the plant is observable and has multiple outputs. In this case, some number of gains can be selected at will by the designer and (2.51) used to solve for the rest.

Equation (2.51) is very similar to (2.44), which is used to find the state feedback gains. In general, the estimator design problem is closely related to the control design problem and similar formulas are obtained in both cases. The similarity between observer design and controller design is embodied in the term *duality*. Duality is discussed more formally in Subsection 7.3.2.

The procedure for observer design is summarized below. Note that the plant must be observable to perform observer design via the given pole placement methodology. A set of n_x (the order of the plant) observer poles is selected that has appropriate damping and meets the requirements for the convergence rate. The observer gain matrix **G** is found using (2.51). Then **F** and **H** are found using (2.47) and (2.49), respectively. The observer is then specified by (2.45). This observer is a filter that can be implemented in either hardware or software (the digital approximation) for performing state estimation.

EXAMPLE 2.11 The ac motor in Example 2.9 is described by the state equation:

$$\dot{x}(t) = \begin{bmatrix} 0 & 1 \\ 0 & -1 \end{bmatrix} x(t) + \begin{bmatrix} 0 \\ 1 \end{bmatrix} u(t).$$

The angular position (but not the angular velocity) of the motor is measured:

$$m(t) = [1 \quad 0] \begin{bmatrix} \theta(t) \\ \dot{\theta}(t) \end{bmatrix}.$$

An observer can be constructed to estimate the plant state from this measurement and the control input. The observer poles are chosen to be

$$p_{o1} = p_{o2} = -8.$$

These poles yield an approximate settling time of 0.5 seconds. The observer gain is found by solving (2.51):

$$\det\left(\begin{bmatrix} s & 0 \\ 0 & s \end{bmatrix} - \begin{bmatrix} 0 & 1 \\ 0 & -1 \end{bmatrix} + \begin{bmatrix} g_1 \\ g_2 \end{bmatrix} [1 \quad 0]\right) = (s + 8)(s + 8);$$

$$s^2 + (1 + g_1)s + (g_1 + g_2) = s^2 + 16s + 64.$$

Equating coefficients in this polynomial and solving the resulting equations yields the observer gain matrix:

$$\mathbf{G} = \begin{bmatrix} 15 \\ 49 \end{bmatrix}.$$

The state matrix of the observer is

$$\mathbf{F} = \mathbf{A} - \mathbf{GC} = \begin{bmatrix} 0 & 1 \\ 0 & -1 \end{bmatrix} - \begin{bmatrix} 15 \\ 49 \end{bmatrix} [1 \quad 0] = \begin{bmatrix} -15 & 1 \\ -49 & -1 \end{bmatrix},$$

and the input matrix associated with the control input is

$$\mathbf{H} = \mathbf{B} - \mathbf{GD} = \begin{bmatrix} 0 \\ 1 \end{bmatrix} - \begin{bmatrix} 15 \\ 49 \end{bmatrix} 0 = \begin{bmatrix} 0 \\ 1 \end{bmatrix}.$$

The observer equation is then

$$\dot{\hat{x}}(t) = \begin{bmatrix} -15 & 1 \\ -49 & 0 \end{bmatrix} \hat{x}(t) + \begin{bmatrix} 15 \\ 49 \end{bmatrix} m(t) + \begin{bmatrix} 0 \\ 1 \end{bmatrix} u(t).$$

This observer equation can be used to generate estimates of the plant state. Note that in order to solve for the estimates, an initial condition for the estimated state must be provided. This initial condition should be the designer's best guess at the initial plant state. An initialization of zero is frequently used when no information on the plant state is available.

The observer and plant are simulated with the initial motor state $x(0) = [1 \quad -2]^T$ and the initial state estimate $\hat{x}(0) = [0 \quad 0]^T$. The results are shown in Figure 2.6. Note that the state estimate converges to the plant state in less than a second, and then

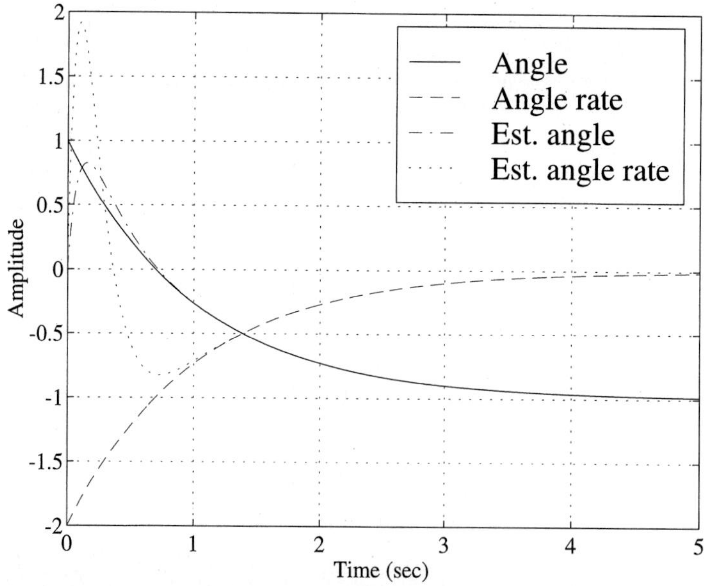

FIGURE 2.6 State estimation using an observer

continues to track the plant state. The plant state does not converge to zero in this example since the plant is unstable. ◆

2.9.3 The Deterministic Separation Principle

The closed-loop poles for the plant (2.41) can be placed at arbitrary locations, if the plant is controllable, by application of the linear, state feedback control law (2.42). This control law requires that the entire state be measured in order to generate the control. This requirement is highly restrictive since it is very costly to measure many plant states, and some plant states may not be measurable. A control law that only requires knowledge of measured outputs is therefore desirable. The plant state can be estimated from these measurements and the control inputs using an observer (2.45). A reasonable approach to controlling a system, given only partial state measurements, is to use the control law (2.42) with the estimated states generated by using the observer. The state model of the resulting controller is

$$\dot{\hat{x}}(t) = \mathbf{F}\hat{x}(t) + \mathbf{G}m(t) + \mathbf{H}u(t); \tag{2.52a}$$

$$u(t) = -\mathbf{K}\hat{x}(t). \tag{2.52b}$$

This design is known as an observer feedback controller. The closed-loop observer feedback system is shown in Figure 2.7.

EXAMPLE 2.12 An observer feedback controller for the ac motor,

$$\dot{x}(t) = \begin{bmatrix} 0 & 1 \\ 0 & -1 \end{bmatrix} x(t) + \begin{bmatrix} 0 \\ 1 \end{bmatrix} u(t);\; m(t) = [1 \quad 0] \begin{bmatrix} \theta(t) \\ \dot{\theta}(t) \end{bmatrix},$$

is obtained by combining the state feedback controller in Example 2.9 with the observer in Example 2.11. The resulting controller is described by the following state model:

$$\dot{\hat{x}}(t) = \begin{bmatrix} -15 & 1 \\ -49 & 0 \end{bmatrix} \hat{x}(t) + \begin{bmatrix} 15 \\ 49 \end{bmatrix} m(t) + \begin{bmatrix} 0 \\ 1 \end{bmatrix} u(t);\; u(t) = -[8 \quad 3]\hat{x}(t).$$

The closed-loop feedback control system is simulated with the initial motor state $x(0) = [1 \quad -2]^T$ and the initial state estimate $\hat{x}(0) = [0 \quad 0]^T$. The results are shown

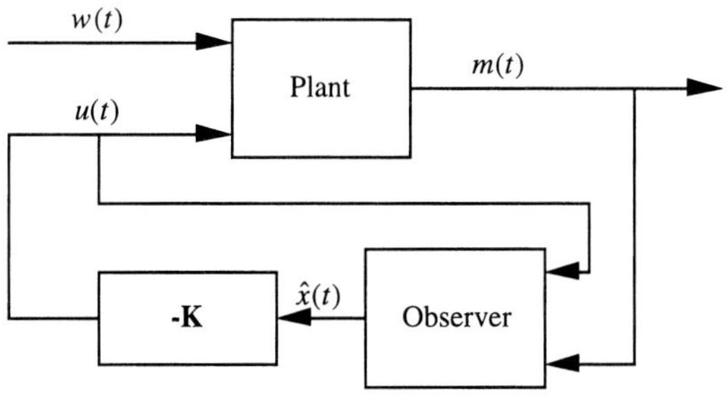

FIGURE 2.7 The observer feedback control system

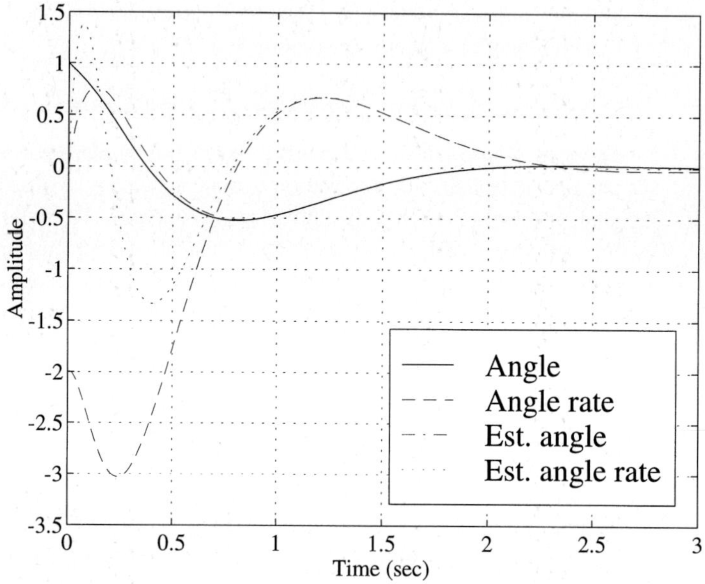

FIGURE 2.8 Observer feedback control of an ac motor

in Figure 2.8. Note that the state estimate converges to the plant state in less than a second, and the plant state converges to zero in roughly 2 seconds. ◆

The closed-loop observer feedback system is described by the plant model (2.41) along with the observer model (2.52a). These equations can be combined into a single state model for analysis:

$$\begin{bmatrix} \dot{x}(t) \\ \hline \dot{\hat{x}}(t) \end{bmatrix} = \begin{bmatrix} \mathbf{A} & \mathbf{0} \\ \hline \mathbf{0} & \mathbf{F} \end{bmatrix} \begin{bmatrix} x(t) \\ \hline \hat{x}(t) \end{bmatrix} + \begin{bmatrix} \mathbf{B} \\ \hline \mathbf{H} \end{bmatrix} u(t) + \begin{bmatrix} \mathbf{0} \\ \hline \mathbf{G} \end{bmatrix} m(t).$$

Substituting (2.41b) and (2.52b) into this equation results in

$$\begin{bmatrix} \dot{x}(t) \\ \hline \dot{\hat{x}}(t) \end{bmatrix} = \begin{bmatrix} \mathbf{A} & -\mathbf{BK} \\ \hline \mathbf{GC} & \mathbf{F} - \mathbf{HK} - \mathbf{GDK} \end{bmatrix} \begin{bmatrix} x(t) \\ \hline \hat{x}(t) \end{bmatrix}.$$

Then, substituting for $\mathbf{H}$ using (2.49) yields the closed-loop state model:

$$\begin{bmatrix} \dot{x}(t) \\ \hline \dot{\hat{x}}(t) \end{bmatrix} = \begin{bmatrix} \mathbf{A} & -\mathbf{BK} \\ \hline \mathbf{GC} & \mathbf{F} - \mathbf{BK} \end{bmatrix} \begin{bmatrix} x(t) \\ \hline \hat{x}(t) \end{bmatrix}. \tag{2.53}$$

The poles of this closed-loop system should be selected to yield good transient performance.

The poles of the observer feedback system can be found directly from (2.53), but these poles can be more simply found after performing the change of variables:

$$\begin{bmatrix} x(t) \\ \hline e(t) \end{bmatrix} = \begin{bmatrix} \mathbf{I} & \mathbf{0} \\ \hline \mathbf{I} & -\mathbf{I} \end{bmatrix} \begin{bmatrix} x(t) \\ \hline \hat{x}(t) \end{bmatrix}; \quad \begin{bmatrix} x(t) \\ \hline \hat{x}(t) \end{bmatrix} = \begin{bmatrix} \mathbf{I} & \mathbf{0} \\ \hline \mathbf{I} & -\mathbf{I} \end{bmatrix} \begin{bmatrix} x(t) \\ \hline e(t) \end{bmatrix}.$$

The transformed state equation is

$$\begin{bmatrix} \dot{x}(t) \\ \dot{e}(t) \end{bmatrix} = \begin{bmatrix} \mathbf{I} & \mathbf{0} \\ \mathbf{I} & -\mathbf{I} \end{bmatrix}^{-1} \begin{bmatrix} \mathbf{A} & -\mathbf{BK} \\ \mathbf{GC} & \mathbf{F} - \mathbf{GBK} \end{bmatrix} \begin{bmatrix} \mathbf{I} & \mathbf{0} \\ \mathbf{I} & -\mathbf{I} \end{bmatrix} \begin{bmatrix} x(t) \\ e(t) \end{bmatrix} = \begin{bmatrix} \mathbf{A} - \mathbf{BK} & \mathbf{BK} \\ \mathbf{0} & \mathbf{F} \end{bmatrix} \begin{bmatrix} x(t) \\ e(t) \end{bmatrix}.$$

Note that the poles remain unchanged when performing this similarity transformation. The closed-loop poles are then the solutions of the following equation (see Appendix equation [A2.5]):

$$\det \begin{bmatrix} s\mathbf{I} - \mathbf{A} + \mathbf{BK} & -\mathbf{BK} \\ \mathbf{0} & s\mathbf{I} - \mathbf{F} \end{bmatrix} = \det(s\mathbf{I} - \mathbf{A} + \mathbf{BK})\det(s\mathbf{I} - \mathbf{F}) = 0.$$

This equation has a solution for all values of s such that either

$$\det(s\mathbf{I} - \mathbf{A} + \mathbf{B}_1\mathbf{K}) = 0 \quad \text{or} \quad \det(s\mathbf{I} - \mathbf{F}) = 0.$$

Therefore, the poles of the closed-loop observer feedback system are equal to the poles of the state feedback system plus the poles of the observer. This result is known as the deterministic separation principle. This principle implies that a state feedback control and an observer can be designed independently. The results are then combined into an observer feedback controller. The closed-loop system that results when using this observer feedback controller will have acceptable transient response provided that the closed-loop state feedback poles and the observer poles individually yield acceptable performance.[21]

The procedure for designing observer feedback systems is summarized as follows. A set of $2n_x$ (2 times the plant order) closed-loop poles is selected that satisfy the transient performance specifications. This set is partitioned into n_x controller poles and n_x observer poles. It is customary to make the observer poles 4 to 10 times faster than the controller poles. For example, the observer poles are often chosen based on the controller poles:

$$p_{oi} = 4p_i.$$

The advantage of making the observer poles faster than the controller poles is that the observer state is a reasonable estimate of the plant state. The estimates can then be used for monitoring the system. The state feedback gain matrix is found using (2.44). The observer gain is found using (2.51), and the remaining observer matrices are found using (2.47) and (2.49). The state model for the observer feedback controller is then given by (2.52).

The state feedback controller, the observer, and the observer feedback controller are all generated using pole placement, which relates most directly to transient performance. In addition, the equations presented all assume that there are no disturbance inputs and no measurement noise. In applications where disturbances and noises are present, faster poles tend to remove more of the plant disturbances but increase the effects of the measurement noises. The pole locations can be selected as a compromise between disturbance rejection and measurement noise filtering in these applications.

---

[21] In general, modifications to settling time and overshoot may occur due to combining of the two sets of poles, but these are typically minor.

2.10 Summary

The state equations in both continuous-time and discrete-time are presented along with the general solution of the continuous-time state equation. The general solution of the state equation is shown to be composed of two terms: an initial condition solution dependent only on the initial condition of the state, and a forced solution consisting of the impulse response convolved with the input. The Laplace transfer function is defined and shown to be simply computed from the state equations. The output of linear, time-invariant systems with sinusoidal inputs is very simply related to the input. For SISO systems, this leads to the concept of frequency response. The concept of frequency response is extended to MIMO systems by the application of the singular value decomposition. This extension leads to the notion of the principal gains, which define a range of frequency-dependent gains for the system. Poles, zeros, and modes are defined for SISO and MIMO systems. Stability and internal stability are also defined and shown to depend on the system poles. A change of basis of the state vector is found to yield a new state model related to the original by a similarity transformation. The concepts of controllability and observability are defined. Practical tests for determining if a system is controllable and/or observable are given. Lastly, state feedback design via pole placement, observer design, and observer feedback design are all presented.

The fundamental equations presented in this chapter are summarized for both continuous-time and discrete-time systems in the following table.

TABLE 2.1 Summary of Formulas for Continuous-Time and Discrete-Time Systems

Formula	Continuous-Time	Discrete-Time
State equations	$\dot{x}(t) = \mathbf{A}x(t) + \mathbf{B}u(t)$ $y(t) = \mathbf{C}x(t) + \mathbf{D}u(t)$	$x(k+1) = \mathbf{\Phi}x(k) + \mathbf{\Gamma}u(k)$ $y(k) = \mathbf{C}x(k) + \mathbf{D}u(k)$ where $\mathbf{\Phi} = e^{\mathbf{A}T};\ \mathbf{\Gamma} = \int_0^T e^{\mathbf{A}\tau}\mathbf{B}d\tau$
Solution of the state equations	$y(t) = \mathbf{C}e^{\mathbf{A}t}x(0) + \int_0^t \mathbf{C}e^{\mathbf{A}(t-\tau)}\mathbf{B}u(\tau)d\tau + \mathbf{D}u(t)$ or $y(t) = \mathbf{C}e^{\mathbf{A}t}x(0) + \int_{-\infty}^{\infty} \mathbf{g}(t-\tau)u(\tau)d\tau$	$y(n) = \mathbf{C\Phi}^n x(0) + \sum_{k=0}^{n-1} \mathbf{C\Phi}^{n-1-k}\mathbf{\Gamma}u(k) + \mathbf{D}u(n)$ or $y(n) = \mathbf{C\Phi}^n x(0) + \sum_{k=-\infty}^{\infty} \mathbf{g}_d(n-k)u(k)$
Impulse Function	$\delta(t) = \begin{cases} \infty & t = 0 \\ 0 & t \neq 0 \end{cases}$	$\delta_d(k) = \begin{cases} 1 & k = 0 \\ 0 & k \neq 0 \end{cases}$
Impulse Response	$\mathbf{g}(t) = \begin{cases} \mathbf{C}e^{\mathbf{A}t}\mathbf{B} + \mathbf{D}\delta(t) & t \geq 0 \\ \mathbf{0} & t < 0 \end{cases}$	$\mathbf{g}_d(k) = \begin{cases} \mathbf{C\Phi}^{k-1}\mathbf{\Gamma} + \mathbf{D}\delta_d(k) & k \geq 0 \\ 0 & k < 0 \end{cases}$ $\mathbf{g}_d(k) \approx T\mathbf{g}(kT)$
Transform analysis	Laplace transform converts differential equations to algebraic equations: $sX(s) - x(0) = \mathbf{A}X(s) + \mathbf{B}U(s)$ $Y(s) = \mathbf{C}X(s) + \mathbf{D}U(s)$	*Z*-transform converts difference equations to algebraic equations: $zX(z) - zx(0) = \mathbf{\Phi}X(z) + \mathbf{\Gamma}U(z)$ $Y(z) = \mathbf{C}X(z) + \mathbf{D}U(z)$
Transfer functions (initial conditions equal zero)	$Y(s) = [\mathbf{C}(s\mathbf{I} - \mathbf{A})^{-1}\mathbf{B} + \mathbf{D}]U(s) = \mathbf{G}(s)U(s)$	$Y(z) = [\mathbf{C}(z\mathbf{I} - \mathbf{\Phi})^{-1}\mathbf{\Gamma} + \mathbf{D}]U(z) = \mathbf{G}_d(z)U(z)$

TABLE 2.1 *(cont.)*

Formula	Continuous-Time	Discrete-Time
Frequency response of SISO systems	$u(t) = \cos(\omega t) \Rightarrow$ $y(t) = \lvert G(j\omega)\rvert \cos(\omega t + \angle G(j\omega))$	$u(k) = \cos(\omega Tk) \Rightarrow$ $y(k) = \lvert G_d(e^{j\omega T})\rvert \cos(\omega Tk + \angle G_d(e^{j\omega T}))$
Principal gains	Singular values of $\mathbf{G}(j\omega)$	Singular values of $\mathbf{G}_d(e^{j\omega T})$
Poles	The eigenvalues of $\mathbf{A}$: $\{p_1, p_2, \ldots, p_{n_x}\}$	The eigenvalues of $\mathbf{\Phi}$: $\{p_{z1}, p_{z2}, \ldots, p_{zn_x}\} = e^{p_1T}, e^{p_2T}, \ldots, e^{p_{n_x}T}\}$
Zeros	$\text{rank}[\mathbf{G}(s_0)] < \min\{n_y, n_u\}$	$\text{rank}[\mathbf{G}_d(z_0)] < \min\{n_y, n_u\}$
Modes	$\{e^{p_1t}, e^{p_2t}, \ldots, e^{p_{n_x}t}\}$	$\{p_{z1}^k, p_{z2}^k, \ldots, p_{zn_x}^k\}$
Stability	System is stable if and only if all poles have negative real parts (are in the left half of the complex plane).	System is stable if and only if all poles have a magnitude of less than 1 (are within the unit circle in the complex plane).
Change of basis	$\mathbf{A} \Rightarrow \mathbf{T}^{-1}\mathbf{AT}$; $\mathbf{B} \Rightarrow \mathbf{T}^{-1}\mathbf{B}$; $\mathbf{C} \Rightarrow \mathbf{CT}$; $\mathbf{D} \Rightarrow \mathbf{D}$	$\mathbf{\Phi} \Rightarrow \mathbf{T}^{-1}\mathbf{\Phi T}$; $\mathbf{\Gamma} \Rightarrow \mathbf{T}^{-1}\mathbf{\Gamma}$; $\mathbf{C} \Rightarrow \mathbf{CT}$; $\mathbf{D} \Rightarrow \mathbf{D}$
Controllability	Controllable if and only if $\mathfrak{Q} = [\mathbf{B} \mid \mathbf{AB} \mid \cdots \mid \mathbf{A}^{(n_x-1)}\mathbf{B}]$ has full rank.	Controllable if and only if $\mathfrak{Q}_d = [\mathbf{\Gamma} \mid \mathbf{\Phi\Gamma} \mid \cdots \mid \mathbf{\Phi}^{(n_x-1)}\mathbf{\Gamma}]$ has full rank.
Observability	Observable if and only if $\mathcal{N} = \begin{bmatrix} \mathbf{C} \\ \mathbf{CA} \\ \vdots \\ \mathbf{CA}^{(n_x-1)} \end{bmatrix}$ has full rank.	Observable if and only if $\mathcal{N}_d = \begin{bmatrix} \mathbf{C} \\ \mathbf{C\Phi} \\ \vdots \\ \mathbf{C\Phi}^{(n_x-1)} \end{bmatrix}$ has full rank.
State feedback via pole placement	$u(t) = -\mathbf{K}x(t)$ where $\det(s\mathbf{I} - \mathbf{A} + \mathbf{BK}) = (s - p_1) \cdots (s - p_{n_x})$	$u(k) = -\mathbf{K}_d x(k)$ where $\det(s\mathbf{I} - \mathbf{\Phi} + \mathbf{\Gamma K}_d) = (s - p_{z1}) \cdots (s - p_{zn_x})$
Observers	$\dot{\hat{x}}(t) = \mathbf{F}\hat{x}(t) + \mathbf{G}m(t) + \mathbf{H}u(t)$ where $\mathbf{F} = \mathbf{A} - \mathbf{GC}$; $\mathbf{H} = \mathbf{B} - \mathbf{GD}$; $\det(s\mathbf{I} - \mathbf{A} + \mathbf{GC}) = (s - p_{o1}) \cdots (s - p_{on_x})$	$\hat{x}(k+1) = \mathbf{F}_d\hat{x}(k) + \mathbf{G}_d m(k) + \mathbf{H}_d u(k)$ where $\mathbf{F}_d = \mathbf{\Phi} - \mathbf{G}_d\mathbf{C}$; $\mathbf{H}_d = \mathbf{\Gamma} - \mathbf{G}_d\mathbf{D}$; $\det(s\mathbf{I} - \mathbf{A} + \mathbf{GC}) = (s - p_{zo1}) \cdots (s - p_{zon_x})$
Observer feedback	$\dot{\hat{x}}(t) = \mathbf{F}\hat{x}(t) + \mathbf{G}m(t) + \mathbf{H}u(t)$ $u(t) = -\mathbf{K}\hat{x}(t)$	$\hat{x}(k+1) = \mathbf{F}_d\hat{x}(k) + \mathbf{G}_d m(k) + \mathbf{H}_d u(k)$ $u(k) = -\mathbf{K}_d x(k)$

REFERENCES

[1] G. F. Franklin, J. D. Powell, and M. L. Workman, *Digital Control of Dynamic Systems,* 2d ed., Addison-Wesley, Menlo Park, CA, 1990.

[2] G. Golub and C. Van Loan, *Matrix Computations,* John Hopkins Press, Baltimore, MD, 1983.

[3] G. Strang, *Linear Algebra and Its Applications,* Academic Press, New York, 1976.

[4] T. Kailath, *Linear Systems,* Prentice-Hall, Englewood Cliffs, NJ, 1980.

[5] J. Kautsky and N. K. Nichols, "Robust pole assignment in linear state feedback," *International Journal of Control* 41 (1985): 1129–55.

[6] Y. S. Hung and A. G. J. MacFarlane, *Multivariable Feedback: A Quasi-Classical Approach,* Lecture Notes in Control and Information Sciences, vol. 40, Springer Verlag, Berlin, 1982.

Some additional references on linear systems and multivariable linear systems follow.

[7] C.-T. Chen, *Linear System Theory and Design,* Holt, Rinehart, and Winston, New York, 1984.

[8] C. A Desoer and F. M. Callier, *Multivariable Feedback Systems,* Springer-Verlag, New York, 1982.

[9] J. M. Maciejowski, *Multivariable Feedback Design,* Addison-Wesley, Reading, MA, 1989.

[10] W. J. Rugh, *Linear System Theory,* Prentice-Hall, Englewood Cliffs, NJ, 1993.

[11] M. Vidyasagar, *Control System Synthesis: A Factorization Approach,* MIT Press, Cambridge, MA, 1985.

[12] W. A. Wolovich, *Linear Multivariable Systems,* Springer-Verlag, New York, 1974.

EXERCISES

2.1 Show that

$$x(t) = \mathbf{\Phi}(t, t_0)x_0 + \int_{t_0}^{t} \mathbf{\Phi}(t, \tau)\mathbf{B}(\tau)u(\tau)d\tau$$

is a solution to the initial value problem,

$$\dot{x}(t) = \mathbf{A}(t)x(t) + \mathbf{B}(t)u(t);\ x(t_0) = x_0,$$

where $\mathbf{\Phi}(t, t_0)$ is the state-transition matrix of the system. *Hint:* Use the properties of the state transition matrix to verify that the solution satisfies the initial condition. Then substitute the solution into the time-varying state equation to verify that it satisfies the differential equation. Assume that the solution of the differential equation exists and is unique.

2.2 Using the power series representation of the state transition matrix, show that

a. $$e^{\mathbf{A}0} = \mathbf{I};$$

b. $$\frac{de^{-\mathbf{A}t}}{dt} = -e^{-\mathbf{A}t}\mathbf{A};$$

c. $$e^{\mathbf{A}t}e^{-\mathbf{A}\tau} = e^{\mathbf{A}(t-\tau)}.$$

2.3 Derive the general solution of the time-invariant state equation:

$$x(t) = e^{\mathbf{A}t}x(0) + \int_{0}^{t} e^{\mathbf{A}(t-\tau)}\mathbf{B}u(\tau)d\tau.$$

Hint: Multiply the state equation by $e^{-\mathbf{A}t}$ and integrate from 0 to t. Use the properties given in Exercise 2.2, and note that a proper combination of terms yields a perfect differential.

2.4 Given the state model in (2.4), do the following.

a. Show that the impulse response of this system is

$$\mathbf{g}(t) = \{\mathbf{C}e^{\mathbf{A}t}\mathbf{B} + \mathbf{D}\delta(t)\}1(t).$$

b. Take the Laplace transform of the impulse response to find the transfer function of the system. Compare this transfer function to that obtained in Section 2.3.

2.5 Given the system described by the state model

$$\begin{bmatrix} \dot{x}_1(t) \\ \dot{x}_2(t) \end{bmatrix} = \begin{bmatrix} 0 & 1 \\ -4 & -5 \end{bmatrix}\begin{bmatrix} x_1(t) \\ x_2(t) \end{bmatrix} + \begin{bmatrix} 0 \\ 1 \end{bmatrix}u(t);$$

$$\begin{bmatrix} y_1(t) \\ y_2(t) \end{bmatrix} = \begin{bmatrix} 1 & 0 \\ 0 & 10 \end{bmatrix}\begin{bmatrix} x_1(t) \\ x_2(t) \end{bmatrix},$$

do the following.

a. Compute the transfer function matrix.

b. Compute the poles, zeros, and modes.

c. Compute the impulse response matrix. How does this relate to the poles and modes?
d. Compute the frequency response matrix and the principal gains. How do the principal gains compare to the gains from the input to each output?

2.6 Given the system described by the state model

$$\begin{bmatrix} \dot{x}_1(t) \\ \dot{x}_2(t) \end{bmatrix} = \begin{bmatrix} -2 & 1 \\ 1 & -2 \end{bmatrix} \begin{bmatrix} x_1(t) \\ x_2(t) \end{bmatrix} + \begin{bmatrix} 2 & 0 \\ 0 & 4 \end{bmatrix} \begin{bmatrix} u_1(t) \\ u_2(t) \end{bmatrix};$$

$$\begin{bmatrix} y_1(t) \\ y_2(t) \end{bmatrix} = \begin{bmatrix} 1 & 1 \\ 1 & -1 \end{bmatrix} \begin{bmatrix} x_1(t) \\ x_2(t) \end{bmatrix},$$

do the following.

a. Compute the transfer function matrix.
b. Compute the poles, zeros, and modes.
c. Compute the impulse response matrix. How does this relate to the poles and modes?
d. Compute the frequency response matrix and the principal gains.
e. Compute the output due to an impulse applied to input 1.
f. Compute the output and the gain when the input is $u(t) = [0 \quad \cos(t)]^T$. How does this compare to the principal gains?
g. Compute the output $y_2(t)$ when the input is $u(t) = [\cos(t) \quad 0]^T$. Compute the gain from this input to the output $y_2(t)$ and compare to the principal gains. Does this make sense?

2.7 Compute and sketch the principal gains of the following systems.

a.
$$\mathbf{G}(s) = \begin{bmatrix} \dfrac{s+2}{s^2+5s+4} \\[2ex] \dfrac{2}{s^2+2s+5} \end{bmatrix}.$$

b.
$$\mathbf{G}(s) = \begin{bmatrix} \dfrac{1}{s+1} & \dfrac{1}{(s+1)(s+2)} \\[2ex] \dfrac{1}{s(s+1)} & \dfrac{1}{s+2} \end{bmatrix}.$$

c.
$$\mathbf{G}(s) = \begin{bmatrix} \dfrac{1}{s+1} & \dfrac{1}{(s+1)(s+2)} & \dfrac{1}{s+1} \\[2ex] \dfrac{1}{s(s+1)} & \dfrac{1}{s+2} & \dfrac{1}{s(s+2)} \end{bmatrix}.$$

For each of these systems, find the input that results in the maximum gain at a frequency of 2 rad/sec.

2.8 The singular value decomposition of the transfer function matrix of a real system is given in (2.26). Show that the gain of this system is σ_1 when the input signal is

$$u(t) = \begin{bmatrix} |V_{11}|\cos(\omega t + \angle V_{11}) \\ |V_{21}|\cos(\omega t + \angle V_{21}) \\ \vdots \\ |V_{n_u 1}|\cos(\omega t + \angle V_{n_u 1}) \end{bmatrix},$$

where

$$V_1(\omega) = \begin{bmatrix} V_{11} \\ V_{21} \\ \vdots \\ V_{n_u 1} \end{bmatrix} = \begin{bmatrix} |V_{11}|e^{j\angle V_{11}} \\ |V_{21}|e^{j\angle V_{21}} \\ \vdots \\ |V_{n_u 1}|e^{j\angle V_{n_u 1}} \end{bmatrix}.$$

2.9 Determine the internal stability of the feedback system in Figure 2.4 with

a.
$$G(s) = \frac{1}{s(s+1)}; \quad K(s) = \frac{s+2}{s+8};$$

b.
$$G(s) = \frac{1}{s(s+1)};\ K(s) = \frac{s}{s+4};$$

c.
$$\mathbf{G}(s) = \begin{bmatrix} \dfrac{1}{s(s+1)} \\ \dfrac{1}{(s+1)} \end{bmatrix};\ \mathbf{K}(s) = [2 \quad 2].$$

2.10 Given the system described by the state equations in Exercise 2.6, perform a similarity transformation with

$$\mathbf{T} = \begin{bmatrix} 1 & 1 \\ -1 & 1 \end{bmatrix}.$$

Is the new system description simpler than the original?

2.11 Show that a linear, time-invariant system is controllable only if

$$\text{rank}(\mathcal{Q}) = n_x.$$

Hint: Write the state using (2.5) and expand the matrix exponential (within the integral) in this expression as a power series. Write the integral as a linear combination of terms involving the powers of **A** with the coefficients being scalar integrals. Then use the Cayley-Hamilton theorem, which states that any power of **A** can be expressed as

$$\mathbf{A}^k = \alpha_0\mathbf{I} + \alpha_1\mathbf{A} + \alpha_2\mathbf{A}^2 + \cdots + \alpha_{n_x-1}\mathbf{A}^{(n_x-1)},$$

to reduce the infinite power series to a finite power series.

2.12 Show that a linear, time-invariant system is observable only if

$$\text{rank}(\mathcal{N}) = n_x.$$

Hint: Write an expression for the output minus the output due to the input:

$$z(t) = y(t) - \int_0^t \mathbf{C}e^{\mathbf{A}(t-\tau)}\mathbf{B}u(\tau)d\tau - \mathbf{D}u(t) = \mathbf{C}e^{\mathbf{A}t}x(0).$$

This expression is valid at all times, specifically at $t = \{t_1, t_2, \cdots, t_{n_x}\}$. Write a set of equations for $z(t)$ at these times. Then, write the matrix exponential in these equations as a finite power series using the Cayley-Hamilton theorem (see Exercise 2.11).

2.13 Given the system described by the following state model,

$$\begin{bmatrix} \dot{x}_1(t) \\ \dot{x}_2(t) \end{bmatrix} = \begin{bmatrix} -2 & 0 \\ 0 & -2 \end{bmatrix}\begin{bmatrix} x_1(t) \\ x_2(t) \end{bmatrix} + \begin{bmatrix} 1 \\ 1 \end{bmatrix}u(t);$$

$$y(t) = [1 \quad 0]\begin{bmatrix} x_1(t) \\ x_2(t) \end{bmatrix},$$

determine the following.

a. Is it controllable?
b. Is it observable?

2.14 Given the system described by the state model,

$$\begin{bmatrix} \dot{x}_1(t) \\ \dot{x}_2(t) \\ \dot{x}_3(t) \end{bmatrix} = \begin{bmatrix} 0 & 1 & 0 \\ 0 & -2 & 1 \\ 0 & 0 & 1 \end{bmatrix}\begin{bmatrix} x_1(t) \\ x_2(t) \\ x_3(t) \end{bmatrix} + \begin{bmatrix} 0 & 0 \\ 1 & 0 \\ 0 & 1 \end{bmatrix}u(t);$$

$$\begin{bmatrix} y_1(t) \\ y_2(t) \end{bmatrix} = \begin{bmatrix} 0 & 1 & 0 \\ 0 & 0 & 1 \end{bmatrix}\begin{bmatrix} x_1(t) \\ x_2(t) \\ x_3(t) \end{bmatrix},$$

determine the following.

a. Is it controllable?
b. Is it observable?

2.15 Given the plant,

$$\dot{x}(t) = \begin{bmatrix} 0 & 1 \\ -1 & -1 \end{bmatrix} x(t) + \begin{bmatrix} 0 \\ 1 \end{bmatrix} u(t); \; m(t) = [1 \quad 0]x(t),$$

generate an observer feedback controller to place the dominant poles at $p_1 = -2 + 2j$ and $p_2 = -2 - 2j$. The observer poles should be chosen to be four times the dominant poles.

2.16 The observer feedback controller of (2.52) is applied to a plant that incorporates a disturbance input and measurement noise:

$$\dot{x}(t) = \mathbf{A}x(t) + \mathbf{B}_u u(t) + \mathbf{B}_w w(t); \; m(t) = \mathbf{C}_m x(t) + v(t).$$

Generate a state equation for the closed-loop system, using the following state:

$$\begin{bmatrix} x(t) \\ \hat{x}(t) \end{bmatrix}.$$

Note that this state equation should have both $w(t)$ and $v(t)$ as inputs.

COMPUTER EXERCISES

2.1 Computer Simulation of an Antenna Altitude Control System
The altitude control system for the antenna in Figure P2.1 is designed to drive the altitude angle error to zero. The antenna is described by the following differential equation:

$$J\ddot{\theta}(t) + G\dot{\theta}(t) = u(t) + w(t),$$

where θ is the altitude angle error in radians, $J = 1000$ kg-m^2 is the moment of inertia of the antenna, $G = 5$ kg-m^2/s is the friction, u is the control torque, and w is a disturbance torque. The control torque is generated from the altitude and altitude rate:

$$u(t) = -2000\theta(t) - 800\dot{\theta}(t).$$

a. Generate the state equations for this system with the disturbance torque as an input. Write a program to simulate the system in discrete-time with a sampling time of 0.01 seconds. Use the discrete-time state model of (2.10) and (2.11).
b. What are the poles and natural modes of the system?
c. Compute the system impulse response, using

$$g(t) = \mathbf{C}e^{\mathbf{A}t}\mathbf{B} + \mathbf{D}\delta(t).$$

Simulate the system to generate the impulse response, plot the result, and compare the result to the calculated impulse response.

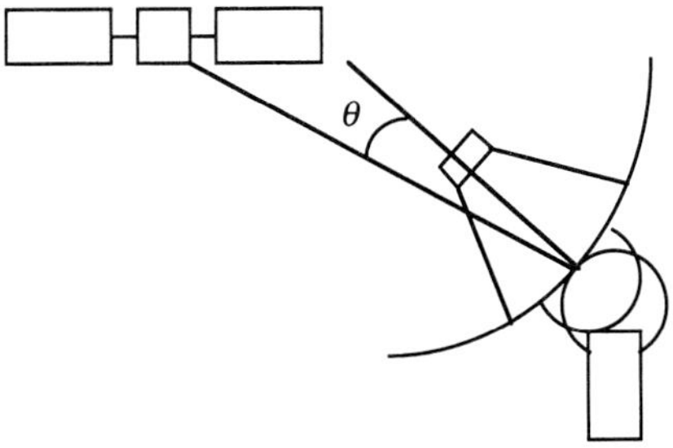

FIGURE P2.1 The antenna in Computer Exercise 2.1

d. Compute the steady-state response of the system to a unit step. Simulate the system to generate the step response and plot.

e. Compute the system's steady-state response to the sinusoidal disturbance torque:

$$w(t) = \cos(2t).$$

Simulate the system to generate this response and plot.

2.2 Principal Gain Computation for an Airplane

An autopilot for the pitch-plane dynamics of an airplane controls the altitude, pitch angle, and speed of the airplane. The autopilot selects the two inputs,

$$u(t) = \begin{bmatrix} u_1(t) \\ u_2(t) \end{bmatrix} = \begin{bmatrix} \textit{Thrust (newtons)} \\ \textit{Elevator angle (deg)} \end{bmatrix},$$

given the three measured outputs:

$$y(t) = \begin{bmatrix} y_1(t) \\ y_2(t) \\ y_3(t) \end{bmatrix} = \begin{bmatrix} \textit{Altitude} \text{ (m)} \\ \textit{Speed} \text{ (m/sec)} \\ \textit{Pitch angle (deg)} \end{bmatrix}.$$

A linearized model of the open-loop aircraft is

$$\dot{x}(t) = \begin{bmatrix} 0 & 0 & 1 & 0 & -1 \\ 0 & -0.05 & -0.2 & 0 & 0.1 \\ 0 & 0 & 0 & 1 & 0 \\ 0 & 0.05 & 0 & -1 & -1 \\ 0 & -0.3 & 0 & 1 & -0.7 \end{bmatrix} x(t) + \begin{bmatrix} 0 & 0 \\ 0.001 & 0 \\ 0 & 0 \\ 0 & -2 \\ 0 & -0.1 \end{bmatrix} u(t);$$

$$y(t) = \begin{bmatrix} 1 & 0 & 0 & 0 & 0 \\ 0 & 1 & 0 & 0 & 0 \\ 0 & 0 & 1 & 0 & 0 \end{bmatrix} x(t).$$

This model is a modification of the model presented in Appendix F of [6].

Compute the principal gains of this open-loop system. Find the maximum gain, the real input that results in this gain, and the real output that results from this input at a frequency of 0.1 rad/sec. Simulate the system with this input and plot the results. Compare the output of the simulation with the expected output. Compute the gain from the simulated output and compare with the expected gain. Find the minimum gain, the real input that results in this minimum gain, and the real output that results from this input at a frequency of 0.1 rad/sec. Simulate the system with this input, plot the results, and compute the gain from these plots. Compare the output of the simulation with the expected output. Compute the gain from the simulated output and compare with the expected gain. *Caution:* This system is unstable since it has a pole at zero. This pole can lead to constant biases in the outputs due to the initial conditions, even if the initial conditions equal zero. Ignore any constant bias when computing the gain due to a sinusoidal input.

2.3 Observer Feedback Design of a Ship Autopilot

An autopilot for a ship is tasked with maintaining the ship's heading. A very simple mathematical model of the ship is given by the following differential equations:

$$M\ddot{\theta}(t) = -d\dot{\theta}(t) - c\alpha(t) + w(t);$$

$$\dot{\alpha}(t) = -0.1\alpha(t) + 0.1\alpha_c(t)$$

where $\theta(t)$ is the heading error (the angle between the ship's true heading and the desired heading), $\alpha(t)$ is the rudder angle, $w(t)$ is a disturbance torque caused by the wind, and $\alpha_c(t)$ is the commanded rudder angle. A sketch of the ship is given in Figure P2.2. Note that all angles are positive as shown. The parameters in the ship model follow: $M = 10^7$ kg-m^2 is the moment of inertia of the ship about a vertical axis through the center of gravity, $d = 10^6$ N-m-sec/rad is the drag coefficient associated with ship rotation, and $c = 5000$ N-m/rad is a coefficient relating rudder angle to the imparted torque.

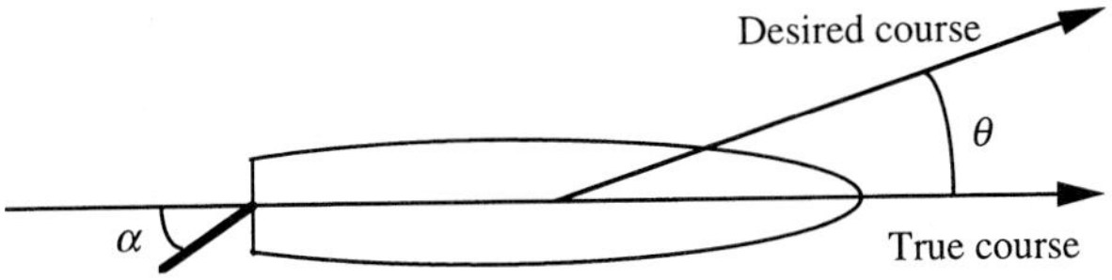

FIGURE P2.2 The ship in Computer Exercise 2.2

The angle $\theta(t)$ is measured using a compass. The autopilot should take this measurement and generate a rudder angle command for steering the ship. The observer feedback methodology is used to design this autopilot. This methodology requires that a state feedback controller be generated, an observer be generated, and then these components combined. Before generating the combination, we test the individual components using computer simulation.

Generate state feedback controllers to place the poles as follows:

a. $$p_1 = p_2 = -0.02,\ p_3 = -0.1;$$

b. $$p_1 = p_2 = -0.04,\ p_3 = -0.1;$$

c. $$p_1 = -0.02 + 0.03j,\ p_2 = -0.02 - 0.03j,\ p_3 = -0.1.$$

Note that the pole at -0.1 is selected to match the plant pole associated with the rudder dynamics. This pole has little effect on the results since the closed-loop system response is dominated by the poles closest to the imaginary axis. Perform two simulations for each set of poles. The first should have an initial heading error $\theta(0) = 0.1$ and no disturbance input. The second should have no initial heading error and a constant disturbance input $w(t) = 1000$. Plot the heading error $\theta(t)$ and the rudder deflection $\alpha(t)$ for each state feedback controller with both simulation conditions. Comment on the effects of changing the pole locations on the controller's performance. Is it practical to use a control system that yields overshoot in the heading error?

Design observers for the ship that have the following pole locations:

a. $$p_1 = p_2 = -0.08,\ p_3 = -0.4;$$

b. $$p_1 = p_2 = -0.16,\ p_3 = -0.4.$$

With no control, simulate of the plant and the observer. The initial condition for the observer is always zero. For the plant, use the two sets of simulation conditions given in the previous paragraph. For each observer and each simulation condition, plot the plant state and the estimated state. Be sure to plot the estimates along with the actual states for ease of comparison. Comment on the effects of changing the pole locations on the observer's performance.

Generate an observer feedback controller by combining the state feedback controller (with the poles in part a) and the observer (with the poles in part a). For each simulation condition, plot the plant state, the estimated state, and the commanded rudder deflection. Comment on the performance of this controller and the quality of the state estimates.

3 Vector Random Processes

Random processes, or signals, appear in control system analysis as models of reference inputs, disturbance inputs, and measurement noise. The angular position of an airplane being tracked by a radar is an example of a random reference input. Note that while the pilot may know where the plane is going, the track is not known *a priori* at the radar and is therefore random. Torque on the radar antenna, generated by a gusty wind, is a random disturbance input. Measurement noise also appears in this radar system due to thermal electrons in the radar receiver. The analysis of this radar tracking system, and many other control systems, therefore, requires the analysis of random signals.

The linear quadratic Gaussian (LQG) controller minimizes the sum of the output variances when the disturbance inputs and measurement errors are random. In particular, the disturbance inputs and measurement errors are assumed to be white noise when optimizing the LQG controller. White noise is used in this application because it provides a good model for many disturbance inputs and for measurement error. In addition, nonwhite random inputs can be generated as the output of a shaping filter driven by white noise. Optimal controls can then be generated for plants with nonwhite random inputs by using LQG results on augmented plants formed by combining shaping filters with the plant. The design and analysis of LQG controllers therefore requires a thorough understanding of the response of linear systems to white noise inputs.

A brief overview of random signal theory is presented in this chapter. This is intended not as a complete portrayal of this theory, but as a development of the tools required to perform the control system design and analysis tasks in the remainder of this book. The chapter begins with a review of random process basics, which includes the definitions of moments, correlation, orthogonality, conditional expectation, stationarity, spectral density, and white noise. These definitions are all provided for the case of vector random processes. A fairly complete treatment of the response of linear systems to white noise inputs is then presented. The chapter concludes with a section on the generation and application of shaping filters.

3.1 The Description of Vector Random Processes

A vector random process, or simply a random process, is defined as a mapping between the outcome of a random experiment and a vector-valued function. For example, the random experiment may consist of monitoring the temperature at several points on the Earth's surface over the course of a year. The individual temperature measurements at given points in time are related random variables, and collectively, these measurements form a vector-valued function of time. Random functions of time are also referred to as random signals.

The collection of all possible vector-valued functions that can result when performing the random experiment is known as the ensemble. A specific outcome of the random experiment is called a sample function. The full probabilistic description of the random process consists of the probability of occurrence of every sample function and every possible event. The events consist of any subset of the ensemble that includes a particular random variable lying within a range of possible values, any pair of random variables lying within a region, any three random variables lying within a region, and so on. The full specification of these probabilities requires that the density function, or the distribution function, for each random variable be specified. In addition, the joint density function for any pair of random variables must be specified, the joint density function for any three random variables must be specified, and so on. This probabilistic description of a random process contains an exhaustive amount of information that can only be fully enumerated in a few special cases.

A simpler description of the random process, which consists of first- and second-order moments is therefore used in practice. The first and second moments are ensemble averages of the individual random variables that make up the random process and pairs of these random variables, respectively. This description is not a complete description of the random process, in general, but still contains a wealth of useful information.

3.1.1 Second-Moment Analysis

The first moments, or the mean, of a random process are the expected values of the random variables that form the process. The second moments, or correlations, of the process are the expected values of these random variables squared and the expected values of products of pairs of these random variables. The correlation function is a function composed of the correlations within a random process. The analysis of a random process, via the specification of the mean and correlation function, is known as second-moment analysis.

The mean and correlation function, along with some additional second-order expectations, are formally defined below. The mean of a random process $x(t)$ is the expected value of the random vector:

$$m_x(t) = E[x(t)] = \int_{-\infty}^{\infty} x f_x(x; t) dx = E\left\{\begin{bmatrix} x_1(t) \\ \vdots \\ x_{n_x}(t) \end{bmatrix}\right\} = \begin{bmatrix} \int_{-\infty}^{\infty} x_1 f_{x_1}(x_1; t) dx_1 \\ \vdots \\ \int_{-\infty}^{\infty} x_{n_x} f_{x_{n_x}}(x_{n_x}; t) dx_{n_x} \end{bmatrix} \tag{3.1}$$

where $E[x(t)]$ is the expected value of $x(t)$, $f_x(x; t)$ is the density function of the random vector $x(t)$, the first integral in the equation represents an nth order multiple integral, and $f_{x_i}(x_i; t)$ is the density function for the ith element of this random vector.[1]

[1] Probability texts typically refer to $f_x(x; t)$ as a joint density function of multiple random variables, and the functions $f_{x_i}(x_i; t)$ as marginal density functions of the individual random variables. This terminology becomes tedious when dealing with vector random processes. Therefore, joint and marginal density functions will be distinguished by the dimensionality of the argument and by context.

All expected values are subsequently written using vector notation, but can be expanded in terms of the individual elements, as in (3.1).

The expected value of the product of a pair of random variables is called the correlation (or the cross-correlation) of the random variables. The correlation provides information on the relationship between the random variables. The correlation function, also known as the autocorrelation function, enumerates all of the correlations of the random process, and is defined as:

$$\mathbf{R}_x(t_1, t_2) = E[x(t_1)x^T(t_2)] = \int_{-\infty}^{\infty}\int_{-\infty}^{\infty} x_1 x_2^T f_{x_1,x_2}(x_1, x_2; t_1, t_2)dx_1 dx_2. \quad (3.2)$$

The correlation function includes both the correlations between random variables at different points in time and the correlations between the elements of the vector random process.

The mean square value of the random process is defined as follows:

$$E[x^T(t)x(t)] = \int_{-\infty}^{\infty} x^T x f_x(x; t)dx,$$

and provides an indication of the "size" of the random process as a function of time. The mean square value can be given in terms of the correlation function:

$$E[x^T(t)x(t)] = \text{tr}\{\mathbf{R}_x(t, t)\},$$

where tr(•) is the trace operator.

The correlation matrix of the random process is defined as follows:

$$\mathbf{\Sigma}_x(t) = E[x(t)x^T(t)] = \int_{-\infty}^{\infty} xx^T f_x(x; t)dx.$$

The diagonal elements of this matrix are the mean square values of each element in the random process. The off-diagonal elements of the correlation matrix are the correlations between the elements of the random process at a given point in time. The correlation matrix can be given in terms of the correlation function:

$$\mathbf{\Sigma}_x(t) = E[x(t)x^T(t)] = \mathbf{R}_x(t, t).$$

The covariance matrix of the random process is defined as:

$$\begin{aligned}\Xi_x(t) &= E[\{x(t) - m_x(t)\}\{x(t) - m_x(t)\}^T] \\ &= \int_{-\infty}^{\infty} \{x(t) - m_x(t)\}\{x(t) - m_x(t)\}^T f_x(x; t)dx.\end{aligned}$$

The diagonal elements of this matrix are the variances $\sigma_{x_i}^2(t)$ of each element in the random process that provides a measure of the spread of that element's density function. The standard deviation $\sigma_x(t)$ is also frequently used in place of the variance to denote the spread of the density function. The off-diagonal elements of the covariance matrix are the cross-covariances of pairs of elements from the random process. The covariance matrix and the correlation matrix are equal when the random process has zero mean. The covariance matrix can be given in terms of the correlation function and the mean:

$$\Xi_x(t) = E[\{x(t) - m_x(t)\}\{x(t) - m_x(t)\}^T] = \mathbf{R}_x(t, t) - m_x(t)m_x^T(t).$$

The covariance function, also known as the autocovariance function, of the random process is defined as follows:

$$\mathbf{C}_x(t_1, t_2) = E[\{x(t_1) - m_x(t_1)\}\{x(t_2) - m_x(t_2)\}^T]$$
$$= \int_{-\infty}^{\infty}\int_{-\infty}^{\infty} \{x(t_1) - m_x(t_1)\}\{x(t_2) - m_x(t_2)\}^T f_{x_1,x_2}(x_1, x_2; t_1, t_2)dx_1 dx_2.$$

The covariance function provides information similar to that provided by the correlation function, but with the effect of the mean removed. The covariance function is equal to the correlation function when the random process has zero mean. In general, the covariance function can be given in terms of the correlation function and the mean:

$$\mathbf{C}_x(t_1, t_2) = E[\{x(t_1) - m_x(t_1)\}\{x(t_2) - m_x(t_2)\}^T] = \mathbf{R}_x(t_1, t_2) - m_x(t_1)m_x^T(t_2).$$

Some additional terminology concerning second-order expectations will prove useful. A pair of random vectors is said to be uncorrelated if

$$E[x(t_1)x^T(t_2)] = m_x(t_1)m_x^T(t_2)$$

or, equivalently, if

$$E[\{x(t_1) - m_x(t_1)\}\{x(t_2) - m_x(t_2)\}^T] = \mathbf{0}.$$

A pair of random variables is said to be orthogonal if

$$E[x(t_1)x^T(t_2)] = \mathbf{0}.$$

The terminology presented above is fairly standard in the engineering literature, but far from intuitive. Uncorrelated means that the correlation equals the product of the means, while orthogonal means that the correlation equals zero. These two definitions are equivalent when either of the random variables has zero mean, which is quite common in estimation problems.[2] The above definitions of uncorrelated and orthogonal also apply when the random vectors are from distinct random processes.

EXAMPLE 3.1 A digital motor controller computes samples of the field voltage based on noisy measurements of the motor shaft position. The field voltage applied to the motor is the computed sample held constant for the sampling interval:

$$u(t) = \sum_{k=-\infty}^{\infty} u_k p_1(t - k),$$

where $p_1(t)$ is the unit pulse function and

$$p_1(t - k) = \begin{cases} 1 & k \le t < k + 1 \\ 0 & \textit{else} \end{cases}.$$

After driving the motor shaft to near the desired position, the noise on the measurements generates control samples $\{u_k\}$ that can be approximated by independent random variables uniformly distributed from -0.3 to 0.3. A sample of the random signal $u(t)$ is

[2] A random variable is often separated into its deterministic mean value and a zero mean random component. For linear systems, these two components can be treated separately without loss of generality.

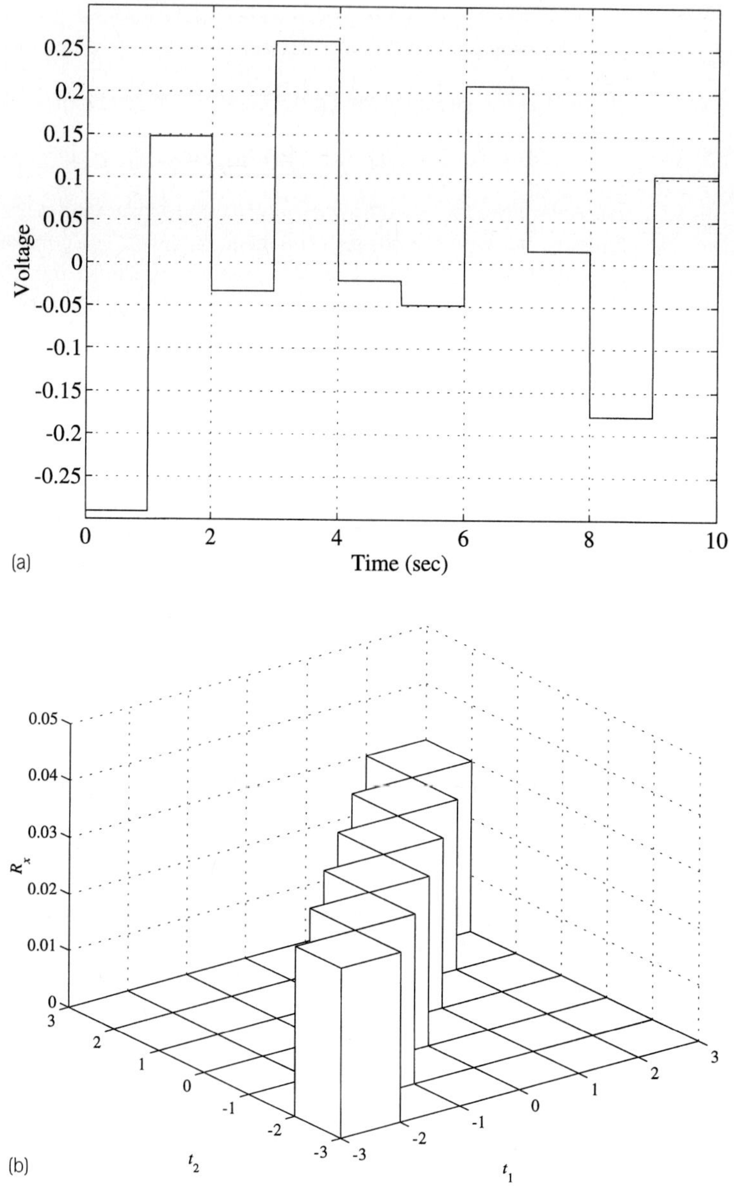

FIGURE 3.1 Motor field voltage in Example 3.1: (a) a sample function; (b) the correlation function

shown in Figure 3.1a. The analysis of the motor's response to this random input requires the mean and correlation function. The mean is

$$m_x(t) = E[x(t)] = E\left[\sum_{k=-\infty}^{\infty} u_k p_1(t - k)\right] = \sum_{k=-\infty}^{\infty} E[u_k]p_1(t - k) = 0.$$

The correlation function (which equals the covariance function) is

$$R_x(t_1, t_2) = E[x(t_1)x(t_2)] = E\left[\sum_{k=-\infty}^{\infty} u_k p_1(t_1 - k) \sum_{l=-\infty}^{\infty} u_l p_1(t_2 - l)\right]$$

$$= \sum_{k=-\infty}^{\infty} \sum_{l=-\infty}^{\infty} E[u_k u_l] p_1(t_1 - k) p_1(t_2 - l)$$

$$= \sum_{k=-\infty}^{\infty} 0.03 p_1(t_1 - k) p_1(t_2 - k),$$

where 0.03 is the variance of the random variables u_k. This correlation function is plotted in Figure 3.1b. The correlation matrix and the covariance matrix are then found:

$$\Sigma_x(t) = \Xi_x(t) = R_x(t, t) = \sum_{k=-\infty}^{\infty} 0.03 p_1^2(t - k) = 0.03. \qquad \blacklozenge$$

◆**EXAMPLE 3.2**   Let $\{u_k\}$ be defined as in Example 3.1. A vector random process can be constructed:

$$\begin{bmatrix} x_1(t) \\ x_2(t) \end{bmatrix} = \begin{bmatrix} u_1 \\ u_1 + u_2 \end{bmatrix}.$$

The correlation matrix of this process is

$$\Sigma_x(t) = E\left\{\begin{bmatrix} x_1(t) \\ x_2(t) \end{bmatrix} [x_1(t) \quad x_2(t)]\right\} = \begin{bmatrix} E[x_1^2(t)] & E[x_1(t)x_2(t)] \\ E[x_1(t)x_2(t)] & E[x_2^2(t)] \end{bmatrix}$$

$$= \begin{bmatrix} E[u_1^2] & E[u_1(u_1 + u_2)] \\ E[u_1(u_1 + u_2)] & E[(u_1 + u_2)^2] \end{bmatrix} = \begin{bmatrix} 0.03 & 0.03 \\ 0.03 & 0.06 \end{bmatrix}. \qquad \blacklozenge$$

3.1.2 Two Random Processes

Multiple random signals are defined in many applications. Consider, for example, a dc motor where the field voltage and the angular velocity of the shaft are related random signals. The relationship of these signals is fully specified by the joint density function, but additional insight can be gained from the conditional density function. The conditional density function of two random vectors is defined as follows:

$$f_{x|y}(x \mid y) = \frac{f_{x,y}(x, y)}{f_x(x)},$$

where $f_{x,y}(x, y)$ is the joint density function of x and y, $f_x(x)$ is the density function (often called the marginal density function) of x, and the notation $(x \mid y)$ is read: x *given* y. These random vectors may be from a single random process at two points in time or from two different random processes.

The conditional mean, the cross-correlation matrix, and the cross-covariance matrix can also be used to quantify the relationship between two random signals. The conditional mean is defined as follows:

$$m_{x|y} = E[x \mid y] = \int_{-\infty}^{\infty} x f_{x|y}(x \mid y) dx,$$

and referred to as *the expected value of x given y.* The cross-correlation matrix and the cross-covariance matrix of a pair of random vectors are defined respectively:

$$\mathbf{\Sigma}_{xy} = E[xy^T] = \int_{-\infty}^{\infty}\int_{-\infty}^{\infty} xy^T f_{x,y}(x, y)dxdy;$$

$$\Xi_{xy} = E[\{x - m_x\}\{y - m_y\}^T] = \int_{-\infty}^{\infty}\int_{-\infty}^{\infty} \{x - m_x\}\{y - m_y\}^T f_{x,y}(x, y)dxdy.$$

These matrices enumerate the correlations and covariances between the random variables within the two vectors.

3.1.3 Wide-Sense Stationarity

The mean value, mean square value, correlation matrix, and covariance matrix of a random process are all functions of time, in general. The description and analysis of the random process is simplified in the case when these expectations are independent of time. In addition, the correlation and covariance functions depend on the time indices of both random vectors being correlated. The analysis of the random process is further simplified when these functions only depend on the difference between the time indices. Collectively, these two attributes are known as wide-sense stationarity.

A random process is termed *wide-sense stationary* if the mean is independent of time:

$$E[x(t)] = m_x,$$

and the correlation function depends only on the time difference $\tau = t_2 - t_1$:

$$\mathbf{R}_x(t_1, t_2) = E[x(t_1)x^T(t_2)] = E[x(t_1)x^T(t_1 + \tau)] = \mathbf{R}_x(\tau).$$

The correlation function of a stationary random process has the symmetry property

$$\mathbf{R}_x(\tau) = \mathbf{R}_x^T(-\tau), \tag{3.3}$$

since the absolute time difference between the vectors being correlated is identical.

The mean square value, the correlation matrix, and the covariance matrix are all constant for a wide-sense stationary random process, and the covariance function, like the correlation function, depends only on the time difference $\tau = t_2 - t_1$:

$$\mathbf{C}_x(t_1, t_2) = \mathbf{C}_x(\tau).$$

The assumption that a random process is stationary is often justified in applications by noting that there is no reason that a particular time is different from other times. Therefore, the probabilistic description of the random process does not change with time, and the process is stationary.

The correlation and covariance functions provide information on the correlation of the random process variables over time. A simpler description of this correlation can be obtained by looking at the spread of these functions. The correlation time is a measure of this spread, and is defined for each element of the random process:

$$\tau_c^{(i)} = \frac{\int_0^{\infty} \mathbf{C}_x^{(i,i)}(\tau)d\tau}{\mathbf{C}_x^{(i,i)}(0)}$$

where $\mathbf{C}_x^{(i,i)}(\tau)$ is the (i, i) component of the covariance function. The correlation time provides an approximate description of the time period over which the random process is strongly correlated. Note that the correlation time is defined with reference to the covariance function in order to remove the effects of the mean on the correlation function's spread.

EXAMPLE 3.1 CONTINUED The random signal in Example 3.1 is not stationary because the correlation function

$$R_x(t_1, t_2) = \sum_{k=-\infty}^{\infty} 0.03 p_1(t_1 - k) p_1(t_2 - k)$$

cannot be written as a function of only the time difference $\tau = t_2 - t_1$ (see Figure 3.1b). ◆

EXAMPLE 3.3 Surface waves on the ocean excite roll motion in a ship equipped with an active roll control system. On a day where the swell is very consistent, these waves can be approximated as sinusoids with a given frequency (determined empirically):

$$x(t) = \alpha \cos(0.5t + \theta),$$

where α is a positive random variable with a mean of 0.63 and a variance of 0.11,[3] θ is uniformly distributed from 0 to 2π, and α and θ are uncorrelated. A sample function from this random process is shown in Figure 3.2a. The mean of this random signal is

$$E[x(t)] = E[\alpha \cos(0.5t + \theta)] = E[\alpha] \int_0^{2\pi} \cos(0.5t + \theta)\frac{1}{2\pi} d\theta = 0.$$

The correlation function of this signal is

$$\begin{aligned} R_x(t_1, t_2) &= E[x(t_1)x(t_2)] \\ &= \frac{1}{2} E[\alpha^2] E[\cos\{0.5(t_2 + t_1) + 2\theta\} + \cos\{0.5(t_2 - t_1)\}] \\ &= \frac{(\mu_\alpha^2 + \sigma_\alpha^2)}{2} \left\{ \int_0^{2\pi} \cos\{0.5(t_2 + t_1) + 2\theta\} \frac{1}{2\pi} d\theta + \cos\{0.5(t_2 - t_1)\} \right\} \\ &= 0.20 \cos\{0.5(t_2 - t_1)\}. \end{aligned}$$

The mean is independent of time, and the correlation function depends only on the time difference $(t_2 - t_1)$, so this random signal is stationary. This result is reasonable since there is no preferred time if the phase is uniformly distributed from zero to 2π. The correlation function is plotted, as a function of this time difference, in Figure 3.2b. ◆

EXAMPLE 3.4 A pilot moves the stick when attempting to maintain a level standard rate (3°/sec) turn. The stick position is a random process: it is not predictable and is different each time the experiment is repeated. Further, the stick position is a vector random process since it is composed of a forward-backward (pitch) position u_p and a side-to-side (roll) position u_r. This random process is zero mean since no stick input is required to maintain

[3] The amplitude of this random sinusoid is modeled as Rayleigh; see [1], page 78.

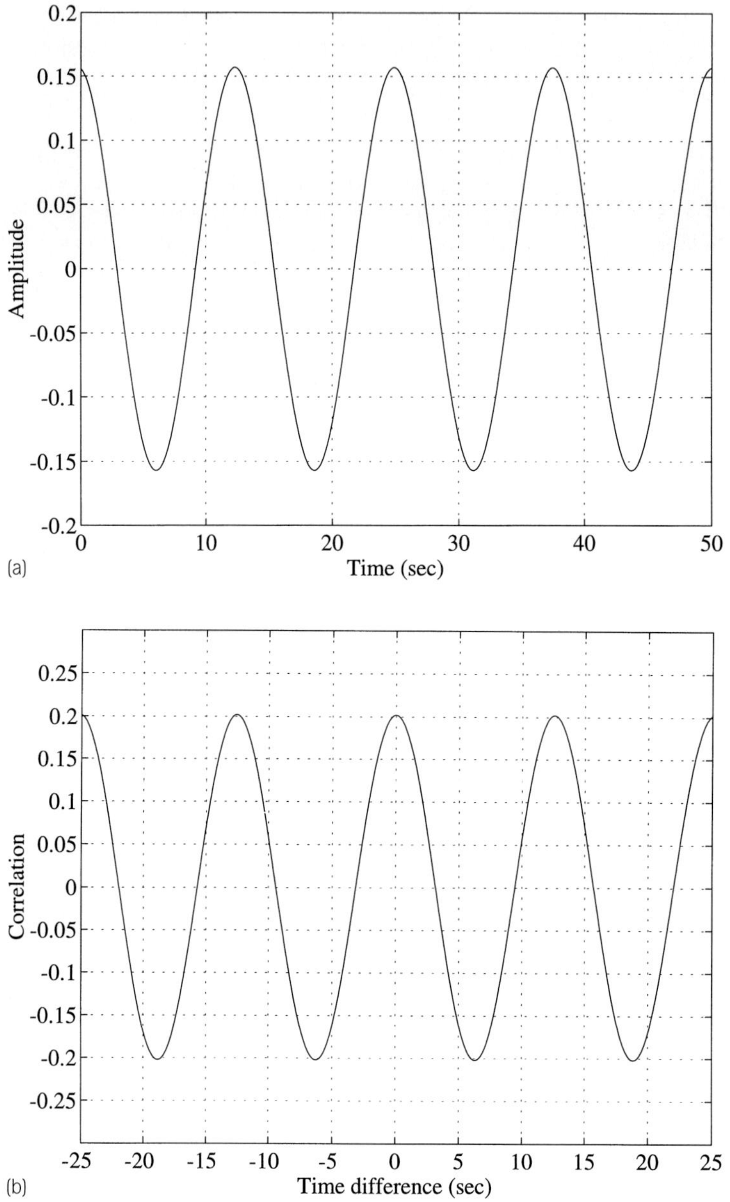

FIGURE 3.2 Surface waves on the ocean: (a) a sample function; (b) the correlation function

the bank of an airplane in a turn, and the plane is assumed to be trimmed for level flight.[4] In addition, this random process is assumed to be stationary since there is no reason to expect the probabilistic description of the random process to change with time. Note

[4] Technically, a small stick input may be required to maintain bank angle. This is ignored since it is typically quite small when performing standard rate turns.

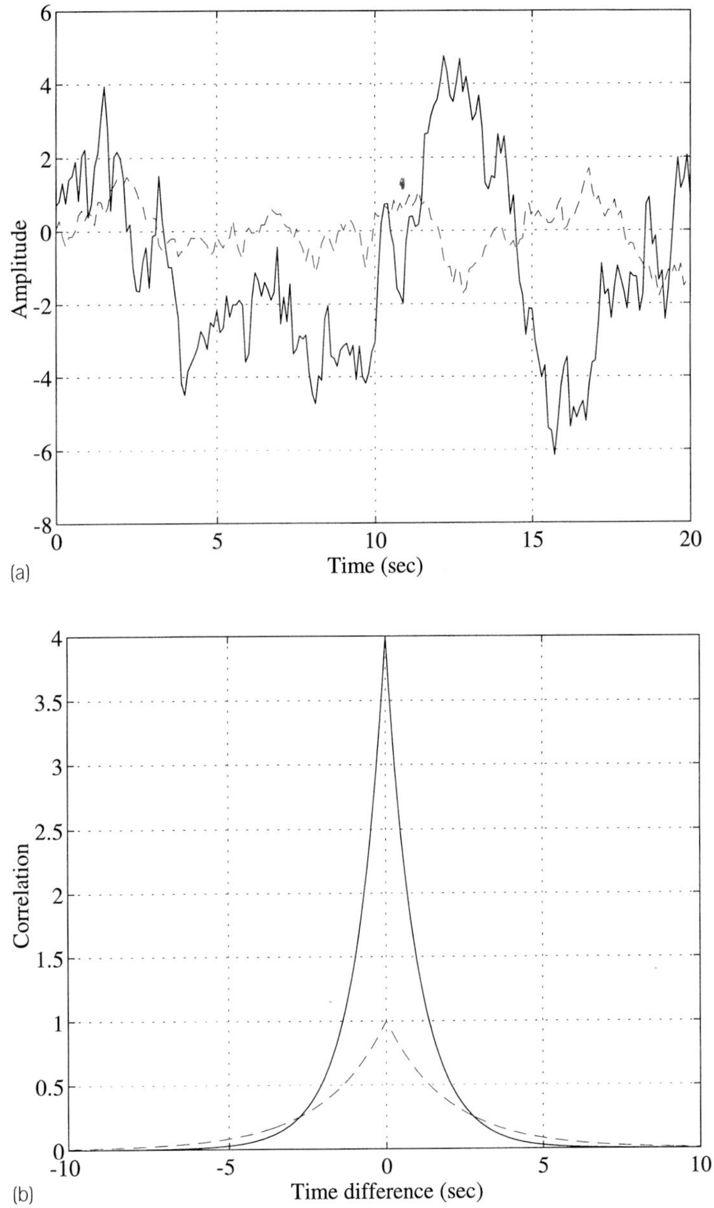

FIGURE 3.3 Stick position during standard rate turn: (a) a sample function; (b) the correlation function

that the random process is only defined during the bank, and does not include roll-in and roll-out. A sample of the stick position is shown in Figure 3.3a. The correlation function, which equals the covariance function since the process is zero mean, is estimated by averaging the results of many trials of this experiment. The resulting data can be approximated by the analytic correlation function:

$$\mathbf{R}_{\left[\begin{smallmatrix}u_p\\u_r\end{smallmatrix}\right]}(\tau) = \mathbf{C}_{\left[\begin{smallmatrix}u_p\\u_r\end{smallmatrix}\right]}(\tau) = \begin{bmatrix} 4e^{-|\tau|} & 0 \\ 0 & e^{-0.5|\tau|} \end{bmatrix}.$$

This correlation function is shown in Figure 3.3b. The fact that the off-diagonal terms in this correlation function are zero indicates that there is no correlation between the pitch and roll positions of the stick. A correlation time can be defined for each of these inputs:

$$\tau_c^{(1)} = \tfrac{1}{4}\int_0^\infty 4e^{-\tau}d\tau = 1 \text{ sec}; \; \tau_c^{(2)} = \tfrac{1}{1}\int_0^\infty e^{-0.5\tau}d\tau = 2 \text{ sec}.$$

Note that these correlation times are a rough measure of the time it takes the sample function (see Figure 3.3a) to change appreciably; that is, the pitch stick position changes significantly about every second, while the roll stick position changes about every two seconds. Further, the component with the longer correlation time is also the component with the broader correlation function. ◆

3.1.4 The Spectral Density

The frequency content of a stationary random process is described by the spectral density (often called the power spectral density):

$$\mathbf{S}_x(\omega) = \int_{-\infty}^{\infty} \mathbf{R}_x(\tau)e^{-j\omega\tau}d\tau.$$

The diagonal elements of the spectral density are strictly real due to the symmetry of the correlation function (3.3) and, therefore, contain only magnitude information on the frequency content of the elements of the random process. This lack of phase information is inherent in the definition of stationarity. For example, consider a random process that is a sinusoid at a specified frequency, but with a random amplitude and phase. This sinusoid is only stationary if the phase angle is uniformly distributed between 0 and 2π. If not, the preferred phase angle yields an increased probability of zero crossings at a particular time, and in turn a decrease in the mean at those times. The off-diagonal elements of the spectral density may be complex, in general. The nonzero argument of these elements indicates a relative phase between the two corresponding signals in the random process. The existence of a relative phase angle between signals in a random process does not violate the tenets of stationarity.

EXAMPLE 3.3 CONTINUED The spectral density of the random signal in Example 3.3 is

$$\mathbf{S}_x(\omega) = \int_{-\infty}^{\infty} 0.20\cos(0.5\tau)e^{-j\omega\tau}d\tau = 0.10\delta(\omega - 0.5) + 0.10\delta(\omega + 0.5).$$

This spectral density (see Figure 3.4) shows that the entire frequency content of the waves is at a single frequency. This fact is quite reasonable (for the mathematical problem posed) since all of the sample functions are at a single frequency but, this is certainly an idealization of a true wave spectrum. ◆

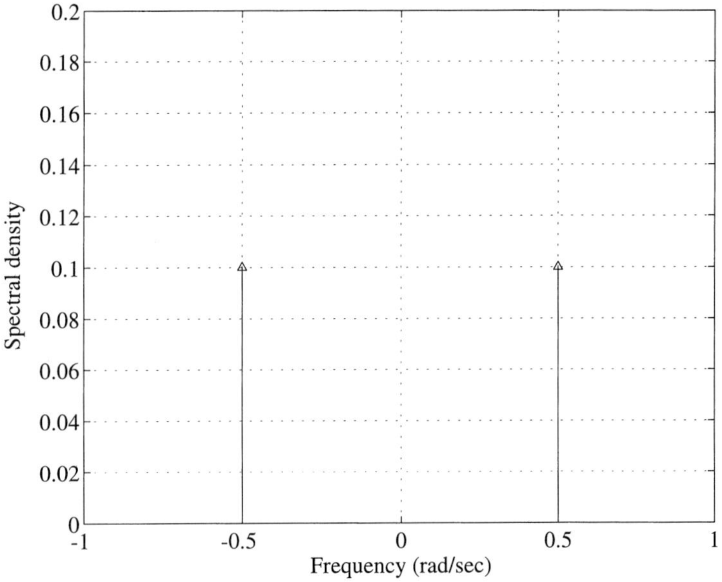

FIGURE 3.4 Surface waves on the ocean: the spectral density

EXAMPLE 3.4 CONTINUED The spectral density of the vector random signal in Example 3.4 is

$$\mathbf{S}_x(\omega) = \int_{-\infty}^{\infty} \begin{bmatrix} 4e^{-|\tau|} & 0 \\ 0 & e^{-0.5|\tau|} \end{bmatrix} e^{-j\omega\tau} d\tau = \begin{bmatrix} \dfrac{8}{\omega^2 + 1} & 0 \\ 0 & \dfrac{1}{\omega^2 + 0.25} \end{bmatrix}.$$

This spectral density (see Figure 3.5) shows the bandwidth of the components of the stick position. As expected, the bandwidth is larger for the pitch component, that is, the component with the shorter correlation time.

3.1.5 Gaussian Random Processes

A Gaussian (or normal) random vector is a random vector whose density function is given:

$$f_x(x) = \frac{1}{(2\pi)^{n_x/2} \det[\Xi_x]^{1/2}} e^{-\frac{1}{2}(x-m_x)^T_{\Xi_x^{-1}}(x-m_x)}.$$

A Gaussian random process is a random process where all random vectors that can be formed as combinations of random variables from the process are Gaussian.[5] A Gaussian random process is fully specified by its mean and covariance function, or equivalently, its mean and correlation function.

[5] A set of random variables that, taken together, form a Gaussian random vector are often termed *jointly Gaussian*.

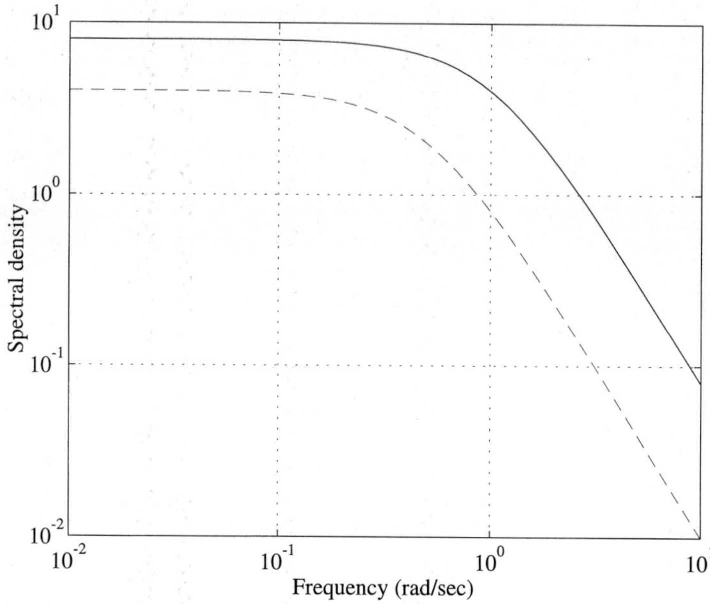

FIGURE 3.5 Stick position during standard rate turn: the spectral density

Gaussian random processes frequently provide excellent models of random signals in applications. This fact results from the central limit theorem: A random variable formed as the sum of a large number of independent random variables is approximately Gaussian. Many random processes are composed of random variables formed as the sum of independent random variables. For example, the convolution integral states that the output of a linear system is generated by integrating the input weighted by the impulse response. Since integration is roughly equivalent to summation, the output of a linear system with a random input is a weighted sum of random variable, and this output is typically Gaussian. The catch in this reasoning is that the random variables in this sum must be independent for the central limit theorem to apply. This condition is satisfied when the correlation time of the input is small compared to the time duration of the impulse response. In this case, the output of the linear system is approximately Gaussian regardless of the density function of the input.

3.2 White Noise

Many random processes that occur in applications have a zero mean and fluctuate rapidly in an unpredictable manner. The static in a radio receiver is an example of such a process. These rapid, unpredictable fluctuations are indicative of a very short correlation time. An idealized version of such random processes, known as white noise, is defined by the following correlation function:

$$\mathbf{R}_w(t_1, t_2) = \mathbf{S}_w(t_1)\delta(t_1 - t_2),$$

where the mean and correlation time are both zero.

The correlation function for a stationary white noise process reduces to

$$\mathbf{R}_w(\tau) = \mathbf{S}_w \delta(\tau),$$

which yields a spectral density that is constant over all frequencies:

$$\mathbf{S}_w(\omega) = \mathbf{S}_w.$$

The term *white noise* results from the analogy with white light, which is composed equally of all colors (frequencies).[6] Continuing this analogy, *colored noise* is used to refer to noise with a bandpass or lowpass spectrum.

The variance of a stationary random process is the correlation function evaluated at zero, and is infinite for white noise. In addition, white noise has an infinite bandwidth. Neither of these attributes is possessed by physical processes. Despite this fact, white noise is a useful approximation for random inputs to linear systems. The moments of the response of a linear system to a white noise input can be simply computed, as shown in the next section. These moments are good approximations to the actual values if the correlation time of the random input is small compared to the time constant of the linear system. The utility of white noise is further extended by noting that physically realizable noise signals can be modeled as the output of a linear system subject to a white noise input. This allows white noise analysis techniques to be used when analyzing linear systems with nonwhite random inputs. Additional comments on the utility of white noise are included in the subsequent sections.

3.3 Linear Systems with White Noise Inputs

Measurement noise and disturbance inputs are often modeled as white noise when evaluating control system performance. Examples include a motor that is controlled using a noisy sensor and an airplane subject to random turbulence. White noise models of these signals often provide an excellent approximation to the real-world signals, and require very little information to enumerate (i.e., only the spectral density). In addition, the computation required to determine the output expectations is typically simplified when employing a white noise input model versus a more general random model.

In Chapter 2, analysis of a system with a deterministic input was accomplished by computing the system output. This approach is not practical for systems with random inputs since the output is also random. Instead, a probabilistic description of the output is required. Computing the full probabilistic description of the output is usually impractical since a full description of the input is only rarely available, and the computations required to yield the full description of the output are often intractable.[7] In addition, the full probabilistic description of the output typically includes more information than necessary for design or analysis. A simpler description of the output is provided by the first and second moments. Formulas for the mean, correlation matrix, correlation function,

---

[6] In reality, the color white may be perceived by the eye in response to light with a wide variety of spectral contents. In addition, the eye is only sensitive to a small range of the electromagnetic spectrum, so this analogy with white light is not perfect.

[7] A notable exception is Gaussian random processes, which are fully described by their first- and second-order moments.

and power spectral density of the output of a linear system subject to a white noise input are developed in the remainder of this section.

A linear system with a random initial state and a stationary white noise input $w(t)$ is modeled:

$$\dot{x}(t) = \mathbf{A}x(t) + \mathbf{B}w(t) \tag{3.4a}$$

$$y(t) = \mathbf{C}x(t). \tag{3.4b}$$

The white noise input has a known correlation function:

$$\mathbf{R}_w(\tau) = E[w(t)w^T(t + \tau)] = \mathbf{S}_w\delta(\tau) \tag{3.5}$$

and a mean of zero. Without loss of generality, the mean of the initial condition is assumed to be zero:[8]

$$m_x(0) = 0. \tag{3.6a}$$

The correlation matrix of the initial condition is assumed to be known:

$$\mathbf{R}_x(0) = E[x(0)x^T(0)]. \tag{3.6b}$$

Random initial conditions can be used to model the effects of random inputs before the initial time, or to model any other uncertainty in the initial condition. The initial conditions are assumed to be uncorrelated with the input:

$$E[x(0)w^T(t)] = E[x(0)]E[w^T(t)] = 0 \quad \text{for all } t \geq 0.$$

This assumption is intuitively justified by noting that there is no reason to believe the initial state is correlated with the input unless the initial state is generated by past inputs. For white noise, past inputs are uncorrelated with current and future inputs, which means the initial state should be uncorrelated with inputs occurring after the initial time.

The mean of the state in (3.4) is found by taking the expected value of the solution of the state equation:

$$\begin{aligned} m_x(t) &= E\left[e^{\mathbf{A}t}x(0) + \int_0^t e^{\mathbf{A}(t-\tau)}\mathbf{B}w(\tau)d\tau\right] \\ &= e^{\mathbf{A}t}E[x(0)] + \int_0^t e^{\mathbf{A}(t-\tau)}\mathbf{B}E[w(\tau)]d\tau = 0. \end{aligned}$$

The mean of the output is also zero:

$$m_y(t) = E[\mathbf{C}x(t)] = \mathbf{C}E[x(t)] = \mathbf{C}m_x(t) = 0.$$

In summary, the means of the state and output are both zero whenever the input is white and the mean of the initial condition is zero.

[8] The initial condition can be segregated into a deterministic mean and a zero mean random component. Superposition allows the separate analysis of these two components, with the methods of Chapter 2 being used to analyze the effects of the deterministic mean.

3.3.1 The Output Correlation Function

The state correlation function in (3.4) is found by substituting the solution of the state equation into the definition of the correlation function:

$$\begin{aligned}\mathbf{R}_x(t_1, t_2) &= E[x(t_1)x^T(t_2)] \\ &= E\left[\left\{e^{\mathbf{A}t_1}x(0) + \int_0^{t_1} e^{\mathbf{A}(t_1-\tau)}\mathbf{B}w(\tau)d\tau\right\}\right. \\ &\qquad \left.\left\{e^{\mathbf{A}t_2}x(0) + \int_0^{t_2} e^{\mathbf{A}(t_2-\gamma)}\mathbf{B}w(\gamma)d\gamma\right\}^T\right].\end{aligned}$$

Rearranging terms and moving the expected value within the integrals yields

$$\begin{aligned}\mathbf{R}_x(t_1, t_2) &= e^{\mathbf{A}t_1}E[x(0)x^T(0)]e^{\mathbf{A}^Tt_2} + \int_0^{t_2} e^{\mathbf{A}t_1}E[x(0)w^T(\gamma)]\mathbf{B}^Te^{\mathbf{A}^T(t_2-\gamma)}d\gamma \\ &+ \int_0^{t_1} e^{\mathbf{A}(t_1-\tau)}\mathbf{B}E[w(\tau)x^T(0)]e^{\mathbf{A}^Tt_2}d\tau \\ &+ \int_0^{t_1}\int_0^{t_2} e^{\mathbf{A}(t_1-\tau)}\mathbf{B}E[w(\tau)w^T(\gamma)]\mathbf{B}^Te^{\mathbf{A}^T(t_2-\gamma)}d\gamma d\tau.\end{aligned}$$

The cross-correlation matrix between the initial condition and the input is zero. Further, the expected values within the first and last terms of this expression are, respectively, the state correlation matrix at time zero and the input correlation function:

$$\mathbf{R}_x(t_1, t_2) = e^{\mathbf{A}t_1}\mathbf{\Sigma}_x(0)e^{\mathbf{A}^Tt_2} + \int_0^{t_1}\int_0^{t_2} e^{\mathbf{A}(t_1-\tau)}\mathbf{B}\mathbf{S}_w\delta(\tau - \gamma)\mathbf{B}^Te^{\mathbf{A}^T(t_2-\gamma)}d\gamma d\tau.$$

Assuming $t_2 > t_1$, and utilizing the definition of the Dirac delta function to integrate with respect to γ, this equation reduces to

$$\mathbf{R}_x(t_1, t_2) = e^{\mathbf{A}t_1}\mathbf{\Sigma}_x(0)e^{\mathbf{A}^Tt_2} + \int_0^{t_1} e^{\mathbf{A}(t_1-\tau)}\mathbf{B}\mathbf{S}_w\mathbf{B}^Te^{\mathbf{A}^T(t_2-\tau)}d\tau.$$

A similar expression can be obtained for $t_2 < t_1$, by integrating first with respect to τ:

$$\mathbf{R}_x(t_1, t_2) = e^{\mathbf{A}t_1}\mathbf{\Sigma}_x(0)e^{\mathbf{A}^Tt_2} + \int_0^{t_2} e^{\mathbf{A}(t_1-\gamma)}\mathbf{B}\mathbf{S}_w\mathbf{B}^Te^{\mathbf{A}^T(t_2-\gamma)}d\gamma.$$

These expressions can be combined by appropriately defining the dummy variable of integration:

$$\boxed{\mathbf{R}_x(t_1, t_2) = e^{\mathbf{A}t_1}\mathbf{\Sigma}_x(0)e^{\mathbf{A}^Tt_2} + \int_0^{\min(t_1,t_2)} e^{\mathbf{A}(t_1-\gamma)}\mathbf{B}\mathbf{S}_w\mathbf{B}^Te^{\mathbf{A}^T(t_2-\gamma)}d\gamma.} \quad (3.7)$$

This formula gives the state correlation function as a function of the initial condition correlation matrix and the input spectral density. The output correlation function can be simply computed from the state correlation function:

$$\boxed{\mathbf{R}_y(t_1, t_2) = E[y(t_1)y^T(t_2)] = \mathbf{C}E[x(t_1)x^T(t_2)]\mathbf{C}^T = \mathbf{C}\mathbf{R}_x(t_1, t_2)\mathbf{C}^T.} \quad (3.8)$$

The Output Correlation Matrix The state correlation function, along with the mean, provides a complete second-moment description of the random state and output. Equation (3.7) is computationally quite tedious, and often yields more information than required. A simpler description of the state is provided by the correlation matrix. The state correlation matrix can be obtained from the state correlation function:

$$\mathbf{\Sigma}_x(t) = \mathbf{R}_x(t, t) = e^{\mathbf{A}t}\mathbf{\Sigma}_x(0)e^{\mathbf{A}^T t} + \int_0^t e^{\mathbf{A}(t-\gamma_1)}\mathbf{B}\mathbf{S}_w\mathbf{B}^T e^{\mathbf{A}^T(t-\gamma_1)}d\gamma_1.$$

Performing the change of variables: $\gamma = t - \gamma_1$ simplifies this expression:

$$\boxed{\mathbf{\Sigma}_x(t) = e^{\mathbf{A}t}\mathbf{\Sigma}_x(0)e^{\mathbf{A}^T t} + \int_0^t e^{\mathbf{A}\gamma}\mathbf{B}\mathbf{S}_w\mathbf{B}^T e^{\mathbf{A}^T\gamma}d\gamma.} \tag{3.9}$$

This formula gives the state correlation matrix as a function of the initial condition correlation matrix and the input spectral density.

The output correlation matrix can be found from the state correlation matrix:

$$\boxed{\mathbf{\Sigma}_y(t) = E[y(t)y^T(t)] = \mathbf{C}E[x(t)x^T(t)]\mathbf{C}^T = \mathbf{C}\mathbf{\Sigma}_x(t)\mathbf{C}^T.} \tag{3.10}$$

EXAMPLE 3.5 A closed-loop motor-speed control system is described by the state model:

$$\dot{x}(t) = -10x(t) + 9w(t);$$

$$y(t) = x(t),$$

where $y(t) = x(t)$ is the angular velocity of the motor shaft, and $w(t)$ is the measurement noise (which is an input to the closed-loop system). The measurement noise is modeled as white noise with the following correlation function:

$$R_w(\tau) = 0.01\delta(\tau).$$

The initial condition on the state is uniformly distributed between $-\pi$ and π, which yields a mean of zero and a variance (correlation matrix) of

$$\Sigma_x(0) = \frac{\pi^2}{3} = 3.29.$$

The state correlation function is given by (3.7):

$$\begin{aligned} R_x(t_1, t_2) &= 3.29e^{-10(t_1+t_2)} + \int_0^{\min(t_1,t_2)} 0.81e^{-10(t_1+t_2)}e^{20\gamma}d\gamma \\ &= 3.25e^{-10(t_1+t_2)} + 0.04e^{-10(t_1+t_2)}e^{20\min(t_1,t_2)}, \end{aligned}$$

and the output correlation function equals the state correlation function because $\mathbf{C} = 1$. The state and output correlation matrices are then

$$\Sigma_x(t) = \Sigma_y(t) = R_x(t, t) = 0.04 + 3.25e^{-20t}.$$ ◆

Propagation of the State Correlation Matrix The state correlation matrix can also be computed as the solution of a differential equation. This differential equation is obtained by differentiating (3.9):

$$\dot{\mathbf{\Sigma}}_x(t) = \mathbf{A}e^{\mathbf{A}t}\mathbf{\Sigma}_x(0)e^{\mathbf{A}^Tt} + e^{\mathbf{A}t}\mathbf{\Sigma}_x(0)e^{\mathbf{A}^Tt}\mathbf{A}^T + \frac{d}{dt}\left\{\int_0^t e^{\mathbf{A}(t-\gamma)}\mathbf{B}\mathbf{S}_w\mathbf{B}^Te^{\mathbf{A}^T(t-\gamma)}d\gamma\right\}.$$

Taking the derivative of the integral yields

$$\begin{aligned}\dot{\mathbf{\Sigma}}_x(t) &= \mathbf{A}e^{\mathbf{A}t}\mathbf{\Sigma}_x(0)e^{\mathbf{A}^Tt} + e^{\mathbf{A}t}\mathbf{\Sigma}_x(0)e^{\mathbf{A}^Tt}\mathbf{A}^T + \mathbf{B}\mathbf{S}_w\mathbf{B}^T \\ &\quad + \int_0^t \frac{d}{dt}\left\{e^{\mathbf{A}(t-\gamma)}\mathbf{B}\mathbf{S}_w\mathbf{B}^Te^{\mathbf{A}^T(t-\gamma)}\right\}d\gamma \\ &= \mathbf{A}e^{\mathbf{A}t}\mathbf{\Sigma}_x(0)e^{\mathbf{A}^Tt} + e^{\mathbf{A}t}\mathbf{\Sigma}_x(0)e^{\mathbf{A}^Tt}\mathbf{A}^T + \mathbf{B}\mathbf{S}_w\mathbf{B}^T \\ &\quad + \int_0^t \mathbf{A}e^{\mathbf{A}(t-\gamma)}\mathbf{B}\mathbf{S}_w\mathbf{B}^Te^{\mathbf{A}^T(t-\gamma)} + e^{\mathbf{A}(t-\gamma)}\mathbf{B}\mathbf{S}_w\mathbf{B}^Te^{\mathbf{A}^T(t-\gamma)}\mathbf{A}^Td\gamma.\end{aligned}$$

Grouping together the terms multiplied by $\mathbf{A}$ and $\mathbf{A}^T$, we have

$$\begin{aligned}\dot{\mathbf{\Sigma}}_x(t) &= \mathbf{A}\left\{e^{\mathbf{A}t}\mathbf{\Sigma}_x(0)e^{\mathbf{A}^Tt} + \int_0^t e^{\mathbf{A}(t-\gamma)}\mathbf{B}\mathbf{S}_w\mathbf{B}^Te^{\mathbf{A}^T(t-\gamma)}d\gamma\right\} \\ &\quad + \left\{e^{\mathbf{A}t}\mathbf{\Sigma}_x(0)e^{\mathbf{A}^Tt} + \int_0^t e^{\mathbf{A}(t-\gamma)}\mathbf{B}\mathbf{S}_w\mathbf{B}^Te^{\mathbf{A}^T(t-\gamma)}d\gamma\right\}\mathbf{A}^T + \mathbf{B}\mathbf{S}_w\mathbf{B}^T.\end{aligned}$$

Recognizing that the terms within the curly brackets are simply the state correlation matrix, the differential equation for the state correlation matrix is

$$\boxed{\dot{\mathbf{\Sigma}}_x(t) = \mathbf{A}\mathbf{\Sigma}_x(t) + \mathbf{\Sigma}_x(t)\mathbf{A}^T + \mathbf{B}\mathbf{S}_w\mathbf{B}^T.} \quad (3.11)$$

The state correlation matrix (or the state covariance matrix) can be found by solving this differential equation subject to an initial condition equal to the initial state correlation matrix. This differential equation is extremely useful in optimal estimation.

EXAMPLE 3.5 CONTINUED

The state correlation matrix in Example 3.5 can be computed using (3.11):

$$\dot{\Sigma}_x(t) = -20\Sigma_x(t) + 0.81.$$

This differential equation is in state form, and the solution is

$$\begin{aligned}\Sigma_x(t) &= \Sigma_x(0)e^{-20t} + \int_0^t 0.81e^{-20\tau}d\tau \\ &= 3.29e^{-20t} - 0.04e^{-20\tau}\big|_0^t = 0.04 + 3.25e^{-20t}.\end{aligned}$$

Note that this is the same answer obtained earlier using (3.9). ◆

Steady-State Correlation Results The state (and the output) of a linear system subject to a random input applied at time zero is nonstationary, in general, even though the input random process is modeled as stationary. This result is reasonable since the input to the system starts at time zero and can be modeled as the product of a unit-step function with a stationary input. This unit-step function can be expected to generate transients that add a time-varying component to the moments of the state. By analogy with deterministic results, these transients are expected to decay to zero with time for stable linear systems. The state is then approximately stationary far from the initial time.

A random process is termed asymptotically stationary in the wide sense if its mean, correlation matrix, and covariance matrix approach constants, and its correlation and covariance functions approach functions of the time difference between samples, as time increases to infinity. The state and output of the system of (3.4), with a random initial condition and subject to the stationary white noise input, are asymptotically stationary in the wide sense provided that the system is stable. This fact is demonstrated by noting that the mean of the state (and also the output) is zero since the means of the initial condition and the input are both zero.[9] The steady-state correlation matrix of the state is obtained by taking the limit of (3.9):

$$\mathbf{\Sigma}_x(\infty) = \lim_{t\to\infty}\left\{ e^{\mathbf{A}t}\mathbf{\Sigma}_x(0)e^{\mathbf{A}^T t} + \int_0^t e^{\mathbf{A}\gamma}\mathbf{B}\mathbf{S}_w\mathbf{B}^T e^{\mathbf{A}^T\gamma}d\gamma \right\}.$$

For a stable system, the matrix exponentials in this expression approach zero as time goes to infinity:

$$\boxed{\mathbf{\Sigma}_x(\infty) = \int_0^\infty e^{\mathbf{A}\gamma}\mathbf{B}\mathbf{S}_w\mathbf{B}^T e^{\mathbf{A}^T\gamma}d\gamma.} \tag{3.12}$$

The integrand in this expression is a matrix whose elements are linear combinations of decaying exponentials, guaranteeing the existence of this improper integral. The output correlation matrix is obtained by substituting (3.12) into (3.8):

$$\boxed{\mathbf{\Sigma}_y(\infty) = \int_0^\infty \mathbf{C}e^{\mathbf{A}\gamma}\mathbf{B}\mathbf{S}_w\mathbf{B}^T e^{\mathbf{A}^T\gamma}\mathbf{C}^T d\gamma = \int_{-\infty}^\infty \mathbf{g}(\gamma)\mathbf{S}_w\mathbf{g}^T(\gamma)d\gamma} \tag{3.13}$$

where $\mathbf{g}(t)$ is the impulse response matrix of the state model (3.4).

The steady-state correlation matrix of the state can also be found from the matrix differential equation (3.11) provided that the system is time-invariant and stable. In steady-state, the derivative of the correlation matrix equals zero:

$$\boxed{\mathbf{A}\mathbf{\Sigma}_x(\infty) + \mathbf{\Sigma}_x(\infty)\mathbf{A}^T + \mathbf{B}\mathbf{S}_x\mathbf{B}^T = \mathbf{0}.} \tag{3.14}$$

This equation is known as a Lyapunov equation, and can be solved to yield the steady-state value of the state correlation matrix. A number of numerical techniques have been developed for the solution of Lyapunov equations, and the interested reader is referred to [2].

[9] The mean value of the state and output approach constant values when the initial condition and the input have nonzero mean values, and the state and output are still asymptotically stationary.

EXAMPLE 3.5 CONTINUED The steady-state correlation matrix of the state in Example 3.5 can be computed using (3.12):

$$\mathbf{\Sigma}_x(\infty) = \int_0^\infty 0.81e^{-20\gamma}d\gamma = -\frac{0.81}{20}e^{-20\gamma}\Big|_0^\infty = 0.04,$$

or by solving the Lyapunov equation (3.14):

$$-10\Sigma_x(\infty) - 10\Sigma_x(\infty) + 0.81 = 0 \Rightarrow \Sigma_x(\infty) = 0.04.$$

◆

EXAMPLE 3.6 A state feedback controller is used to position a high-gain antenna. The plant model is

$$\begin{bmatrix} \dot{x}_1(t) \\ \dot{x}_2(t) \end{bmatrix} = \begin{bmatrix} 0 & 1 \\ 0 & -1 \end{bmatrix}\begin{bmatrix} x_1(t) \\ x_2(t) \end{bmatrix} + \begin{bmatrix} 0 \\ 0.1 \end{bmatrix}u(t) \ ; \ y(t) = [1 \quad 0]\begin{bmatrix} x_1(t) \\ x_2(t) \end{bmatrix}.$$

The measurements used in the state feedback controller are corrupted by noise:

$$u(t) = -[20 \quad 10]\left(\begin{bmatrix} x_1(t) \\ x_2(t) \end{bmatrix} + \begin{bmatrix} w_1(t) \\ w_2(t) \end{bmatrix}\right),$$

where the measurement noise is white with the following spectral density matrix:

$$\mathbf{S}_w = \begin{bmatrix} 0.0025 \text{ rad}^2/\text{Hz} & 0 \\ 0 & 0.05 \text{ rad}^2/\text{sec}^2/\text{Hz} \end{bmatrix}.$$

The off-diagonal zeros in this spectral density matrix indicate that the two measurement noises are uncorrelated.

The closed-loop state equation is then

$$\begin{bmatrix} \dot{x}_1(t) \\ \dot{x}_2(t) \end{bmatrix} = \begin{bmatrix} 0 & 1 \\ -2 & -2 \end{bmatrix}\begin{bmatrix} x_1(t) \\ x_2(t) \end{bmatrix} + \begin{bmatrix} 0 & 0 \\ -2 & -1 \end{bmatrix}\begin{bmatrix} w_1(t) \\ w_2(t) \end{bmatrix}; \ y(t) = [1 \quad 0]\begin{bmatrix} x_1(t) \\ x_2(t) \end{bmatrix}.$$

Solving the Lyapunov equation (3.14),

$$\begin{bmatrix} 0 & 1 \\ -2 & -2 \end{bmatrix}\mathbf{\Sigma}_x(\infty) + \mathbf{\Sigma}_x(\infty)\begin{bmatrix} 0 & -2 \\ 1 & -2 \end{bmatrix} + \begin{bmatrix} 0 & 0 \\ -2 & -1 \end{bmatrix}\begin{bmatrix} 0.01 & 0 \\ 0 & 0.05 \end{bmatrix}\begin{bmatrix} 0 & -2 \\ 0 & -1 \end{bmatrix} = \begin{bmatrix} 0 & 0 \\ 0 & 0 \end{bmatrix},$$

yields the steady-state correlation matrix of the state:

$$\mathbf{\Sigma}_x(\infty) = \begin{bmatrix} 0.0075 & 0 \\ 0 & 0.015 \end{bmatrix}.$$

Using (3.10), the variance and the standard deviation of the output are found to be $\sigma_y^2 = 0.0075 \text{ rad}^2$ and $\sigma_y = 0.087 \text{ rad} = 5.0°$, respectively. Therefore, this system maintains the antenna within 5° of the desired position roughly 67% of the time.[10] ◆

The limiting state correlation function can be obtained by letting $t_2 = t_1 + \tau$ in (3.7), and taking the limit as t_1 approaches infinity (see Exercise 3.10):

---

[10] The probability of being within one standard deviation of the mean is 0.67 for a Gaussian random variable. While the density functions of the noise and the output are not specified in this example, assuming the output is Gaussian is usually reasonable.

$$\mathbf{R}_x(\tau) = \int_{\max(0,-\tau)}^{\infty} e^{\mathbf{A}\gamma}\mathbf{B}\mathbf{S}_w\mathbf{B}^T e^{\mathbf{A}^T(\gamma+\tau)}d\gamma. \tag{3.15}$$

In practice, the above expressions for the state correlation matrix and state correlation function can be used after the initial transients have decayed to much less than the diagonal terms in the steady-state correlation matrix. The settling time is also used to indicate when steady state is obtained, but this can yield erroneous results when the initial state covariance matrix is large compared to the steady state value. The steady-state output correlation function is obtained by substituting (3.15) into (3.8):

$$\begin{aligned}\mathbf{R}_y(\tau) &= \int_{\max(0,-\tau)}^{\infty} \mathbf{C}e^{\mathbf{A}\gamma}\mathbf{B}\mathbf{S}_w\mathbf{B}^T e^{\mathbf{A}^T(\gamma+\tau)}\mathbf{C}^T d\gamma \\ &= \int_{-\infty}^{\infty} \mathbf{g}(\gamma)\mathbf{S}_w\mathbf{g}^T(\gamma+\tau)d\gamma.\end{aligned} \tag{3.16}$$

3.3.2 The Output Spectral Density

The state and the output of a stable linear system, subject to a stationary white noise input, is asymptotically stationary. The spectral density of the steady-state output is computed by taking the Fourier transform of the correlation function (3.16):

$$\mathbf{S}_y(\omega) = \int_{-\infty}^{\infty} \mathbf{R}_y(\tau)e^{-j\omega\tau}d\tau = \int_{-\infty}^{\infty}\left[\int_{-\infty}^{\infty} \mathbf{g}(\gamma)\mathbf{S}_w\mathbf{g}^T(\gamma+\tau)d\gamma\right]e^{-j\omega\tau}d\tau.$$

Interchanging the order of integration yields

$$\mathbf{S}_y(\omega) = \int_{-\infty}^{\infty} \mathbf{g}(\gamma)\mathbf{S}_w\left[\int_{-\infty}^{\infty} \mathbf{g}^T(\gamma+\tau)e^{-j\omega\tau}d\tau\right]d\gamma. \tag{3.17}$$

The integral within the brackets can be evaluated by performing the change of variables $\gamma_1 = \gamma + \tau$:

$$\int_{-\infty}^{\infty} \mathbf{g}^T(\gamma+\tau)e^{-j\omega\tau}d\tau = \int_{-\infty}^{\infty} \mathbf{g}^T(\gamma_1)e^{-j\omega(\gamma_1-\gamma)}d\gamma_1 = \mathbf{G}^T(j\omega)e^{j\omega\gamma}.$$

Substituting this result into (3.17) yields

$$\mathbf{S}_y(\omega) = \int_{-\infty}^{\infty} \mathbf{g}(\gamma)e^{j\omega\gamma}d\gamma\,\mathbf{S}_w\mathbf{G}^T(j\omega);$$

$$\mathbf{S}_y(\omega) = \mathbf{G}(-j\omega)\mathbf{S}_w\mathbf{G}^T(j\omega). \tag{3.18}$$

This expression gives the spectral density of the steady-state output of a stable linear system subject to a stationary white noise input. The spectral density of the state can also be computed by utilizing the transfer function matrix from the random input to the state in (3.18).

EXAMPLE 3.5 CONTINUED The transfer function of the closed-loop motor-speed control system in Example 3.5 is

$$G(s) = \frac{9}{s + 10}.$$

The output spectral density is given by (3.18):

$$S_y(\omega) = \frac{9}{-j\omega + 10} 0.01 \frac{9}{j\omega + 10} = \frac{0.81}{\omega^2 + 100}.$$

EXAMPLE 3.6 CONTINUED The closed-loop system transfer function of the antenna-position control system in Example 3.6 is

$$\mathbf{G}(s) = \begin{bmatrix} \frac{-2}{s^2 + 2s + 2} & \frac{-1}{s^2 + 2s + 2} \end{bmatrix}.$$

The output spectral density is given by (3.18):

$$S_y(\omega) = \frac{1}{(-j\omega)^2 - 2j\omega + 2}[-2 \quad -1]\begin{bmatrix} 0.0025 & 0 \\ 0 & 0.05 \end{bmatrix}\begin{bmatrix} -2 \\ -1 \end{bmatrix}\frac{1}{(j\omega)^2 + 2j\omega + 2}$$
$$= \frac{0.06}{\omega^4 + 4}.$$

3.3.3 Approximation of Real Inputs by White Noise

White noise can be used as an approximation to a nonwhite noise process that is input to a linear system. This approximation provides accurate results when the bandwidth of the noise is large compared to the system bandwidth. Equivalently, white noise provides a good idealization of the input whenever the correlation time is sufficiently small.[11] These facts can be demonstrated by considering the output spectral density of the linear system.

The output spectral density of a stable linear system subject to a stationary white noise input is given by (3.18), reproduced below for convenience:

$$\mathbf{S}_y(\omega) = \mathbf{G}(-j\omega)\mathbf{S}_w\mathbf{G}^T(j\omega). \tag{3.18}$$

The output spectral density of this system subject to a nonwhite stationary random input is given by substituting the frequency dependent input spectral density into (3.18):[12]

$$\mathbf{S}_y(\omega) = \mathbf{G}(-j\omega)\mathbf{S}_w(\omega)\mathbf{G}^T(j\omega). \tag{3.19}$$

Comparing (3.18) and (3.19), we see that the spectral densities of the outputs are approximately equal when the frequency dependent input spectral density in (3.19) is approximately equal to the constant spectral density in (3.18). This equality is required

[11] The correlation time of the input must be small compared to the system time constants, which are defined in Chapter 4.

[12] This result can be derived in a straightforward manner by including the correlation function of the nonwhite noise in the derivation of (3.16) and then following the steps used in the derivation of (3.18).

to hold only over the range of frequencies where $|\mathbf{G}(j\omega)|$ has elements that differ significantly from zero (i.e., over the bandwidth of the system). Therefore, a nonwhite noise process can be approximated by white noise whenever the nonwhite noise process is input to a linear system, and the spectral density of the nonwhite process is approximately flat over the bandwidth of the linear system. Note that this approximation only makes sense when the noise is input to a strictly proper system; that is, there is no direct feedthrough between the input and the output. When a direct feedthrough exists, the bandwidth of the system is infinite.

Random inputs are frequently characterized by a variance and a correlation time. For example, consider a control system where sensor noise is an input to the closed-loop system. Sensor manufacturers often provide variance and correlation-time specifications on their products. For analysis purposes, the variance and correlation time of the input can be translated into a spectral density for an equivalent white noise input. Let $w(t)$ be a random input with the correlation time τ_c and the variance σ_w^2. A correlation function for this input can be given:[13]

$$R_w(\tau) = \begin{cases} \sigma_w^2 & |\tau| \le \tau_c \\ 0 & \text{else} \end{cases}.$$

The spectral density is obtained by taking the Fourier transform of this function:

$$S_w(\omega) = 2\tau_c\sigma_w^2\frac{\sin(\omega\tau_c)}{\omega\tau_c}.$$

At low frequencies, this spectral density is approximately constant:

$$S_w \approx 2\tau_c\sigma_w^2. \tag{3.20}$$

The random input can then be modeled as white noise with the spectral density (3.20) provided the bandwidth of the system is small compared to the bandwidth of the input.

3.3.4 Simulation of Systems with White Noise Inputs

Computer simulation is often used during control system analysis both as a primary analysis tool and as a means of verifying analytic results. Simulations can be used to evaluate performance of systems with nonlinearities, time variations, and highly complex dynamics. Unfortunately, the infinite variance and the zero correlation time of white noise makes it impossible to simulate. A good approximation to the response of a system with a white noise input can be obtained using simulation, provided that the white noise is approximated by an appropriate discrete-time noise process. The approximation of white noise by a discrete-time noise process is developed below.

The best discrete-time approximation to a continuous-time white noise process is obtained by making the samples uncorrelated:

$$\mathbf{R}_w^{(d)}(n) = E[x(k)x^T(k+n)] = \mathbf{\Sigma}_w^{(d)}\delta_d(n),$$

[13] Note that while this is a reasonable correlation function, other correlation functions can be constructed that have the same correlation time and variance. The use of an alternate (but still reasonable) correlation function has only a minor effect on the spectral density of the equivalent white noise.

where

$$\delta_d(n) = \begin{cases} 1 & n = 0 \\ 0 & n \neq 0 \end{cases}$$

is the unit sample function, and the (d) is used to denote discrete-time. A discrete-time random process with this correlation matrix is termed discrete-time white noise. The correlation matrix of discrete-time white noise is finite, and the correlation time is on the order of the sampling interval and is therefore nonzero.

Simulation of the system with a discrete-time white noise input yields accurate results provided that the correlation time of the discrete-time process (approximately the sample time) is much less than the fastest time constant of the system. Standard simulation practice requires that the sampling time be chosen to be much less than the fastest time constant of the system, so this assumption is reasonable. Further, the correlation matrix of the white noise must be selected such that the output correlation matrix matches that resulting from the original white noise process. The output correlation matrix of a discrete-time system, subject to a discrete-time white noise input, is given:[14]

$$\mathbf{\Sigma}_y^{(d)}(\infty) = \sum_{k=0}^{\infty} \mathbf{g}_d(k)\mathbf{\Sigma}_w^{(d)}\mathbf{g}_d^T(k). \tag{3.21}$$

The output correlation matrix of a continuous-time system, subject to a white noise input with spectral density $\mathbf{S}_w$, is given by (3.13):

$$\mathbf{\Sigma}_y(\infty) = \int_0^{\infty} \mathbf{g}(\tau)\mathbf{S}_w\mathbf{g}^T(\tau)d\tau.$$

This result can be approximated using the Reimann sum:

$$\mathbf{\Sigma}_y(\infty) \approx T\sum_{k=0}^{\infty} \mathbf{g}(kT)\mathbf{S}_w\mathbf{g}^T(kT). \tag{3.22}$$

Samples of the continuous-time impulse response are related to the discrete-time impulse response (see Table 2.1):

$$\mathbf{g}(kT) \approx \frac{1}{T}\mathbf{g}_d(k),$$

where the factor $1/T$ equalizes the area of the continuous-time impulse with that of the discrete-time impulse. Substituting this expression into (3.22) yields

$$\mathbf{\Sigma}_y(\infty) \approx \sum_{k=0}^{\infty} \mathbf{g}_d(k)\frac{\mathbf{S}_w}{T}\mathbf{g}_d^T(k). \tag{3.23}$$

Equating the output correlation matrix computed in discrete time (3.21) with the approximate output correlation matrix computed in continuous time (3.23) yields

$$\mathbf{\Sigma}_w^{(d)} = \frac{\mathbf{S}_w}{T}.$$

[14] This result can be obtained by substituting the convolution summation (see Table 2.1) into the definition of the correlation matrix. The subsequent manipulations required to yield this result are analogous to those performed in the continuous-time case.

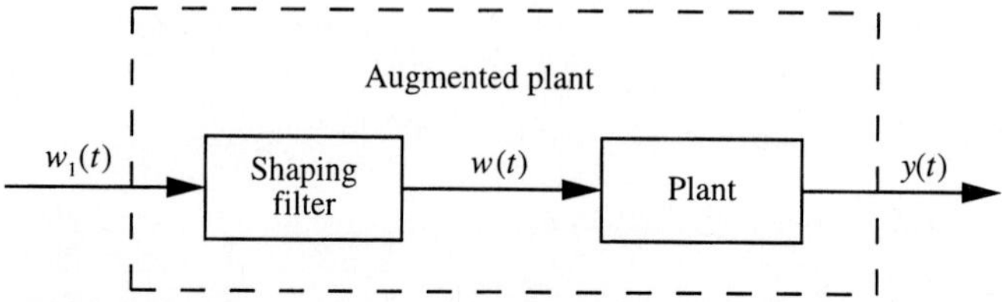

FIGURE 3.6 Using shaping filters to generate nonwhite noise

This equation gives the variance of the discrete-time white noise that yields an output variance equal to that obtained for the continuous-time system with a white noise input. This variance should be used when performing discrete-time simulations of continuous-time systems with white noise inputs. Note that this approximation is only valid when the continuous-time white noise is the input to a system whose bandwidth is much less than the sampling frequency.

3.4 Colored Noise Via Shaping Filters

A number of control systems are best modeled as linear systems subject to nonwhite noise. The results obtained for white noise can be simply extended for use in these cases. The key to this extension is to recognize that any reasonable nonwhite noise process can be generated as the output of a linear filter, called a shaping filter, with a white noise input.[15] The response of the original system to the nonwhite noise input is then equivalent to the response of the series combination of the original system and the shaping filter to a white noise input, as shown in Figure 3.6.

The fact that any nonwhite noise process can be generated from white noise can be simply understood by considering the spectral density. The spectral density contains all of the second-moment information on the process, and the two processes are referred to as second-moment equivalent if they have identical spectral density functions. The spectral density function $\mathbf{S}_w(\omega)$ of a given colored noise process can be obtained by putting white noise with spectral density $\mathbf{S}_{w1} = \mathbf{I}$ into a filter with transfer function $\mathbf{G}(j\omega)$:

$$\mathbf{S}_w(\omega) = \mathbf{G}(-j\omega)\mathbf{G}^T(j\omega).$$

The process of generating a linear filter that is both stable and causal and that yields the given spectral density $\mathbf{S}_w(\omega)$, is known as spectral factorization. The estimation of the spectral density function of a random process from measurements of the process is known as spectral estimation. Many spectral estimation algorithms directly estimate the parameters in $\mathbf{G}(j\omega)$ and therefore avoid the need for spectral factorization. The study of spectral estimation and spectral factorization is quite involved and only peripherally related to the material in this book. Therefore, only the simplest case of a scalar, nonwhite noise process generated by a first-order filter is included in this discussion. Spec-

[15] Note that only the second-moment properties of the output are required to match those obtained with the nonwhite noise.

tral estimation and spectral factorization are covered in a number of excellent texts including [3], [4], and [5].

Probabilistic information on random disturbance inputs is often very difficult to obtain. In many applications, only the variance and correlation time of the input are available. The spectral density of an equivalent white noise input can be generated from these parameters if the correlation time is sufficiently small (i.e., the bandwidth of the noise is sufficiently large). When the correlation time is large, the disturbance input (or any random process specified only by correlation time and variance) can be modeled as the output of a first-order shaping filter,

$$G(s) = \frac{b}{s + a}, \tag{3.24}$$

driven by white noise with a unit spectral density. The spectral density of the shaping filter output is

$$S_w(\omega) = \frac{b^2}{\omega^2 + a^2},$$

and the output correlation function is

$$R_w(\tau) = \frac{1}{2\pi} \int_{-\infty}^{\infty} S_w(\omega) e^{j\omega\tau} d\omega = \frac{b^2}{2a} e^{-a|\tau|}.$$

The shaping filter output then has the variance

$$\Sigma_w = R_w(0) = \frac{b^2}{2a},$$

and the correlation time

$$\tau_c = \frac{\int_0^{\infty} R_w(\tau) d\tau}{R_w(0)} = \frac{S_w(0)}{2R_w(0)} = \frac{1}{a}.$$

The filter (3.24) is fully specified by the following parameters:

$$a = \frac{1}{\tau_c} \quad \text{and} \quad b = \sqrt{\frac{2\Sigma_w}{\tau_c}}, \tag{3.25}$$

which can be found from the variance and correlation time of the disturbance input. Note that higher order filters also exist that yield the same variance and correlation time, but they are typically only used when required to match additional information on the spectral density.

EXAMPLE 3.7 Consider a control system for an aircraft tasked with providing a given rate of turn in response to a command from the pilot. The rate-of-turn command can be modeled as a random process since it is unknown *a priori*. For commercial aircraft, the maximum rate of turn is the standard rate of 3 degrees per second (one-half standard rate turns are also common for commercial carriers). The rate-of-turn command can therefore be assumed to be between $-3°$/s and $+3°$/s. Further, turning maneuvers typically take less than one minute since a 180-degree turn at the standard rate takes one minute. Therefore, it is expected that this random process will remain correlated only over time intervals of less than one minute. The rate-of-turn command can therefore be modeled as a random process with a correlation time[16] of 60 sec and a variance of 3 deg^2/sec^2. This variance is computed assuming that the acceleration is uniformly distributed between the limits. Inserting this variance and correlation time into (3.25) yields the shaping filter parameters:

$$a = \frac{1}{60} \quad \text{and} \quad b = \frac{1}{\sqrt{10}}.$$

The performance of the aircraft control system can be evaluated by analyzing this airplane model with the above-specified nonwhite noise input. Alternatively, this performance can be found by analyzing the series combination of the airplane model and the first-order shaping filter (3.24) with a unit spectral density white noise input. ◆

◆EXAMPLE 3.8 When manually implementing a standard rate turn, the pilot moves the stick (in both the pitch and yaw directions) as discussed in Example 3.4. The correlation function for this stick motion (as given in Example 3.4) is

$$\mathbf{R}_{\left[\begin{smallmatrix} u_p \\ u_r \end{smallmatrix}\right]}(\tau) = \begin{bmatrix} 4e^{-|\tau|} & 0 \\ 0 & e^{-0.5|\tau|} \end{bmatrix}.$$

Since the two stick motions are uncorrelated, they can be generated from two uncorrelated white noise inputs by application of two shaping filters. The coefficients of these filters are

$$a_1 = \frac{1}{\tau_c^{(1)}} = \frac{1}{1} = 1 \quad \text{and} \quad b_1 = \sqrt{\frac{2\Sigma_w^{(1)}}{\tau_c^{(1)}}} = \sqrt{\frac{2 \cdot 4}{1}} = \sqrt{8}$$

and

$$a_2 = \frac{1}{\tau_c^{(2)}} = \frac{1}{2} = 0.5 \quad \text{and} \quad b_2 = \sqrt{\frac{2\Sigma_w^{(2)}}{\tau_c^{(2)}}} = \sqrt{\frac{2 \cdot 1}{2}} = 1,$$

where the superscripts (1) and (2) indicate the pitch and roll component, respectively. The turn rate and altitude of the aircraft during a turn can then be found by analyzing the system in Figure 3.7. Note that the pitch and roll dynamics block in this diagram represents the transfer function from the stick inputs to the turn rate and altitude. ◆

3.5 Summary

Random signals are used to model reference inputs, disturbance inputs, and measurement noise in control systems. The complete description of a random signal consists of the joint probability density function of all combinations of random variables comprising the signal. This description is impossible to enumerate except in a few special cases,

[16] This is only an approximate correlation time. Detailed measurements can be used to provide a more accurate correlation time than obtained via the given thought experiment.

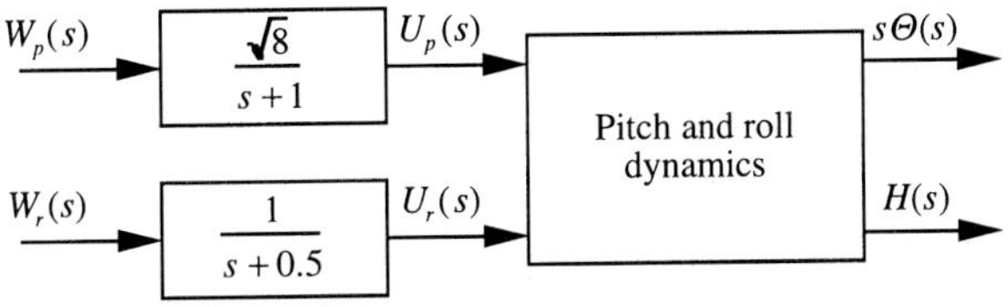

FIGURE 3.7 The augmented plant for evaluating the effects of nonwhite stick motion on turn rate and altitude

and contains more information than typically necessary for design and analysis. A more compact description of a random signal is provided by the moments of the signal.

The mean, correlation matrix, covariance matrix, correlation function, and covariance function of a vector random signal are defined. The correlation matrix, the covariance matrix, and the covariance function can all be computed from the mean and the correlation function, and these relationships are specified. The mean and correlation function serve as the fundamental second-moment description of a random signal in this book.

The conditional density function, conditional mean, cross-correlation matrix, and cross-covariance matrix are also defined. These quantities are used to enumerate the relationship between pairs of random vectors.

A random signal is said to be stationary if the statistics of the random signal are independent of time. In particular, a signal whose mean is constant and whose correlation function depends only on the difference between the time indices of the vectors being correlated is termed wide-sense stationary. Wide-sense stationary random signals can be described by the spectral density, which provides information on the frequency content of the random signal.

White noise is defined, and the response of linear systems to white noise inputs is fully explored. Formulas for the mean, correlation matrix, correlation function, and spectral density of the state and the output of a time-invariant linear system with a white noise input are derived. The state correlation matrix can also be computed as the solution of a differential equation. This solution methodology can be simply extended to time-varying systems, and is especially important for future applications. The correlation matrix of the state and the output approach steady-state values when the system is time-invariant and stable. The steady-state correlation matrix of the state can be found by solving an algebraic equation known as a Lyapunov equation, which is derived by setting the derivative equal to zero in the differential equation for the correlation matrix.

Shaping filters are used to generate colored noise from white noise. The development of a scalar, first-order shaping filter for generating colored noise described by a variance and a correlation time is presented. Appending a shaping filter to a plant allows us to extend the results on linear system response with white noise inputs generated in this chapter for use with nonwhite inputs.

The definitions of expectations and moments presented in this chapter can be directly applied to discrete-time signals. The relationships between continuous-time and discrete-time white noise are summarized in the following table along with the fundamental equations governing the response of linear systems to these inputs.

TABLE 3.1 Summary of Formulas for Continuous-Time and Discrete-Time Systems

Formula	Continuous-Time	Discrete-Time
White noise	$\mathbf{R}_w(\tau) = \mathbf{S}_w\delta(\tau)$	$\mathbf{R}_w(n) = \mathbf{\Sigma}_w\delta_d(n)\,;\ \mathbf{\Sigma}_w = \dfrac{\mathbf{S}_w}{T}$
State model	$\dot{x}(t) = \mathbf{A}x(t) + \mathbf{B}w(t)\,;\ y(t) = \mathbf{C}x(t)$	$x(k+1) = \mathbf{\Phi}x(k) + \mathbf{\Gamma}w(k)\,;\ y(k) = \mathbf{C}x(k)$
State correlation matrix of a linear system with white noise input	$\mathbf{\Sigma}_x(t) = e^{\mathbf{A}t}\mathbf{\Sigma}_x(0)e^{\mathbf{A}^Tt} + \int_0^t e^{\mathbf{A}\gamma}\mathbf{BS}_w\mathbf{B}^Te^{\mathbf{A}^T\gamma}d\gamma$	$\mathbf{\Sigma}_x(n) = \mathbf{\Phi}^n\mathbf{\Sigma}_x(0)(\mathbf{\Phi}^n)^T + \sum_{k=0}^{n}\mathbf{\Phi}^k\mathbf{\Gamma}\mathbf{\Sigma}_w\mathbf{\Gamma}^T(\mathbf{\Phi}^k)^T$
	$\dot{\mathbf{\Sigma}}_x(t) = \mathbf{A}\mathbf{\Sigma}_x(t) + \mathbf{\Sigma}_x(t)\mathbf{A}^T + \mathbf{BS}_w\mathbf{B}^T$	$\mathbf{\Sigma}_x(k+1) = \mathbf{\Phi}\mathbf{\Sigma}_x(k)\mathbf{\Phi}^T + \mathbf{\Gamma}\mathbf{\Sigma}_w\mathbf{\Gamma}^T$
	$\mathbf{\Sigma}_x(\infty) = \int_0^\infty e^{\mathbf{A}\gamma}\mathbf{BS}_w\mathbf{B}^Te^{\mathbf{A}^T\gamma}d\gamma$	$\mathbf{\Sigma}_x(\infty) = \sum_{k=0}^{\infty}\mathbf{\Phi}^k\mathbf{\Gamma}\mathbf{\Sigma}_w\mathbf{\Gamma}^T(\mathbf{\Phi}^k)^T$
	$\mathbf{A}\mathbf{\Sigma}_x(\infty) + \mathbf{\Sigma}_x(\infty)\mathbf{A}^T + \mathbf{BS}_w\mathbf{B}^T = \mathbf{0}$	$\mathbf{\Sigma}_x(\infty) = \mathbf{\Phi}\mathbf{\Sigma}_x(\infty)\mathbf{\Phi}^T + \mathbf{\Gamma}\mathbf{\Sigma}_w\mathbf{\Gamma}^T$
Output correlation matrix of a linear system with white noise input	$\mathbf{\Sigma}_y(t) = \mathbf{C}\mathbf{\Sigma}_x(t)\mathbf{C}^T$	$\mathbf{\Sigma}_y(n) = \mathbf{C}\mathbf{\Sigma}_x(n)\mathbf{C}^T$
Output spectral density of a linear system with white noise input	$\mathbf{S}_y(\omega) = \mathbf{G}(-j\omega)\mathbf{S}_w\mathbf{G}^T(j\omega)$	$\mathbf{S}_y(\omega) = \mathbf{G}_d(e^{-j\omega T})\mathbf{\Sigma}_w\mathbf{G}_d^T(e^{j\omega T})$

REFERENCES

[1] A. Papoulis, *Probability, Random Variables, and Stochastic Processes,* 3d ed., McGraw-Hill, New York, 1991.

[2] P. Hagander, "Numerical solution of $\mathbf{A}^T\mathrm{S} + \mathbf{SA} + \mathbf{Q} = \mathbf{0}$," *Information Science* 4 (1972): 35–50.

[3] S. M. Kay, *Modern Spectral Estimation: Theory and Application,* Prentice-Hall, Englewood Cliffs, NJ, 1988.

[4] S. L. Marple, Jr., *Digital Spectral Analysis with Applications,* Prentice-Hall, Englewood Cliffs, NJ, 1987.

[5] C. W. Therrien, *Discrete Random Signals and Statistical Signal Processing,* Prentice-Hall, Englewood Cliffs, NJ, 1992.

Some additional references on stochastic processes follow:

[6] R. G. Brown and P. Y. C. Hwang, *Introduction to Random Signals and Applied Kalman Filtering,* 2d ed., John Wiley and Sons, New York, 1992.

[7] W. A. Gardner, *Introduction to Random Processes with Applications to Signals and Systems,* 2d ed., McGraw-Hill, New York, 1990.

[8] P. Z. Peebles, Jr., *Probability, Random Variables, and Random Signal Principles,* 3d ed., McGraw-Hill, New York, 1993.

[9] G. R. Cooper and C. D. McGillem, *Probabilistic Methods of Signal and System Analysis,* 2d ed., Holt, Rinehart, and Winston, New York, 1971.

Some estimation and control books that include sections on linear system response to white noise inputs follow:

[10] B. Friedland, *Control System Design: An Introduction to State Space Methods,* McGraw-Hill, New York, 1986.

[11] H. Kwakernaak and R. Sivan, *Linear Optimal Control Systems,* Wiley-Interscience, New York, 1972.

[12] M. S. Grewal and A. P. Andrews, *Kalman Filtering: Theory and Practice,* Prentice-Hall, Englewood Cliffs, NJ, 1993.

[13] G. E. Franklin, J. D. Powell, and M. L. Workman, *Digital Control of Dynamic Systems,* 2d ed., Addison-Wesley, Reading, MA, 1990.

EXERCISES

3.1 Given two uncorrelated random variables, z_1 and z_2 with the density functions,

$$f(z_1) = \begin{Bmatrix} 1 & z_1 \in [0 \quad 1] \\ 0 & z_1 \notin [0 \quad 1] \end{Bmatrix}; f(z_2) = \begin{Bmatrix} 1 & z_2 \in [0 \quad 1] \\ 0 & z_2 \notin [0 \quad 1] \end{Bmatrix}.$$

For

$$x = \begin{bmatrix} x_1 \\ x_2 \end{bmatrix} = \begin{bmatrix} z_1 + z_2 \\ z_1 - z_2 \end{bmatrix},$$

a. Compute the mean value of x.
b. Compute the covariance matrix of x.
c. What is the variance of x_1?

3.2 Show that the covariance matrix and the correlation matrix are related:

$$\Xi_x(t) = E[\{x(t) - m_x(t)\}\{x(t) - m_x(t)\}^T] = \mathbf{\Sigma}_x(t) - m_x(t)m_x^T(t).$$

3.3 Given a stationary, zero mean, random process, $w(t)$ has the correlation function:

$$R_w(\tau) = \sigma_w^2 e^{-a|\tau|}.$$

a. Generate an expression for the spectral density of $w(t)$.
b. Generate an expression for the correlation time of $w(t)$.

3.4 Given the system

$$\dot{x}(t) = -2x(t) + 3w(t);\ y(t) = 4x(t),$$

where $w(t)$ is white noise with a spectral density of 16, and $x(0) = 0$, do the following:

a. Compute the variance of $x(t)$ using (3.9).
b. Compute the variance of $x(t)$ using (3.11).
c. Compute the variance of $y(t)$.
d. Compute the steady-state variance of $x(t)$ using (3.14).

3.5 Derive the expression for the state correlation matrix of a linear, time-invariant system driven by white noise:

$$\mathbf{\Sigma}_x(t) = e^{\mathbf{A}t}\mathbf{\Sigma}_x(0)e^{\mathbf{A}^Tt} + \int_0^t e^{\mathbf{A}\gamma}\mathbf{B}\mathbf{S}_w\mathbf{B}^T e^{\mathbf{A}^T\gamma}d\gamma.$$

Note that the white noise input is uncorrelated with the state initial condition. Derive this result directly; that is, do not use the expression given in the text for the state correlation function.

3.6 An ac motor control system is shown in Figure P3.1. The actuator noise $w(t)$ is white with a spectral density of 100, the measurement noise $v(t)$ is white with a spectral density of 0.01, and the reference input $r(t)$ is a nonrandom constant. In addition, the actuator noise and the measurement noise are independent.

a. Find the steady-state covariance matrix of the state of this system. Define the states to be the output of the motor and the derivative of the motor output.
b. Find the steady-state variances of the error $e(t)$ due to the measurement noise, due to the actuator noise, and due to the combination of both noises. How is the combined variance related to the variance due to each individual noise?

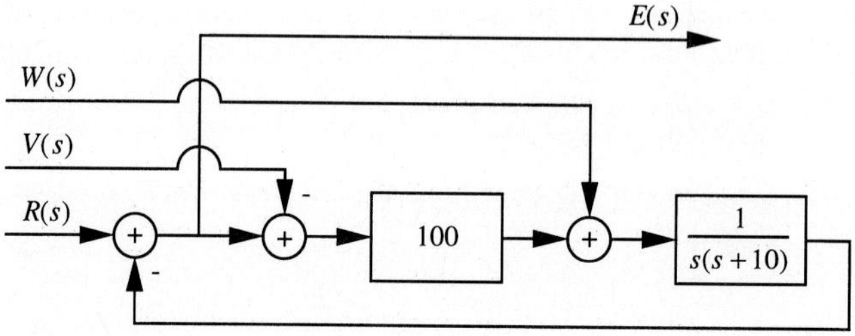

FIGURE P3.1 An ac motor control system with random inputs

3.7 The plant described by the state model

$$\dot{x}(t) = \begin{bmatrix} 0 & 1 & 0 \\ 0 & 0 & 1 \\ 0 & -8 & -4 \end{bmatrix} x(t) + \begin{bmatrix} 0 \\ 0 \\ 2 \end{bmatrix} u(t);\ y(t) = [1 \quad 0 \quad 0]x(t)$$

is being controlled using a proportional controller,

$$u(t) = -3\{y(t) + v(t)\},$$

where $v(t)$ is measurement noise. The measurement noise is assumed to be white with a spectral density of 5. For the closed-loop system, find the steady-state covariance matrices of the state and the output.

3.8 A time-varying linear plant,

$$\dot{x}(t) = \mathbf{A}(t)x(t) + \mathbf{B}(t)w(t);\ y(t) = \mathbf{C}(t)x(t),$$

has zero initial conditions, and is subject to an input with the general correlation matrix $\mathbf{R}_w(t_1, t_2)$. Show that the state correlation function of this plant is

$$\mathbf{R}_x(t_1, t_2) = \int_0^{t_1}\int_0^{t_2} \mathbf{\Phi}(t_1, \tau)\mathbf{B}(\tau)\mathbf{R}_w(\tau, \gamma)\mathbf{B}^T(\gamma)\mathbf{\Phi}^T(t_2, \gamma)d\gamma d\tau.$$

3.9 The discrete-time state equation

$$x(k + 1) = \mathbf{\Phi}x(k) + \mathbf{\Gamma}w(k)$$

is driven by stationary, discrete-time white noise:

$$E[w(k)w^T(l)] = \mathbf{\Sigma}_w\delta_{kl} = \begin{cases} \mathbf{\Sigma}_w & k = l \\ 0 & k \neq l \end{cases}.$$

Derive a difference equation that describes the evolution of the state covariance matrix. *Hint:* Substitute the state equation into the state covariance matrix:

$$\mathbf{\Sigma}_x(k + 1) = E[x(k + 1)x^T(k + 1)],$$

and simplify the result.

3.10 Derive the following limiting state correlation function:

$$\mathbf{R}_x(\tau) = \int_{\max(0,-\tau)}^{\infty} e^{\mathbf{A}\gamma}\mathbf{B}\mathbf{S}_w\mathbf{B}^T e^{\mathbf{A}^T(\gamma+\tau)}d\gamma.$$

Hint: The limiting state correlation function can be obtained by letting $t_2 = t_1 + \tau$ in (3.7), and taking the limit as t_1 approaches infinity. A change of variables is also required to obtain this form of the result.

3.11 Let $x(t)$ be a random signal with zero mean, a variance of 4, and a correlation time of 0.5 seconds.

a. Generate a reasonable correlation function for this random process.
b. What is the spectral density of a reasonable white noise approximation of this signal?

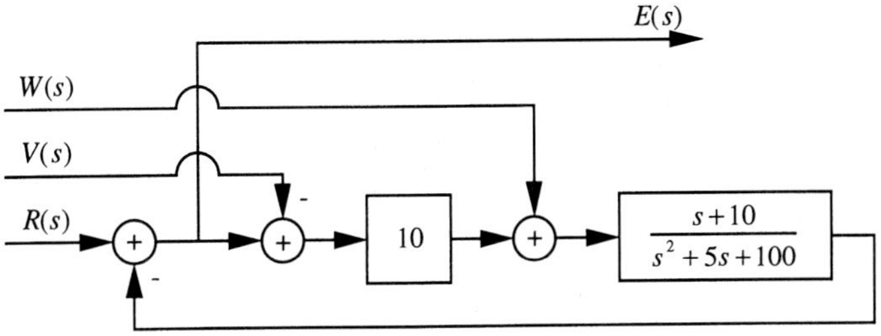

FIGURE P3.2 A unity feedback control system with measurement and actuator noise

3.12 You are given the control system in Figure P3.2. The measurement noise $v(t)$ is white with a spectral density of 0.1, and the reference input $r(t)$ is a nonrandom constant. In addition, the actuator noise $w(t)$ and the measurement noise are independent. Find the steady-state covariance matrix of the error given the following:

a. The actuator noise is white with a spectral density of 10.
b. The actuator noise has a correlation time of 0.1 second and a variance of 100.
c. The actuator noise has a correlation time of 0.001 second and a variance of 10,000.
d. The actuator noise has a correlation time of 10^{-5} second and a variance of 10^6.

COMPUTER EXERCISES

3.1 Optimal Angular Velocity Estimation for an AC Motor

A tachometer is used to measure the angular velocity of an ac motor. These tachometer measurements include noise. Filtering may be used to reduce this noise and therefore generate improved angular velocity estimates. The accuracy of these estimates can be computed using the random process theory presented in this chapter. Filter parameters can then be adjusted to optimize the accuracy of the estimates.

The variance of the estimation error,

$$e(t) = \omega(t) - \hat{\omega}(t),$$

is used as a measure of estimator accuracy. Estimation errors are generated by both random plant inputs (plant noise) and measurement noise. Computation of the estimation error variance then requires a model of the ac motor, a description of the filter, and a probabilistic description of both the plant and the measurement noises.

A mathematical model for the ac motor is

$$\dot{\omega}(t) = -\tfrac{1}{10}\omega(t) + 2\{u(t) + w(t)\},$$

where $\omega(t)$ is the angular velocity of the motor in deg/sec. The ac voltage applied to the motor $\{u(t) + w(t)\}$ is the sum of a deterministic voltage $u(t)$ and random voltage $w(t)$, both in volts. The random applied voltage is assumed to be white noise with the following spectral density:

$$S_w = 1\frac{\text{volts}^2}{\text{Hz}}.$$

The noisy angular velocity measurements are

$$m(t) = \omega(t) + v(t).$$

The measurement noise $v(t)$ is assumed to be white noise with the following spectral density:

$$S_v = 100\frac{\text{deg}^2}{\text{sec}^2 - \text{Hz}},$$

and to be uncorrelated with the plant noise.

A Luenberger observer is used to estimate angular velocity. This observer is described by the differential equation

$$\dot{\hat{\omega}}(t) = -\tfrac{1}{10}\hat{\omega}(t) + 2u(t) + G[m(t) - \hat{\omega}(t)],$$

where G is the observer gain.

The estimation error variance can be computed from the state model formed by appending the observer equation to the plant dynamics. This state model has three inputs: the input voltage $u(t)$, the plant noise $w(t)$, and the measurement noise $v(t)$. The estimation error is the output of this model. The steady-state estimation error variance can be given as follows:

$$\sigma_e^2 = S_w \int_0^\infty g_{ew}^2(t)dt + S_v \int_0^\infty g_{ev}^2(t)dt,$$

where $g_{ew}(t)$ is the impulse response from the plant noise to the estimation error, and $g_{ev}(t)$ is the impulse response from the measurement noise to the estimation error. Note that the two terms in this expression are the variance of the estimation error due to the plant and measurement noises, respectively. The impulse response from the deterministic input $u(t)$ does not appear in this expression since this input does not contain randomness.

For various values of G, generate the error variances due to plant noise, measurement noise, and both noises by numerical integration of the simulated impulse response. As a check, solve the Lyapunov equation for the state covariance matrix, and compute the estimation error variance from this matrix. This result should match the total error variance obtained using the impulse response method.

Plot the variance of the estimation error due to the plant noise, the measurement noise, and both noises as a function of G. Find the value of G that minimizes the estimation error variance. Comment on the effects of G on the estimation error variances due to the individual noises.

3.2 Conveyer Belt Positioning System Performance Evaluation

A dc motor is driving a conveyor belt used to transport materials on an assembly line from station to station. This motor is modeled as follows:

$$\begin{bmatrix} \dot{x}_1(t) \\ \dot{x}_2(t) \\ \dot{x}_3(t) \end{bmatrix} = \begin{bmatrix} -50 & 0 & 0 \\ 10 & -350 & 0 \\ 0 & 1 & 0 \end{bmatrix} \begin{bmatrix} x_1(t) \\ x_2(t) \\ x_3(t) \end{bmatrix} + \begin{bmatrix} 10{,}000 \\ 0 \\ 0 \end{bmatrix} u(t) + \begin{bmatrix} 0 \\ 100 \\ 0 \end{bmatrix} w(t);$$

$$m(t) = \begin{bmatrix} 0 & 0 & 1 \end{bmatrix} \begin{bmatrix} x_1(t) \\ x_2(t) \\ x_3(t) \end{bmatrix} + v(t)$$

where $u(t)$ is the motor field voltage in volts (the control input), $w(t)$ is the load torque in N-m, $m(t)$ is the measured motor shaft angle in degrees, and $v(t)$ is the measurement noise in degrees. The load torque varies slowly as material is added and removed from the conveyor belt, and can be modeled as a random process with a variance of 1000 (N^2-m^2) and a correlation time of 2 seconds. The measurement noise is white with a spectral density of 0.01 deg^2/Hz. The measurement noise and load torque are assumed to be uncorrelated.

This motor is being controlled with the lead compensator (see Figure P3.3):

$$\dot{x}_c(t) = -300x_c(t) - 14{,}400\{r(t) - m(t)\};$$

$$u(t) = x_c(t) + 60\{r(t) - m(t)\}$$

where $r(t)$ is the desired position of the motor shaft in degrees. The desired position of the motor shaft is not random.

Generate a state model driven by white noise (and deterministic inputs) for this system; that is, the nonwhite load torque should be generated using a shaping filter. Compute the

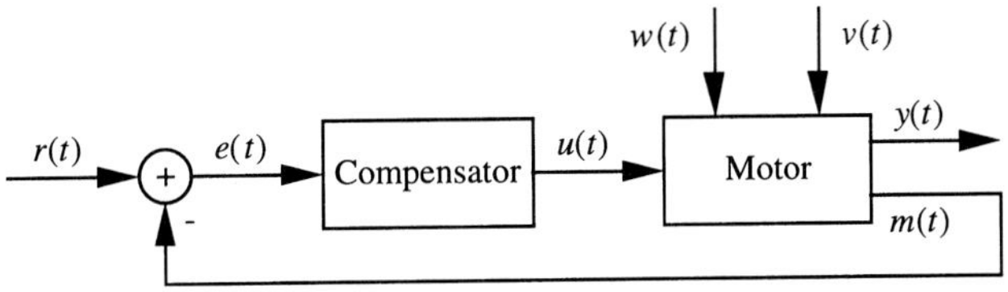

FIGURE P3.3 Conveyor belt control system

standard deviation of the motor shaft angle due to the load torque, due to the measurement noise, and due to the combination of the load torque and the measurement noise. What percentage of the time is the motor shaft angle within 1° of the desired angle when this system is operating?

4 Performance

Control systems are designed to maintain the system output at a desired value. The output is a vector, in general, which consists of those linear combinations of the plant states and inputs that are important in terms of system performance. A number of factors influence the control system's ability to maintain the desired output, including initial conditions, the desired output, and external disturbance inputs.

The control system must remove the effects of unknown initial conditions on the output. The initial condition response of the system also describes the transient response of the system subject to inputs applied at a given time, such as steps, ramps, and partial sinusoids. The term transient performance is, therefore, used to describe the decay of initial conditions in a control system.

The error between the desired output and the actual output is dependent on the desired output. For example, a control system is better able to match the output to a constant desired value than to a desired output that varies rapidly. A control system designed to force the output to follow a nonzero desired output is termed a tracking system, and its performance is termed tracking performance.

The control system must maintain the actual output close to the desired output in the presence of disturbances. Disturbances are inputs to the plant that are not generated by the controller. For example, an airplane autopilot, tasked with maintaining a constant altitude, must compensate for the effects of gravity, which is a disturbance input. The ability of a control system to limit the effects of disturbances is termed disturbance rejection.

Quantitative performance criteria are required to evaluate control systems and to compare the merits of competing control system designs. These criteria can include closed-loop system parameters, the response of the closed-loop system subject to specific test conditions, and the gain of the closed-loop system. For example, the closed-loop poles and the damping ratio are system parameters that can be used as performance criteria. The percent overshoot is an example of a performance criterion based on the response of the closed-loop system to a specific test condition (in this case, the step input). The closed-loop gain from a disturbance input to an output of interest can also be used as a performance criterion.

An important type of performance criterion is the cost function. The cost function is a real, non-negative, quantitative measure of performance, where superior performance is indicated by a smaller numerical value. Evaluating performance with a cost function allows a direct comparison between competing control system designs, since the real numbers are ordered. The use of cost functions also enables control system optimization, that is, the design of a control system to minimize the cost.

This chapter begins by defining a very general model of a feedback control system that explicitly includes the desired outputs and disturbance inputs. Transient performance, tracking performance, and disturbance rejection are then discussed. A number of cost functions, which find application in control system design, are also presented and discussed, along with methods for computing the various performance criteria and costs.

4.1 General Models of Feedback Control Systems

A block diagram of a general feedback control system is shown in Figure 4.1. The disturbance input $w_0(t)$ is the vector of inputs to the plant that are not generated by the control system. Measurement noise is included as a disturbance input since the measurement process is included as part of the plant in this model. The reference input $r(t)$ is an input to the control system that specifies the desired behavior of some or all of the plant outputs, and is equal to the desired value of these outputs. Note that reference inputs are only included for those outputs that have nonzero desired values. The reference input and the disturbance input are both external inputs to the plant and can be combined into a single input:

$$w(t) = \begin{bmatrix} r(t) \\ \hdashline w_0(t) \end{bmatrix}, \tag{4.1}$$

where $w(t)$ is called a generalized disturbance input [or simply a disturbance input when this does not lead to confusion with $w_0(t)$]. The control input $u(t)$ is the vector of inputs to the plant that are generated by the control system. The reference output $y(t)$ is the vector of plant outputs that are of interest. These outputs include the errors between plant states and the desired values of these states (or linear combinations of the plant states), which are termed the output errors. Additionally, the control input can be incorporated into the reference output when desired. The measured output $m(t)$ is the vector of plant outputs that are directly measured and, therefore, available for feedback to the controller. In general, the measured output is distinct from the reference output, although they may be identical in some applications.

A well-designed control system should keep the output errors small in the presence of both changing reference inputs and disturbance inputs. Intuitively, disturbances are inputs that tend to generate undesirable behavior in the plant. The reference input, on the other hand, is an input that tends to generate desirable behavior in the plant. These

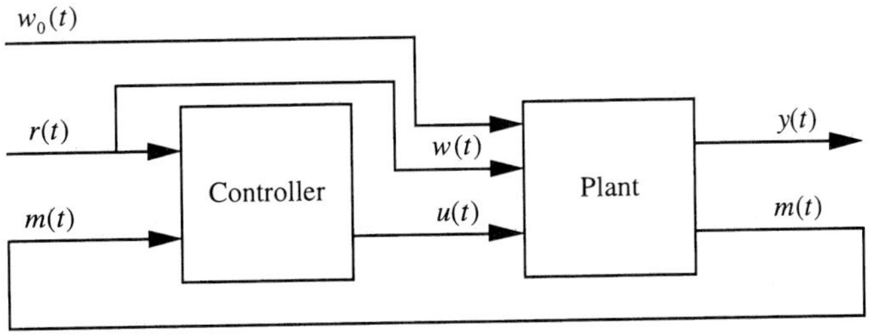

FIGURE 4.1 A general closed-loop control system

inputs are lumped together as generalized disturbances since they both tend to generate nonzero output errors: the disturbances by changing the plant state, and the reference input by changing the desired behavior of the plant state. Further, a well-designed control system should use a reasonable amount of control, that is, maintain the control inputs at sufficiently small levels that the actuators are not saturated and do not utilize excessive amounts of energy, fuel, and so on. Therefore, it may be desirable to include the control input as part of the reference output in some applications. The reference output can then be used exclusively to evaluate control system performance.

The state model and the transfer function model for the closed-loop system in Figure 4.1 are developed below. The state model of the plant is

$$\dot{x}(t) = \mathbf{A}x(t) + [\mathbf{B}_u \mid \mathbf{B}_w]\begin{bmatrix} u(t) \\ \hline w(t) \end{bmatrix}; \tag{4.2a}$$

$$\begin{bmatrix} m(t) \\ \hline y(t) \end{bmatrix} = \begin{bmatrix} \mathbf{C}_m \\ \hline \mathbf{C}_y \end{bmatrix} x(t) + \begin{bmatrix} \mathbf{0} & \mathbf{D}_{mw} \\ \hline \mathbf{D}_{yu} & \mathbf{D}_{yw} \end{bmatrix}\begin{bmatrix} u(t) \\ \hline w(t) \end{bmatrix}. \tag{4.2b}$$

In general, $w(t) = [r(t) \quad w_0(t)]^T$ includes disturbance inputs to the plant that enter through $\mathbf{B}_w$, errors in the measurement that enter through $\mathbf{D}_{mw}$, and reference inputs that enter through $\mathbf{D}_{yw}$. Note that any number of these disturbances may be absent. The term $\mathbf{D}_{yu}$ results in input-to-output coupling between the control input and the reference output. This coupling is typically used to incorporate the control input into the reference output. It is assumed that there is no input-to-output coupling between the control input and the measured output. This assumption is valid in almost all applications and simplifies the algebra in deriving future results. The transfer function model of the plant is

$$\begin{bmatrix} M(s) \\ \hline Y(s) \end{bmatrix} = \mathbf{G}(s)\begin{bmatrix} U(s) \\ \hline W(s) \end{bmatrix} = \begin{bmatrix} \mathbf{G}_{mu}(s) & \mathbf{G}_{mw}(s) \\ \hline \mathbf{G}_{yu}(s) & \mathbf{G}_{yw}(s) \end{bmatrix}\begin{bmatrix} U(s) \\ \hline W(s) \end{bmatrix} \tag{4.3a}$$

where

$$\mathbf{G}_{mu}(s) = \mathbf{C}_m(s\mathbf{I} - \mathbf{A})^{-1}\mathbf{B}_u; \tag{4.3b}$$

$$\mathbf{G}_{mw}(s) = \mathbf{C}_m(s\mathbf{I} - \mathbf{A})^{-1}\mathbf{B}_w + \mathbf{D}_{mw}; \tag{4.3c}$$

$$\mathbf{G}_{yu}(s) = \mathbf{C}_y(s\mathbf{I} - \mathbf{A})^{-1}\mathbf{B}_u + \mathbf{D}_{yu}; \tag{4.3d}$$

$$\mathbf{G}_{yw}(s) = \mathbf{C}_y(s\mathbf{I} - \mathbf{A})^{-1}\mathbf{B}_w + \mathbf{D}_{yw}. \tag{4.3e}$$

The state model of the controller in Figure 4.1 is

$$\dot{x}_c(t) = \mathbf{A}_c x_c(t) + [\mathbf{B}_{cr} \mid \mathbf{B}_{cm}]\begin{bmatrix} r(t) \\ \hline m(t) \end{bmatrix}; \tag{4.4a}$$

$$u(t) = \mathbf{C}_c x_c(t) + [\mathbf{D}_{cr} \mid \mathbf{D}_{cm}]\begin{bmatrix} r(t) \\ \hline m(t) \end{bmatrix}. \tag{4.4b}$$

The transfer function model of the controller is

$$U(s) = \mathbf{G}_c(s)\begin{bmatrix} R(s) \\ \hdashline M(s) \end{bmatrix} = [\mathbf{G}_{cr}(s) \;\vdots\; \mathbf{G}_{cm}(s)]\begin{bmatrix} R(s) \\ \hdashline M(s) \end{bmatrix} \tag{4.5a}$$

where

$$\mathbf{G}_{cr}(s) = \mathbf{C}_c(s\mathbf{I} - \mathbf{A}_c)^{-1}\mathbf{B}_{cr} + \mathbf{D}_{cr}; \tag{4.5b}$$

$$\mathbf{G}_{cm}(s) = \mathbf{C}_c(s\mathbf{I} - \mathbf{A}_c)^{-1}\mathbf{B}_{cm} + \mathbf{D}_{cm}. \tag{4.5c}$$

The difference, $[r(t) - m(t)]$, is the only input to the controller in many applications. This is an error feedback system and is a special case of the controller described above with

$$\mathbf{B}_{cm} = -\mathbf{B}_{cr}\ ;\ \mathbf{D}_{cm} = -\mathbf{D}_{cr}\ ;\ \mathbf{G}_{cm}(s) = -\mathbf{G}_{cr}(s).$$

A state model of the closed-loop system is obtained by combining the model of the plant with the model of the controller. Appending (4.4a) to (4.2a) yields

$$\begin{bmatrix} \dot{x}(t) \\ \hdashline \dot{x}_c(t) \end{bmatrix} = \begin{bmatrix} \mathbf{A} & \mathbf{0} \\ \hdashline \mathbf{0} & \mathbf{A}_c \end{bmatrix}\begin{bmatrix} x(t) \\ \hdashline x_c(t) \end{bmatrix} + \begin{bmatrix} \mathbf{B}_u \\ \hdashline \mathbf{0} \end{bmatrix} u(t) + \begin{bmatrix} \mathbf{B}_w \\ \hdashline \mathbf{0} \end{bmatrix} w(t)$$
$$+ \begin{bmatrix} \mathbf{0} \\ \hdashline \mathbf{B}_{cr} \end{bmatrix} r(t) + \begin{bmatrix} \mathbf{0} \\ \hdashline \mathbf{B}_{cm} \end{bmatrix} m(t),$$

where all of the inputs are displayed as individual terms. The control input is the output of the controller (4.4b):

$$\begin{bmatrix} \dot{x}(t) \\ \hdashline \dot{x}_c(t) \end{bmatrix} = \begin{bmatrix} \mathbf{A} & \mathbf{0} \\ \hdashline \mathbf{0} & \mathbf{A}_c \end{bmatrix}\begin{bmatrix} x(t) \\ \hdashline x_c(t) \end{bmatrix} + \begin{bmatrix} \mathbf{B}_u \\ \hdashline \mathbf{0} \end{bmatrix}\{\mathbf{C}_c x_c(t) + \mathbf{D}_{cr} r(t) + \mathbf{D}_{cm} m(t)\}$$
$$+ \begin{bmatrix} \mathbf{B}_w \\ \hdashline \mathbf{0} \end{bmatrix} w(t) + \begin{bmatrix} \mathbf{0} \\ \hdashline \mathbf{B}_{cr} \end{bmatrix} r(t) + \begin{bmatrix} \mathbf{0} \\ \hdashline \mathbf{B}_{cm} \end{bmatrix} m(t).$$

Combining like terms yields

$$\begin{bmatrix} \dot{x}(t) \\ \hdashline \dot{x}_c(t) \end{bmatrix} = \begin{bmatrix} \mathbf{A} & \mathbf{B}_u\mathbf{C}_c \\ \hdashline \mathbf{0} & \mathbf{A}_c \end{bmatrix}\begin{bmatrix} x(t) \\ \hdashline x_c(t) \end{bmatrix} + \begin{bmatrix} \mathbf{B}_w \\ \hdashline \mathbf{0} \end{bmatrix} w(t)$$
$$+ \begin{bmatrix} \mathbf{B}_u\mathbf{D}_{cr} \\ \hdashline \mathbf{B}_{cr} \end{bmatrix} r(t) + \begin{bmatrix} \mathbf{B}_u\mathbf{D}_{cm} \\ \hdashline \mathbf{B}_{cm} \end{bmatrix} m(t).$$

The measured output is given by (4.2b):

$$\begin{bmatrix} \dot{x}(t) \\ \hdashline \dot{x}_c(t) \end{bmatrix} = \begin{bmatrix} \mathbf{A} & \mathbf{B}_u\mathbf{C}_c \\ \hdashline \mathbf{0} & \mathbf{A}_c \end{bmatrix}\begin{bmatrix} x(t) \\ \hdashline x_c(t) \end{bmatrix} + \begin{bmatrix} \mathbf{B}_w \\ \hdashline \mathbf{0} \end{bmatrix} w(t) + \begin{bmatrix} \mathbf{B}_u\mathbf{D}_{cr} \\ \hdashline \mathbf{B}_{cr} \end{bmatrix} r(t)$$
$$+ \begin{bmatrix} \mathbf{B}_u\mathbf{D}_{cm} \\ \hdashline \mathbf{B}_{cm} \end{bmatrix} [\mathbf{C}_m x(t) + \mathbf{D}_{mw} w(t)].$$

Again, combining like terms,

$$\begin{bmatrix} \dot{x}(t) \\ \hdashline \dot{x}_c(t) \end{bmatrix} = \begin{bmatrix} \mathbf{A} + \mathbf{B}_u\mathbf{D}_{cm}\mathbf{C}_m & \mathbf{B}_u\mathbf{C}_c \\ \hdashline \mathbf{B}_{cm}\mathbf{C}_m & \mathbf{A}_c \end{bmatrix} \begin{bmatrix} x(t) \\ \hdashline x_c(t) \end{bmatrix} + \begin{bmatrix} \mathbf{B}_w + \mathbf{B}_u\mathbf{D}_{cm}\mathbf{D}_{mw} \\ \hdashline \mathbf{B}_{cm}\mathbf{D}_{mw} \end{bmatrix} w(t) + \begin{bmatrix} \mathbf{B}_u\mathbf{D}_{cr} \\ \hdashline \mathbf{B}_{cr} \end{bmatrix} r(t).$$

Noting that the reference input forms the first part of the generalized disturbance input, the state equation of the closed-loop system is

$$\begin{aligned} \begin{bmatrix} \dot{x}(t) \\ \hdashline \dot{x}_c(t) \end{bmatrix} &= \begin{bmatrix} \mathbf{A} + \mathbf{B}_u\mathbf{D}_{cm}\mathbf{C}_m & \mathbf{B}_u\mathbf{C}_c \\ \hdashline \mathbf{B}_{cm}\mathbf{C}_m & \mathbf{A}_c \end{bmatrix} \begin{bmatrix} x(t) \\ \hdashline x_c(t) \end{bmatrix} \\ &\quad + \begin{bmatrix} \mathbf{B}_w + \mathbf{B}_u\mathbf{D}_{cm}\mathbf{D}_{mw} + [\mathbf{B}_u\mathbf{D}_{cr} \;\vdots\; \mathbf{0}] \\ \hdashline \mathbf{B}_{cm}\mathbf{D}_{mw} + [\mathbf{B}_{cr} \;\vdots\; \mathbf{0}] \end{bmatrix} w(t) \\ &= \mathbf{A}_{cl} \begin{bmatrix} x(t) \\ \hdashline x_c(t) \end{bmatrix} + \mathbf{B}_{cl} w(t), \end{aligned} \tag{4.6}$$

where $\mathbf{A}_{cl}$ and $\mathbf{B}_{cl}$ are defined above, and the partitioning of the matrices within $\mathbf{B}_{cl}$ is defined by the partitioning of the generalized disturbance input (4.1).

The output of the closed-loop system is the reference output as given in (4.2b):

$$y(t) = \mathbf{C}_y x(t) + \mathbf{D}_{yu} u(t) + \mathbf{D}_{yw} w(t).$$

Noting that the control input is the output of the controller (4.4b):

$$y(t) = \mathbf{C}_y x(t) + \mathbf{D}_{yu}\{\mathbf{C}_c x_c(t) + \mathbf{D}_{cr} r(t) + \mathbf{D}_{cm} m(t)\} + \mathbf{D}_{yw} w(t).$$

Rearranging and substituting for $m(t)$ using (4.2b) gives:

$$\begin{aligned} y(t) &= \mathbf{C}_y x(t) + \mathbf{D}_{yu}\mathbf{C}_c x_c(t) + \mathbf{D}_{yu}\mathbf{D}_{cr} r(t) \\ &\quad + \mathbf{D}_{yu}\mathbf{D}_{cm}\{\mathbf{C}_m x(t) + \mathbf{D}_{mw} w(t)\} + \mathbf{D}_{yw} w(t). \end{aligned}$$

Combining terms yields the measurement equation for the closed-loop system:

$$\begin{aligned} y(t) &= [\mathbf{C}_y + \mathbf{D}_{yu}\mathbf{D}_{cm}\mathbf{C}_m \;\vdots\; \mathbf{D}_{yu}\mathbf{C}_c] \begin{bmatrix} x(t) \\ \hdashline x_c(t) \end{bmatrix} \\ &\quad + \{\mathbf{D}_{yw} + \mathbf{D}_{yu}\mathbf{D}_{cm}\mathbf{D}_{mw} + [\mathbf{D}_{yu}\mathbf{D}_{cr} \;\vdots\; \mathbf{0}]\} w(t) \\ &= \mathbf{C}_{cl} \begin{bmatrix} x(t) \\ \hdashline x_c(t) \end{bmatrix} + \mathbf{D}_{cl} w(t) \end{aligned} \tag{4.7}$$

where $\mathbf{C}_{cl}$ and $\mathbf{D}_{cl}$ are defined above, and the partitioning of the matrix within $\mathbf{D}_{cl}$ is defined by the partitioning of the generalized disturbance input (4.1). Equations (4.6) and (4.7) form the state model of the closed-loop system. The input to this system is the generalized disturbance input, which consists of both the true disturbance inputs and the reference inputs. The output of this system is the reference output of the plant, which may contain the output errors, the control input, and other linear combinations of the plant states.

The transfer function of the closed-loop system can be found in a straightforward manner from the state model in (4.6) and (4.7). Alternatively, the transfer function of the closed-loop system can be found from the plant and controller transfer functions as shown below. The measured output and the control input are given by (4.3a) and (4.5a), respectively:

$$M = [\mathbf{G}_{mu} \;\vdots\; \mathbf{G}_{mw}]\begin{bmatrix} U \\ \text{----} \\ W \end{bmatrix} = \mathbf{G}_{mu}U + \mathbf{G}_{mw}W;$$

$$U = [\mathbf{G}_{cr} \;\vdots\; \mathbf{G}_{cm}]\begin{bmatrix} R \\ \text{----} \\ M \end{bmatrix} = \mathbf{G}_{cr}R + \mathbf{G}_{cm}M.$$

Note that the argument s has been dropped to simplify the notation in the following development. Substituting for the measurement, the control input becomes

$$U = (\mathbf{I} - \mathbf{G}_{cm}\mathbf{G}_{mu})^{-1}\mathbf{G}_{cr}R + (\mathbf{I} - \mathbf{G}_{cm}\mathbf{G}_{mu})^{-1}\mathbf{G}_{cm}\mathbf{G}_{mw}W.$$

Substituting for the control input in (4.3a) yields

$$Y = \mathbf{G}_{yu}\{(\mathbf{I} - \mathbf{G}_{cm}\mathbf{G}_{mu})^{-1}\mathbf{G}_{cr}R + (\mathbf{I} - \mathbf{G}_{cm}\mathbf{G}_{mu})^{-1}\mathbf{G}_{cm}\mathbf{G}_{mw}W\} + \mathbf{G}_{yw}W.$$

Recognizing that the reference input is part of the generalized disturbance input and combining terms results in

$$\begin{aligned} Y = \{&[\mathbf{G}_{yu}(\mathbf{I} - \mathbf{G}_{cm}\mathbf{G}_{mu})^{-1}\mathbf{G}_{cr} \;\vdots\; \mathbf{0}] \\ &+ \mathbf{G}_{yu}(\mathbf{I} - \mathbf{G}_{cm}\mathbf{G}_{mu})^{-1}\mathbf{G}_{cm}\mathbf{G}_{mw} + \mathbf{G}_{yw}\}W = \mathbf{G}_{cl}W \end{aligned} \tag{4.8}$$

where $\mathbf{G}_{cl}$ is the transfer function model of the closed-loop control system, and is defined above. Note that the partitioning of the matrix within this transfer function model is defined by (4.1).

The state model and the transfer function model of the closed-loop system are fairly complex, but very general. These models provide a framework for doing all of the performance analysis used in linear control system design. Both models can be simplified for a number of special cases that are of interest in various applications. These include the following cases:

- The generalized disturbance is equal to the reference input; that is, there are no true disturbance inputs.
- The generalized disturbance is equal to the true disturbance input; that is, there are no reference inputs.
- The reference output is composed of only output errors.
- The reference output is only equal to linear combinations of the plant states.
- The reference output is only equal to the control input.

Combinations of the above special cases may also occur in applications. While these special cases result in a simplification of the closed-loop model, the determination of these models may be very involved, and may even require the use of a digital computer.

EXAMPLE 4.1 The block diagram of a control system for a field-controlled dc motor is given in Figure 4.2. Disturbance inputs are included, and a distinction is made between the actual position of the motor shaft and the measured position of the motor shaft. The plant (the motor) has four inputs: The control input, which is the field voltage $u(t)$; two true disturbance inputs, which are the load torque $d(t)$ and the measurement noise $v(t)$; and the reference input $r(t)$, which specifies the desired motor-shaft angle. The measured output is the measured angular position of the motor shaft. This output is available for feedback to the controller and is denoted $m(t)$. The reference output consists of both the error $e(t)$ between the actual position of the motor shaft and the desired position of the motor shaft, and the control input. A state model of this plant is

$$\begin{bmatrix} \dot{x}_1(t) \\ \dot{x}_2(t) \\ \dot{x}_3(t) \end{bmatrix} = \begin{bmatrix} -p_1 & 0 & 0 \\ k_2 & -p_2 & 0 \\ 0 & 1 & 0 \end{bmatrix} \begin{bmatrix} x_1(t) \\ x_2(t) \\ x_3(t) \end{bmatrix} + \left[\begin{array}{c|ccc} k_1 & 0 & 0 & 0 \\ 0 & 0 & k_2 k_3 & 0 \\ 0 & 0 & 0 & 0 \end{array}\right] \begin{bmatrix} u(t) \\ \hline r(t) \\ d(t) \\ v(t) \end{bmatrix};$$

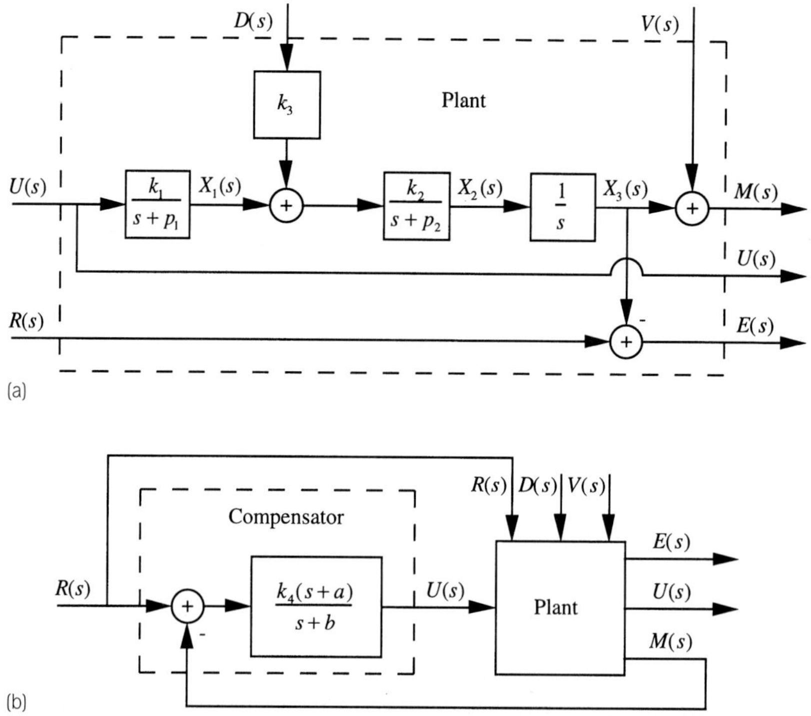

FIGURE 4.2 Block diagram of a control system for a field-controlled dc motor: (a) plant model; (b) control system

$$\begin{bmatrix} m(t) \\ \hdashline e(t) \\ u(t) \end{bmatrix} = \begin{bmatrix} 0 & 0 & 1 \\ \hdashline 0 & 0 & -1 \\ 0 & 0 & 0 \end{bmatrix} \begin{bmatrix} x_1(t) \\ x_2(t) \\ x_3(t) \end{bmatrix} + \begin{bmatrix} 0 & 0 & 0 & 1 \\ \hdashline 0 & 1 & 0 & 0 \\ 1 & 0 & 0 & 0 \end{bmatrix} \begin{bmatrix} u(t) \\ \hdashline r(t) \\ d(t) \\ v(t) \end{bmatrix},$$

where the matrices are partitioned as in (4.2). The controller is a lead network with the following state model:

$$\dot{x}_c(t) = -bx_c(t) + [k_4(a - b) \;\vdots\; -k_4(a - b)] \begin{bmatrix} r(t) \\ \hdashline m(t) \end{bmatrix};$$

$$u(t) = x_c(t) + [k_4 \;\vdots\; -k_4] \begin{bmatrix} r(t) \\ \hdashline m(t) \end{bmatrix}.$$

The state model of the closed-loop system is given by (4.6) and (4.7):

$$\begin{bmatrix} \dot{x}_1(t) \\ \dot{x}_2(t) \\ \dot{x}_3(t) \\ \hdashline \dot{x}_c(t) \end{bmatrix} = \begin{bmatrix} -p_1 & 0 & -k_1 k_4 & k_1 \\ k_2 & -p_2 & 0 & 0 \\ 0 & 1 & 0 & 0 \\ \hdashline 0 & 0 & k_4(b - a) & -b \end{bmatrix} \begin{bmatrix} x_1(t) \\ x_2(t) \\ x_3(t) \\ \hdashline x_c(t) \end{bmatrix} + \begin{bmatrix} k_1 k_4 & 0 & -k_1 k_4 \\ 0 & k_2 k_3 & 0 \\ 0 & 0 & 0 \\ \hdashline k_4(a - b) & 0 & k_4(b - a) \end{bmatrix} \begin{bmatrix} r(t) \\ d(t) \\ v(t) \end{bmatrix};$$

$$\begin{bmatrix} e(t) \\ u(t) \end{bmatrix} = \begin{bmatrix} 0 & 0 & -1 & 0 \\ 0 & 0 & -k_4 & 1 \end{bmatrix} \begin{bmatrix} x_1(t) \\ x_2(t) \\ x_3(t) \\ \hdashline x_c(t) \end{bmatrix} + \begin{bmatrix} 1 & 0 & 0 \\ k_4 & 0 & -k_4 \end{bmatrix} \begin{bmatrix} r(t) \\ d(t) \\ v(t) \end{bmatrix}.$$

Note that the closed-loop output is composed of both the output error and the control input. Including the control input as a reference output allows the designer to use this model to analyze the effects of the disturbances on the control. For example, this model can be used to identify and correct a control system design that results in excessive field voltages when subject to expected disturbances.

The plant transfer function is

$$\begin{bmatrix} M(s) \\ \hdashline E(s) \\ U(s) \end{bmatrix} = \begin{bmatrix} \dfrac{k_1 k_2}{s(s + p_1)(s + p_2)} & 0 & \dfrac{k_2 k_3}{s(s + p_2)} & 1 \\ \hdashline \dfrac{-k_1 k_2}{s(s + p_1)(s + p_2)} & 1 & \dfrac{-k_2 k_3}{s(s + p_2)} & 0 \\ 1 & 0 & 0 & 0 \end{bmatrix} \begin{bmatrix} U(s) \\ \hdashline R(s) \\ D(s) \\ V(s) \end{bmatrix}.$$

The controller transfer function is

$$U(s) = \left[\frac{k_4(s + a)}{s + b} \;\vdots\; -\frac{k_4(s + a)}{s + b} \right] \begin{bmatrix} R(s) \\ \hdashline M(s) \end{bmatrix}.$$

The transfer function model of the closed-loop system is given by (4.8):

$$\begin{bmatrix} E \\ U \end{bmatrix} = \begin{bmatrix} \dfrac{s(s+p_1)(s+p_2)(s+b)}{d(s)} & \dfrac{-k_2k_3(s+p_1)(s+b)}{d(s)} & \dfrac{k_1k_2k_4(s+a)}{d(s)} \\ \dfrac{k_4s(s+p_1)(s+p_2)(s+a)}{\alpha(s)} & \dfrac{-k_2k_3k_4(s+p_1)(s+a)}{\alpha(s)} & \dfrac{-k_4s(s+p_1)(s+p_2)(s+a)}{\alpha(s)} \end{bmatrix} \begin{bmatrix} R \\ D \\ V \end{bmatrix},$$

where the denominator is

$$\begin{aligned} \alpha(s) &= s^4 + (p_1 + p_2 + b)s^3 + (p_1p_2 + p_1b + p_2b)s^2 \\ &\quad + (p_1p_2b + k_1k_2k_4)s + k_1k_2k_4a. \end{aligned}$$

The determination of the transfer function matrix is quite tedious in this example. Fortunately, the transfer function matrix can be generated using a computer. Symbolic math software can be used to find the transfer function matrix using (4.8); alternatively, the state model (4.6) and (4.7) can be generated and used to find this matrix. ◆

In summary, state and transfer function models for the general closed-loop control system in Figure 4.1 are given in (4.6) and (4.7), and in (4.8), respectively. These models can be used to analyze the effects of disturbances and reference inputs on the states, output errors, and control inputs of this system.

4.2 Transient Performance Analysis

The general response of a linear system consists of the sum of the initial condition response and the forced response. The forced response is generated, in part, by the reference input, which specifies the desired behavior of the plant. Therefore, control of the plant should be accomplished via the forced response. The initial condition response is extraneous and should go to zero in a timely manner to avoid corrupting the output errors. Transient performance criteria describe the response of the closed-loop system to initial conditions.

The settling time, rise time, and percent overshoot are all measures of transient performance that are introduced in elementary control systems courses for application to SISO systems subject to a step input. These performance criteria can be extended to MIMO systems, but unfortunately, the meaning is somewhat watered down in the process. For example, when moving a motor shaft from one angle to another, overshoot in angular velocity is often desirable, while overshoot in angle is undesirable. Incorporating both of these overshoots into a single number may not be useful. An alternative approach to transient analysis is to evaluate these performance criteria for each of the relevant individual transfer functions of a MIMO system. These performance criteria can be easily generated using analytic methods or computer simulation, and cumulatively, they provide a clear picture of the transient performance of the system.

The natural frequency and damping ratio are measures of transient performance that are based on the analysis of a second-order, all-pole system. These measures of performance can be used for MIMO systems where the response is dominated by a pair of poles. Since MIMO systems are often of relatively high order, and zeros are typically present in some of the individual transfer functions, the system performance may depart appreciably from that expected given the natural frequency and damping ratio of the dominant poles.

The transient performance measures given above can all be simply related to the pole locations when the plant is a first- or second-order, all-pole system. For MIMO systems, an indication of the transient performance is still provided by the closed-loop poles:

$$p = -\alpha + j\omega$$

where α is assumed to be positive; that is, the system is assumed to be stable. The approximate settling time of each mode,[1]

$$T_s = \frac{4}{|\alpha|}, \tag{4.9}$$

provides an indication of the time it takes that mode to decay to zero. The damping ratio of each mode,[2]

$$\zeta = \frac{\alpha}{\sqrt{\omega^2 + \alpha^2}}, \tag{4.10}$$

provides an indication of how much oscillation is experienced during the decay of that mode. Cumulatively, the settling times and damping ratios of the modes provide a rough understanding of the total transient response. This understanding is very useful, especially in evaluating the effects of changing design parameters. But caution must be exercised in assuming that pole information alone can yield precise quantitative values for settling time and overshoot. The final decision on the acceptability of a control system design, for a high-order system or for a MIMO system, should be based on the complete analytic solution of the system or a computer simulation of the complete system.

EXAMPLE 4.2 Consider the fifth-order MIMO system described by the following transfer function:

$$\mathbf{G}(s) = \frac{1}{(s + 10)(s - 10e^{j(4/5)\pi})(s - 10e^{-j(4/5)\pi})(s - 10e^{j(3/5)\pi})(s - 10e^{-j(3/5)\pi})} \begin{bmatrix} 100{,}000 \\ 50{,}000(s + 1) \end{bmatrix}$$

where the poles are arranged in a Butterworth filter pattern. The settling time and damping ratio of each of the modes are

[1] This result is an approximation of the 2% settling time.

[2] This formula yields a damping of one for all poles on the real axis. The concept of an overdamped system (damping greater than 1) is not very useful for systems of order greater than 2.

TABLE 4.1 Settling Times and Damping Ratios for the Individual Modes

Pole	Settling Time (seconds)	Damping Ratio
-10	0.40	1
$10e^{j(4/5)\pi}$	0.49	0.81
$10e^{-j(4/5)\pi}$	0.49	0.81
$10e^{j(3/5)\pi}$	1.29	0.31
$10e^{-j(3/5)\pi}$	1.29	0.31

The step response of this system is given in Figure 4.3. Note that the settling time is approximately the maximum settling time of the individual modes. The step response for the first output $y_1(t)$, the all-pole transfer function, is less oscillatory than expected from the damping ratio of the dominant pair of poles (i.e., the poles with the longest settling times). This decrease in oscillation is due to the presence of the three additional, and more highly damped, poles in the transfer function. The step response for the second output $y_2(t)$ is more oscillatory than expected from the damping ratio of the dominant pair of poles. The increase in oscillation is due to the presence of the zero in the transfer function for this output. The settling time is also increased slightly by the zero in this transfer function. Note that zeros in the transfer function often produce much larger oscillations than expected solely on the basis of the pole locations. ◆

From this example, it is clear that the system poles do not fully specify the transient response of the system, but they do provide a general indication of the settling time and the magnitude of oscillation within the transient response.

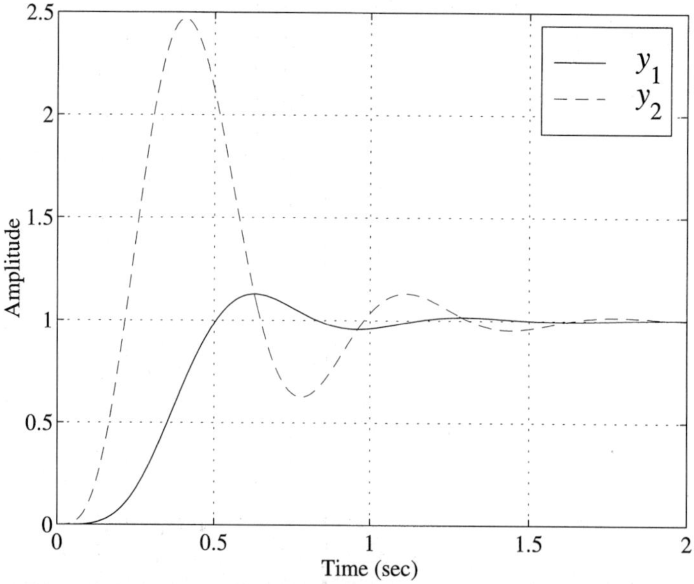

FIGURE 4.3 Step response for Example 4.2

4.3 Tracking Performance Analysis

The tracking error and the required control can be computed for a specific reference input using either the state or transfer function models of the closed-loop system. Linearity insures that the response due to the reference input and the response due to any possible disturbances can be computed separately. The response due to the reference input is generated by setting all initial conditions to zero and all disturbance inputs to zero. A state model can be constructed that gives only the response due to the reference input, by deleting the columns associated with the disturbance inputs $w_0(t)$ in both the input matrix and the input-to-output coupling matrix of the closed-loop state model (4.6) and (4.7). A transfer function matrix from the reference input to the reference output can also be obtained by deleting the columns associated with the disturbance inputs in the transfer function model (4.8). Note that reference inputs generate both steady-state and transient responses that can be computed using these models.

The specification of realistic reference inputs depends on the application. Many applications require that the output be driven to a constant reference input. For example, an airplane autopilot is typically required to maintain the airplane's heading and altitude at desired constant values (reference inputs). These reference inputs are occasionally changed upon encountering waypoints. Abrupt changes in reference inputs can be described mathematically as step functions.

Other applications may have reference inputs that lie within a frequency band. For example, a fire control radar is required to follow the motion of an aircraft that may be maneuvering. The reference signal in this application is the angle to the aircraft. The variability in this angle is limited by the maximum speed and acceleration of the aircraft. Note that the variability also depends on the distance to the aircraft. The reference input is therefore constrained to be slowly varying or, equivalently, to be bandlimited.

Reference inputs can also be modeled as colored noise processes. For example, the angle of a maneuvering aircraft can be modeled as a random process since the aircraft moves in an unpredictable manner. This random process is colored noise since it is constrained to vary slowly. Reference inputs described as colored noise can always be modeled as the output of a shaping filter driven by white noise. The analysis of colored-noise reference inputs can then be accomplished by analyzing the combination of the plant and the shaping filter subject to a white noise input.

The tracking performance of a control system can be evaluated by applying a representative reference input to the system and finding the resulting tracking error and control input. The frequency response of the closed-loop system can also be used to quantify the tracking performance.

4.4 Disturbance Rejection Analysis

The tracking error and the required control can be computed for a specific disturbance input. The response due to the disturbance inputs is generated by setting all initial conditions to zero and all reference inputs to zero. A state model can be constructed that gives the response due to the disturbance input by deleting the columns associated with the reference inputs in both the input matrix and the input-to-output coupling matrix of the closed-loop state model (4.6) and (4.7). A transfer function matrix from the disturbance input to the reference output can also be obtained by deleting the columns associated with the reference inputs in the transfer function model (4.8). Note that disturbance inputs generate both steady-state and transient responses that can be computed using these models.

The control system must maintain the output close to the desired value in the presence of disturbances. Disturbances are inputs beyond the control of the designer and are usually inputs that tend to drive the output away from its desired value. Disturbance inputs consist of an infinite variety of types, which complicates the analysis of disturbance rejection. A set of "typical" disturbances are therefore defined. The system response, subject to these disturbances, is used to characterize the disturbance rejection of the system.

Disturbance inputs often exist for short periods of time. Wind gusts on antennas, meteor strikes on spacecraft, and the sticking of a motor shaft are all examples of short-duration disturbances. Short-duration disturbances can be approximated by impulse functions.

Constant and step disturbances are also commonly encountered. Gravity on an airplane, engine torque on a helicopter, and solar pressure on a geosynchronous satellite are examples of constant or nearly constant disturbances. Step disturbances are encountered when a load is placed on a motor, a robot picks up an object, and when a satellite experiences solar pressure upon departing Earth's shadow. The steady-state analysis of step and constant disturbances is identical. The analysis of step disturbances requires, in addition, the computation of the transient response.

Sinusoidal disturbances such as waves acting on a ship, acoustic waves acting on a structure (earthquakes acting on a building), and vibrations caused by rotating machinery frequently appear in control applications. The Fourier transform can be used to decompose more complex disturbances into sinusoidal disturbances. The simplicity of evaluating the effects of sinusoidal disturbances makes this model very attractive in applications. Disturbances can often be characterized as existing only within a given frequency band. The frequency response (from the disturbance input to the output) provides a very useful tool for evaluating the effects of sinusoidal or bandlimited disturbances.

Disturbances are often best modeled as random processes. A simple random process that is often employed in disturbance rejection analysis is white noise. Examples of white noise disturbances are turbulence acting on a jetliner, choppy seas acting on a ship, and measurement noise in a closed-loop control system. White noise is an idealization of a zero-mean random input with a short correlation time (true white noise does not exist in nature). When colored noise disturbances are more appropriate, this colored noise can always be modeled as the output of a shaping filter driven by white noise. The analysis of colored noise disturbances can then be accomplished by analyzing the combination of the plant and the shaping filter subject to a white noise disturbance.

The specification of disturbances in a particular application proves to be one of the more difficult tasks in control design. The disturbance types discussed above are the most commonly used, but many other types are possible. The disturbance types presented are selected both for their applicability to a wide range of control systems, and for the ease with which the disturbance response is computed. The ignorance of the control designer with respect to the true disturbances often results in one of the given types providing a reasonable fit to the available information.

The disturbance rejection of a control system can be evaluated by applying a representative disturbance input to the system and finding the resulting tracking error and control input. The frequency response of the closed-loop system can also be used to quantify the disturbance rejection.

4.5 Cost Functions and Norms

The performance of a control system can be quantified in many applications by a cost function. A cost function is, in general, a real-valued, non-negative function of the system, or of the time histories of the state, reference output, and control input, subject to a given set of initial conditions and inputs. The cost (the real number resulting from application of the cost function) can be used to evaluate the performance of a system, where superior performance is indicated by a smaller cost. For example, the cost associated with a control system may be required to be less than some maximum allowable value. The cost can also be used to compare the performance of multiple controller designs; that is, the decision on which of several alternative designs is superior can be made by comparing their respective costs. The controller that minimizes the cost, over all possible designs or a set of possible candidate designs, is known as an optimal controller. The synthesis of optimal controllers is discussed in detail later in this book. For now, cost functions are presented as a means of quantifying control system performance.

The selection of a cost function for a particular application is a useful art in control system design. To illustrate this selection process, cost functions used for typical classes of problems are presented in the remainder of this section. The cost functions given are all based on mathematical objects called norms. The properties of norms are outlined prior to continuing the discussion of cost functions.

4.5.1 Norms

A general cost function is a non-negative, real-valued, nonlinear function of the closed-loop system, or of the time histories of the state, the reference output, and the control input. This definition is very broad and allows the specification of a wide variety of cost functions. The majority of cost functions used in applications have additional properties related to mathematical objects called norms.

The norm, denoted $\|\bullet\|_p$, is a real-valued function of the elements of a linear space $\mathcal{B}$.[3] A linear space is a set where any linear combination of elements is also an element of the set, and can be composed of vectors, signals, systems, or other possible collections of elements. A norm has the following properties:

$$\|x\|_p \geq 0; \tag{4.11a}$$

$$\|x\|_p = 0 \text{ if and only if } x = 0; \tag{4.11b}$$

$$\|\alpha x\|_p = |\alpha| \|x\|_p; \tag{4.11c}$$

$$\|x + y\|_p \leq \|x\|_p + \|y\|_p. \tag{4.11d}$$

[3] A variety of subscripts will be used, in addition to the generic p, to distinguish between different norms.

where $x, y \in \mathscr{B}$, and α is a scalar. Note that (4.11a) is a property assumed for general cost functions, while (4.11b), (4.11c), and (4.11d) are additional properties possessed by norms. Intuitively, the norm provides a measure of the size of the vector, signal, or system. In practice, norms, like size, can be defined in a variety of ways. Norms can also be used to denote the distance between two vectors, two signals, or two systems:

$$\|y - x\|_p.$$

Again, this distance can be defined in a number of different ways.

Examples of vector and signal norms are given below. Norms of systems are then discussed after first presenting some general cost functions.

Vector Norms A familiar example of a norm is the Euclidean vector norm, or the vector 2-norm, which appears in elementary geometry and vector analysis. The Euclidean norm is defined as follows:

$$\|x\|_2 = \sqrt{x^T x} = \sqrt{\sum_{k=1}^{n_x} x_k^2}$$

on the space of real vectors. This norm can be generalized to operate on the space of complex vectors:

$$\|x\|_2 = \sqrt{x^\dagger x} = \sqrt{\sum_{k=1}^{n_x} |x_k|^2}. \tag{4.12}$$

Note that this definition is identical to the previous definition over the space of real vectors.

A modification of the Euclidean vector norm can be obtained by adding a positive definite weighting matrix $\mathbf{W}$:

$$\|x\|_W = \sqrt{x^\dagger \mathbf{W} x}. \tag{4.13}$$

Note that the weighting matrix in this expression must be positive definite to ensure that this function satisfies (4.11a) and (4.11b).

The vector ∞-norm is defined as follows:

$$\|x\|_\infty = \max_i |x_i|. \tag{4.14}$$

Additional vector norms can be defined as required in applications. Note that the above norms all specify the length of a vector (in some sense), and satisfy the properties given in (4.11).

Signal Norms The Euclidean norm can be generalized to operate on signals:

$$\|x(t)\|_2 = \sqrt{\int_{-\infty}^{\infty} x^\dagger(t) x(t) dt}. \tag{4.15}$$

This norm is referred to as the signal 2-norm. Note that all signals do not have finite 2-norms. When using norms, the signal will be assumed to be an element of the linear space over which the norm is defined (in this case, the space $\mathscr{L}_2$) unless otherwise specified.

Weighted signal 2-norms are defined as follows:

$$\|x(t)\|_{\mathbf{W}(t)} = \sqrt{\int_{-\infty}^{\infty} x^{\dagger}(t)\mathbf{W}(t)x(t)dt}, \tag{4.16}$$

where $\mathbf{W}(t)$ is positive definite at all times. An additional generalization of the signal 2-norm is obtained by defining this norm over a finite time interval:

$$\|x(t)\|_{2,[t_0,t_f]} = \sqrt{\int_{t_0}^{t_f} x^{\dagger}(t)x(t)dt}. \tag{4.17}$$

The signal ∞-norm is defined as follows:

$$\|x(t)\|_{\infty} = \sup_t \max_i |x_i(t)|. \tag{4.18}$$

The supremum is used in this expression since the set of times is infinite. In this case, $x_i(t)$ may asymptotically approach a value but never actually achieve the maximum. The signal ∞-norm can also be defined on a finite time interval:

$$\|x(t)\|_{\infty,[t_0,t_f]} = \max_{t\in[t_0,t_f]} \max_i |x_i(t)|. \tag{4.19}$$

The same notation is used for the vector, signal, and system (after defining this norm) 2-norms (and ∞-norms). The distinction between the given norms is provided by the set over which they operate. This should be apparent when these norms are used and, we hope, will not lead to confusion.

4.5.2 Quadratic Cost Functions

The goals of the control system are to drive the output errors to zero,[4] and to do this while using a reasonable amount of control. These goals are typically at odds. The tighter the control of the output errors, the more control is required. The more reasonable the control used (i.e., the less control used), the larger the output errors. A typical control design represents a compromise between keeping the output errors small and keeping the controls small. The cost function should, therefore, include a measure of both the size of the output errors and the size of the control. One such cost function is

$$\boxed{J = \int_0^{t_f} y^T(t)\mathbf{Y}(t)y(t)dt = \|y(t)\|^2_{\mathbf{Y}(t)},} \tag{4.20}$$

where the reference output is assumed to include both the output errors and the control inputs.

The cost function (4.20) is called quadratic since it is a quadratic function of the reference output. The weighting function $\mathbf{Y}(t)$ is a positive definite matrix, selected to quantify the relative importance of the various output errors and control inputs. This weighting function is also time-dependent which allows this relative importance to

[4] The output errors include both the errors between the outputs and the reference inputs, and any states or linear combinations of states that the control system is tasked with driving to zero.

change with time. The parameter t_f is the final time, which can be infinity if the control system is intended to operate indefinitely. The norm in (4.20) operates on the set of real signals, and is a finite-time, weighted signal 2-norm.

The cost function (4.20) can be expanded to yield a more detailed description of the cost:

$$J = \int_0^{t_f} [e^T(t) \mid u^T(t)] \left[\begin{array}{c|c} \mathbf{Z}(t) & \mathbf{0} \\ \hline \mathbf{0} & \mathbf{R}(t) \end{array}\right] \left[\begin{array}{c} e(t) \\ \hline u(t) \end{array}\right] dt \tag{4.21}$$

$$= \int_0^{t_f} e^T(t)\mathbf{Z}(t)e(t) + u^T(t)\mathbf{R}(t)u(t)dt,$$

where the weighting functions $\mathbf{Z}(t)$ and $\mathbf{R}(t)$ are positive definite matrix functions of time. Note that this form of the cost function is not as general as the one given in (4.20), but is sufficiently general to be useful in an abundance of applications. The output error can be written

$$e(t) = r(t) - \mathbf{C}_e x(t),$$

where $\mathbf{C}_e$ is the portion of $\mathbf{C}_y$ that generates the output error, and the parts of the $\mathbf{D}$ matrices that contribute to the error are assumed to be zero. The cost (4.21) can then be written in terms of the system state:

$$J = \int_0^{t_f} \{r(t) - \mathbf{C}_e x(t)\}^T \mathbf{Z}(t)\{r(t) - \mathbf{C}_e x(t)\} + u^T(t)\mathbf{R}(t)u(t)dt. \tag{4.22}$$

This version of the quadratic cost function can be readily simplified in a number of special cases.

The state can often be defined in such a manner that good control is synonymous with linear combinations of the states being close to zero. For example, the state of a servomotor may include the angular displacement of the shaft from the desired position. The goal of the control system is then to drive this element of the state to zero while using a small amount of control. A control system designed to drive the state, or linear combinations of the state, to zero is termed a *regulator*. The quadratic cost function for a regulator is

$$J = \int_0^{t_f} x^T(t)\mathbf{Q}(t)x(t) + u^T(t)\mathbf{R}(t)u(t)dt, \tag{4.23}$$

where $\mathbf{Q}(t)$ is a positive semidefinite matrix,

$$\mathbf{Q}(t) = \mathbf{C}_e^T \mathbf{Z}(t)\mathbf{C}_e,$$

selected to weight the appropriate states. The weighting on the control $\mathbf{R}(t)$ is used to impose a penalty on the use of excessive amounts of control. Note that in many applications, the cost function is generated by directly selecting $\mathbf{Q}$ as opposed to both defining $\mathbf{C}_e$ and defining the weighting matrix on the output $\mathbf{Z}(t)$.

The weighting matrices in (4.22) or (4.23) are constant in many applications. Constant weighting matrices are used whenever output errors (and nonzero control inputs) at any point in time are equally undesirable. For example, an autopilot for an airplane is required to maintain a heading during the entire flight, and the suppression of heading errors is equally important at all times. Constant weighting matrices are also typically used when the control system is designed to operate indefinitely or for extended periods of time.

The output error may be important only at the final time in some applications. For example, a missile autopilot is designed to position the missile as close as possible to a desired target at impact (the final time). The cost function (4.22) then simplifies to

$$J = \{r(t_f) - \mathbf{C}_e x(t_f)\}^T \mathbf{V}\{r(t_f) - \mathbf{C}_e x(t_f)\} + \int_0^{t_f} u^T(t)\mathbf{R}(t)u(t)dt.$$

Note that excessive control is undesirable at all times, so the term involving the control input is still integrated over all time.

The cost functions given above all represent compromises between the output and the control input. Performance can also be evaluated using only one of these vectors while the other is subject to a constraint. For example, a satellite may be required to change from one orbit to another orbit while using the minimum amount of fuel. The performance of a candidate system can be evaluated with the following cost function:

$$J = \int_0^{t_f} u^T(t)\mathbf{R}u(t)dt$$

as long as this system yields the desired final state:

$$x(t_f) = x_d.$$

As a second example, consider the missile that is designed to hit a target. A reasonable cost function is the square of the miss distance (the nearest approach of the missile to the target):

$$J = [x(t_f) - x_d]^T[x(t_f) - x_d],$$

subject to the constraints that each of the control inputs are bounded:

$$|u_i(t)| \leq \bar{u}_i.$$

These constraints on the control inputs quantify the limitations of the control actuators.

The quadratic cost is dependent on the reference input applied, the disturbance input applied, the initial conditions, the final conditions, and/or constraints on the state and control. Collectively these inputs, conditions, and constraints are known as the test conditions. The performance of a system can be quantified by the cost when the system is subject to worst-case test conditions or nominal test conditions. *Nominal* refers to test conditions that are representative of normal operating conditions, and *worst-case* is self explanatory. The test conditions may be simplified to yield information on the effects of one initial condition or the effects of one disturbance input. Performance is also often evaluated for simple test conditions, which include step inputs, impulse inputs, initial conditions, and so on. Quadratic cost functions can be used in a wide range of engineering design applications due to the flexibility provided by the specification of weighting matrices and test conditions.

The Computation of Quadratic Cost The quadratic cost can be computed by direct integration of the output error and control trajectories that result when the test conditions are applied to the system. The state and control trajectories can be found analytically (for simple systems) or via computer simulation. Note that direct computer simulation cannot be applied when the test conditions include constraints on the final state. A reverse-time simulation called the adjoint simulation can be applied when only final conditions are given [1], page 622. When both initial and final conditions are given, the state and control trajectories are found by solution of a two-point boundary-value problem [2], page 212. The solution of two-point boundary-value problems is difficult and typically requires extensive numerical iteration. The determination of quadratic cost for test conditions that involve constraints on the final state of the system is beyond the scope of this book but is addressed in [2], page 55.

EXAMPLE 4.3 A field-controlled dc motor is being used to position an antenna that is required to track a satellite in low earth orbit. A lead compensator being used to control is motor. Models of this motor and controller are given in Example 4.1. The closed-loop state model of this combination also given in Example 4.1 is:

$$\begin{bmatrix} \dot{x}_1(t) \\ \dot{x}_2(t) \\ \dot{x}_3(t) \\ \hline \dot{x}_c(t) \end{bmatrix} = \left[\begin{array}{ccc|c} -p_1 & 0 & -k_1k_4 & k_1 \\ k_2 & -p_2 & 0 & 0 \\ 0 & 1 & 0 & 0 \\ \hline 0 & 0 & k_4(b-a) & -b \end{array}\right] \begin{bmatrix} x_1(t) \\ x_2(t) \\ x_3(t) \\ \hline x_c(t) \end{bmatrix} + \begin{bmatrix} k_1k_4 & 0 & -k_1k_4 \\ 0 & k_2k_3 & 0 \\ 0 & 0 & 0 \\ \hline k_4(a-b) & 0 & k_4(b-a) \end{bmatrix} \begin{bmatrix} r(t) \\ d(t) \\ v(t) \end{bmatrix}$$

$$\begin{bmatrix} e(t) \\ u(t) \end{bmatrix} = \left[\begin{array}{ccc|c} 0 & 0 & -1 & 0 \\ 0 & 0 & -k_4 & 1 \end{array}\right] \begin{bmatrix} x_1(t) \\ x_2(t) \\ x_3(t) \\ \hline x_c(t) \end{bmatrix} + \begin{bmatrix} 1 & 0 & 0 \\ k_4 & 0 & -k_4 \end{bmatrix} \begin{bmatrix} r(t) \\ d(t) \\ v(t) \end{bmatrix}$$

where $r(t)$ is the reference input, $d(t)$ is the disturbance torque, $v(t)$ is the measurement noise, $e(t)$ is the difference between the motor shaft angle and the reference input, and $u(t)$ is the field voltage, which is the control input. Further, the parameters are defined as follows:

$$p_1 = 4 \; ; \; p_2 = 0.1 \; ; \; k_1 = 10 \; ; \; k_2 = 0.01 \; ; \; k_3 = 10 \; ; \; k_4 = 150 \; ; \; a = 0.4 \; ; \; b = 4.$$

It is desirable to keep the output error less than 1 degree, while using an input of less than 15 volts. A reasonable cost function is then

$$J = \frac{1}{300} \int_0^{150} e^2(t) + \frac{1}{15^2} u^2(t) dt.$$

Note that a factor of 1/150 is used to normalize the cost with respect to the integration time, a factor of 1/2 is used to normalize the cost with respect to the number of terms within the integral, and a factor of $1/15^2$ is used to normalize the desired size of the

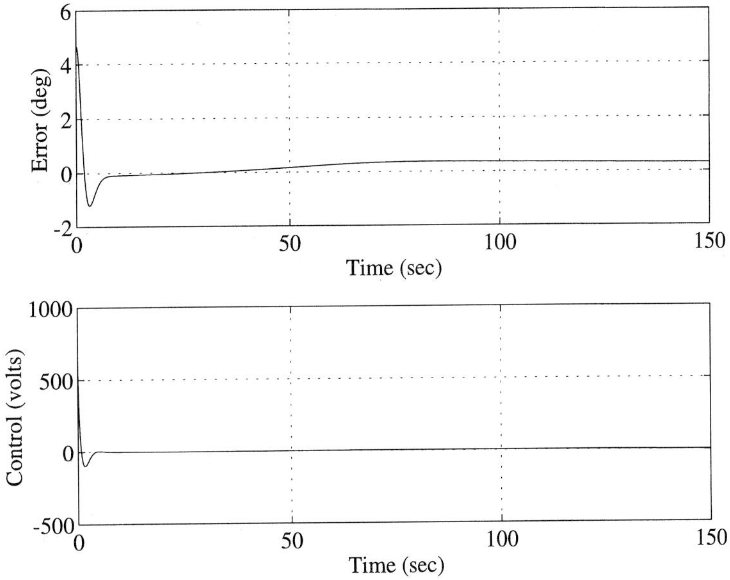

FIGURE 4.4 The reference output for Example 4.3

control to 1 (which equals the desired size of the error). The test conditions consist of a nominal path for the satellite:

$$r(t) = \tan^{-1}\left(\frac{h}{562.5 - \nu t}\right),$$

where $h = 320$ km and $\nu = 7.5$ km/s are the height of the satellite and the velocity of the satellite, respectively.[5] The initial conditions of the closed-loop plant are

$$x(0) = [0 \quad 0 \quad 25 \quad 0]^T.$$

The initial elevation angle of the antenna is 25 degrees, which is selected by estimating the initial satellite elevation angle. The true initial elevation angle of the satellite (the angle when the satellite is first detected) is 30 degrees. This yields a 5 degree error in the initial angle estimate. The disturbance torque is generated by gravity, and is due to imperfect balancing of the antenna. This torque depends on the elevation angle, but can be approximated by the following time-varying function:

$$d(t) = \cos\left[\tan^{-1}\left(\frac{h}{562.5 - \nu t}\right)\right]$$

The measurement noise is small enough that it can be ignored, that is, set equal to zero. These test conditions represent normal operating conditions. The resulting error trajectory and control trajectory are plotted in Figure 4.4. The cost is

$$J = 1.74,$$

[5] The satellite is assumed to be going directly overhead, and the curvature of the orbit is ignored in generating this trajectory.

which is greater than 1, indicating that the specifications are not achieved with this controller. Looking at the plots, the control input is seen to exceed the specifications during the initial transient, but meets that specification after this transient decays to zero. Care must be used in interpreting the meaning of the cost when the test conditions include both initial conditions and inputs. ◆

4.5.3 Cost Functions for Systems with Random Inputs

The state and control trajectories become random processes when the system is subject to random disturbance inputs. The quadratic cost functions, as defined above, are then random variables instead of the desired real values. The expected value of the random cost can be used to provide a real measure of performance:

$$J = E\left[\int_0^{t_f} y^T(t)\mathbf{Y}(t)y(t)dt\right] = \int_0^{t_f} E[y^T(t)\mathbf{Y}(t)y(t)]dt. \tag{4.24}$$

In applications where the random inputs are stationary, the system operates long enough that the initial transient can be ignored, the weighting matrix is time-invariant, and the closed-loop system is stable, the cost,

$$\boxed{J = E[y^T(t)\mathbf{Y}y(t)],} \tag{4.25}$$

is proportional to the cost in (4.24).[6] This cost function can be expanded in terms of the state and the control inputs as was done in the previous subsection.

The Computation of Cost for Systems with Random Inputs The cost (4.25) can be computed from the correlation matrix of the reference output. Taking the trace of (4.25) leaves the cost (a scalar) unchanged:

$$J = \text{tr}\{E[y^T(t)\mathbf{Y}y(t)]\}.$$

Using the fact that the trace is invariant under cyclic permutation (see the Appendix, equation A2.3), the cost can be written as follows:

$$J = \text{tr}\{\mathbf{Y}E[y(t)y^T(t)]\} = \text{tr}\{\mathbf{Y}\boldsymbol{\Sigma}_y\}.$$

The correlation matrix of the reference output can be computed from the correlation function of the disturbance input:

$$\boldsymbol{\Sigma}_y = \int_0^{\infty}\int_0^{\infty} \mathbf{g}_{cl}(\tau_1)\mathbf{R}_w(\tau_1 - \tau_2)\mathbf{g}_{cl}^T(\tau_2)d\tau_1 d\tau_2.$$

where $\mathbf{g}_{cl}(t)$ is the impulse response of the closed-loop system. Note that the disturbance input and reference output are both assumed to be stationary, and that the closed-loop system is assumed to be stable. Combining the two expressions above yields

[6] The cost in (4.25) is proportional to the cost in (4.24) provided that the final time is finite. For infinite final times, (4.24) becomes infinite and is unusable as a cost function.

the cost in terms of both the impulse response matrix of the closed-loop system and the correlation function of the disturbance input:

$$J = \operatorname{tr}\left\{\mathbf{Y} \int_0^\infty \int_0^\infty \mathbf{g}_{cl}(\tau_1)\mathbf{R}_w(\tau_1 - \tau_2)\mathbf{g}_{cl}^T(\tau_2)d\tau_1 d\tau_2\right\}. \tag{4.26}$$

This cost simplifies to

$$\boxed{J = \operatorname{tr}\left\{\mathbf{Y} \int_0^\infty \mathbf{g}_{cl}(\tau)\mathbf{S}_w\mathbf{g}_{cl}^T(\tau)d\tau\right\}} \tag{4.27}$$

when the disturbance input is white noise with spectral density $\mathbf{S}_w$. The impulse response can be evaluated numerically, and the cost evaluated numerically using (4.26) or (4.27) provided that the correlation function of the disturbance input is known.

The output correlation matrix can also be computed from the correlation matrix of the state,

$$\mathbf{\Sigma}_y = \mathbf{C}_y E[x(t)x^T(t)]\mathbf{C}_y^T = \mathbf{C}_y\mathbf{\Sigma}_x\mathbf{C}_y^T,$$

and the cost written as

$$\boxed{J = \operatorname{tr}\{\mathbf{Y}\mathbf{C}_y\mathbf{\Sigma}_x\mathbf{C}_y^T\}.} \tag{4.28}$$

In addition, the correlation matrix of the state can be computed from the Lyapunov equation:

$$\mathbf{A}_{cl}\mathbf{\Sigma}_x + \mathbf{\Sigma}_x\mathbf{A}_{cl}^T + \mathbf{B}_{cl}\mathbf{S}_w\mathbf{B}_{cl}^T = \mathbf{0}. \tag{4.29}$$

Note that this equation is only valid when the disturbance input is white noise, and the closed-loop system is stable. Reliable numerical algorithms for solving this Lyapunov equation can be found in many computer-aided control system design packages. Numerical solution of this Lyapunov equation and substitution into (4.28) typically yields a faster and more accurate cost estimate than numerical evaluation of (4.27). A discussion of the solution of Lyapunov equations is included in Section A7 of the Appendix.

EXAMPLE 4.4 An autopilot is used to control the altitude of a helicopter while hovering. The altitude dynamics of the helicopter are

$$\begin{bmatrix} \dot{x}_1(t) \\ \dot{x}_2(t) \end{bmatrix} = \begin{bmatrix} 0 & 1 \\ 0 & 0 \end{bmatrix}\begin{bmatrix} x_1(t) \\ x_2(t) \end{bmatrix} + \left[\begin{array}{c|cc} 0 & 0 & 0 \\ 1 & 1 & 0 \end{array}\right]\begin{bmatrix} u(t) \\ \hline w_0(t) \\ v(t) \end{bmatrix};$$

$$\begin{bmatrix} m(t) \\ \hline e(t) \\ u(t) \end{bmatrix} = \begin{bmatrix} 1 & 0 \\ \hline 1 & 0 \\ 0 & 0 \end{bmatrix}\begin{bmatrix} x_1(t) \\ x_2(t) \end{bmatrix} + \left[\begin{array}{c|cc} 0 & 0 & 1 \\ \hline 0 & 0 & 0 \\ 1 & 0 & 0 \end{array}\right]\begin{bmatrix} u(t) \\ \hline w_0(t) \\ v(t) \end{bmatrix},$$

where $e(t)$ is the altitude error (the zero on the coordinate system is set at the desired altitude, so the reference input is zero) and $u(t)$ is the vertical acceleration, which is proportional to the throttle setting. The disturbance $w_0(t)$ is a vertical acceleration caused by wind gusts and $v(t)$ is the measurement noise. Note that the delay due to engine spin-up is ignored in this example. An observer feedback controller for this plant is described by the following state model:

$$\begin{bmatrix}\dot{\hat{x}}_1(t)\\ \dot{\hat{x}}_2(t)\end{bmatrix} = \begin{bmatrix}-1.6 & 1\\ -0.68 & -0.4\end{bmatrix}\begin{bmatrix}\hat{x}_1(t)\\ \hat{x}_2(t)\end{bmatrix} + \begin{bmatrix}1.6\\ 0.64\end{bmatrix} m(t);$$

$$u(t) = [-0.04 \quad -0.4]\begin{bmatrix}\hat{x}_1(t)\\ \hat{x}_2(t)\end{bmatrix},$$

where $\hat{x}(t)$ is an estimate of $x(t)$. The closed-loop system is then described by the following state model:

$$\begin{bmatrix}\dot{x}_1(t)\\ \dot{x}_2(t)\\ \hline \dot{\hat{x}}_1(t)\\ \dot{\hat{x}}_2(t)\end{bmatrix} = \left[\begin{array}{cc|cc}0 & 1 & 0 & 0\\ 0 & 0 & -0.04 & -0.4\\ \hline 1.6 & 0 & -1.6 & 1\\ 0.64 & 0 & -0.68 & -0.4\end{array}\right]\begin{bmatrix}x_1(t)\\ x_2(t)\\ \hline \hat{x}_1(t)\\ \hat{x}_2(t)\end{bmatrix} + \begin{bmatrix}0 & 0\\ 1 & 0\\ \hline 0 & 1.6\\ 0 & 0.64\end{bmatrix}\begin{bmatrix}w_0(t)\\ v(t)\end{bmatrix};$$

$$\begin{bmatrix}e(t)\\ u(t)\end{bmatrix} = \left[\begin{array}{cc|cc}1 & 0 & 0 & 0\\ 0 & 0 & -0.04 & -0.4\end{array}\right]\begin{bmatrix}x_1(t)\\ x_2(t)\\ \hline \hat{x}_1(t)\\ \hat{x}_2(t)\end{bmatrix} + \begin{bmatrix}0 & 0\\ 0 & 0\end{bmatrix}\begin{bmatrix}w_0(t)\\ v(t)\end{bmatrix}.$$

The two disturbance inputs, $w_0(t)$ and $v(t)$, are assumed to be independent and are modeled as white noise with the following spectral density:

$$S_w = \begin{bmatrix}S_{w_0} & 0\\ 0 & S_v\end{bmatrix} = \begin{bmatrix}1\dfrac{m^2}{\sec^2 - Hz} & 0\\ 0 & 400\dfrac{m^2}{Hz}\end{bmatrix}.$$

The control input should remain within ± 10 m/sec^2, and the desired altitude should be maintained to within ± 20 m on average during steady-state operation. A reasonable cost function is then

$$J = \frac{1}{2}E\left[\frac{1}{20^2}e^2(t) + \frac{1}{10^2}u^2(t)\right],$$

where the factor of 1/2 normalizes the cost function to the number of terms.

The resulting trajectories of the error and the control input are given in Figure 4.5, after the decay of the transient. The cost found by solving the Lyapunov equation is

$$J = 0.321,$$

while the cost found by statistically averaging the trajectory is

$$J = 0.275.$$

The latter cost is a random variable and is only an approximation to the actual cost. The cost is less than 1, indicating that the specifications are achieved with this controller.

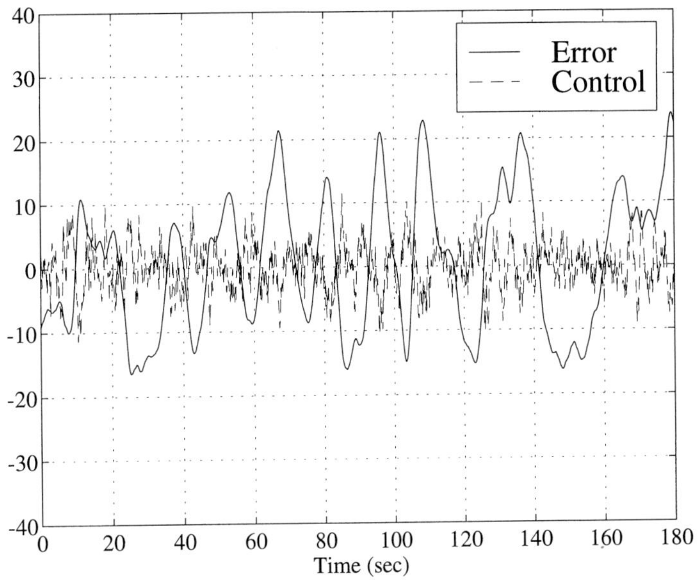

FIGURE 4.5 The reference output for Example 4.4

This is the case, as can be seen by observing that both trajectories stay within the desired bounds most of the time. The altitude error does occasionally exceed the bound. This behavior is typical of systems with Gaussian random inputs, since there is always a finite probability (often very small) of exceeding any finite bound. ◆

4.5.4 The System 2-Norm Cost Function

The computation of the cost functions presented above requires that the excitation (initial conditions, final conditions, reference inputs, and disturbance inputs) be known. In many applications, neither the inputs, the initial conditions, nor the final conditions are known *a priori*. An alternative type of cost function focuses on the gain between the inputs and the outputs, as defined by the Fourier transfer function of the closed-loop system.

The system 2-norm is proportional to the root mean square gain of the system, and can be thought of as an average gain for the system. This average is performed both over all the elements of the matrix transfer function and over all frequencies:

$$\boxed{\|\mathbf{G}\|_2 = \sqrt{\frac{1}{2\pi}\int_{-\infty}^{\infty} \operatorname{tr}\{\mathbf{G}^\dagger(j\omega)\mathbf{G}(j\omega)\}d\omega}.} \tag{4.30a}$$

Note that the $\operatorname{tr}\{\mathbf{G}^\dagger(j\omega)\mathbf{G}(j\omega)\}$ is the sum of the magnitudes squared of all of the elements of $\mathbf{G}(j\omega)$. The system 2-norm is actually only proportional to the true average since it is not normalized to either the number of elements in the transfer function matrix or to the frequency range. The system 2-norm can also be written in terms of the impulse response matrix by using Parseval's theorem:[7]

$$\boxed{\|\mathbf{G}\|_2 = \sqrt{\int_0^\infty \operatorname{tr}\{\mathbf{g}^T(t)\mathbf{g}(t)\}dt}.} \tag{4.30b}$$

The indefinite integral in (4.30b) only exists when the system is stable. In fact, the whole concept of gain is not very useful for an unstable system whose transient persists indefinitely. Therefore, the system 2-norm is only defined for stable systems.

The system 2-norm can be interpreted as the gain of the system subject to a stationary white noise input. For example, let the input to the system be a white noise vector with the following correlation function:

$$E[w(t)w^T(t+\tau)] = S_w\mathbf{I}\delta(\tau),$$

where S_w is the scalar spectral density of each element of the white noise input. The mean square power of the output is

$$E[y^T(t)y(t)] = \operatorname{tr}\{\Sigma_y\},$$

where Σ_y is the output covariance matrix, which is given as

$$\Sigma_y = S_w \int_0^\infty \mathbf{g}(t)\mathbf{g}^T(t)dt.$$

Combining these two equations, we have

$$E[y^T(t)y(t)] = S_w \int_{-\infty}^\infty \operatorname{tr}\{\mathbf{g}^T(t)\mathbf{g}(t)\}dt = S_w \int_{-\infty}^\infty \operatorname{tr}\{\mathbf{g}(t)\mathbf{g}^T(t)\}dt = S_w\|\mathbf{G}\|_2^2.$$

Taking the square root,

$$\sqrt{E[y^T(t)y(t)]} = \|\mathbf{G}\|_2\sqrt{S_w},$$

the system 2-norm is seen to be the gain between the square root of the spectral density of a white noise input and the root mean square value of the output. Note that the individual inputs are assumed to be white, uncorrelated, and to have identical spectral densities in this interpretation.

Computation of the System 2-Norm The 2-norm of a generic stable system is a function of the impulse response:

$$\mathbf{g}(t) = \{\mathbf{C}e^{\mathbf{A}t}\mathbf{B} + \mathbf{D}\delta(t)\}1(t).$$

Substituting this expression into (4.30b) yields

$$\|\mathbf{G}\|_2 = \sqrt{\int_0^\infty \operatorname{tr}\{[\mathbf{C}e^{\mathbf{A}t}\mathbf{B} + \mathbf{D}\delta(t)]^T[\mathbf{C}e^{\mathbf{A}t}\mathbf{B} + \mathbf{D}\delta(t)]\}dt}$$

$$= \sqrt{\int_0^\infty \operatorname{tr}\{\mathbf{B}^Te^{\mathbf{A}^Tt}\mathbf{C}^T\mathbf{C}e^{\mathbf{A}t}\mathbf{B} + \mathbf{B}^Te^{\mathbf{A}^Tt}\mathbf{C}^T\mathbf{D}\delta(t) + \mathbf{D}^T\mathbf{C}e^{\mathbf{A}t}\mathbf{B}\delta(t) + \mathbf{D}^T\mathbf{D}\delta^2(t)\}dt}.$$

[7] System norms are denoted by the norm symbol operating on the transfer function matrix, even though the system may be characterized in either the frequency-domain by transfer function multiplication or in the time-domain by convolution with the impulse response.

This expression includes an integral of the square of the Dirac delta function. This integral is infinite, and the system 2-norm is therefore not defined, except when the input-to-output coupling matrix is zero.

For the closed-loop system of (4.6) and (4.7), the input-to-output coupling matrix is

$$\mathbf{D}_{cl} = \mathbf{D}_{yw} + \mathbf{D}_{yu}\mathbf{D}_{cm}\mathbf{D}_{mw} + [\mathbf{D}_{yu}\mathbf{D}_{cr} \mid \mathbf{0}].$$

The term $\mathbf{D}_{yw}$ in this matrix is typically zero when the reference input is zero and can be deleted from the general control system model. Note that the reference input is part of the generalized disturbance $w(t)$. The second term in this input-to-output coupling matrix is zero when at least one of the matrices $\mathbf{D}_{yu}$, $\mathbf{D}_{cm}$, or $\mathbf{D}_{mw}$ are zero; that is, when the reference output does not contain a control term, the controller is strictly proper, or there is no measurement noise. The final term in the input-to-output coupling matrix is zero when either $\mathbf{D}_{yu}$ or $\mathbf{D}_{cr}$ are zero, that is, when either the reference output does not contain a control term, there is no reference input, or the controller is strictly proper. Therefore, the closed-loop, system 2-norm is defined when this system is stable, there is no reference input, and either the reference output does not contain a control term, the controller is strictly proper, or there is no measurement noise.[8]

The 2-norm of a generic system reduces to

$$\|\mathbf{G}\|_2 = \sqrt{\int_0^\infty \mathrm{tr}\{\mathbf{B}^T e^{\mathbf{A}^T t}\mathbf{C}^T\mathbf{C}e^{\mathbf{A}t}\mathbf{B}\}dt} = \sqrt{\int_0^\infty \mathrm{tr}\{\mathbf{C}e^{\mathbf{A}t}\mathbf{B}\mathbf{B}^T e^{\mathbf{A}^T t}\mathbf{C}^T\}dt}$$

when the input-to-output coupling matrix equals zero. Interchanging the order of the trace and the integration operators yields

$$\|\mathbf{G}\|_2 = \sqrt{\mathrm{tr}\left\{\mathbf{B}^T \int_0^\infty e^{\mathbf{A}^T t}\mathbf{C}^T\mathbf{C}e^{\mathbf{A}t}dt\mathbf{B}\right\}} = \sqrt{\mathrm{tr}\left\{\mathbf{C} \int_0^\infty e^{\mathbf{A}t}\mathbf{B}\mathbf{B}^T e^{\mathbf{A}^T t}dt\mathbf{C}^T\right\}}. \tag{4.31}$$

Defining the observability grammian,

$$\mathbf{L}_o = \int_0^\infty e^{\mathbf{A}^T t}\mathbf{C}^T\mathbf{C}e^{\mathbf{A}t}dt, \tag{4.32}$$

and the controllability grammian,

$$\mathbf{L}_c = \int_0^\infty e^{\mathbf{A}t}\mathbf{B}\mathbf{B}^T e^{\mathbf{A}^T t}dt, \tag{4.33}$$

the 2-norm can be computed using either expression:

$$\|\mathbf{G}\|_2 = \sqrt{\mathrm{tr}\{\mathbf{B}^T\mathbf{L}_o\mathbf{B}\}} = \sqrt{\mathrm{tr}\{\mathbf{C}\mathbf{L}_c\mathbf{C}^T\}}. \tag{4.34}$$

[8] Special cases may exist where the system 2-norm is defined even when these conditions are not satisfied.

The observability and controllability grammians can be found by solving the following Lyapunov equations:[9]

$$\mathbf{A}^T\mathbf{L}_o + \mathbf{L}_o\mathbf{A} = -\mathbf{C}^T\mathbf{C}; \tag{4.35}$$

$$\mathbf{A}\mathbf{L}_c + \mathbf{L}_c\mathbf{A}^T = -\mathbf{B}\mathbf{B}^T. \tag{4.36}$$

The equation for the system 2-norm as a function of the controllability grammian can be intuitively understood based on the interpretation of this norm as the system gain subject to a white noise input. Comparing (4.36) and (4.29), the controllability grammian is seen to be the correlation matrix of the state when the inputs are uncorrelated white noise with unity spectral densities. The system 2-norm is then given by the square root of (4.28), with the weighting matrix $\mathbf{Y}$ set equal to the identity, which equals the second expression in (4.34). Formal proofs that the grammians can be computed using (4.35) and (4.36) are given in Section A7 of the Appendix along with additional information on the solution of these Lyapunov equations. A numerically efficient method for computing the system 2-norm is first to solve one of the Lyapunov equations (4.35) or (4.36) and then use (4.34) to compute the norm.

EXAMPLE 4.5 The 2-norm of the closed-loop system in Example 4.4 is

$$\|\mathbf{G}_{cl}\|_2 = 11.2.$$

This number provides an indication of the average system gain. Note that this average gain is dominated by the largest gain, that is, the gain from the disturbance input to the control input. Weighting matrices and weighting functions will be subsequently incorporated within the framework of the system 2-norm to allow greater control over the definition of this average.

4.5.5 The System ∞-Norm Cost Function

The maximum gain of a generic system over all frequencies is given by the system ∞-norm:

$$\boxed{\|\mathbf{G}\|_\infty = \sup_\omega \bar{\sigma}[\mathbf{G}(j\omega)].} \tag{4.37}$$

The fact that this cost function is a norm, that is, possesses the properties given in (4.11), can be easily verified. This cost function is particularly applicable to the design of systems where the performance is specified by bounds on the output error and the control, and reasonable bounds can be generated for sinusoidal disturbance inputs. This is a very intuitive way of specifying the cost, since most specifications for control systems take the form of bounds on the errors and controls. The ∞-norm also finds application in robustness analysis, as will be seen in the next chapter.

The ∞-norm of a system provides a bound on the maximum system gain, where the gain is defined in terms of the signal 2-norm:

$$\|\mathbf{g}(t) \otimes w(t)\|_2 \leq \|\mathbf{G}\|_\infty \|w(t)\|_2 \tag{4.38}$$

[9] Software is available for solving these equations in the MATLAB® Control Systems Toolbox [3].

where $\mathbf{g}(t) \otimes w(t)$ is the input convolved with the impulse response matrix, which yields the time-domain system output. This result is surprising at first glance, since the gain used in the definition of the ∞-norm is based on the transfer function gain. This transfer function gain, in turn, is based on the gain of the system with a sinusoidal input, but the signal 2-norm (over an infinite time interval) of a sinusoidal input does not exist. The bound (4.38) can be intuitively justified by noting that the gain of the system with a sinusoidal input is defined by the signal 2-norm gain when the norm is evaluated over a finite time period. The bound then makes sense for truncated sinusoids, which can be used to construct more complex functions with finite signal 2-norms via the Fourier series. A formal proof of the bound (4.38) is included in the Appendix; see equation (A9.2).

The bound (4.38) is a tight bound, that is, the equality is nearly achieved for some input signal. In fact, the ∞-norm equals

$$\|\mathbf{G}\|_\infty = \sup_{w \neq 0} \frac{\|\mathbf{g}(t) \otimes w(t)\|_2}{\|w(t)\|_2}. \tag{4.39}$$

A formal proof of this property is included in the Appendix; see equation (A9.3).

Equation (4.39) is often used as a definition of the system ∞-norm. This definition can be generalized to finite time intervals:

$$\|\mathbf{G}\|_{\infty,[t_0,t_f]} = \sup_{w \neq 0} \frac{\|\mathbf{g}(t) \otimes w(t)\|_{2,[t_0,t_f]}}{\|w(t)\|_{2,[t_0,t_f]}}. \tag{4.40}$$

This ∞-norm is interpreted as the maximum system gain over the given time interval.

The ∞-norm (both finite time and infinite time) of a series combination of subsystems is bounded:

$$\boxed{\|\mathbf{G}_1\mathbf{G}_2\|_\infty \leq \|\mathbf{G}_1\|_\infty\|\mathbf{G}_2\|_\infty.} \tag{4.41}$$

This result is reasonable since the maximum overall gain of a series combination of subsystems cannot exceed the product of the maximum gains of each subsystem. More formally, (4.42) can be demonstrated by noting that for any nonzero input,

$$\|\mathbf{g}_1(t) \otimes \mathbf{g}_2(t) \otimes w(t)\|_2 \leq \|\mathbf{G}_1\|_\infty\|\mathbf{g}_2(t) \otimes w(t)\|_2 \leq \|\mathbf{G}_1\|_\infty\|\mathbf{G}_2\|_\infty\|w(t)\|_2$$

Dividing by the 2-norm of the input and taking the supremum of the result yields (4.41):

$$\|\mathbf{G}_1\mathbf{G}_2\|_\infty = \sup_{w \neq 0} \frac{\|\mathbf{g}_1(t) \otimes \mathbf{g}_2(t) \otimes w(t)\|_2}{\|w(t)\|_2} \leq \|\mathbf{G}_1\|_\infty\|\mathbf{G}_2\|_\infty.$$

This property, known as the submultiplicative property, will prove very useful in subsequent chapters.

Computation of the System ∞-Norm The ∞-norm of a system described by a state model is

$$\sup_\omega \bar{\sigma}[\mathbf{G}(j\omega)] = \sup_\omega \bar{\sigma}[\mathbf{C}(j\omega\mathbf{I} - \mathbf{A})^{-1}\mathbf{B} + \mathbf{D}]. \tag{4.42}$$

The ∞-norm can be computed by iterating over frequency to find the maximum. Alternatively, the principal gains can be plotted verses frequency, and the maximum found from the plot. A third method of computing the ∞-norm is presented in the Appendix, Section A10.

EXAMPLE 4.6 The ∞-norm of the closed-loop system in Example 4.3 is computed by plotting the principal gains of the system, as shown in Figure 4.6. The maximum in Figure 4.6 is

$$\|\mathbf{G}_{cl}\|_\infty = 212.$$

This number provides an indication of the maximum system gain. Note that this maximum gain is dominated by the gain from the disturbance input to the control input. Weighting matrices and weighting functions are subsequently incorporated within the plant to allow control over which inputs and outputs are most significant. ◆

4.5.6 Weighting Functions for System Norms

The system 2-norm and the system ∞-norm treat all inputs and outputs, and all frequencies equally. In many applications, the inputs are dissimilar and may be of varying size. In addition, inputs may be limited to specific frequency bands. The outputs may also be dissimilar. For example, the output errors may be required to be several orders of magnitude smaller than the control inputs, and the various output errors may have specifications with dissimilar magnitudes. The 2-norm and the ∞-norm cost functions can be modified to accommodate these nonuniform specifications by the addition of weighting functions. In this section, the discussion is limited to the application of weighting func-

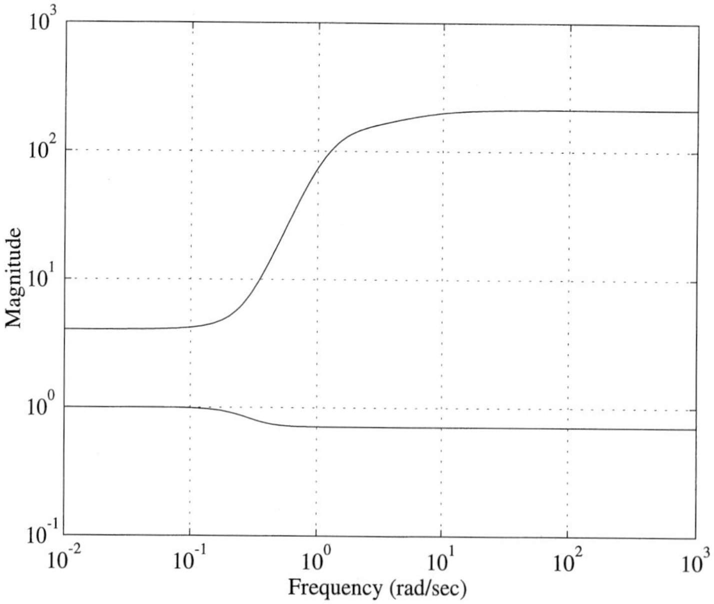

FIGURE 4.6 The principal gains for Example 4.3

tions for the ∞-norm cost function, but similar weighting functions can also be applied for use with the 2-norm cost function.

The disturbance input may be band-limited. The ∞-norm, in this case, is not a good measure of the gain from the disturbance input to the output since the maximum gain may be outside of the frequency band of interest. A more appropriate measure of system gain would be

$$\sup_{|\omega| \leq \mathcal{BW}} \bar{\sigma}[\mathbf{G}(j\omega)],$$

where $\mathcal{BW}$ is the bandwidth of the disturbance. A very general way of implementing this restriction is by application of an input or an output weighting function to the system:

$$\|\mathbf{GW}_i\|_\infty = \sup_\omega \bar{\sigma}[\mathbf{G}(j\omega)\mathbf{W}_i(j\omega)]; \tag{4.43a}$$

$$\|\mathbf{W}_o\mathbf{G}\|_\infty = \sup_\omega \bar{\sigma}[\mathbf{W}_o(j\omega)\mathbf{G}(j\omega)], \tag{4.43b}$$

where either

$$W_i(j\omega) = \begin{cases} \mathbf{I} & |\omega| \leq \mathcal{BW} \\ \mathbf{0} & |\omega| > \mathcal{BW} \end{cases} \quad \text{or} \quad W_o(j\omega) = \begin{cases} \mathbf{I} & |\omega| \leq \mathcal{BW} \\ \mathbf{0} & |\omega| > \mathcal{BW} \end{cases}. \tag{4.43c}$$

In many applications, different inputs and/or outputs may have differing magnitudes, frequency ranges, and significance, which can be included in the cost function by utilizing differing weights on the various inputs and outputs. The specification of weighting functions is illustrated in the following examples.

EXAMPLE 4.7 For a given servomotor, the two inputs are the reference input, which has an amplitude on the order of 1 and the measurement noise, which has an amplitude on the order of 10^{-3}. Further, the measurement noise may be white, that is, have a flat spectrum, while the reference input is assumed to have a bandwidth of 1 rad/sec. The system infinity norm reflects the worst-case system gain. If this gain is from the measurement noise to the output, it is less significant than if this gain is from the reference input to the output since the reference input is significantly larger than the measurement noise. In addition, if the gain is for the reference input, but outside of its frequency band, then the gain is of no significance since there is no input at this frequency. An input weighting function can be selected that reflects these considerations:

$$\mathbf{W}_i(j\omega) = \begin{bmatrix} W_1(j\omega) & 0 \\ 0 & W_2(j\omega) \end{bmatrix};$$

$$W_1(j\omega) = \begin{cases} 1 & |\omega| \leq 1 \\ 0 & |\omega| > 1 \end{cases};$$

$$W_2(j\omega) = 10^{-3},$$

where the reference input is the first element of the input vector, and the measurement noise is the second element of the input vector. The ∞-norm (4.43a) incorporating this weighting function provides an accurate representation of the size of the output. ◆

EXAMPLE 4.8 For an airplane on an instrument final approach (flying totally with reference to instruments), the outputs are the two angular position errors in altitude and azimuth. The azimuth must be maintained within 10°, and the elevation angle must be maintained within 1° on final approach to assure obstacle avoidance. This information can be collected into an output weighting function:

$$\mathbf{W}_o(j\omega) = \begin{bmatrix} 1/10 & 0 \\ 0 & 1 \end{bmatrix}.$$

The ∞-norm in (4.43b), with this weighting function, provides an accurate representation of whether or not the specifications are met for the worst case input. ◆

The weighting functions given above are filters that can be approximated by realizable transfer functions. The advantages of performing this approximation are (1) Computer-aided design software may be more user friendly when applied to systems with weighting functions that are rational transfer functions. (2) The $\mathcal{H}_2$ and $\mathcal{H}_\infty$ design methodologies, which are subsequently presented, operate on the system-plus-weighting functions described by rational transfer functions. (3) In real applications, a disturbance input (for example) may not be perfectly band-limited, but instead rolls off in a manner closer to that given by a rational transfer function than that given by an ideal lowpass filter.

Weighting functions are typically approximated by low-order filters (usually zero through third order) when evaluating performance. The uncertainty in the frequency dependence of the inputs is usually sufficiently large that the use of higher-order weighting functions is not justified. The approximation of a weighting function by a rational transfer function is illustrated in the continuation of Example 4.7.

EXAMPLE 4.7 CONTINUED The ideal lowpass filter $W_1(j\omega)$ can be approximated by a first-order transfer function:

$$W_1(s) \approx \frac{1}{s+1}$$

The magnitude of the frequency response of this system is plotted in Figure 4.7 along with the ideal lowpass filter's frequency response. ◆

The use of weighting matrices and weighting functions allows the quadratic, system 2-norm, and system ∞-norm cost functions to be tailored for use in a wide range of applications. Weighting matrices can be added to increase (or decrease) the contribution of any given input or output to the cost. Weighting functions can be added to select frequency bands of interest and to select times of interest. Input weighting functions are typically selected equal to the amplitude spectrum of disturbance inputs. Output weighting functions are typically selected equal to the inverse of the specifications on the outputs (often frequency-dependent). Time weighting functions are selected to increase the weighting during critical phases of operation. In summary, the use of weighting provides great flexibility when using the given cost functions.[10]

An optimal control system is designed to minimize a cost function that typically incorporates weighting. In the design case, the weighting function is selected both to

[10] The use of a cost function different from those presented may be desirable in some applications. The specification of these alternative cost functions is left to the imagination of the control system designer.

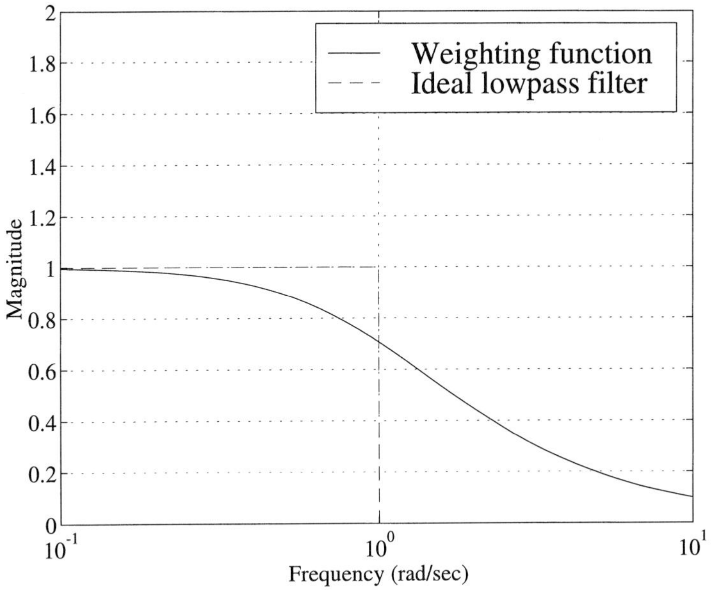

FIGURE 4.7 Magnitude frequency response of a first-order weighting function

specify performance and to tailor the controller to the given application. Understanding the effects of the weighting function on the optimal control system is vital for design and also helps clarify the role of the weighting function in performance evaluation. Therefore, many additional examples of weighting function selection are given in the controller design chapters.

4.6 Summary

The state-space and transfer function models of a general feedback control system are presented. The inputs to this general system consist of disturbances (inputs that are beyond the control of the control system designer) and reference inputs (inputs that typify the desired plant behavior). The outputs of this general system consist of the output error (the error between the reference input and the corresponding output), linear combinations of the state (those linear combinations that should be kept near zero), and the control input (the commands for the plant actuators). The control input is of interest since the actuators usually have a limited range.

The transient performance criteria of settling time and damping ratio are reviewed. The closed-loop pole locations are presented as a measure of the transient performance of the system. We note that the pole locations cannot be used to provide exact quantitative predictions of settling time and overshoot.

Tracking performance is defined as the ability of the control system to match the output to a desired value (the reference input). The evaluation of tracking performance is accomplished by computing the error for representative reference inputs, or by computing the gain between the reference input and the error. A number of reference input types that are frequently encountered in applications are discussed.

Disturbance rejection is defined as the ability of the control system to keep the output from being influenced by disturbance inputs. The evaluation of disturbance rejection is accomplished by computing the error for representative disturbance inputs, or by computing the gain between the disturbance input and the error. A number of disturbance input types that are frequently encountered in applications are discussed.

Cost functions provide a means of comparing the performance of competing control system designs. The quadratic, system 2-norm, and system ∞-norm cost functions are presented since they are frequently used in the design and analysis of control systems. The general properties of norms are also given since most useful cost functions are based on norms. The application of weighting matrices for quadratic cost functions and weighting functions for system norm cost functions are described. The inclusion of weighting in these cost functions greatly expands their utility in the analysis and design of control systems. Methods for practical computation of the various costs are also presented.

The closed-loop models (state model and transfer function model) can be applied to discrete-time systems by simply replacing continuous-time plant and controller models with their discrete-time analogues. The cost functions presented in this chapter are summarized for both continuous-time and discrete-time in Table 4.2. Note that all trans-

TABLE 4.2 Summary of Formulas for Continuous-Time and Discrete-Time Systems

Formula	Continuous-Time	Discrete-Time
Quadratic cost	$J = \int_0^{t_f} y^T(t)\mathbf{Y}(t)y(t)dt$	$J = \sum_{k=0}^{k_f} y^T(k)\mathbf{Y}_d(k)y(k)$ $\mathbf{Y}_d(k) = T\mathbf{Y}(kT)$
Quadratic cost for systems with white noise random inputs	$J = E[y^T(t)\mathbf{Y}y(t)]$	$J = E[y^T(k)\mathbf{Y}y(k)]$
	$J = \text{tr}\left\{\mathbf{Y}\int_0^{\infty} \mathbf{g}(\tau)\mathbf{S}_w\mathbf{g}^T(\tau)d\tau\right\}$	$J = \text{tr}\left\{\mathbf{Y}\sum_{k=0}^{\infty} \mathbf{g}(k)\boldsymbol{\Sigma}_w\mathbf{g}^T(k)\right\}$ where $\boldsymbol{\Sigma}_w = E[w(k)w^T(k)]$
	$J = \text{tr}\{\mathbf{Y}\mathbf{C}_y\boldsymbol{\Sigma}_x\mathbf{C}_y^T\}$ where $\boldsymbol{\Sigma}_x$ is found by solving $\mathbf{A}\boldsymbol{\Sigma}_x + \boldsymbol{\Sigma}_x\mathbf{A}^T + \mathbf{B}\mathbf{S}_w\mathbf{B}^T = \mathbf{0}$	$J = \text{tr}\{\mathbf{Y}\mathbf{C}_y\boldsymbol{\Sigma}_x\mathbf{C}_y^T\}$ where $\boldsymbol{\Sigma}_x$ is found by solving $\boldsymbol{\Sigma}_x = \boldsymbol{\Phi}\boldsymbol{\Sigma}_x\boldsymbol{\Phi}^T + \boldsymbol{\Gamma}\boldsymbol{\Sigma}_w\boldsymbol{\Gamma}^T$
The system 2-norm	$\|\mathbf{G}\|_2^2 = \frac{1}{2\pi}\int_{-\infty}^{\infty} \text{tr}\{\mathbf{G}^\dagger(j\omega)\mathbf{G}(j\omega)\}\,d\omega$	$\|\mathbf{G}\|_2^2 = \frac{1}{2\pi}\int_{-\pi}^{\pi} \text{tr}\{\mathbf{G}^\dagger(e^{j\omega})\mathbf{G}(e^{j\omega})\}\,d\omega$
	$\|\mathbf{G}\|_2^2 = \int_0^{\infty} \text{tr}\{\mathbf{g}^T(t)\mathbf{g}(t)\}\,dt$	$\|\mathbf{G}\|_2^2 = \sum_{k=0}^{\infty} \text{tr}\{\mathbf{g}^T(k)\mathbf{g}(k)\}$
	$\|\mathbf{G}\|_2^2 = \text{tr}\{\mathbf{C}\mathbf{L}_c\mathbf{C}^T\}$ where $\mathbf{L}_c = \int_{-\infty}^{\infty} e^{\mathbf{A}t}\mathbf{B}\mathbf{B}^Te^{\mathbf{A}^Tt}dt$ which can be found by solving: $\mathbf{A}\mathbf{L}_c + \mathbf{L}_c\mathbf{A}^T = -\mathbf{B}\mathbf{B}^T$	$\|\mathbf{G}\|_2^2 = \text{tr}\{\mathbf{C}\mathbf{L}_c\mathbf{C}^T\}$ where $\mathbf{L}_c = \sum_{k=0}^{\infty} \boldsymbol{\Phi}^k\boldsymbol{\Gamma}\boldsymbol{\Gamma}^T(\boldsymbol{\Phi}^T)^k$ which can be found by solving: $\mathbf{L}_c = \boldsymbol{\Phi}\mathbf{L}_c\boldsymbol{\Phi}^T + \boldsymbol{\Gamma}\boldsymbol{\Gamma}^T$
The system ∞-norm	$\|\mathbf{G}\|_\infty = \sup_{\omega} \bar{\sigma}[\mathbf{G}(j\omega)]$	$\|\mathbf{G}\|_\infty = \sup_{\omega\in[-\pi,\pi]} \bar{\sigma}[\mathbf{G}(e^{j\omega})]$

fer functions in the discrete-time column are z-transfer functions. In addition, discrete-time white noise inputs are assumed in deriving the applicable discrete-time results. The relationship between continuous-time and discrete-time white noise is discussed in detail in Chapter 3.

REFERENCES

[1] T. Kailath, *Linear Systems,* Prentice-Hall, Englewood Cliffs, NJ, 1980.

[2] A. E. Bryson, Jr. and Y. C. Ho, *Applied Optimal Control: Optimization, Estimation and Control,* Hemisphere, New York, 1975.

[3] A. Grace, A. J. Laub, J. N. Little, and C. Thompson, *Control Systems Toolbox For Use with MATLAB®,* The Math Works, Natick, MA, 1990.

Some additional references on general performance follow:

[4] R. C. Dorf, *Modern Control Systems,* 5th ed., Addison-Wesley, Reading, MA, 1990.

[5] J. M. Maciejowski, *Multivariable Feedback Design,* Addison-Wesley, Reading, MA, 1989.

[6] N. S. Nise, *Control Systems Engineering,* Benjamin-Cummings, Redwood City, CA, 1992.

Some additional references on cost functions and norms follow:

[7] J. C. Doyle, B. A. Francis, and A. R. Tannenbaum, *Feedback Control Theory,* Macmillan, New York, 1992.

[8] B. Friedland, *Control System Design: An Introduction to State Space Methods,* McGraw-Hill, New York, 1986.

[9] D. Kirk, *Optimal Control Theory: An Introduction,* Prentice-Hall, Englewood Cliffs, NJ, 1970.

[10] M. Visyasagar, *Control System Synthesis: A Factorization Approach,* MIT Press, Boston, MA, 1985.

EXERCISES

4.1 The measured elevation angle of a satellite-tracking antenna is described by the following state model:

$$\begin{bmatrix} \dot{x}_1(t) \\ \dot{x}_2(t) \end{bmatrix} = \begin{bmatrix} 0 & 1 \\ 0 & -0.1 \end{bmatrix} \begin{bmatrix} x_1(t) \\ x_2(t) \end{bmatrix} + \begin{bmatrix} 0 \\ 1 \end{bmatrix} u(t) + \begin{bmatrix} 0 \\ 1 \end{bmatrix} w(t);\ m(t) = [1 \quad 0] \begin{bmatrix} x_1(t) \\ x_2(t) \end{bmatrix} + v(t),$$

where $w(t)$ is a disturbance torque generated by wind, and $m(t)$ is the measured elevation angle, which is equal to the elevation angle plus the measurement noise $v(t)$. A unity feedback lead compensator is used to control the elevation angle. This controller is described by the following transfer function model:

$$U(s) = \frac{1.5(s + .4)}{(s + 2)}\{R(s) - M(s)\},$$

where $r(t)$ is the reference input.

a. Generate a closed-loop state model of this system. The input to this state model should include $w(t)$, $v(t)$, and $r(t)$. The output of this state model should be the error between the reference input and the elevation angle.

b. Generate a closed-loop transfer function model of this system. The input and output for this model are as given in part *a*.

c. Generate state and transfer function models for the closed-loop system, where the input is restricted to be only $r(t)$.

4.2 Generate a state model for the system in Figure P4.1.

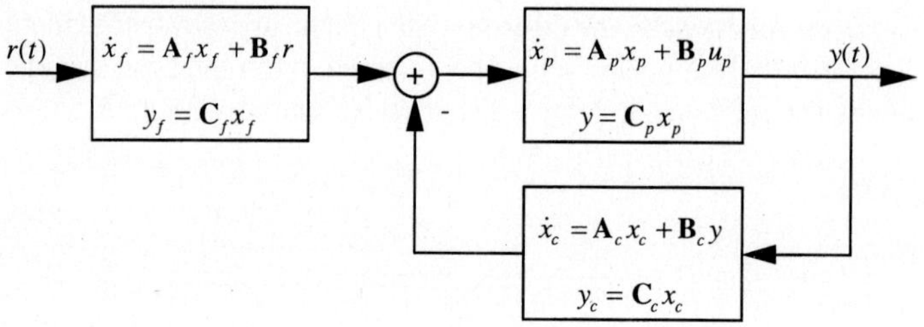

FIGURE P4.1 Combined feedback and feedforward control

4.3 Show that the following are norms:

a. $$\|x\|_\infty = \sup_t |x(t)| \quad \text{where } x(t) \in \Re \text{ for all } t;$$

b. $$\|\mathbf{M}\| = \sup_{\|x\|_2=1} \|\mathbf{M}x\|_2 \quad \text{where } \mathbf{M} \in \Re^{n_x \times n_x}.$$

Feel free to use the norm properties for the vector 2-norm, which is a true norm.

4.4 Given the plant

$$\dot{x}(t) = u(t);\ y(t) = x(t)$$

and the proportional control

$$u(t) = -Ky(t),$$

generate an analytic expression for the following cost function:

$$J = \int_0^\infty \left\{ y^2(t) + \frac{1}{10}u^2(t) \right\} dt$$

as a function of K. Assume the initial condition $x(0) = 1$ and K is positive. What value of K minimizes this expression?

4.5 Given the plant

$$\dot{x}(t) = u(t) + 3w(t);\ y(t) = x(t)$$

and the proportional controller

$$u(t) = -2y(t),$$

let $w(t)$ be a white noise input with unit spectral density. Generate an analytic expression for the following cost function:

$$J = E\left\{ y^2(t) + \frac{1}{10}u^2(t) \right\}.$$

Assume that the system has reached steady state. *Hint:* Generate the closed-loop system and solve a scalar Lyapunov equation to find the variance of $x(t)$. Then use this to generate the terms in the cost function.

4.6 Given the system

$$\dot{x}(t) = -3x(t) + 2w(t);\ y(t) = x(t),$$

do the following.

a. Generate the 2-norm of this system.
b. Generate the ∞-norm of this system.

4.7 Given the system

$$\begin{bmatrix} \dot{x}_1(t) \\ \dot{x}_2(t) \end{bmatrix} = \begin{bmatrix} -1 & 2 \\ 0 & -3 \end{bmatrix} \begin{bmatrix} x_1(t) \\ x_2(t) \end{bmatrix} + \begin{bmatrix} 0 \\ 1 \end{bmatrix} w(t);\ y(t) = [1 \quad 0] \begin{bmatrix} x_1(t) \\ x_2(t) \end{bmatrix},$$

do the following.

a. Generate the 2-norm of this system.
b. Generate the ∞-norm of this system.

4.8 Given the system

$$\begin{bmatrix} \dot{x}_1(t) \\ \dot{x}_2(t) \end{bmatrix} = \begin{bmatrix} -1 & 2 \\ 0 & -3 \end{bmatrix} \begin{bmatrix} x_1(t) \\ x_2(t) \end{bmatrix} + \begin{bmatrix} 0 \\ 1 \end{bmatrix} w(t); \; y(t) = [1 \quad 0] \begin{bmatrix} x_1(t) \\ x_2(t) \end{bmatrix},$$

where $w(t)$ is a band-limited disturbance input with a magnitude less than 10 and a bandwidth of 5 rad/sec, do the following.

a. Generate a rational weighting function for describing $w(t)$.
b. Generate an augmented system using this weighting function such that the system ∞-norm yields a bound on the output.

4.9 Given the system

$$\begin{bmatrix} \dot{x}_1(t) \\ \dot{x}_2(t) \end{bmatrix} = \begin{bmatrix} -1 & 2 \\ 0 & -3 \end{bmatrix} \begin{bmatrix} x_1(t) \\ x_2(t) \end{bmatrix} + \begin{bmatrix} 1 & 0 \\ 0 & 1 \end{bmatrix} \begin{bmatrix} w_1(t) \\ w_2(t) \end{bmatrix}; \; y(t) = [1 \quad 0] \begin{bmatrix} x_1(t) \\ x_2(t) \end{bmatrix},$$

where the disturbance inputs $w_1(t)$ and $w_2(t)$ are assumed to be uncorrelated white noises with spectral densities of 2 and 10, respectively, generate an augmented system such that the 2-norm of this system equals the steady-state standard deviation of the output.

4.10 You are given the system

$$\begin{bmatrix} \dot{x}_1(t) \\ \dot{x}_2(t) \end{bmatrix} = \begin{bmatrix} -1 & 2 \\ 0 & -3 \end{bmatrix} \begin{bmatrix} x_1(t) \\ x_2(t) \end{bmatrix} + \begin{bmatrix} 0 & 1 \\ 1 & 0 \end{bmatrix} \begin{bmatrix} w_1(t) \\ w_2(t) \end{bmatrix}; \; y(t) = [1 \quad 0] \begin{bmatrix} x_1(t) \\ x_2(t) \end{bmatrix}.$$

The disturbance input $w_1(t)$ is band-limited with a magnitude less than 10 and a bandwidth of 5 rad/sec. The disturbance input $w_2(t)$ is band-limited with a magnitude less than 1 and a bandwidth of 10 rad/sec. Assuming the disturbance inputs are sinusoids with frequencies in the given ranges, generate an augmented system using rational weighting functions such that the ∞-norm of this system yields a bound on the output.

COMPUTER EXERCISES

4.1 Performance Analysis for a DC Motor

The dc motor in Example 4.1 (see Figure 4.2) is modeled as follows:

$$\begin{bmatrix} \dot{x}_1(t) \\ \dot{x}_2(t) \\ \dot{x}_3(t) \end{bmatrix} = \begin{bmatrix} -50 & 0 & 0 \\ 10 & -350 & 0 \\ 0 & 1 & 0 \end{bmatrix} \begin{bmatrix} x_1(t) \\ x_2(t) \\ x_3(t) \end{bmatrix} + \left[\begin{array}{c|ccc} 10{,}000 & 0 & 0 & 0 \\ 0 & 0 & 100 & 0 \\ 0 & 0 & 0 & 0 \end{array}\right] \begin{bmatrix} u(t) \\ \hline r(t) \\ d(t) \\ v(t) \end{bmatrix};$$

$$\begin{bmatrix} m(t) \\ \hline e(t) \\ V(t) \end{bmatrix} = \begin{bmatrix} 0 & 0 & 1 \\ \hline 0 & 0 & -1 \\ 0 & 0 & 0 \end{bmatrix} \begin{bmatrix} x_1(t) \\ x_2(t) \\ x_3(t) \end{bmatrix} + \left[\begin{array}{c|ccc} 0 & 0 & 0 & 1 \\ \hline 0 & 1 & 0 & 0 \\ 1 & 0 & 0 & 0 \end{array}\right] \begin{bmatrix} u(t) \\ \hline r(t) \\ d(t) \\ v(t) \end{bmatrix},$$

where numerical values have been given for the poles and gains. This motor is being controlled with the following lead network:

$$\dot{x}_c(t) = -300x_c(t) + [-14{,}400 \mid 14{,}400] \begin{bmatrix} r(t) \\ \hline m(t) \end{bmatrix};$$

$$u(t) = x_c(t) + [60 \mid -60] \begin{bmatrix} r(t) \\ \hline m(t) \end{bmatrix}.$$

Compute the settling time and damping ratio of each closed-loop pole. Based on this data, estimate the settling time and percent overshoot of the closed-loop system. The percent overshoot can be estimated using the formula for a second-order, all-pole system:

$$\%Overshoot = 100e^{-\zeta\pi/\sqrt{1-\zeta^2}}$$

applied to the dominant poles. Simulate the system with a step reference input and plot the error and control input. Assume initial conditions and disturbance torques are zero. How do the simulated results compare to those estimated from the poles?

Compute the dc gain between the reference input and both the tracking error and the control input. How do these results compare to the simulated results obtained for the step input? The steady-state value of the control input is zero in this case and does not provide useful information on the size of the transient control. An approximate magnitude for the transient control can be obtained by looking at the closed-loop frequency response at a frequency that approximates the rise of the step response. This frequency response approximates the transient control magnitude (due to a unit step reference input) since both inputs require similar tracking rates. Note that this is only a rough approximation. Compare this frequency-domain result to that obtained from the simulation with the unit step input.

Compute the Fourier transfer function between the disturbance input (both the measurement noise and the disturbance torque) and the reference output (both the tracking error and the control input). Use this transfer function to predict the effects of a dc disturbance torque of unit magnitude on the reference output. Use this transfer function to predict the effects of a 60 Hz measurement noise of magnitude 0.01 on the reference output. Simulate the system with each of these inputs and plot the results. Assume that all other inputs and initial conditions are zero in this simulation. Use the results to verify your calculations.

4.2 Performance Analysis for a Rotorcraft with Autopilot

An autopilot is used to control the pitch and yaw of a rotorcraft. The available sensors consist of gyros for measuring pitch and yaw and rate gyros for measuring pitch rate and yaw rate (the entire state). The control inputs consist of vanes that supply torques in the pitch and yaw axes. Disturbances consist of torques in both pitch and yaw (measurement noise is ignored). A model of the plant is

$$\begin{bmatrix} \dot{\theta} \\ \ddot{\theta} \\ \dot{\Psi} \\ \ddot{\Psi} \end{bmatrix} = \begin{bmatrix} 0 & 1 & 0 & 1 \\ 0 & 0 & 0 & -7 \\ 0 & 0 & 0 & 1 \\ 0 & 7 & 0 & 0 \end{bmatrix} \begin{bmatrix} \theta \\ \dot{\theta} \\ \Psi \\ \dot{\Psi} \end{bmatrix} + \begin{bmatrix} 0 & 0 \\ -15 & 0 \\ 0 & 0 \\ 0 & -15 \end{bmatrix} \begin{bmatrix} \delta_\theta \\ \delta_\Psi \end{bmatrix} + \begin{bmatrix} 0 & 0 \\ 1 & 0 \\ 0 & 0 \\ 0 & 1 \end{bmatrix} \begin{bmatrix} w_\theta \\ w_\Psi \end{bmatrix},$$

where θ is the pitch angle, Ψ is the yaw angle, δ_θ is the pitch vane deflection, δ_Ψ is the yaw vane deflection, w_θ is the disturbance torque in pitch, and w_Ψ is the disturbance torque in yaw. The outputs of interest are the pitch and yaw angles. The autopilot generates control-vane deflections from the measured state:

$$\begin{bmatrix} \delta_\theta \\ \delta_\Psi \end{bmatrix} = -\begin{bmatrix} -1.33 & -0.27 & 0 & 0.47 \\ -0.47 & 0 & -1.33 & -0.27 \end{bmatrix} \begin{bmatrix} \theta \\ \dot{\theta} \\ \Psi \\ \dot{\Psi} \end{bmatrix}$$

Find the principal gains from the disturbance inputs to the pitch and yaw angles, and from the disturbance inputs to the control inputs and plot. What is the ∞-norm of each of the above transfer functions (no weighting functions).

Assume that the disturbance inputs lie in a band of frequency from 0 Hz to 5 Hz. What are reasonable weighting functions to use when evaluating performance? What are the weighted ∞-norms of the given transfer functions? Simulate the system with representative disturbance torques and plot the pitch angle, yaw angle, and vane deflections. Compare the results to the principal gains and ∞-norms.

5 Robustness

Most control system designers have synthesized a controller with good performance, simulated the controller to verify this performance, and then implemented the controller, only to discover that the performance is totally unacceptable. The problem encountered by these designers is that the mathematical model used for design and simulation is significantly different from the true model. This discrepancy can often lead to poor performance and even instability of the control system.

Mathematical modeling uses physics, chemistry, aerodynamics, and so on, in order to produce equations that describe the plant.[1] A number of assumptions are typically made during this process in order to yield a simple model. Examples include using Newtonian gravitation as opposed to general relativity, ignoring friction between moving parts, and ignoring vibration of a motor shaft. These assumptions are justified by the need for simple design models and the difficulty encountered in generating ever more accurate models. For example, using general relativity may result in an extremely complex, nonlinear model that makes design nearly impossible. As another example, the friction between moving parts may be difficult to determine and may change episodically, creating errors in an initially accurate model. For these reasons, a mathematical model is never a perfect representation of the physical object.

The control system engineer should be assured that a design will function acceptably before committing to implementation. Such assurance can be obtained by analyzing control system stability and performance with respect to a range of plant models that is expected to encompass the actual plant. This type of analysis is termed robustness analysis.

The analysis of robustness requires that the discrepancy between the mathematical model of the plant and the actual plant be quantified. Since a perfect mathematical model of the plant is not available, this discrepancy can not be uniquely defined. Instead, a set of mathematical models is defined which includes the actual plant dynamics. This set is specified by a nominal plant and a set of perturbations termed admissible perturbations. The admissible perturbations are typically assumed to be bounded, where the bound is dependent on the uncertainty in the model.

A controller that functions adequately for all admissible perturbations is termed robust. Two types of robustness, robust stability and robust performance, are considered in this chapter. A control system is said to be robustly stable if it is stable for all admissible perturbations. A control system is said to perform robustly if it satisfies the

[1] Models can also be generated empirically by matching observed plant behavior. This process, known as system identification, is typically used when theoretical modeling is impossible or overly complicated.

performance specifications for all admissible perturbations. Note that stability and performance robustness depend on the controller, the nominal model, and the set of perturbations. Performance robustness also depends on the performance specifications.

This chapter begins by developing some results on the internal stability of feedback systems that are used in robustness analysis. The stability robustness of a SISO plant subject to gain and phase perturbations is then addressed. This analysis leads to the gain and phase margins of classical control. Two more general types of perturbations are then presented: transfer functions (unstructured uncertainty); and multiple transfer functions and/or parameters (structured uncertainty). The stability and performance robustness of general systems subject to these perturbations is then addressed.

5.1 Internal Stability of Feedback Systems

The internal stability of the feedback systems shown in Figure 5.1a and 5.1b is determined by considering the closed-loop system in Figure 5.1c. This feedback system is internally stable provided that all four of the transfer functions between each input and each error,

$$\left[\begin{array}{c:c} \mathbf{G}_{e_1u_1}(s) & \mathbf{G}_{e_1u_2}(s) \\ \hdashline \mathbf{G}_{e_2u_1}(s) & \mathbf{G}_{e_2u_2}(s) \end{array}\right] = \left[\begin{array}{c:c} \{\mathbf{I} + \mathbf{K}(s)\mathbf{G}(s)\}^{-1} & -\{\mathbf{I} + \mathbf{K}(s)\mathbf{G}(s)\}^{-1}\mathbf{K}(s) \\ \hdashline \{\mathbf{I} + \mathbf{G}(s)\mathbf{K}(s)\}^{-1}\mathbf{G}(s) & \{\mathbf{I} + \mathbf{G}(s)\mathbf{K}(s)\}^{-1} \end{array}\right], \tag{5.1}$$

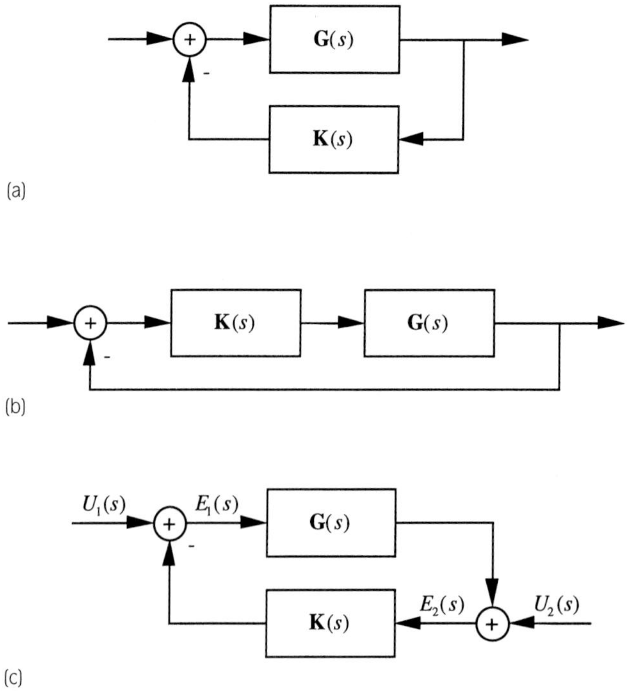

FIGURE 5.1 Feedback system block diagrams: (a) standard feedback; (b) unity feedback; (c) general feedback used to evaluate internal stability

are stable (see Subsection 2.6.1 for details). Expanding the matrix inverses in these transfer functions yields

$$\left[\begin{array}{c:c} \mathbf{G}_{e_1u_1}(s) & \mathbf{G}_{e_1u_2}(s) \\ \hdashline \mathbf{G}_{e_2u_1}(s) & \mathbf{G}_{e_2u_2}(s) \end{array}\right] = \left[\begin{array}{c:c} \dfrac{\operatorname{adj}\{\mathbf{I}+\mathbf{K}(s)\mathbf{G}(s)\}}{\det\{\mathbf{I}+\mathbf{K}(s)\mathbf{G}(s)\}} & -\dfrac{\operatorname{adj}\{\mathbf{I}+\mathbf{K}(s)\mathbf{G}(s)\}\mathbf{K}(s)}{\det\{\mathbf{I}+\mathbf{K}(s)\mathbf{G}(s)\}} \\ \hdashline \dfrac{\operatorname{adj}\{\mathbf{I}+\mathbf{G}(s)\mathbf{K}(s)\}\mathbf{G}(s)}{\det\{\mathbf{I}+\mathbf{G}(s)\mathbf{K}(s)\}} & \dfrac{\operatorname{adj}\{\mathbf{I}+\mathbf{G}(s)\mathbf{K}(s)\}}{\det\{\mathbf{I}+\mathbf{G}(s)\mathbf{K}(s)\}} \end{array}\right],$$

where adj(•) denotes the adjugate[2] and det(•) denotes the determinant. Note that the division above is valid since $\det\{\mathbf{I}+\mathbf{K}(s)\mathbf{G}(s)\}$ is a scalar quantity. The poles of these four transfer functions (denoted simply as the poles of the feedback system) must satisfy one of the following conditions:

$$\det\{\mathbf{I}+\mathbf{G}(s)\mathbf{K}(s)\} = 0, \tag{5.2a}$$

$$\det\{\mathbf{I}+\mathbf{K}(s)\mathbf{G}(s)\} = 0, \tag{5.2b}$$

$$s \text{ is a pole of } \operatorname{adj}\{\mathbf{I}+\mathbf{G}(s)\mathbf{K}(s)\}, \tag{5.2c}$$

$$s \text{ is a pole of } \operatorname{adj}\{\mathbf{I}+\mathbf{K}(s)\mathbf{G}(s)\}, \tag{5.2d}$$

$$s \text{ is a pole of } \mathbf{G}(s), \tag{5.2e}$$

$$s \text{ is a pole of } \mathbf{K}(s). \tag{5.2f}$$

These conditions for the poles of the feedback system can be condensed by utilizing some properties of the adjugate and the determinant operations. The adjugate of a matrix takes sums of the products of individual terms in the matrix. This is equivalent to taking series and parallel combinations of these transfer functions, operations that do not create new poles. The poles of the $\operatorname{adj}\{\mathbf{I}+\mathbf{G}(s)\mathbf{K}(s)\}$ are therefore a subset of the poles of $\{\mathbf{I}+\mathbf{G}(s)\mathbf{K}(s)\}$. Each pole of $\operatorname{adj}\{\mathbf{I}+\mathbf{G}(s)\mathbf{K}(s)\}$ must be a pole of the plant $\mathbf{G}(s)$ or a pole of the controller $\mathbf{K}(s)$ since the identity matrix has no poles, and all the poles of $\mathbf{G}(s)\mathbf{K}(s)$ are either plant poles or controller poles. Similarly, each pole of $\operatorname{adj}\{\mathbf{I}+\mathbf{K}(s)\mathbf{G}(s)\}$ is a plant pole or a controller pole. The conditions (5.2c) and (5.2d) for the poles of the feedback system are therefore superfluous. Also note that [see the Appendix, equation (A8.1)],

$$\det\{\mathbf{I}+\mathbf{G}(s)\mathbf{K}(s)\} = \det\{\mathbf{I}+\mathbf{K}(s)\mathbf{G}(s)\}.$$

Thus, the poles of the four transfer functions in (5.1) must satisfy one of the following three conditions:

$$\boxed{\begin{aligned} \det\{\mathbf{I}+\mathbf{G}(s)\mathbf{K}(s)\} &= 0, && (5.3a)\\ s \text{ is a pole of } \mathbf{G}(s), & && (5.3b)\\ s \text{ is a pole of } \mathbf{K}(s). & && (5.3c) \end{aligned}}$$

[2] The adjugate matrix appearing in the inverse is called the adjoint matrix by many authors. I prefer the designation *adjugate* because *adjoint,* as defined in functional analysis, leads to an alternative definition of the adjoint of a matrix.

The feedback system is then internally stable provided that the solutions of (5.3a), the plant poles, and the controller poles, all have negative real parts.[3]

These conditions for the poles of a feedback system can be further simplified for the case of SISO systems where there are no pole-zero cancellations between the plant and the controller. In this case, the $\det\{\mathbf{I} + \mathbf{G}(s)\mathbf{K}(s)\}$ is equal to $\{1 + G(s)K(s)\}$. Additionally, the poles of $G(s)$ and $K(s)$ are canceled by the denominator of $\{1 + G(s)K(s)\}$ and do not appear as poles of any of the four transfer functions in (5.1).

For SISO systems with no pole-zero cancellations between $G(s)$ and $K(s)$, the poles of all four of the transfer functions in (5.1) are the solutions of

$$\boxed{1 + G(s)K(s) = 0.} \tag{5.4}$$

Further, the systems in Figure 5.1a, b, and c are internally stable if and only if all the solutions of (5.4) have negative real parts.

This result states that internal stability of a SISO system with no pole-zero cancellations between the plant and the controller depends only on the solution of (5.4).

An important special case for MIMO systems is when both the plant and the controller are stable. The MIMO feedback system, with a stable plant and a stable controller is internally stable if all the solutions of (5.3a) have negative real parts. This test for internal stability is both necessary and sufficient, as demonstrated below. If the system is internally stable, then

$$\det[\mathbf{G}_{e_2u_2}(s)] = \det[\{\mathbf{I} + \mathbf{G}(s)\mathbf{K}(s)\}^{-1}] = \frac{1}{\det[\mathbf{I} + \mathbf{G}(s)\mathbf{K}(s)]}$$

is bounded for all s with non-negative real parts. Therefore, the determinant in this equation is nonzero for all non-negative values of s.

A MIMO feedback system, as in Figure 5.1a, b, or c with both $\mathbf{G}(s)$ and $\mathbf{K}(s)$ stable, is internally stable if and only if all of the solutions of

$$\boxed{\det\{\mathbf{I} + \mathbf{G}(s)\mathbf{K}(s)\} = 0} \tag{5.5}$$

have negative real parts.

This result provides a convenient test for stability of a MIMO feedback system with a stable plant and a stable controller.

5.2 The SISO Nyquist Stability Criterion

The Nyquist stability criterion is a graphical method of determining the stability of a feedback system. The Nyquist criterion was developed originally as a method of determining the stability of a feedback system without actually having to solve for the closed-loop poles (a job that was difficult in 1932 when the Nyquist criterion was

[3] These conditions guarantee internal stability, but are not required for internal stability. For example, controllers are frequently used to internally stabilize unstable plants.

developed [1]). The digital computer is now the tool of choice for determining poles and hence stability, but the Nyquist criterion is still useful in several respects:

1. The phase and gain margins, which provide measures of the stability robustness of a feedback system, are readily determined from a Nyquist plot.
2. The Nyquist stability criterion can be applied to systems that include time delays. The transfer function of such a system is not a ratio of finite polynomials; therefore, finding poles of the closed-loop system is not possible using polynomial methods.
3. Modifications to the controller frequency response that increase gain and/or phase margin are readily obtained from the Nyquist plot.

Of primary interest in this book is applying the Nyquist stability criterion for the first reason cited above, namely, to determine phase and gain margins, since they are indicators of stability robustness.

A SISO feedback system is internally stable if all the zeros of $\{1 + G(s)K(s)\}$ have negative real parts, or equivalently, there are no zeros of $\{1 + G(s)K(s)\}$ in the closed right half-plane.[4] The number of zeros of $\{1 + G(s)K(s)\}$ in a region of the complex plane can be found using the argument principle (see [2], page 298). This principle states that if a closed contour encircling (in the clockwise direction) the right half-plane is mapped through $\{1 + G(s)K(s)\}$, the resulting contour will encircle the origin $N_c = N_p - N_z$ times in the counterclockwise direction. Here, N_p is the number of poles of $\{1 + G(s)K(s)\}$ in the right half-plane, and N_z is the number of zeros of $\{1 + G(s)K(s)\}$ in the right half-plane. The contour generated by the loop transfer function $G_l(s) = G(s)K(s)$ is identical to the contour generated by $\{1 + G(s)K(s)\}$ except it is shifted one unit to the right. In addition, the poles of $\{1 + G(s)K(s)\}$ equal the poles of $G_l(s)$ since adding the value 1 does not change the pole locations. These facts, along with the argument principle, lead to the Nyquist criterion (see [3], page 517), for determining stability based on the loop transfer function:

> A feedback system is stable if and only if the image of a closed contour encircling (in the clockwise direction) the right half-plane as mapped through $G_l(s)$ encircles the point minus one N_p times in the counterclockwise direction, where N_p is the number of poles of $G_l(s)$ in the right half-plane.

The Nyquist criterion is illustrated by application to an example.

EXAMPLE 5.1 We are given

$$G(s) = \frac{1}{s-1};\ K(s) = 2;\ \text{then } G_l(s) = \frac{2}{s-1}.$$

The closed contour encircling the right half-plane is given in Figure 5.2a. The Nyquist plot is given in Figure 5.2b. There is one counterclockwise encirclement (check that it

[4] This is true provided there are no pole-zero cancellations between the plant and the controller. It will be assumed that no such cancellations exist throughout the remainder of this book unless otherwise noted.

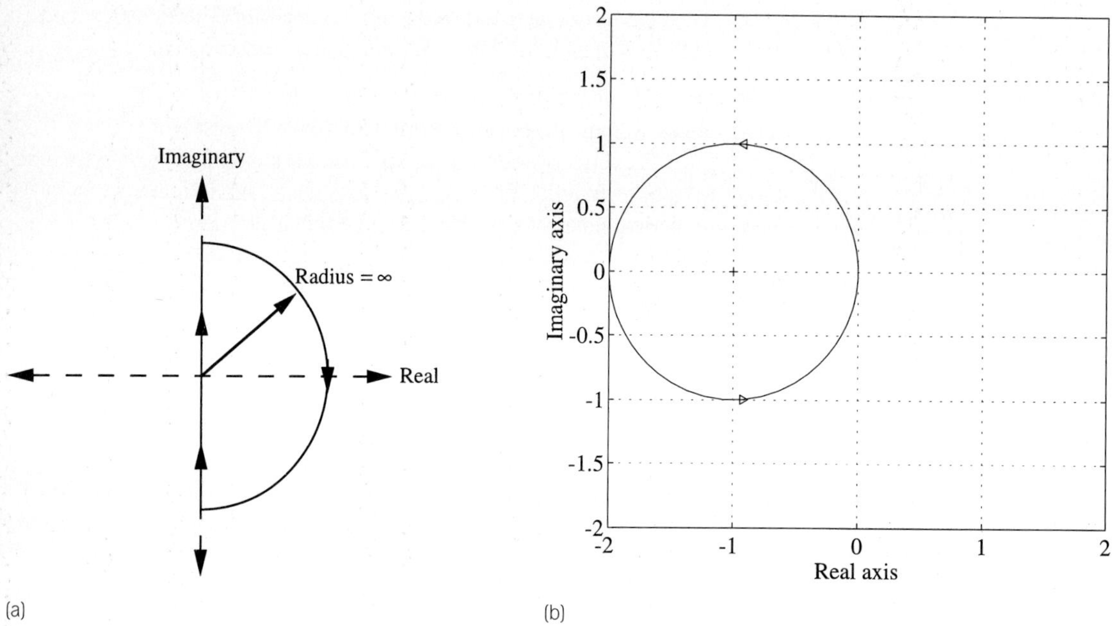

FIGURE 5.2 The Nyquist stability criterion applied to Example 5.1: (a) the contour encircling the right half-plane; (b) the Nyquist locus or Nyquist plot

is counterclockwise) of the point minus one, and $G_l(s)$ has one pole with a positive real part. Therefore, the system is stable. ◆

The contour encircling the right half-plane must not pass through any poles of the loop transfer function. If poles of the loop transfer function lie on the imaginary axis (the boundary of the right half-plane), a detour can be added to the contour encircling the right half-plane, as shown in Example 5.2.

EXAMPLE 5.2 We are given

$$G(s) = \frac{1}{s(s+1)}\ ;\ K(s) = 2\ ;\ \text{then } G_l(s) = \frac{2}{s(s+1)}.$$

The closed contour encircling the right half-plane is given in Figure 5.3a. The Nyquist plot is given in Figure 5.3b. Note that the circle of radius ε on the closed contour yields the large circle on the Nyquist plot. The radius of this large circle approaches infinity as ε goes to zero. There are no encirclements of the point minus one, and $G_l(s)$ has no poles with positive real parts. The number of encirclements equals the number of right half-plane poles, and the system is stable. ◆

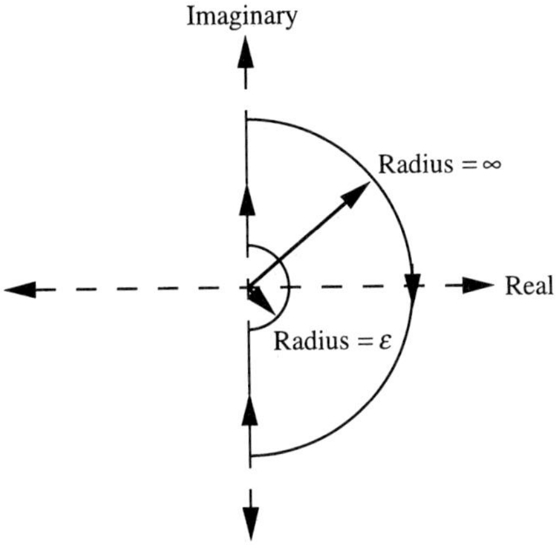

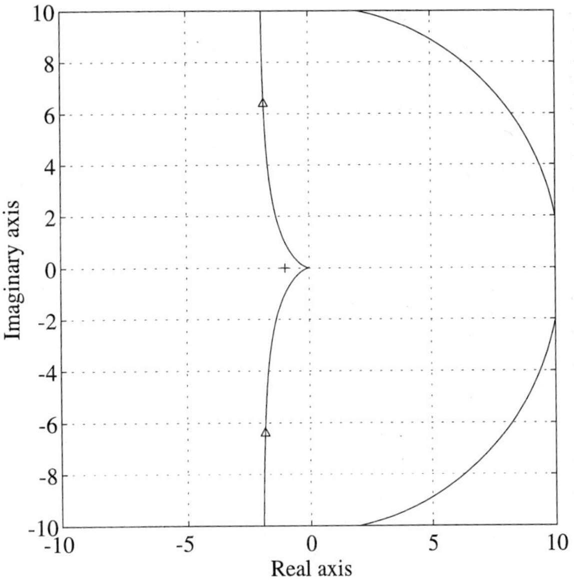

FIGURE 5.3 The Nyquist stability criterion applied to Example 5.2: (a) the contour encircling the right half-plane; (b) the Nyquist locus or Nyquist plot

5.3 Gain and Phase Margins for SISO Systems

Uncertainty in the plant can be modeled by the inclusion of a variable gain:

$$G(s) = gG_0(s)$$

where $G_0(s)$ is the nominal plant, and the uncertain gain g lies in an admissible region:

$$g_{\min} \leq g \leq g_{\max}.$$

The admissible region includes a gain of 1 since this results in the actual plant equaling the nominal plant. A block diagram of a closed-loop control system with this plant uncertainty is given in Figure 5.4.

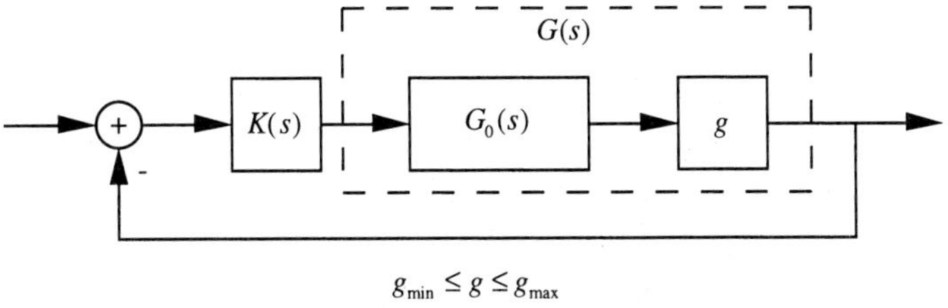

FIGURE 5.4 SISO plant with uncertain gain

The Nyquist stability criterion can be applied to this uncertain system to evaluate stability robustness. The loop transfer function of the system with gain uncertainty is

$$G_1(s) = gG_0(s)K(s).$$

The direct approach to determining robustness is to evaluate stability for each g within the possible range. Fortunately, this is unnecessary since the effect of a gain on the Nyquist plot is quite simple. The gain merely multiplies the loop transfer function of the nominal system; that is, each point on the Nyquist plot of the actual system is g times the corresponding point on the Nyquist plot of the nominal system. The effect of the uncertain gain is then merely to expand or contract the nominal Nyquist plot.

The nominal system is assumed to be stable since the controller is designed for the nominal plant and, of course, a stable system is designed. As the gain is changed, the system goes from stable to unstable when the Nyquist plot touches the point minus one. At this point, one or more poles of the closed-loop system lie on the imaginary axis. Therefore, the range of possible gains that maintain stability of the closed-loop system can be immediately determined from the Nyquist plot of the nominal system. The resulting design is termed robust if this stability range includes the range of admissible gains.

EXAMPLE 5.3 We are given the plant and the controller

$$G(s) = \frac{g}{s^2 - 2s + 1}\,; \; K(s) = \frac{1500s - 100}{s^2 + 30s + 400},$$

where the uncertain plant gain is bounded:

$$0.8 \leq g \leq 1.2.$$

The stability robustness of this system is determined from the Nyquist plot of the nominal loop transfer function, which is shown in Figure 5.5. The Nyquist locus is seen to cross the real axis adjacent the minus one point at -0.25 and -1.8. The uncertain gain can then range between $1/1.8 = 0.56$ and $1/0.25 = 4.0$ before the system goes unstable. Since the actual uncertain gain only varies between 0.8 and 1.2, the system is stable for all gains in the prescribed range and is robustly stable. ◆

The range of gains that result in a stable system is readily computed from the Nyquist plot. This range of gains is often presented instead of explicit information on stability robustness since the true range of admissible gains is often not well specified. The range of gains that guarantee stability is specified using the gain margin and the downside gain margin:

DEFINITION: The gain margin GM^+ is defined as the minimum gain greater than 1 that results in the feedback system (Figure 5.4) becoming unstable.

DEFINITION: The downside gain margin GM^- is defined as the maximum positive gain less than 1 that results in the feedback system (Figure 5.4) becoming unstable.

Note that the gain margin is defined as infinite if there is no gain increase that generates an unstable closed-loop system. Similarly, the downside gain margin is defined as zero if

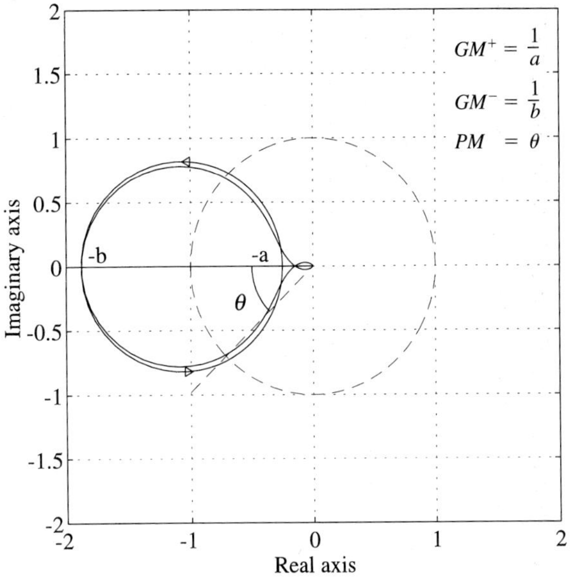

FIGURE 5.5 Nyquist locus for Examples 5.3 and 5.4

there is no decrease in gain that results in instability. The largest range of admissible gains that guarantees a stable system is

$$(GM^-, GM^+).$$

The loop gain is the product of the plant gain and the controller gain. Any uncertainty in the gain of the controller, therefore, has the same effect on stability as uncertainty in the plant gain. The gain margins then provide limits on the allowable uncertainty in the combination of the plant and the controller gains.

A second simple type of uncertainty is often employed when considering the robustness of SISO systems. The plant is assumed to have an uncertain phase shift,

$$G(s) = e^{-j\phi}G_0(s), \tag{5.6}$$

where the uncertain phase ϕ lies within an admissible region:

$$\phi_{\min} \leq \phi \leq \phi_{\max}.$$

A phase shift, as given in (5.6), is not a realistic type of uncertainty, since ideal phase shifters cannot be constructed. A more realistic type of uncertainty would be an uncertain time delay or the addition of unmodeled dynamics, both of which add a frequency-dependent phase shift to the system. The phase uncertainty given above is an idealized representation of these more general uncertainties that cause phase shifts. It is assumed that zero is an admissible phase shift, since this results in the actual plant equaling the nominal plant. A block diagram of a control system with a phase uncertainty is shown in Figure 5.6.

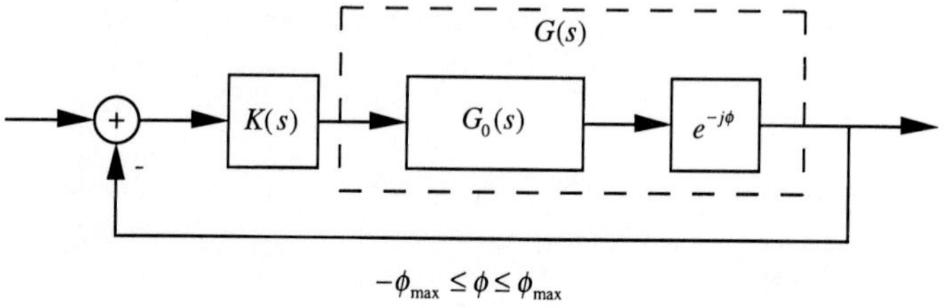

FIGURE 5.6 SISO plant with uncertain phase

The Nyquist stability criterion can be applied to evaluate robustness. Stability results based on polynomial methods cannot be applied in this case, since the phase shift cannot be represented as a ratio of polynomials in s. The loop transfer function of the system in Figure 5.6 is

$$G_l(s) = e^{-j\phi}K(s)G_0(s).$$

Robustness can be evaluated from the nominal Nyquist plot, since the phase perturbation has a simple effect on the Nyquist locus. The phase shift $e^{-j\phi}$ multiplies each point on the Nyquist plot, causing a rotation through an angle ϕ around the origin. Assuming the nominal system is stable, the system becomes unstable when the Nyquist locus is rotated far enough to touch the point minus one. Due to the symmetry of the Nyquist locus, the negative of this phase shift also results in instability.

The range of phase shifts that guarantees stability is given by specifying the phase margin:

DEFINITION: The phase margin PM is defined as the minimum amount of phase shift that, when added into the feedback loop, results in the system becoming unstable.

The phase margin is defined as infinite if the gain of the loop transfer function is always less than one and no phase shift results in instability.

EXAMPLE 5.4 Consider the system described in Example 5.3. The gain and phase margins can be computed from the Nyquist plot given in Figure 5.5. The gain and phase margins are

$$GM^+ = 4; \; GM^- = 0.56; \; PM = 45°,$$

and the system is stable for a range of gain perturbations between 0.56 and 4. The system is also stable for a range of phase perturbations between $-45°$ and $+45°$. Note that each perturbation is considered separately when defining the margins. Given only these phase and gain margins, there is no guarantee that the system will remain stable if a gain perturbation of 2 is combined with a phase perturbation of 30°. ◆

The gain margin, the downside gain margin, and the phase margin (collectively known as the stability margins) combine to yield information on the robustness of the closed-loop system. In many applications, explicit information on the uncertainty of the model is not available. For SISO systems, experience has shown that a gain margin greater than 2 (6 dB), a downside gain margin less than 0.5 (-6 dB), and a phase margin greater than 30° provide sufficient stability robustness in most applications. Additionally, feedback systems with these gain and phase margins typically yield performance close to that computed for the nominal system.

In summary, the stability margins provide information on stability robustness when the plant is subject to gain and phase perturbations. These margins are simply computed from the nominal Nyquist plot. The stability margins can also be computed from other representations of the system frequency response, for example, Bode plots and Nichol's plots. Additionally, the gain margins can be found by solving for the roots of the characteristic equation as a function of gain, or by using the Routh stability test.[5]

5.4 Unstructured Uncertainty

The gain and phase perturbations presented above are special cases of uncertainty in the mathematical model of the plant. Many other types of perturbations exist and could be used when evaluating system robustness. In this section, uncertainty is modeled as a perturbation to the nominal plant. This perturbation is a bounded transfer function, where *bounded* is defined in terms of the system ∞-norm. This type of plant uncertainty is termed unstructured since no detailed model of the perturbation (the unknown transfer function) is employed.

5.4.1 Unstructured Uncertainty Models

An unstructured perturbation can be connected to the plant in a number of ways, each generating a unique set of possible plant models. Five basic connections of the perturbation to the nominal plant model are presented: additive perturbation, input-multiplicative perturbation, output-multiplicative perturbation, input feedback perturbation, and output feedback perturbation. An additive unstructured uncertainty models the actual plant as equal to the nominal plant plus a perturbation:

$$\mathbf{G}(s) = \mathbf{G}_0(s) + \mathbf{\Delta}_a(s)$$

where $\mathbf{\Delta}_a(s)$ denotes the additive perturbation. An input-multiplicative uncertainty models the actual plant as the nominal plant plus a series combination of the perturbation and the nominal plant (the perturbation appears on the input to the nominal plant):

$$\mathbf{G}(s) = \mathbf{G}_0(s)[\mathbf{I} + \mathbf{\Delta}_i(s)],$$

where $\mathbf{\Delta}_i(s)$ denotes the input-multiplicative perturbation. An output-multiplicative uncertainty models the actual plant as the nominal plant plus a series combination of the nominal plant and the perturbation (the perturbation appears on the output to the nominal plant):

$$\mathbf{G}(s) = [\mathbf{I} + \mathbf{\Delta}_o(s)]\mathbf{G}_0(s),$$

where $\mathbf{\Delta}_o(s)$ denotes the output-multiplicative perturbation. An input feedback uncertainty models the actual plant as the nominal plant in series with the perturbation in a feedback loop (the feedback loop appears on the input to the nominal plant):

$$\mathbf{G}(s) = \mathbf{G}_0(s)[\mathbf{I} + \mathbf{\Delta}_{fi}(s)]^{-1},$$

[5] The Routh stability test is described in most introductory control texts; for example, see Dorf [4], page 183.

where $\boldsymbol{\Delta}_{fi}(s)$ denotes the input feedback perturbation. An output feedback uncertainty models the actual plant as the nominal plant in series with the perturbation in a feedback loop (the feedback loop appears on the output to the nominal plant):

$$\mathbf{G}(s) = [\mathbf{I} + \boldsymbol{\Delta}_{fo}(s)]^{-1}\mathbf{G}_0(s),$$

where $\boldsymbol{\Delta}_{fo}(s)$ denotes the output feedback perturbation. Block diagrams of these five uncertainty models, appearing in a feedback system, are given in Figure 5.7. Note that for SISO systems, blocks can be interchanged without affecting the system response, and there is no difference between the input and output uncertainty models, but these models can yield substantially different results for MIMO systems. The actual plant equals the nominal plant in all cases when the perturbation equals zero.

The uncertainty models are used to represent various types of uncertainty in the

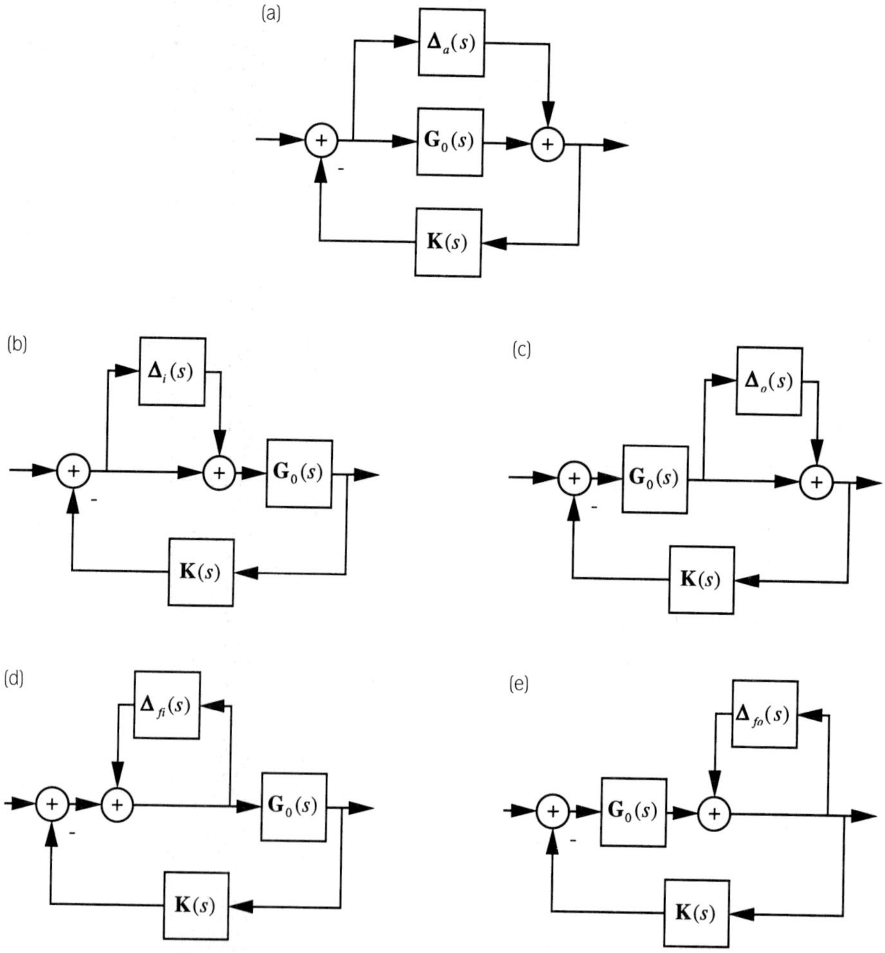

FIGURE 5.7 Unstructured uncertainties in the plant model: (a) additive uncertainty; (b) input-multiplicative uncertainty; (c) output-multiplicative uncertainty; (d) input feedback uncertainty; (e) output feedback uncertainty

plant. The additive perturbation represents unknown dynamics operating in parallel with the plant. The multiplicative perturbations represent unknown dynamics operating in series with the plant. The feedback perturbations are used primarily to represent uncertainty in the gain and phase of the plant (or the control loop if a feedback control is applied to the plant). As the magnitude of the bound on a scalar feedback perturbation approaches 1, the allowable gain increase approaches infinity, the gain decrease approaches 1/2, and the maximum phase shift approaches 60 degrees. A feedback system that is tolerant of bounded feedback perturbations (with a bound near 1) therefore has a large gain margin, a reasonable downside gain margin, and a large phase margin, that is, good stability margins.

Stability robustness or performance robustness can be evaluated when the perturbations in these models are bounded:

$$\bar{\sigma}\{\mathbf{\Delta}'(j\omega)\} \le \Delta_{\max}(j\omega) \tag{5.7}$$

where $\bar{\sigma}(\bullet)$ is the maximum singular value, and $\mathbf{\Delta}'$ can be any of the perturbations described above. The bound given for the perturbation is in general frequency-dependent, allowing the specification of plant uncertainty to vary over frequency. In many applications the plant model is accurate at low frequencies but less accurate at high frequencies. The frequency-dependent bound allows this information to be incorporated into robustness analysis. Note that the perturbation transfer function is always stable since it possesses a bounded gain.

The unstructured uncertainty models presented above can all be analyzed in a similar manner by placing them within a common framework (called the standard form for robustness analysis or simply the standard form). This common framework has the perturbation normalized and in a feedback loop, as shown in Figure 5.8. The plant $\mathbf{P}(s)$ has three inputs and three outputs (in general, each of these inputs and outputs can be a vector):

$$\begin{bmatrix} Y_d(s) \\ \hline Y(s) \\ \hline M(s) \end{bmatrix} = \left[\begin{array}{c:c:c} \mathbf{P}_{y_d w_d}(s) & \mathbf{P}_{y_d w}(s) & \mathbf{P}_{y_d u}(s) \\ \hdashline \mathbf{P}_{y w_d}(s) & \mathbf{P}_{yw}(s) & \mathbf{P}_{yu}(s) \\ \hdashline \mathbf{P}_{m w_d}(s) & \mathbf{P}_{mw}(s) & \mathbf{P}_{mu}(s) \end{array}\right] \begin{bmatrix} W_d(s) \\ \hline W(s) \\ \hline U(s) \end{bmatrix} = \mathbf{P}(s) \begin{bmatrix} W_d(s) \\ \hline W(s) \\ \hline U(s) \end{bmatrix}. \tag{5.8}$$

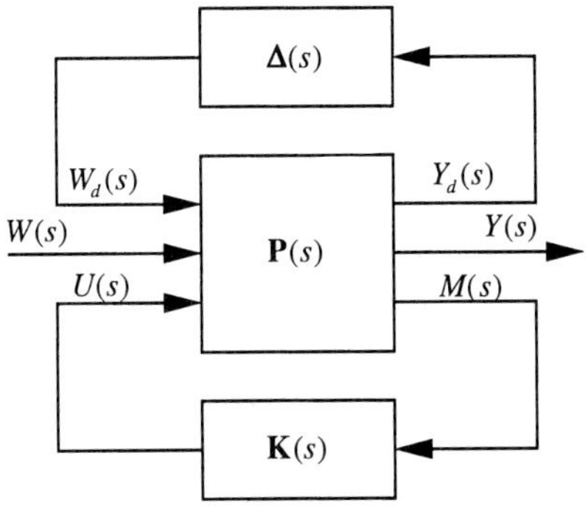

FIGURE 5.8 Standard form of a general feedback system with uncertainty

The inputs consist of the perturbation input W_d (the output of the feedback perturbation), the disturbance input W, and the control input U. The outputs consist of the perturbation output Y_d (the input to the feedback perturbation), the reference output Y, and the measured output M.

The perturbation is normalized by incorporating the actual perturbation bound into the plant transfer function. The perturbation bound (5.7) can be written as

$$\frac{1}{\Delta_{\max}(j\omega)}\bar{\sigma}\{\mathbf{\Delta}'(j\omega)\} \leq 1,$$

and the normalized perturbation defined as follows:

$$\mathbf{\Delta}(j\omega) = \frac{1}{\Delta_{\max}(j\omega)}\mathbf{\Delta}'(j\omega).$$

The maximum singular value of the normalized perturbation is then

$$\bar{\sigma}\{\mathbf{\Delta}(j\omega)\} = \frac{1}{\Delta_{\max}(j\omega)}\bar{\sigma}\{\mathbf{\Delta}'(j\omega)\} \leq 1.$$

Taking the maximum over all frequencies, the set of perturbations $\mathbf{\Delta}(j\omega)$ that satisfies this bound is defined by

$$\|\mathbf{\Delta}(j\omega)\|_{\infty} \leq 1.$$

The normalized perturbation is incorporated into the system model by making the following substitution:

$$\mathbf{\Delta}'(j\omega) = \Delta_{\max}(j\omega)\mathbf{\Delta}(j\omega).$$

The use of the normalized perturbation simplifies robustness analysis by both normalizing the perturbation's magnitude and, more important, removing the frequency dependence of the bound. To simplify the notation, the term *perturbation* may be used to refer to either a normalized or an unnormalized perturbation. This practice is followed when the normalization of the perturbation is unimportant, or when the normalization of the perturbation is specified by the bound.

The perturbation bound $\Delta_{\max}(j\omega)$ is, in general, a scalar transfer function that is not necessarily rational (a ratio of polynomials in $j\omega$). The normalized plant may then be nonrational. Many design methods and analysis tools (especially computer-aided design tools) require that the plant be modeled as a rational transfer function. This can be accomplished by using a rational transfer function approximation of the perturbation bound. The rational approximation should have a magnitude that closely matches the perturbation bound. The phase of the approximation can be arbitrary, since the perturbation bound appears in series with the normalized perturbation, which has a totally uncertain phase. The generation of the perturbation bound is illustrated in the following example.

EXAMPLE 5.5 Consider the input multiplicative uncertainty model given in Figure 5.7b. The nominal plant model is fairly accurate (within about 0.5%) at frequencies below 10 rad/sec, but fairly inaccurate (within about 50%) at frequencies above 1000 rad/sec. The accuracy of the model should transition between these two extremes at intermediate frequencies.

The frequency-dependent uncertainty bound is given in Figure 5.9a. This bound can be approximated by a first-order transfer function with a zero at 10 rad/sec and a pole at 1000 rad/sec:

$$\Delta'_{\max}(j\omega) = g\frac{(j\omega + 10)}{(j\omega + 1000)}.$$

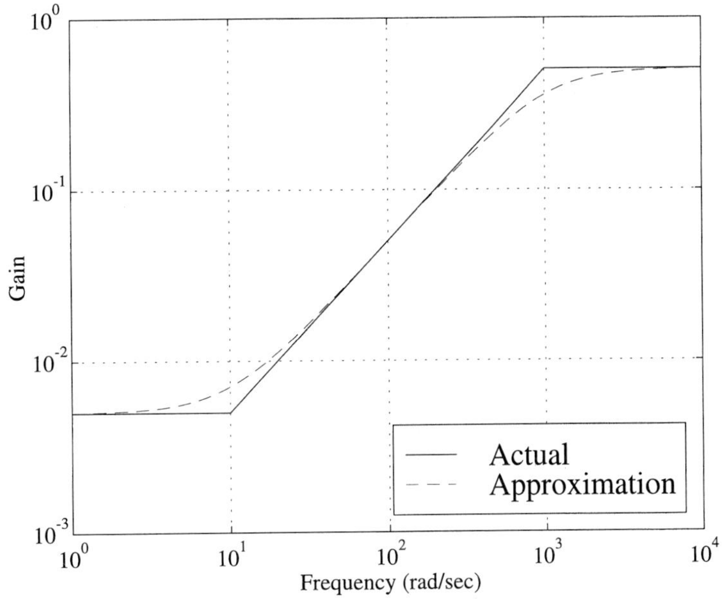

(a)

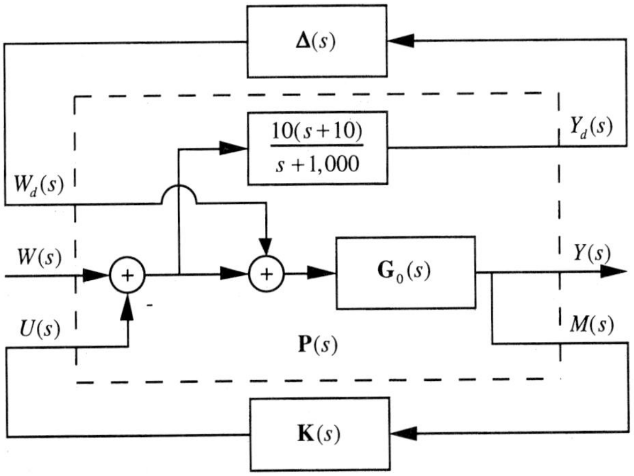

(b)

FIGURE 5.9 The uncertainty model for Example 5.5: (a) the bound on the perturbation; (b) the uncertainty model in standard form

Note that the zero at 10 rad/sec generates an increase in the magnitude response with a slope of 20 dB/decade. This generates the desired rise in the uncertainty over the two-decade frequency range. The rise is then canceled by the pole at 1000 rad/sec. The transfer function gain is found by matching the bound at zero frequency:

$$g = 0.005\frac{(0 + 1000)}{(0 + 10)} = 0.5.$$

The magnitude response of the rational approximation to the uncertainty bound is plotted in Figure 5.9a. Assuming that this approximation is sufficiently accurate, the system can be placed in standard form by normalizing the perturbation and placing it in a feedback loop, as shown in Figure 5.9b. ◆

A few comments concerning this example are in order. First, the specification of uncertainty is rather vague. This is often the case in applications where model accuracy is difficult to determine. Model uncertainty can be overstated to guarantee robustness in applications with a high cost of failure. This typically results in a loss of performance since the controller is required to accommodate a wider range of possible plant models and is therefore less optimized for the actual plant. A considerable amount of judgment is required to both assess uncertainty and to generate a reasonable compromise between robustness and performance. This is part of the art of controller design and, unfortunately for the student (but fortunately for the engineer seeking job security), it is application-specific and most easily learned from experience. Secondly, a more accurate rational transfer function approximation to the uncertainty bound can be generated by using a higher order filter.[6] Note that high-order approximations may be overkill when the specification of uncertainty is vague. In this case, high-order filters complicate the analysis needlessly by yielding very accurate approximations to ballpark guesses and are probably not useful.

5.4.2 Stability Robustness Analysis

The stability robustness of systems with unstructured uncertainty is addressed by analysis of the standard model. Combining the nominal plant $\mathbf{P}(s)$ with the feedback $\mathbf{K}(s)$ results in a system consisting of the nominal closed-loop system $\mathbf{N}(s)$ with the perturbation $\boldsymbol{\Delta}(s)$ in a feedback loop, as shown in Figure 5.10a. The transfer function $\mathbf{N}(s)$ is found from (5.8) by noting that

$$U(s) = \mathbf{K}(s)M(s) = \mathbf{K}(s)\mathbf{P}_{mw_d}(s)W_d(s) + \mathbf{K}(s)\mathbf{P}_{mw}(s)W(s) + \mathbf{K}(s)\mathbf{P}_{mu}(s)U(s)$$

and solving for $U(s)$:

$$\begin{aligned} U(s) &= \{\mathbf{I} - \mathbf{K}(s)\mathbf{P}_{mu}(s)\}^{-1}\mathbf{K}(s)\mathbf{P}_{mw_d}(s)W_d(s) \\ &\quad + \{\mathbf{I} - \mathbf{K}(s)\mathbf{P}_{mu}(s)\}^{-1}\mathbf{K}(s)\mathbf{P}_{mw}(s)W(s). \end{aligned}$$

[6] The subject of filter design is treated in introductory signal processing courses and is not discussed further, except tc note that excellent filter design CAD tools are available from a number of sources.

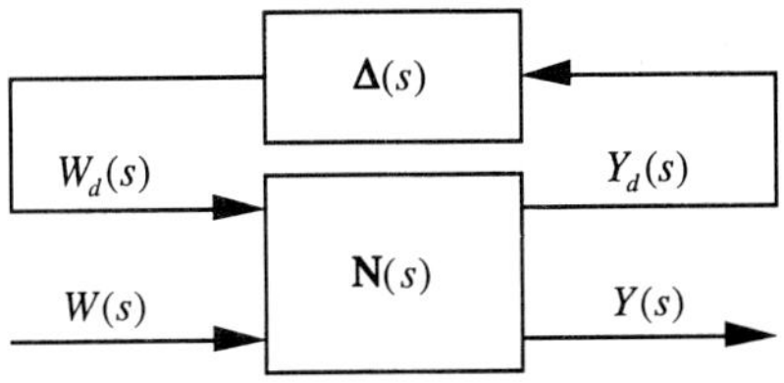

(a)

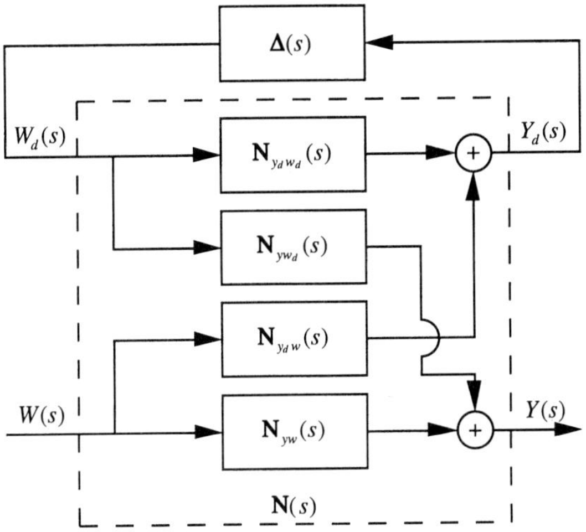

(b)

FIGURE 5.10 The unstructured uncertainty model for robustness analysis: (a) the basic model; (b) expanded to show the subsystems of **N**(s)

Substituting $U(s)$ into the equations for $Y_d(s)$ and $Y(s)$ in (5.8) yields the following closed-loop system:

$$\begin{bmatrix} Y_d \\ \hdashline Y \end{bmatrix} = \left[\begin{array}{c:c} \mathbf{P}_{y_d w_d} + \mathbf{P}_{y_d u}\{\mathbf{I} - \mathbf{KP}_{mu}\}^{-1}\mathbf{KP}_{m w_d} & \mathbf{P}_{y_d w} + \mathbf{P}_{y_d u}\{\mathbf{I} - \mathbf{KP}_{mu}\}^{-1}\mathbf{KP}_{mw} \\ \hdashline \mathbf{P}_{y w_d} + \mathbf{P}_{yu}\{\mathbf{I} - \mathbf{KP}_{mu}\}^{-1}\mathbf{KP}_{m w_d} & \mathbf{P}_{yw} + \mathbf{P}_{yu}\{\mathbf{I} - \mathbf{KP}_{mu}\}^{-1}\mathbf{KP}_{mw} \end{array}\right] \begin{bmatrix} W_d \\ \hdashline W \end{bmatrix},$$

where all Laplace variables s have been dropped to simplify the notation. The nominal closed-loop transfer function $\mathbf{N}(s)$ is then[7]

[7] Transformations resulting from the closing of a single loop around a MIMO system are referred to as *linear fractional transformations* in the literature.

$$\mathbf{N} = \begin{bmatrix} \mathbf{N}_{y_d w_d} & \mathbf{N}_{y_d w} \\ \mathbf{N}_{y w_d} & \mathbf{N}_{yw} \end{bmatrix}$$
$$= \begin{bmatrix} \mathbf{P}_{y_d w_d} + \mathbf{P}_{y_d u}\{\mathbf{I} - \mathbf{K}\mathbf{P}_{mu}\}^{-1}\mathbf{K}\mathbf{P}_{m w_d} & \mathbf{P}_{y_d w} + \mathbf{P}_{y_d u}\{\mathbf{I} - \mathbf{K}\mathbf{P}_{mu}\}^{-1}\mathbf{K}\mathbf{P}_{mw} \\ \mathbf{P}_{y w_d} + \mathbf{P}_{yu}\{I - \mathbf{K}\mathbf{P}_{mu}\}^{-1}\mathbf{K}\mathbf{P}_{m w_d} & \mathbf{P}_{yw} + \mathbf{P}_{yu}\{\mathbf{I} - \mathbf{K}\mathbf{P}_{mu}\}^{-1}\mathbf{K}\mathbf{P}_{mw} \end{bmatrix}.$$

A considerable amount of time can be saved in practice by using a computer to evaluate this rather complex expression for the nominal closed-loop transfer function.

The nominal closed-loop system is assumed to be stable since the controller has been designed for the nominal system. The perturbation is also stable since it has a bounded gain. The combined system in Figure 5.10b is then internally stable provided that the feedback loop containing the perturbation is internally stable, this feedback loop being the only possible source of instability.

The internal stability of this feedback loop (shown in Figure 5.11) is evaluated by considering the following system:

$$\begin{bmatrix} E_1 \\ E_2 \end{bmatrix} = \begin{bmatrix} (\mathbf{I} - \boldsymbol{\Delta}\mathbf{N}_{y_d w_d})^{-1} & (\mathbf{I} - \boldsymbol{\Delta}\mathbf{N}_{y_d w_d})^{-1}\boldsymbol{\Delta} \\ (\mathbf{I} - \mathbf{N}_{y_d w_d}\boldsymbol{\Delta})^{-1}\mathbf{N}_{y_d w_d} & (\mathbf{I} - \mathbf{N}_{y_d w_d}\boldsymbol{\Delta})^{-1} \end{bmatrix} \begin{bmatrix} U_1 \\ U_2 \end{bmatrix} \tag{5.9}$$
$$= \begin{bmatrix} \mathbf{G}_{e_1 u_1} & \mathbf{G}_{e_1 u_2} \\ \mathbf{G}_{e_2 u_1} & \mathbf{G}_{e_2 u_2} \end{bmatrix} \begin{bmatrix} U_1 \\ U_2 \end{bmatrix}.$$

A condition for internal stability of this system is generated by finding a bound on the signal 2-norm for the output of each of the four transfer functions in (5.9). Consider one of these transfer functions, for example $\mathbf{G}_{e_1 u_2}(s)$ which describes the relationship between $U_2(s)$ and $E_1(s)$:

$$E_1(s) = [\mathbf{I} - \boldsymbol{\Delta}(s)\mathbf{N}_{y_d w_d}(s)]^{-1}\boldsymbol{\Delta}(s)U_2(s) = \mathbf{G}_{e_1 u_2}(s)U_2(s).$$

This transfer function implies that

$$E_1(s) = \boldsymbol{\Delta}(s)\mathbf{N}_{y_d w_d}(s)E_1(s) + \boldsymbol{\Delta}(s)U_2(s).$$

Converting this expression into the time domain yields

$$e_1(t) = \boldsymbol{\delta}(t) \otimes \mathbf{n}_{y_d w_d}(t) \otimes e_1(t) + \boldsymbol{\delta}(t) \otimes u_2(t).$$

where $\mathbf{n}_{y_d w_d}(t)$ and $\boldsymbol{\delta}(t)$ are the impulse responses of $\mathbf{N}_{y_d w_d}(s)$ and $\boldsymbol{\Delta}(s)$, respectively, and $\otimes$ denotes convolution. Taking the signal 2-norm of both sides of this expression and employing the triangle inequality yields

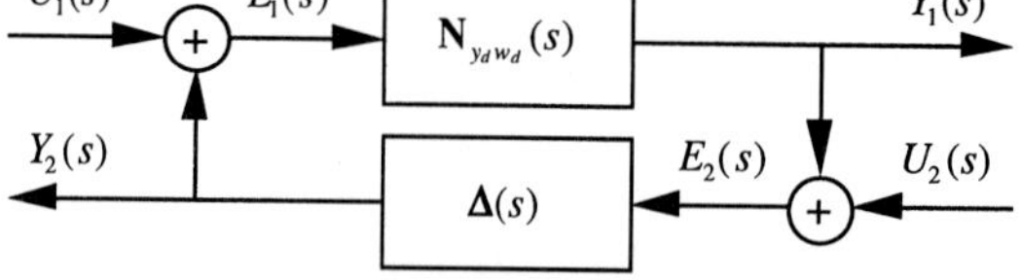

FIGURE 5.11 Internal stability of the uncertain system

$$\begin{aligned}\|e_1(t)\|_2 &= \|\boldsymbol{\delta}(t) \otimes \mathbf{n}_{y_d w_d}(t) \otimes e_1(t) + \boldsymbol{\delta}(t) \otimes u_2(t)\|_2 \\ &\le \|\boldsymbol{\delta}(t) \otimes \mathbf{n}_{y_d w_d}(t) \otimes e_1(t)\|_2 + \|\boldsymbol{\delta}(t) \otimes u_2(t)\|_2 .\end{aligned}$$

The ∞-norm can be used to bound the terms on the right side of this inequality:

$$\|e_1(t)\|_2 \le \|\boldsymbol{\Delta}\mathbf{N}_{y_d w_d}\|_\infty \|e_1(t)\|_2 + \|\boldsymbol{\Delta}\|_\infty \|u_2(t)\|_2 .$$

Further, the ∞-norm of a product is less than or equal to the product of the ∞-norms:

$$\|e_1(t)\|_2 \le \|\boldsymbol{\Delta}\|_\infty \|\mathbf{N}_{y_d w_d}\|_\infty \|e_1(t)\|_2 + \|\boldsymbol{\Delta}\|_\infty \|u_2(t)\|_2 .$$

Solving for the signal 2-norm of $e_1(t)$ yields

$$\|e_1(t)\|_2 \le (1 - \|\boldsymbol{\Delta}\|_\infty \|\mathbf{N}_{y_d w_d}\|_\infty)^{-1} \|\boldsymbol{\Delta}\|_\infty \|u_2(t)\|_2 .$$

The system described by the transfer function $\mathbf{G}_{e_1 u_2}$ has a bounded gain and is therefore $\mathcal{L}_2$ stable and BIBO stable provided the inverse above is finite.[8] Noting that $\|\boldsymbol{\Delta}\|_\infty \le 1$, this inverse is finite if

$$\|\mathbf{N}_{y_d w_d}\|_\infty < 1. \tag{5.10}$$

A similar analysis can be applied to the other three transfer functions in (5.9). All of these transfer functions are stable provided condition (5.10) is satisfied. The following result is then obtained:

> A general feedback system, as given in Figure 5.10, where the perturbation is bounded,
>
> $$\|\boldsymbol{\Delta}\|_\infty \le 1, \tag{5.11}$$
>
> is internally stable for all possible perturbations provided the nominal closed-loop system is stable and
>
> $$\boxed{\|\mathbf{N}_{y_d w_d}\|_\infty = \sup_\omega \{\bar{\sigma}[\mathbf{N}_{y_d w_d}(j\omega)]\} < 1.} \tag{5.12}$$

This result is known as the small-gain theorem and provides a test for robust stability with respect to bounded perturbations. An alternative way of presenting the above results is to state that the system is internally stable if the ∞-norm of the loop transfer function is less than 1. This result makes sense, in terms of the Nyquist stability criterion for SISO systems, since the phase of the unstructured perturbation is arbitrary.

The robust stability condition (5.12) is presented as a sufficient condition for internal stability. In fact, this condition is also necessary for robust stability with respect to unstructured perturbations; that is, a destabilizing perturbation can be found that satisfies the bound (5.11) whenever (5.12) does not hold. A necessary and sufficient stability robustness condition is subsequently derived for structured perturbations. The robustness condition (5.12) is a special case of this more general result, since an unstructured perturbation is a special case of a structured perturbation. The fact that (5.12) is necessary for internal stability follows immediately from the necessity results presented for structured perturbations.

---

[8] $\mathcal{L}_2$ stability is defined in Section A5 of the Appendix. The fact that $\mathcal{L}_2$ stability is equivalent to the BIBO stability is also shown in A5.

EXAMPLE 5.6 We are given the nominal plant

$$G_0(s) = \frac{10}{s - 1},$$

with an unstructured additive uncertainty,

$$\|\Delta(s)\|_\infty \le 1,$$

and a controller,

$$K(s) = \tfrac{1}{2}.$$

The block diagram of this system in standard form is shown in Figure 5.12. Note that reference inputs and outputs are not specified for this example since they have no effect on robust stability analysis. From the block diagram, the transfer function from W_d to Y_d is computed with the loop closed through the controller:

$$N_{y_d w_d}(s) = \frac{-K(s)}{1 + K(s)G_0(s)} = \frac{-\frac{1}{2}(s - 1)}{s + 4}.$$

Note that the nominal system is stable. The maximum singular value of the scalar transfer function $N_{y_y w_d}(j\omega)$ equals the magnitude of this transfer function:

$$\bar{\sigma}\{N_{y_d w_d}(j\omega)\} = |N_{y_d w_d}(j\omega)| = \frac{1}{2}\frac{\sqrt{\omega^2 + 1}}{\sqrt{\omega^2 + 16}}$$

The infinity norm is then

$$\|N_{y_d w_d}(j\omega)\|_\infty = \sup_\omega \frac{1}{2}\frac{\sqrt{\omega^2 + 1}}{\sqrt{\omega^2 + 16}} = \frac{1}{2},$$

which implies the system is robustly stable.

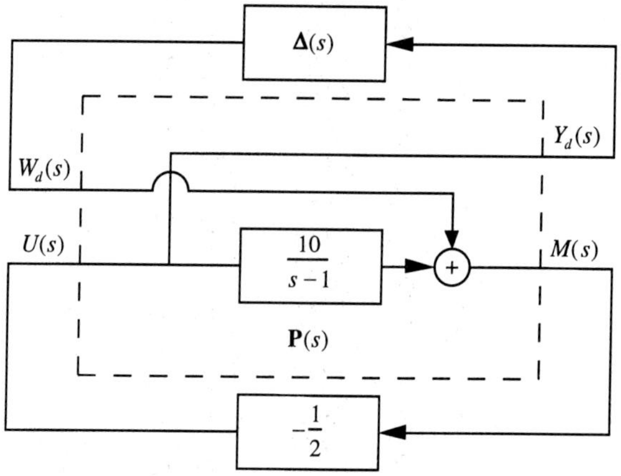

FIGURE 5.12 Standard form for the system in Example 5.6

EXAMPLE 5.7 The hover mode pitch and yaw dynamics of a vertical-takeoff-and-landing remotely piloted vehicle are given by the following state equations:

$$\begin{bmatrix} \dot{p} \\ \ddot{p} \\ \dot{q} \\ \ddot{q} \end{bmatrix} = \begin{bmatrix} 0 & 1 & 0 & 0 \\ 0 & 0 & 0 & -1.6 \\ 0 & 0 & 0 & 1 \\ 0 & 1.8 & 0 & 0 \end{bmatrix} \begin{bmatrix} p \\ \dot{p} \\ q \\ \dot{q} \end{bmatrix} + \left[\begin{array}{cc:cc} 0 & 0 & 0 & 0 \\ 11 & 0 & 0 & 0 \\ 0 & 0 & 0 & 0 \\ 0 & 12 & 0 & 0 \end{array}\right] \begin{bmatrix} \delta_p \\ \delta_q \\ \hdashline p_c \\ q_c \end{bmatrix} = \mathbf{A}x + [\mathbf{B}_u \;\vdots\; \mathbf{0}] \begin{bmatrix} u \\ \hdashline w \end{bmatrix};$$

$$\begin{bmatrix} m \\ \hdashline y \end{bmatrix} = \begin{bmatrix} p \\ \dot{p} \\ q \\ \dot{q} \\ \hdashline e_p \\ e_q \end{bmatrix} = \begin{bmatrix} 1 & 0 & 0 & 0 \\ 0 & 1 & 0 & 0 \\ 0 & 0 & 1 & 0 \\ 0 & 0 & 0 & 1 \\ \hdashline -1 & 0 & 0 & 0 \\ 0 & 0 & -1 & 0 \end{bmatrix} \begin{bmatrix} p \\ \dot{p} \\ q \\ \dot{q} \end{bmatrix} + \left[\begin{array}{cc:cc} 0 & 0 & 0 & 0 \\ 0 & 0 & 0 & 0 \\ 0 & 0 & 0 & 0 \\ 0 & 0 & 0 & 0 \\ \hdashline 0 & 0 & 1 & 0 \\ 0 & 0 & 0 & 1 \end{array}\right] \begin{bmatrix} \delta_p \\ \delta_q \\ \hdashline p_c \\ q_c \end{bmatrix}$$

$$= \begin{bmatrix} \mathbf{C}_m \\ \hdashline \mathbf{C}_y \end{bmatrix} x + \left[\begin{array}{c:c} \mathbf{0} & \mathbf{0} \\ \hdashline \mathbf{0} & \mathbf{D}_{yw} \end{array}\right] \begin{bmatrix} u \\ \hdashline w \end{bmatrix},$$

where these equations are partitioned as in (4.2). The control inputs are elevator deflection δ_p and rudder deflection δ_q. The measured outputs are pitch angle p, yaw angle q, pitch rate, and yaw rate. For completeness, the disturbance inputs (commanded pitch angle p_c and commanded yaw angle q_c) and the reference outputs (pitch error and yaw error) are included in the model. The plant is subject to an unstructured input multiplicative uncertainty on the control:

$$\|\Delta(s)\|_\infty \leq \frac{1}{2}.$$

The controller is specified by the following equation:

$$u(t) = -\mathbf{K}\mathbf{C}_m(x - \mathbf{K}_r w),$$

where

$$\mathbf{K} = \begin{bmatrix} 15.92 & 1.08 & 0.91 & -0.18 \\ 1.00 & 0.25 & 10.76 & 0.84 \end{bmatrix}; \; \mathbf{K}_r = \begin{bmatrix} 1 & 0 \\ 0 & 0 \\ 0 & 1 \\ 0 & 0 \end{bmatrix}.$$

Note that $\mathbf{C}_m$ is the identity matrix, so this is state feedback plus a feedforward control from the reference input. This controller is designed to place the poles at

$$\{-5 + 10j, -5 - 10j, -6 + 12j, -6 - 12j\}$$

and drive the pitch and yaw errors to zero. A block diagram of this system in standard form appears in Figure 5.13a. The transfer function from W_d to Y_d is found to be

$$\mathbf{N}_{y_d w_d}(s) = -\frac{1}{2}\mathbf{K}\mathbf{C}_m(s\mathbf{I} - \mathbf{A} + \mathbf{B}_u\mathbf{K}\mathbf{C}_m)^{-1}\mathbf{B}_u.$$

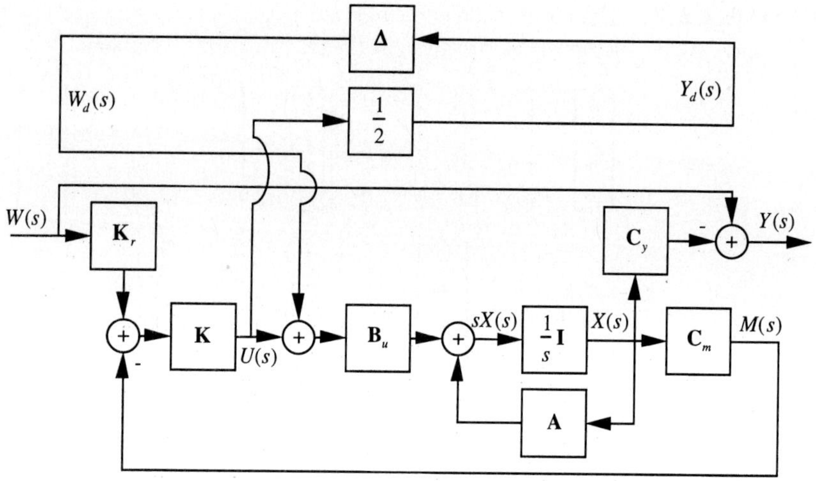

(a)

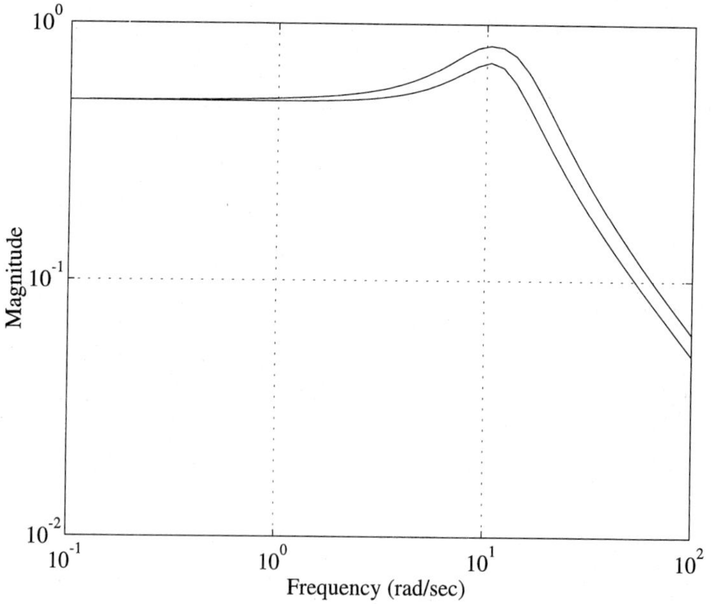

(b)

FIGURE 5.13 Stability robustness analysis for Example 5.7: (a) block diagram in standard form; (b) principal gains

The principal gains of this transfer function are computed numerically and plotted in Figure 5.13b. The infinity norm is the maximum value of the maximum principal gain,

$$\|\mathbf{N}_{y_d w_d}(s)\|_\infty = 0.84,$$

which implies that the system is robustly stable. ◆

5.5 Structured Uncertainty

Unstructured uncertainty is modeled by connecting an unknown but bounded perturbation to the plant. This type of uncertainty model requires very little information, that is, only the bound and how the perturbation is connected. In many applications, additional constraints on the set of admissible perturbations are available. These constraints add "structure" to the set of admissible perturbations, and the uncertainty is termed structured. Structured uncertainty is therefore a more general form of uncertainty than unstructured uncertainty.

Structured uncertainty arises when the plant is subject to multiple perturbations. Multiple perturbations occur when the plant contains a number of uncertain parameters, or when the plant contains multiple unstructured uncertainties. For example, the plant model may be well specified except for two uncertain time constants, which are modeled as a nominal value plus a perturbation. Another example of structured uncertainty is when the plant contains both an input multiplicative perturbation and an additive perturbation. Clearly, structured uncertainty is a very general way of modeling uncertainty.[9]

5.5.1 The Structured Uncertainty Model

A plant subject to structured uncertainty can be placed in a standard form analogous to that used for unstructured uncertainty. The standard form of the structured uncertainty model has the individual perturbations normalized to 1 and placed in a feedback loop around the nominal plant. The standard form of the structured uncertainty model is shown in Figure 5.14. The structured perturbation $\mathbf{\Delta}(s)$ is a block diagonal transfer function:

$$\mathbf{\Delta}(s) = \begin{bmatrix} \mathbf{\Delta}_1(s) & \mathbf{0} & \cdots & \mathbf{0} \\ \mathbf{0} & \mathbf{\Delta}_2(s) & \cdots & \mathbf{0} \\ \vdots & \vdots & \ddots & \vdots \\ \mathbf{0} & \mathbf{0} & \cdots & \mathbf{\Delta}_n(s) \end{bmatrix}, \tag{5.13}$$

where n is the number of perturbations and the blocks $\mathbf{\Delta}_i(s) \in \mathscr{C}^{l_i x n_i}$ represent the individual perturbations applied to the plant. An individual block can represent an uncertainty in a parameter (a scalar perturbation) or an unstructured uncertainty. The set of all transfer function matrices with this block diagonal form is denoted $\bar{\mathbf{\Delta}}$. The structured perturbation is normalized so that its infinity norm is bounded by 1:

$$\|\mathbf{\Delta}\|_\infty \leq 1. \tag{5.14a}$$

[9] Structured uncertainties also arise when the perturbation is restricted to be purely real or when other constraints on the perturbation are present. In this book, we concentrate on the modeling and analysis of structured uncertainties that arise from multiple perturbations to the plant and ignore other sources of structure in the perturbation. This is done because plants with multiple perturbations form a very general class of systems that are useful in a wide range of applications. It should be noted that many applications also yield purely real perturbations. Some useful results on the analysis of systems with real perturbations are given in [5, 6].

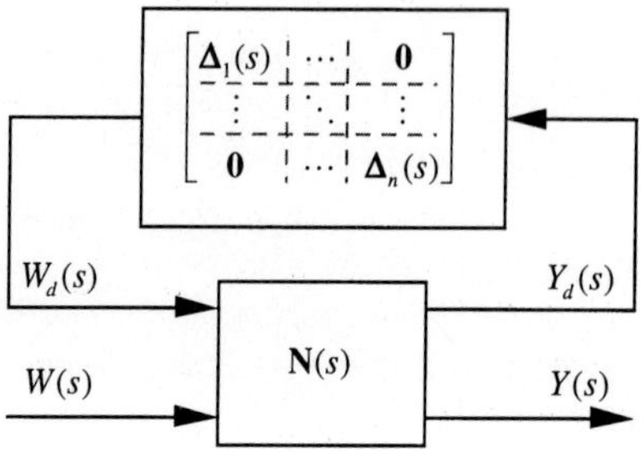

FIGURE 5.14 Standard form of the structured uncertainty model

Additionally, all the blocks in the perturbation are scaled so that their infinity norms are bounded by 1:

$$\|\mathbf{\Delta}_1\|_\infty \le 1;\ \|\mathbf{\Delta}_2\|_\infty \le 1;\ \cdots;\ \|\mathbf{\Delta}_n\|_\infty \le 1. \tag{5.14b}$$

Note that the bounds in (5.14b) imply the bound in (5.14a). The perturbation is normalized by incorporating the actual bound (which may be frequency-dependent) into the plant, as discussed in Section 5.4.1. The subset of $\bar{\mathbf{\Delta}}$ that satisfies the bounds in (5.14) is termed the set of admissible perturbations.

EXAMPLE 5.8 Consider the following transfer function:

$$G(s) = \frac{1}{(s + p_1)(s + p_2)}.$$

The two poles are uncertain, as given by

$$p_1 \in [0.9,\ 1.1];\ p_2 \in [3,\ 5].$$

These poles can be modeled as a nominal value plus a perturbation:

$$p_1 = 1 + \delta_1;\ p_2 = 4 + \delta_2,$$

where δ_1 ranges from -0.1 to 0.1 and δ_2 ranges from -1 to 1. These perturbations can be placed in a feedback loop around the nominal plant, as shown in Figure 5.15. Figure 5.15a shows the basic block diagram of this system. The perturbations are isolated in separate feedback loops in Figure 5.15b. The result is put in standard form, as shown in Figure 5.15c, by normalizing the perturbations. The resulting perturbation has the property that

$$\|\Delta_1\|_\infty \le 1\ ;\ \|\Delta_2\|_\infty \le 1\ ;\ \|\mathbf{\Delta}\|_\infty \le 1,$$

where both $\Delta_1(s)$ and $\Delta_2(s)$ are real perturbations. The transfer function $\mathbf{N}(s)$ is

$$\mathbf{N}(s) = \left[\begin{array}{c|c} \mathbf{N}_{y_d w_d}(s) & \mathbf{N}_{y_d w}(s) \\ \hline \mathbf{N}_{y w_d}(s) & \mathbf{N}_{y_d}(s) \end{array}\right] = \left[\begin{array}{cc|c} \dfrac{-0.1}{(s+1)} & 0 & \dfrac{0.1}{(s+1)} \\ \dfrac{-1}{(s+1)(s+4)} & \dfrac{-1}{(s+4)} & \dfrac{1}{(s+1)(s+4)} \\ \hline \dfrac{-1}{(s+1)(s+4)} & \dfrac{-1}{(s+4)} & \dfrac{1}{(s+1)(s+4)} \end{array}\right],$$

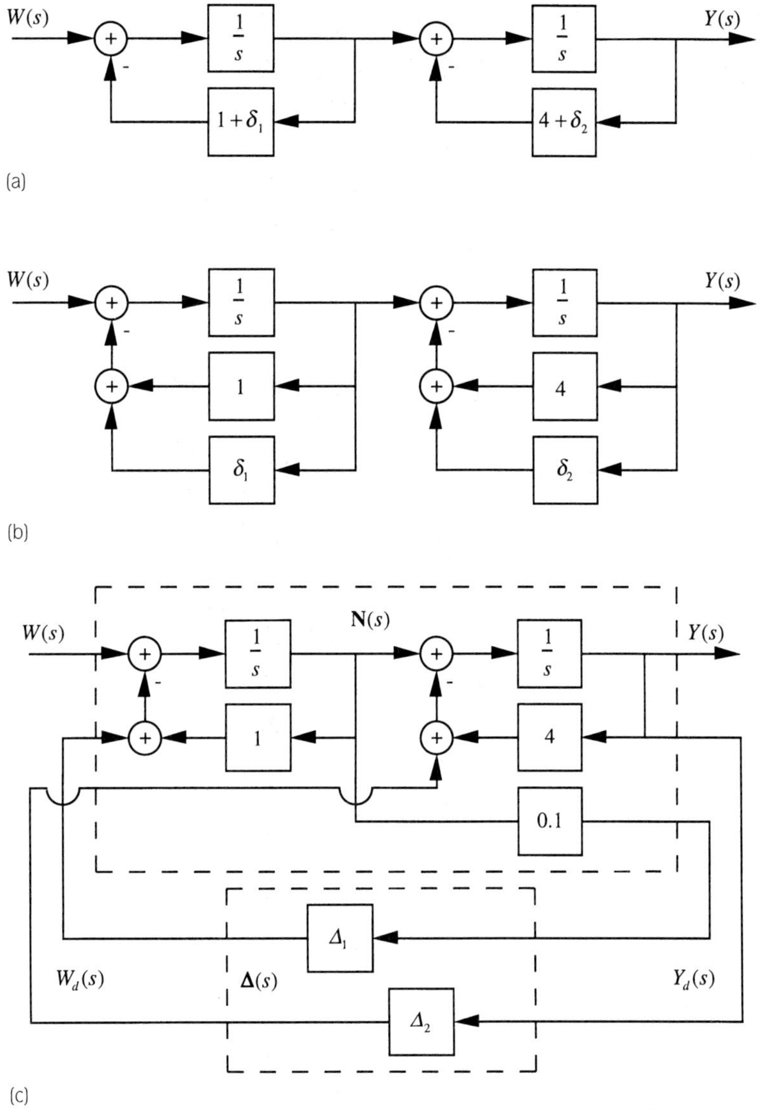

FIGURE 5.15 An example of modeling parameter uncertainties as feedback perturbations: (a) the model with uncertainty; (b) the model with parameter perturbations separated from the nominal values; (c) the model with feedback perturbations in standard form

where the block structure separates the inputs and outputs going to the perturbation from the reference inputs and outputs. ◆

The structured uncertainty model can include unstructured uncertainty blocks, as shown in the following example.

EXAMPLE 5.9 Consider the transfer function below:

$$G(s) = G_0(s)[1 + \Delta_o(s)] = \frac{1}{(s + 4 + \delta)}[1 + \Delta_o(s)],$$

where $\Delta_o(s)$ is an unstructured uncertainty. The perturbations $\Delta_o(s)$ and δ are bounded:

$$\bar{\sigma}[\Delta_o(j\omega)] \leq \left| \frac{10(j\omega + 10)}{j\omega + 1000} \right|; \ |\delta| \leq 5,$$

where δ is assumed to be real, and the frequency-dependent bound on $\Delta_o(s)$ is given as a rational transfer function of $j\omega$ (see Section 5.4.1). These perturbations can be placed in a feedback loop around the nominal plant, as shown in Figure 5.16, where

$$\Delta_o(s) = \frac{10(s + 10)}{s + 1000}\Delta_1(s); \ \delta = 5\Delta_2(s).$$

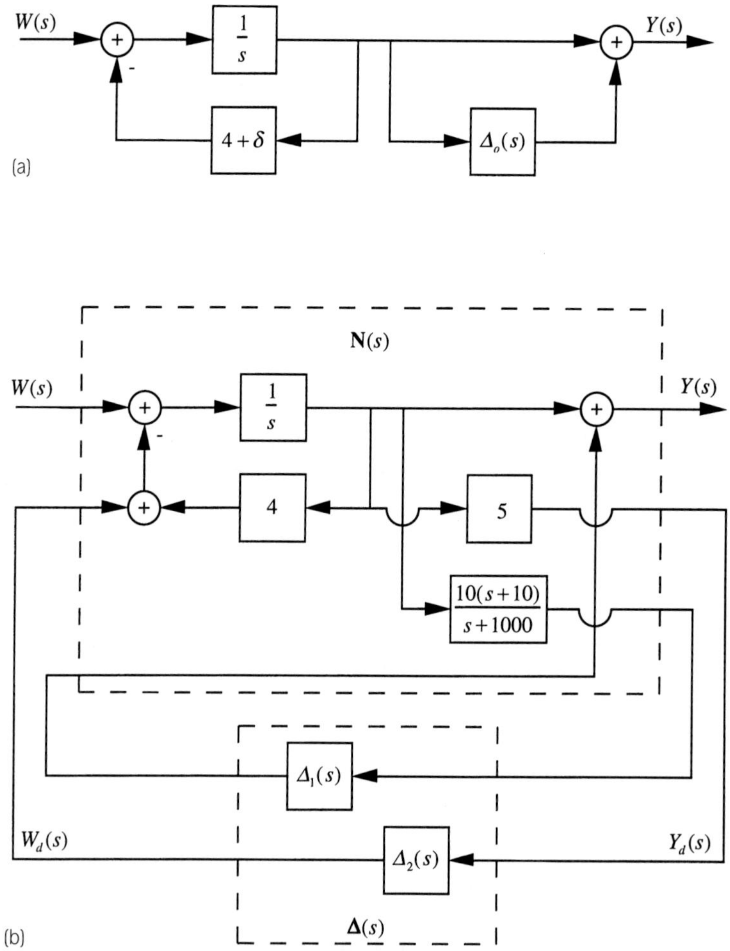

FIGURE 5.16 An example of modeling uncertainties as feedback perturbations: (a) the model with uncertainties; (b) the model with feedback perturbations

The resulting perturbation has the property that

$$\|\Delta_1\|_\infty \leq 1;\ \|\Delta_2\|_\infty \leq 1;\ \|\boldsymbol{\Delta}\|_\infty \leq 1,$$

$\Delta_1(s)$ is a complex perturbation, and $\Delta_2(s)$ is a real perturbation. ◆

A more general description of uncertainty can be obtained by allowing blocks to be purely real or adding the constraint that two or more blocks are proportional. The analysis of robustness for these cases is not treated in order to keep the notation and derivations simple. Additionally, the standard structured uncertainty model is applicable to a wide variety of systems even without these cases. For robustness analysis of systems with real perturbations, see [5, 6] and for repeated or proportional perturbations, see [7]. The remainder of this section will discuss how to analyze the stability robustness of systems with structured complex perturbations in standard form and no repeated blocks.

5.5.2 The Structured Singular Value and Stability Robustness

The stability of a system subject to a structured uncertainty is determined by analyzing the feedback system in Figure 5.14. The nominal closed-loop system is assumed to be stable. Any unstable poles of this system are therefore caused by closing the loop through the perturbation and are the solutions of

$$\det\{\mathbf{I} + \mathbf{N}_{y_d w_d}(s)\boldsymbol{\Delta}(s)\} = 0. \tag{5.15}$$

Stability robustness may be evaluated by determining the "size" of the smallest perturbation that results in a pole—a solution of (5.15)—with a non-negative real part. A perturbation that results in such a pole is termed a destabilizing perturbation.

The locus of the solutions of (5.15) is a continuous function of $\boldsymbol{\Delta}(s)$. Therefore, the smallest destabilizing perturbation has one or more poles on the imaginary axis. The "size" of the smallest destabilizing perturbation is defined as follows:

$$\inf_{\omega}\left\{\min_{\boldsymbol{\Delta}(j\omega)\in\bar{\boldsymbol{\Delta}}}\{\bar{\sigma}[\boldsymbol{\Delta}(j\omega)] \text{ such that } \det\{\mathbf{I} + \mathbf{N}_{y_d w_d}(j\omega)\boldsymbol{\Delta}(j\omega)\} = 0\}\right\}. \tag{5.16}$$

Note that $\boldsymbol{\Delta}(j\omega)$ in (5.16) is any perturbation (with the appropriate block structure) that places a pole at a specific point $j\omega$ on the imaginary axis. The maximum singular value is a measure of the size of this perturbation. The minimization over all appropriate perturbations results in the size of the smallest perturbation that places a pole at $j\omega$. The minimization over frequency then yields the size of the smallest perturbation that places a pole anywhere on the imaginary axis, that is, the size of the smallest destabilizing perturbation. A system in standard form is robustly stable if and only if the smallest destabilizing perturbation is greater than 1 (the infinity norm of the largest admissible perturbation):

$$\inf_{\omega}\left\{\min_{\boldsymbol{\Delta}(j\omega)\in\bar{\boldsymbol{\Delta}}}\{\bar{\sigma}[\boldsymbol{\Delta}(j\omega)] \text{ such that } \det\{\mathbf{I} + \mathbf{N}_{y_d w_d}(j\omega)\boldsymbol{\Delta}(j\omega)\} = 0\}\right\} > 1. \tag{5.17}$$

Unfortunately, finding the size of the smallest destabilizing perturbation using (5.16) is not a trivial matter. In fact, this problem is intractable in all but the simplest of cases. Therefore, bounds on the size of the smallest destabilizing perturbation are developed. Before generating these bounds, the stability robustness condition (5.17) is put in a form that is more useful for both application and computation.

The stability robustness condition is placed in an alternate form by inverting (5.17):

$$\sup_{\omega}\left\{\frac{1}{\min\limits_{\Delta(j\omega)\in\bar{\Delta}}\{\bar{\sigma}[\Delta(j\omega)] \text{ such that } \det\{\mathbf{I}+\mathbf{N}_{y_d w_d}(j\omega)\Delta(j\omega)\}=0\}}\right\}<1. \tag{5.18}$$

Note that the supremum of the ratio in this equation equals the inverse of the infimum of the denominator. This alternate stability robustness condition is very similar in form to the result obtained for unstructured uncertainty in (5.12). In addition, this form of the stability robustness condition can be simply generalized to include performance robustness.

The term within the brackets in (5.18) is called the structured singular value (SSV) and is formally defined as follows:

$$\mu_{\bar{\Delta}}(\mathbf{N}) = \frac{1}{\min\limits_{\Delta\in\bar{\Delta}}[\bar{\sigma}(\Delta) \text{ such that } \det(\mathbf{I}+\mathbf{N}\Delta)=0]}; \tag{5.19a}$$

$$\mu_{\bar{\Delta}}(\mathbf{N}) = 0 \quad \text{if } \det(\mathbf{I}+\mathbf{N}\Delta)\neq 0 \text{ for all } \Delta\in\bar{\Delta}. \tag{5.19b}$$

The structured singular value is, in general, a real-valued function of a complex matrix $\mathbf{N}$, which depends on the structure of the perturbations as defined by $\bar{\Delta}$.

The stability robustness criteria for a system with unstructured uncertainty is summarized below.

A general feedback system, as given in Figure 5.14, is internally stable for all possible perturbations:

$$\Delta(j\omega)\in\bar{\Delta} \quad \text{and} \quad \|\Delta(j\omega)\|_{\infty}\leq 1, \tag{5.20}$$

if and only if the nominal closed-loop system is internally stable and

$$\boxed{\sup_{\omega}\{\mu_{\bar{\Delta}}[\mathbf{N}_{y_d w_d}(j\omega)]\}<1.} \tag{5.21}$$

Note that satisfying this condition is both necessary and sufficient for robust stability. This test for stability robustness is very general and can be used in a wide range of applications, provided the structured singular value can be computed.

5.5.3 Bounds on the Structured Singular Value

The determination of robust stability is dependent on the computation of the structured singular value. The direct computation of the SSV by a search over all $\bar{\Delta}$ is impractical since this set has an infinite number of elements. Quantized searches over $\bar{\Delta}$ may yield unreliable results and are often hindered by the high dimensionality of $\bar{\Delta}$. Therefore, bounds on the SSV, which can be generated using a moderate amount of computation, are presented. These bounds are often tight; that is, they provide good estimates of the SSV.

Upper Bounds The structured singular value is bounded from above by the maximum singular value:

$$\mu_{\bar{\Delta}}(\mathbf{N})\leq\bar{\sigma}(\mathbf{N}). \tag{5.22}$$

To demonstrate this bound, first note that if $\mu_{\bar{\mathbf{\Delta}}}(\mathbf{N}) = 0$, the bound is valid since $\bar{\sigma}(\mathbf{N}) \geq 0$. Now, when $\mu_{\bar{\mathbf{\Delta}}}(\mathbf{N}) \neq 0$, the definition of $\mu_{\bar{\mathbf{\Delta}}}(\mathbf{N})$ states that

$$\frac{1}{\mu_{\bar{\mathbf{\Delta}}}(\mathbf{N})} = \min_{\mathbf{\Delta} \in \bar{\mathbf{\Delta}}} [\bar{\sigma}(\mathbf{\Delta}) \text{ such that } \det(\mathbf{I} + \mathbf{N\Delta}) = 0]. \tag{5.23}$$

Let $\tilde{\mathbf{\Delta}}$ be the value of $\mathbf{\Delta} \in \bar{\mathbf{\Delta}}$ that yields the minimum in (5.23). The fact that the determinant in (5.23) equals zero implies that for some nonzero vector v,

$$(\mathbf{I} + \mathbf{N}\tilde{\mathbf{\Delta}})v = 0; \tag{5.24}$$

$$v = -\mathbf{N}\tilde{\mathbf{\Delta}}v.$$

Taking the vector 2-norm of this equation, we have

$$\|v\|_2 = \|\mathbf{N}\tilde{\mathbf{\Delta}}v\|_2.$$

The gain of the matrix in this expression can be bounded by the maximum singular value:

$$\|v\|_2 \leq \bar{\sigma}(\mathbf{N}\tilde{\mathbf{\Delta}})\|v\|_2;$$

$$\|v\|_2 \leq \bar{\sigma}(\mathbf{N})\bar{\sigma}(\tilde{\mathbf{\Delta}})\|v\|_2.$$

Dividing by $\|v\|_2$ and $\bar{\sigma}(\tilde{\mathbf{\Delta}})$ yields the result given in (5.22):

$$\mu_{\bar{\mathbf{\Delta}}}(\mathbf{N}) = \frac{1}{\bar{\sigma}(\tilde{\mathbf{\Delta}})} \leq \bar{\sigma}(\mathbf{N}).$$

In summary, the structured singular value of a matrix is bounded from above by the maximum singular value.

The bound in (5.22) is easy to compute but tends to be overly conservative; that is, the structured singular value can be appreciably less than the maximum singular value. Additional bounds can be generated by returning to the system interpretation of the SSV and considering the block diagrams in Figure 5.17.

The SSV of a transfer function $\mathbf{N}(s)$ is the inverse of the smallest perturbation that, when placed in the feedback loop, yields a closed-loop pole located at s. The closed-loop poles of this system are not changed by the inclusion (as shown in Figure 5.17b) of the diagonal scaling matrices $\mathfrak{D}_L(s)$ and $\mathfrak{D}_R(s)$ and their inverses:

$$\mathfrak{D}_L(s) = \left[\begin{array}{c:c:c:c} d_1(s)\mathbf{I}_{l_1} & \mathbf{0} & \cdots & \mathbf{0} \\ \hdashline \mathbf{0} & d_2(s)\mathbf{I}_{l_2} & \cdots & \mathbf{0} \\ \hdashline \vdots & \vdots & \ddots & \vdots \\ \hdashline \mathbf{0} & \mathbf{0} & \cdots & d_n(s)\mathbf{I}_{l_n} \end{array}\right]; \tag{5.25a}$$

$$\mathfrak{D}_R(s) = \left[\begin{array}{c:c:c:c} d_1(s)\mathbf{I}_{n_1} & \mathbf{0} & \cdots & \mathbf{0} \\ \hdashline \mathbf{0} & d_2(s)\mathbf{I}_{n_2} & \cdots & \mathbf{0} \\ \hdashline \vdots & \vdots & \ddots & \vdots \\ \hdashline \mathbf{0} & \mathbf{0} & \cdots & d_n(s)\mathbf{I}_{n_n} \end{array}\right]. \tag{5.25b}$$

The uncertainty blocks have the dimensions $\mathbf{\Delta}_i(s) \in \mathscr{C}^{l_i \times n_i}$, and the identity matrices have the dimensions $\mathbf{I}_{n_i} \in \Re^{n_i \times n_i}$. These dimensions match up with the perturbation blocks to yield

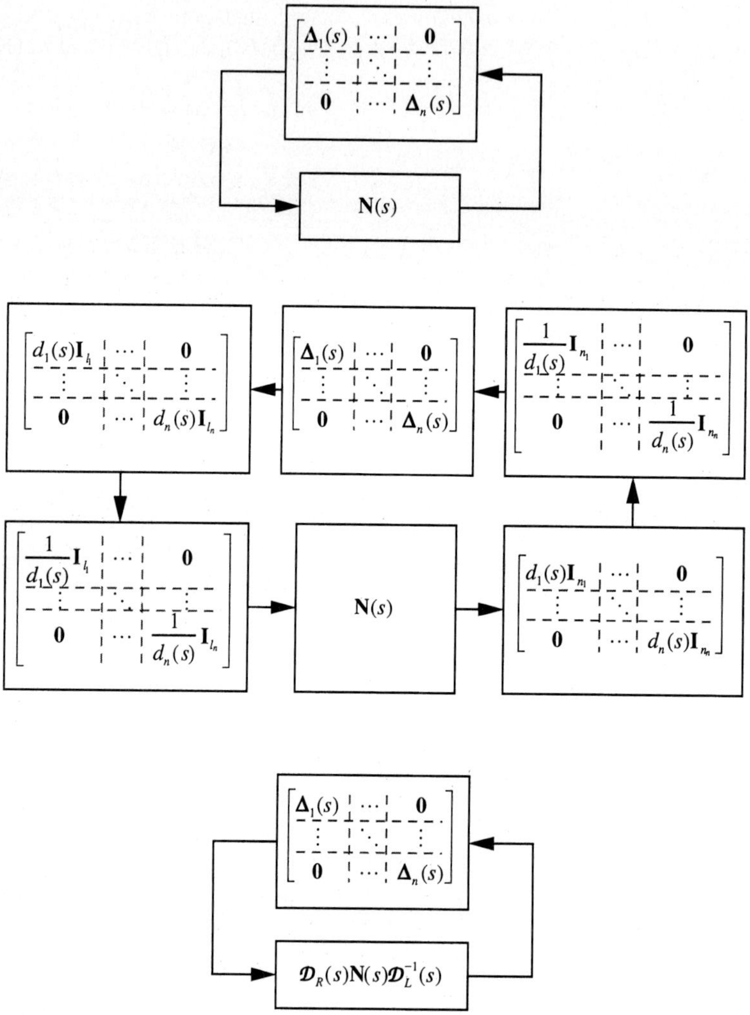

FIGURE 5.17 Diagonal scaling of the plant: (a) the feedback perturbation; (b) diagonal scaling added to the plant and the perturbation; (c) diagonal scaling leaves the diagonal perturbation unchanged

$$
\begin{aligned}
&\mathfrak{D}_L(s)\boldsymbol{\Delta}(s)\mathfrak{D}_R^{-1}(s)\\
&= \left[\begin{array}{c|c|c|c}
d_1(s)\boldsymbol{\Delta}_1(s)\dfrac{1}{d_1(s)} & \mathbf{0} & \cdots & \mathbf{0} \\
\hdashline
\mathbf{0} & d_2(s)\boldsymbol{\Delta}_2(s)\dfrac{1}{d_2(s)} & \cdots & \mathbf{0} \\
\hdashline
\vdots & \vdots & \ddots & \vdots \\
\hdashline
\mathbf{0} & \mathbf{0} & \cdots & d_n(s)\boldsymbol{\Delta}_n(s)\dfrac{1}{d_n(s)}
\end{array}\right] = \boldsymbol{\Delta}(s).
\end{aligned}
$$

Therefore, the system in Figure 5.17c is equivalent to the system in Figure 5.17a and

$$\mu_{\bar{\Delta}}[\mathbf{N}(s)] = \mu_{\bar{\Delta}}[\mathcal{D}_R(s)\mathbf{N}(s)\mathcal{D}_L^{-1}(s)].$$

An additional bound on the structured singular value,

$$\mu_{\bar{\Delta}}[\mathbf{N}(s)] = \mu_{\bar{\Delta}}[\mathcal{D}_R(s)\mathbf{N}(s)\mathcal{D}_L^{-1}(s)] \le \bar{\sigma}[\mathcal{D}_R(s)\mathbf{N}(s)\mathcal{D}_L^{-1}(s)],$$

is then obtained, since the maximum singular value is changed by inclusion of the diagonal scaling matrices. Since this result is valid for all diagonal scaling matrices (with the given block structure) and for all s, then

$$\mu_{\bar{\Delta}}(\mathbf{N}) \le \min_{\substack{\{d_1,d_2,\cdots d_n\} \\ d_i \in \mathscr{C}}} \bar{\sigma}(\mathcal{D}_R\mathbf{N}\mathcal{D}_L^{-1}). \tag{5.26}$$

The parameters d_i are called $\mathcal{D}$-scales. The bound (5.26) is valid for all complex $\mathcal{D}$-scales, and as a special case, for all real $\mathcal{D}$-scales. For the case of complex perturbations, the phase shift of the perturbation is arbitrary, and any phase shift (including sign changes) imparted by the $\mathcal{D}$-scales has no effect on the bound. Therefore, the minimization in (5.26) can be performed over the set of positive real $\mathcal{D}$-scales without loss of generality:

$$\boxed{\mu_{\bar{\Delta}}(\mathbf{N}) \le \min_{\substack{\{d_1,d_2,\cdots d_n\} \\ d_i \in (0,\infty)}} \bar{\sigma}(\mathcal{D}_R\mathbf{N}\mathcal{D}_L^{-1}).} \tag{5.27}$$

The following example illustrates the computation of this upper bound for the SSV.

EXAMPLE 5.10 Consider the matrix

$$\mathbf{N} = \begin{bmatrix} -1 & 1 \\ 2 & 0 \end{bmatrix}.$$

The set of perturbations $\bar{\Delta}$ is the set of diagonal matrices in $\mathscr{C}^{2\times 2}$; that is, $n = 2$, and $n_1 = n_2 = l_1 = l_2 = 1$. The SSV of $\mu_{\bar{\Delta}}(\mathbf{N})$ is bounded:

$$\mu_{\bar{\Delta}}(\mathbf{N}) \le \min_{d_1,d_2} \bar{\sigma}\left\{\begin{bmatrix} d_1 & 0 \\ 0 & d_2 \end{bmatrix}\begin{bmatrix} -1 & 1 \\ 2 & 0 \end{bmatrix}\begin{bmatrix} 1/d_1 & 0 \\ 0 & 1/d_2 \end{bmatrix}\right\} = \min_{d_1,d_2} \bar{\sigma}\begin{bmatrix} -1 & \dfrac{d_1}{d_2} \\ \dfrac{2d_2}{d_1} & 0 \end{bmatrix}.$$

Note that diagonal scaling only affects the off-diagonal terms of a matrix and therefore has no effect on the diagonal perturbation matrix. Only ratios of the diagonal scaling parameters appear in $\mathcal{D}_R\mathbf{N}\mathcal{D}_L^{-1}$. Therefore, one of these parameters can be chosen arbitrarily. In this case, let $d_2 = 1$. In general, one of the diagonal scaling parameters can be arbitrarily selected and is usually set equal to 1. The SSV is then

$$\mu_{\bar{\Delta}}[\mathbf{N}] \le \min_{d_1} \bar{\sigma}\begin{bmatrix} -1 & d_1 \\ \dfrac{2}{d_1} & 0 \end{bmatrix}.$$

The analytic solution of even this simple example is very difficult due to the complexity of the analytic expression for the singular value of a matrix. Numerical methods are therefore used to find the minimum:

$$d_1 = 1.4142;\ \mu_{\bar{\Delta}}(\mathbf{N}) \leq 2.$$

Note that this is a smaller upper bound than is provided by the maximum singular value:

$$\bar{\sigma}(\mathbf{N}) = 2.2882.$$

The actual SSV can be computed for this simple example by finding the perturbation matrix that has the smallest spectral norm and satisfies the following equation:

$$\det(\mathbf{I} - \mathbf{\Delta N}) = \det\begin{bmatrix} 1+\Delta_1 & -\Delta_1 \\ -2\Delta_2 & 1 \end{bmatrix} = 1 + \Delta_1 - 2\Delta_1\Delta_2 = 0.$$

Using this equation to solve for Δ_2 in terms of Δ_1 yields

$$\Delta_2 = \frac{1+\Delta_1}{2\Delta_1}.$$

The perturbation matrix is then

$$\mathbf{\Delta} = \begin{bmatrix} \Delta_1 & 0 \\ 0 & \dfrac{1+\Delta_1}{2\Delta_1} \end{bmatrix}.$$

The solution that minimizes the spectral norm (the maximum singular value) of this matrix is found by direct computation. The spectral norm of a diagonal matrix is the maximum diagonal element (in absolute value):

$$\bar{\sigma}(\mathbf{\Delta}) = \max\left(|\Delta_1|, \left|\frac{1+\Delta_1}{2\Delta_1}\right|\right).$$

The value of Δ_1 that minimizes this expression is found by equating the two absolute values and solving. The resulting $\mathbf{\Delta}$ is

$$\mathbf{\Delta} = \begin{bmatrix} -0.5 & 0 \\ 0 & -0.5 \end{bmatrix}.$$

The SSV is the inverse of the spectral norm of this perturbation:

$$\mu_{\bar{\Delta}}(\mathbf{N}) = 2.$$

In this example, the bound computed using (5.27) is equal to the actual SSV. ◆

The expression $\bar{\sigma}(\mathscr{D}_R \mathbf{N} \mathscr{D}_L^{-1})$ is a convex function of $\{d_1, d_2, \ldots, d_{n-1}\}$ for $d_i \in (0, \infty)$. In this case, $\bar{\sigma}(\mathscr{D}_R \mathbf{N} \mathscr{D}_L^{-1})$ has a single global minimum, no local minima, and the minimum value can be computed reliably using any good gradient-decent algorithm (see [8], page 197, for a good summary of decent algorithms).

Computing the upper bound for the SSV using a brute force gradient-decent algorithm is often computationally quite intensive. This computational intensity is compounded by the fact that robustness analysis requires that the SSV be computed over a range of frequencies. These computational difficulties have led to the development of a number of algorithms for optimizing the upper bound for the SSV that exploit the structure inherent in this problem. Two of the more efficient of these algorithms are described in [6] and [9]. A very readable treatment of the basics of these algorithms can be found in [10], page 119.

The upper bound in (5.27) has been found to lie close to the true SSV in a number of applications. This bound is an equality for perturbations consisting of three or fewer complex blocks, typically within 5% of the true SSV for larger numbers of complex blocks, and rarely worse than within 15% of the true SSV. The exception is when some of the perturbations are real.

The bound in (5.27) may be overly conservative for mixed real and complex perturbations, although even for these mixed perturbations, (5.27) often yields acceptable results. The extension of (5.27) to the case of mixed real and complex perturbations is treated in [5, 6].

A bound similar to (5.27) can be developed for the case where repeated perturbations are present. Repeated perturbations occur when multiple system parameters are dependent on a single external factor. For example, the Mach number (speed) of an aircraft influences both the lift produced by a given angle of attack and the torque produced by a given angle of attack. Therefore, the two parameters in the model relating lift and moment to angle of attack are related. These parameters can be modeled as nominal values plus scaled versions of the same perturbation. An extension of (5.27) to the case of repeated perturbations is given in [7].

Lower Bounds The structured singular value is bounded from below by the spectral radius of $\mathbf{N}_s$:

$$\mu_{\bar{\boldsymbol{\Delta}}}(\mathbf{N}) \geq \rho(\mathbf{N}_s) \tag{5.28}$$

where $\mathbf{N}_s$ is generated from $\mathbf{N}$ by the addition of zero columns and/or rows. These zero columns and rows are added to make $\mathbf{N}_s$ square and to make the perturbation blocks square, and have no effect on the structured singular value. The spectral radius of a matrix is defined as the largest of the eigenvalue magnitudes:

$$\rho(\mathbf{N}) = \max_i |\lambda_i(\mathbf{N})|.$$

The following example is used to illustrate how $\mathbf{N}_s$ is generated and to demonstrate that

$$\mu_{\bar{\boldsymbol{\Delta}}}(\mathbf{N}) = \mu_{\bar{\boldsymbol{\Delta}}_s}(\mathbf{N}_s).$$

Note that $\bar{\boldsymbol{\Delta}}_s$ is the set of block diagonal perturbations with appropriately sized square blocks.

EXAMPLE 5.11 Let $\mathbf{N} \in \mathscr{C}^{2\times 3}$, and $\boldsymbol{\Delta} \in \mathscr{C}^{3\times 2}$. The block structure of $\bar{\boldsymbol{\Delta}}$ is defined by

$$\boldsymbol{\Delta} = \begin{bmatrix} \boldsymbol{\Delta}_1 & \mathbf{0} \\ \mathbf{0} & \boldsymbol{\Delta}_2 \end{bmatrix},$$

where $\boldsymbol{\Delta}_1 \in \mathscr{C}^{2\times 1}$, and $\boldsymbol{\Delta}_2 \in \mathscr{C}^{1\times 1}$. The determinant used in computing $\mu_{\bar{\boldsymbol{\Delta}}}(\mathbf{N})$ is

$$\begin{aligned} \det(\mathbf{I} + \mathbf{N}\boldsymbol{\Delta}) &= \det\left(\begin{bmatrix} 1 & 0 \\ 0 & 1 \end{bmatrix} + \begin{bmatrix} N_{11} & N_{12} & N_{13} \\ N_{21} & N_{22} & N_{23} \end{bmatrix}\begin{bmatrix} \Delta_{11} & 0 \\ \Delta_{21} & 0 \\ 0 & \Delta_{32} \end{bmatrix}\right) \\ &= \det\left(\begin{bmatrix} 1 + N_{11}\Delta_{11} + N_{12}\Delta_{21} & N_{13}\Delta_{32} \\ N_{21}\Delta_{11} + N_{22}\Delta_{21} & 1 + N_{23}\Delta_{32} \end{bmatrix}\right), \end{aligned}$$

and is equivalent to the determinant used in computing $\mu_{\bar{\Delta}_s}(\mathbf{N}_s)$:

$$\det(\mathbf{I}+\mathbf{N}_s\mathbf{\Delta}_s) = \det\left(\begin{bmatrix}1&0&0\\0&1&0\\0&0&1\end{bmatrix} + \begin{bmatrix}N_{11}&N_{12}&N_{13}\\0&0&0\\N_{21}&N_{22}&N_{23}\end{bmatrix}\begin{bmatrix}\Delta_{11}&\Delta_{12}&0\\\Delta_{21}&\Delta_{22}&0\\0&0&\Delta_{32}\end{bmatrix}\right)$$
$$= \det\left(\begin{bmatrix}1+N_{11}\Delta_{11}+N_{12}\Delta_{21} & N_{11}\Delta_{12}+N_{12}\Delta_{22} & N_{13}\Delta_{32}\\ 0 & 1 & 0\\ N_{21}\Delta_{11}+N_{22}\Delta_{21} & N_{21}\Delta_{12}+N_{22}\Delta_{22} & 1+N_{23}\Delta_{32}\end{bmatrix}\right)$$
$$= \det\left(\begin{bmatrix}1+N_{11}\Delta_{11}+N_{12}\Delta_{21} & N_{13}\Delta_{32}\\ N_{21}\Delta_{11}+N_{22}\Delta_{21} & 1+N_{23}\Delta_{32}\end{bmatrix}\right).$$

The structured singular value of $\mathbf{N}_s$ is therefore equal to the structured singular value of $\mathbf{N}$. ◆

To show that the SSV is bounded as in (5.28), the equivalent bound

$$\mu_{\bar{\Delta}_s}(\mathbf{N}_s) \geq \rho(\mathbf{N}_s)$$

is demonstrated by considering (5.23):

$$\frac{1}{\mu_{\bar{\Delta}_s}(\mathbf{N}_s)} = \min_{\mathbf{\Delta}_s\in\bar{\Delta}_s}\,[\bar{\sigma}(\mathbf{\Delta}_s) \text{ such that } \det(\mathbf{I}+\mathbf{N}_s\mathbf{\Delta}_s)=0]. \tag{5.29}$$

A particular perturbation matrix is

$$\mathbf{\Delta}_s = -\frac{1}{\lambda_1}\mathbf{I} \in \bar{\mathbf{\Delta}}_s,$$

where λ_1 is the eigenvalue of $\mathbf{N}_s$ with the largest absolute value. For this perturbation, the matrix within the determinant in (5.29) becomes

$$\mathbf{I}+\mathbf{N}_s\mathbf{\Delta}_s = \mathbf{I} - \frac{1}{\lambda_1}\mathbf{N}_s.$$

This matrix is singular since

$$(\mathbf{I}+\mathbf{N}_s\mathbf{\Delta}_s)\phi_1 = \left(\mathbf{I}-\frac{1}{\lambda_1}\mathbf{N}_s\right)\phi_1 = \left(1-\frac{1}{\lambda_1}\lambda_1\right)\phi_1 = 0$$

where ϕ_1 is the eigenvector of $\mathbf{N}_s$ associated with the eigenvalue λ_1. Therefore, the $\det(\mathbf{I}+\mathbf{N}_s\mathbf{\Delta}_s)=0$ for the given perturbation, and the SSV is bounded:

$$\mu_{\bar{\Delta}_s}(\mathbf{N}_s) = \frac{1}{\min_{\mathbf{\Delta}_s\in\bar{\Delta}_s}\,[\bar{\sigma}(\mathbf{\Delta}_s) \text{ such that } \det(\mathbf{I}+\mathbf{N}_s\mathbf{\Delta}_s)=0]}$$
$$\geq \frac{1}{\bar{\sigma}(-\frac{1}{\lambda_1}\mathbf{I})} = |\lambda_1| = \rho(\mathbf{N}_s).$$

The bound in (5.28) is easy to compute, but tends to be overly conservative; that is, the SSV can be appreciably greater than the spectral radius. Additional bounds can be generated by noting that

$$\mu_{\bar{\Delta}_s}[\mathbf{N}_s\mathbf{U}] = \mu_{\bar{\Delta}_s}[\mathbf{N}_s] = \mu_{\bar{\Delta}}[\mathbf{N}], \tag{5.30}$$

where $\mathbf{U} \in \bar{\mathbf{\Delta}}_s$, and $\mathbf{U}$ is a unitary matrix; that is,

$$\mathbf{U}^{\dagger}\mathbf{U} = \mathbf{I}.$$

Combining (5.30) with the bound (5.28) yields a lower bound for the SSV (see Exercise 5.12):

$$\boxed{\mu_{\bar{\Delta}}[\mathbf{N}] \geq \max_{\substack{\mathbf{U}\in\bar{\mathbf{\Delta}}_s \\ \mathbf{U}^{\dagger}\mathbf{U}=\mathbf{I}}} \rho[\mathbf{N}_s\mathbf{U}].} \tag{5.31}$$

The lower bound in (5.31) is identically equal to the SSV. Unfortunately, the expression $\rho[\mathbf{N}_s\mathbf{U}]$ is not a convex function of the elements of $\mathbf{U}$, and may have multiple local maxima. Numerical optimization algorithms may fail to find the true maximum in (5.31). Therefore, it is best to treat (5.31) as a lower bound and not as an equality, since numerical algorithms may not yield the equality.

EXAMPLE 5.12 Consider a motor with the following transfer function:

$$G(s) = \frac{g}{s(s + \tau_1)(s + \tau_2)}$$

where the forward path gain and each of the time constants are uncertain:

$$g \in [2.7 \times 10^6, 3.3 \times 10^6] \ ; \ \tau_1 \in [50, 60] \ ; \ \tau_2 \in [300, 400].$$

These parameters can be modeled as nominal values plus perturbations:

$$g = 3 \times 10^6 + 3 \times 10^5\Delta_1 \ ; \ \tau_1 = 55 + 5\Delta_2 \ ; \ \tau_2 = 350 + 50\Delta_3$$

where Δ_1, Δ_2, and Δ_3 are real perturbations bounded by 1. A unity feedback controller is designed for this plant:

$$K(s) = 1.$$

A block diagram of the closed-loop system is given in Figure 5.18a. The closed-loop state model for this system is

$$\dot{x}(t) = \begin{bmatrix} 0 & 1 & 0 \\ 0 & -350 & 1 \\ -3 \times 10^6 & 0 & -55 \end{bmatrix} x(t) + \begin{bmatrix} 0 & 0 & 0 \\ 0 & 0 & 1 \\ 1 & 1 & 0 \end{bmatrix} w_d(t);$$

$$y_d(t) = \begin{bmatrix} -3 \times 10^5 & 0 & 0 \\ 0 & 0 & 5 \\ 0 & 50 & 0 \end{bmatrix} x(t),$$

and the closed-loop transfer function from W_d to Y_d is

$$\mathbf{N}_{y_d w_d}(s) = \begin{bmatrix} -3 \times 10^5 & 0 & 0 \\ 0 & 0 & 5 \\ 0 & 50 & 0 \end{bmatrix} \left(s\mathbf{I} - \begin{bmatrix} 0 & 1 & 0 \\ 0 & -350 & 1 \\ -3 \times 10^6 & 0 & -55 \end{bmatrix} \right)^{-1} \begin{bmatrix} 0 & 0 & 0 \\ 0 & 0 & 1 \\ 1 & 1 & 0 \end{bmatrix}.$$

The bounds on the SSV for this system are computed numerically. The diagonal scales used in optimizing the upper bound and the unitary matrix used in optimizing the lower bound must be computed at every frequency. The SSV is plotted in Figure 5.18b. Note

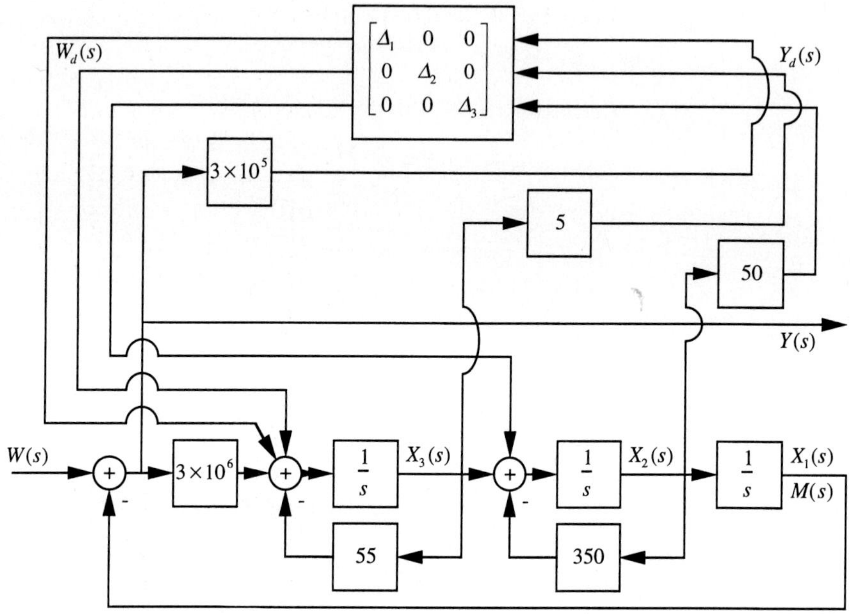

(a)

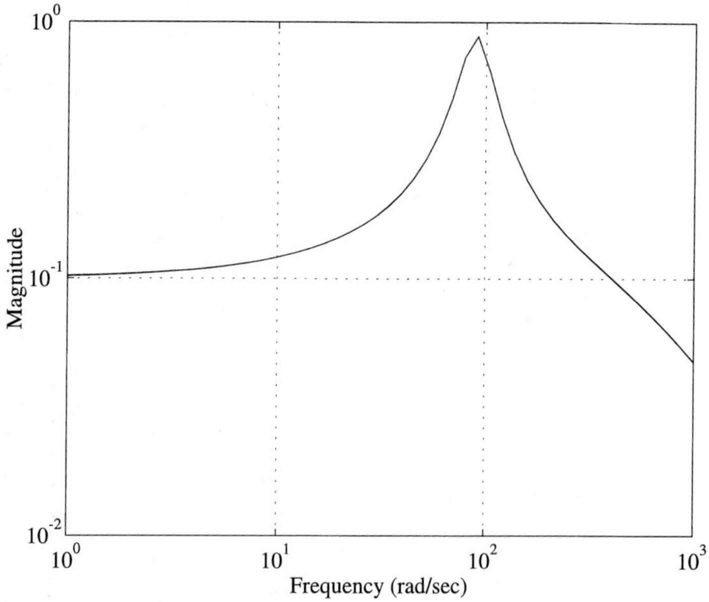

(b)

FIGURE 5.18 Stability robustness of Example 5.12: (a) block diagram; (b) the structured singular value that the lower bound and the upper bound coincide, indicating that the true SSV is computed exactly. The supremum of $\mu_{\bar{\Delta}}[\mathbf{N}_{y_d w_d}(j\omega)]$ is 0.88, which indicates that the system is robustly stable. ◆

5.5.4 Additional Properties of the Structured Singular Value

The structured singular value is central to the analysis of robustness. A number of properties of the SSV have been generated in the literature. In this section, two of these properties are developed. These properties are selected because they add insight into the meaning of the SSV, and they are used in subsequent developments.

A consequence of the definition of the structured singular value is that

$$\mu_{\bar{\Delta}}(\mathbf{N}) < 1 \tag{5.32}$$

if and only if

$$\det(\mathbf{I} - \mathbf{N}\boldsymbol{\Delta}) > 0 \tag{5.33}$$

for all admissible perturbations:

$$\boldsymbol{\Delta} \in \bar{\boldsymbol{\Delta}} \quad \text{and} \quad \|\boldsymbol{\Delta}\|_\infty \leq 1. \tag{5.34}$$

The condition (5.33) results from the fact that the determinant is a continuous function of the perturbation. The determinant is positive (indeed, equal to 1) for the admissible perturbation $\boldsymbol{\Delta} = \mathbf{0}$. If for some other admissible perturbation the determinant is nonpositive, the determinant must equal zero for an admissible perturbation. The singular value of this perturbation is less than 1, implying that $\mu_{\bar{\Delta}}(\mathbf{N})$ is greater than 1.

The SSV is bounded by the maximum singular value (5.22), in general. When the set of perturbations is the set of all complex matrices (i.e., $\bar{\boldsymbol{\Delta}}$ is the set of unstructured uncertainties), this bound becomes an equality. In addition, the bound is achieved when the perturbation is

$$\boldsymbol{\Delta} = -\frac{1}{\sigma_1} V_1 U_1^\dagger, \tag{5.35}$$

where σ_1, V_1, and U_1 are from the singular value decomposition of $\mathbf{N}$:

$$\mathbf{N} = \sum_{i=1}^{p} \sigma_i U_i V_i^\dagger .$$

To demonstrate that the SSV equals the maximum singular value, note that

$$(\mathbf{I} + \mathbf{N}\boldsymbol{\Delta})U_1 = \left\{\mathbf{I} - \sum_{i=1}^{p} \sigma_i U_i V_i^\dagger \frac{1}{\sigma_1} V_1 U_1^\dagger \right\} U_1 = 0.$$

Therefore, this matrix must be singular:

$$\det(\mathbf{I} + \mathbf{N}\boldsymbol{\Delta}) = 0.$$

The bound (5.22) on the SSV is then achieved for the given perturbation, and

$$\mu_{\bar{\Delta}}(\mathbf{N}) = \bar{\sigma}(\mathbf{N}) \tag{5.36}$$

when the perturbations are unstructured.

The result (5.36) shows that the maximum singular value of a matrix can be defined in a manner analogous to the structured singular value:

$$\bar{\sigma}(\mathbf{N}) = \frac{1}{\min\limits_{\Delta \in \mathscr{C}^{l x n}} [\bar{\sigma}(\mathbf{\Delta}) \text{ such that } \det(\mathbf{I} + \mathbf{N\Delta}) = 0]}. \tag{5.37}$$

The structured singular value is then a generalization of the maximum singular value. This generalization is accomplished by placing restrictions on the set over which the minimization is performed in (5.37). This relationship explains the similarity between the robust stability conditions for unstructured and structured uncertainty.

5.6 Performance Robustness Analysis Using the SSV

The stability of systems subject to gain, phase, unstructured, and structured perturbations was addressed in the previous sections of this chapter. Stability is typically required of feedback control systems, but stability alone does not insure suitability of the system. Suitability of the controller is dependent on both stability and the meeting of certain performance specifications.

Performance can be described in a number of ways, as discussed in Chapter 4. A particular method of specifying performance is to bound the ∞-norm of the closed-loop transfer function. This form of performance specification is quite general and can be simply incorporated in SSV robustness analysis. The performance specification is given as

$$\|\mathbf{H}(s)\|_{\infty} < 1, \tag{5.38}$$

where $\mathbf{H}(s)$ is the perturbed closed-loop transfer function. This transfer function is obtained by closing the loop containing the perturbation in Figure 5.19a:

$$\mathbf{H}(s) = \mathbf{N}_{yw_d}(s)[\mathbf{I} - \mathbf{\Delta}(s)\mathbf{N}_{y_dw_d}(s)]^{-1}\mathbf{\Delta}(s)\mathbf{N}_{y_dw}(s) + \mathbf{N}_{yw}(s).$$

A bound of 1 can be used to specify any frequency-dependent bound on this gain by incorporating weighting functions into the plant. A system is said to possess robust performance if the system remains internally stable and the specification in (5.38) is satisfied for all admissible perturbations. Robust performance can also be defined using other performance criteria or cost functions. The ∞-norm cost function is typically used to specify performance robustness because it yields a robustness test that is easily applied in practice.

The perturbed closed-loop transfer function is dependent on both the nominal closed-loop transfer function and the perturbation. The conditions for performance robustness can be precisely stated in terms of these transfer functions:

$$\boxed{\sup_{\omega}\{\mu_{\bar{\Delta}}[\mathbf{N}_{y_dw_d}(s)]\} < 1;} \tag{5.39a}$$

$$\boxed{\|\mathbf{H}(s)\|_{\infty} = \|\mathbf{N}_{yw_d}(s)[\mathbf{I} - \mathbf{\Delta}(s)\mathbf{N}_{y_dw_d}(s)]^{-1}\mathbf{\Delta}(s)\mathbf{N}_{y_dw}(s) + \mathbf{N}_{yw}(s)\|_{\infty} < 1} \tag{5.39b}$$

for all admissible $\mathbf{\Delta}$.

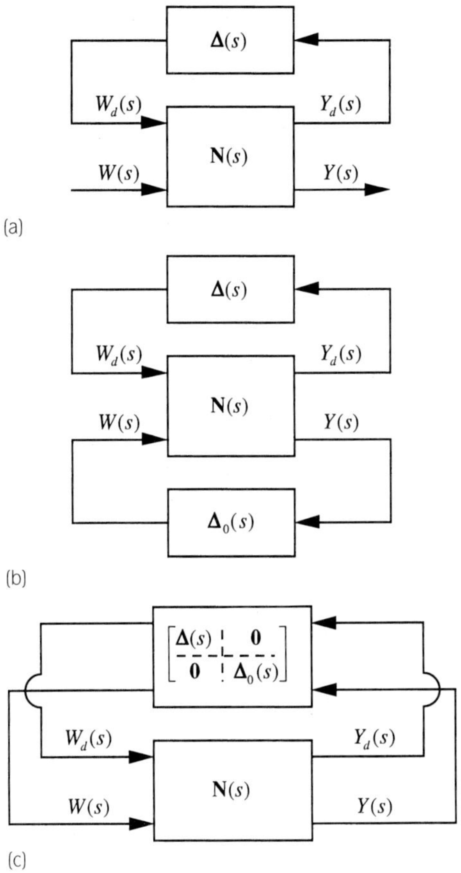

FIGURE 5.19 Performance robustness analysis using the SSV: (a) the system with uncertainty; (b) adding a performance block; (c) structured uncertainty incorporating the performance block

This robust performance problem can be converted into an equivalent robust stability problem by appending an uncertainty block to the system. This uncertainty block (known as a performance block) connects the performance output with the disturbance input, as shown in Figure 5.19. The system in Figure 5.19a meets the performance robustness objectives if and only if the system in Figure 5.19c is robustly stable. This fact can be intuitively understood by noting that for robust stability, the gain must be less than 1. This requirement is the same as the performance requirement. Therefore, requiring robust stability with respect to the perturbation $\mathbf{\Delta}_0(s)$, as shown in Figure 5.19b, is equivalent to the performance requirement. Adding this new perturbation to the original perturbations yields the block diagram in Figure 5.19c. A more mathematical demonstration of the connection between robust performance of the system in Figure 5.19a and robust stability of the system in Figure 5.19c is presented below.

Robust stability of the system with the performance block (Figure 5.19c) implies that the system is stable for all perturbations of the following kind:

$$\boldsymbol{\Delta}_p(s) = \left[\begin{array}{c:c} \boldsymbol{\Delta}(s) & \mathbf{0} \\ \hdashline \mathbf{0} & \boldsymbol{\Delta}_0(s) \end{array}\right] \text{ such that } \|\boldsymbol{\Delta}_p(s)\|_\infty \le 1. \tag{5.40}$$

The system is also stable for all perturbations of the following kind:

$$\boldsymbol{\Delta}_p(s) = \left[\begin{array}{c:c} \boldsymbol{\Delta}(s) & \mathbf{0} \\ \hdashline \mathbf{0} & \mathbf{0} \end{array}\right] \text{ such that } \|\boldsymbol{\Delta}(s)\|_\infty \le 1, \tag{5.41}$$

since this is a subset of the perturbations in (5.40). The feedback loop of the system in Figure 5.19c, with this set of perturbations, is identical to the feedback loop of the system in Figure 5.19a. This can be seen either from the block diagram or by looking at the closed-loop poles, which are the solutions of

$$\det\left\{\mathbf{I} - \left[\begin{array}{c:c} \mathbf{N}_{y_d w_d}(s) & \mathbf{N}_{y_d w}(s) \\ \hdashline \mathbf{N}_{y w_d}(s) & \mathbf{N}_{yw}(s) \end{array}\right]\left[\begin{array}{c:c} \boldsymbol{\Delta}(s) & \mathbf{0} \\ \hdashline \mathbf{0} & \mathbf{0} \end{array}\right]\right\} = \det\left\{\mathbf{I} - \left[\begin{array}{c:c} \mathbf{N}_{y_d w_d}(s)\boldsymbol{\Delta}(s) & \mathbf{0} \\ \hdashline \mathbf{N}_{y w_d}(s)\boldsymbol{\Delta}(s) & \mathbf{0} \end{array}\right]\right\}$$
$$= \det\{\mathbf{I} - \mathbf{N}_{y_d w_d}(s)\boldsymbol{\Delta}(s)\} = 0.$$

The system is robustly stable when subject to the perturbations in (5.41) if and only if

$$\sup_\omega\{\mu_{\bar{\boldsymbol{\Delta}}}[\mathbf{N}_{y_d w_d}(j\omega)]\} < 1,$$

which satisfies the first condition for robust performance (5.39a).

The system with the performance block is robustly stable provided that

$$\sup_\omega\{\mu_{\bar{\boldsymbol{\Delta}}_p}[\mathbf{N}(j\omega)]\} < 1. \tag{5.42}$$

A consequence of the definition of the structured singular value [see (5.33)] is that (5.42) is true if and only if

$$\det[\mathbf{I} - \mathbf{N}(j\omega)\boldsymbol{\Delta}_p(j\omega)] > 0 \tag{5.43}$$

for all frequencies and all admissible perturbations:

$$\boldsymbol{\Delta}_p(j\omega) \in \bar{\boldsymbol{\Delta}}_p \text{ such that } \|\boldsymbol{\Delta}_p(j\omega)\|_\infty \le 1.$$

The determinant in (5.43) can be expanded using equation (A2.6) in the Appendix:

$$\begin{aligned} &\det[\mathbf{I} - \mathbf{N}(j\omega)\boldsymbol{\Delta}_p(j\omega)] \\ &\quad = \det\left[\begin{array}{c:c} \mathbf{I} - \mathbf{N}_{y_d w_d}\boldsymbol{\Delta} & -\mathbf{N}_{y_d w}\boldsymbol{\Delta}_0 \\ \hdashline -\mathbf{N}_{y w_d}\boldsymbol{\Delta} & \mathbf{I} - \mathbf{N}_{yw}\boldsymbol{\Delta}_0 \end{array}\right] \\ &\quad = \det[\mathbf{I} - \mathbf{N}_{y_d w_d}\boldsymbol{\Delta}]\det\{[\mathbf{I} - \mathbf{N}_{yw}\boldsymbol{\Delta}_0] - \mathbf{N}_{y w_d}\boldsymbol{\Delta}[\mathbf{I} - \mathbf{N}_{y_d w_d}\boldsymbol{\Delta}]^{-1}\mathbf{N}_{y_d w}\boldsymbol{\Delta}_0\} \\ &\quad = \det(\mathbf{I} - \mathbf{N}_{y_d w_d}\boldsymbol{\Delta})\det[\mathbf{I} - \{\mathbf{N}_{yw} + \mathbf{N}_{y w_d}\boldsymbol{\Delta}(\mathbf{I} - \mathbf{N}_{y_d w_d}\boldsymbol{\Delta})^{-1}\mathbf{N}_{y_d w}\}\boldsymbol{\Delta}_0] > 0, \end{aligned} \tag{5.44}$$

where the frequency designation has been dropped to simplify this expression. Since the system is robustly stable,

$$\det(\mathbf{I} - \mathbf{N}_{y_d w_d}\boldsymbol{\Delta}) > 0.$$

Equation (5.44) is then satisfied if and only if

$$\det[\mathbf{I} - \{\mathbf{N}_{yw}(j\omega) + \mathbf{N}_{yw_d}(j\omega)\boldsymbol{\Delta}(j\omega) \times (\mathbf{I} - \mathbf{N}_{y_dw_d}(j\omega)\boldsymbol{\Delta}(j\omega))^{-1}\mathbf{N}_{y_dw}(j\omega)\}\boldsymbol{\Delta}_0(j\omega)] > 0 \quad (5.45)$$

for all frequencies and all admissible perturbations. The condition (5.45) holds both for all admissible $\boldsymbol{\Delta}$ and all $\boldsymbol{\Delta}_0$ such that

$$\|\boldsymbol{\Delta}_0(j\omega)\|_\infty \leq 1. \quad (5.46)$$

Equation (5.45) is satisfied for all unstructured perturbations $\boldsymbol{\Delta}_0$ that satisfy (5.46) if and only if (by the small-gain theorem)

$$\|\mathbf{N}_{yw}(s) + \mathbf{N}_{yw_d}(s)\boldsymbol{\Delta}(s)[\mathbf{I} - \mathbf{N}_{y_dw_d}(s)\boldsymbol{\Delta}(s)]^{-1}\mathbf{N}_{y_dw}(s)\|_\infty < 1. \quad (5.47)$$

Applying the push-through theorem [see equation (A8.2) in the Appendix] to this result yields the condition for robust performance (5.39b).

In summary, the structured singular value can be applied to evaluate robust performance. The only limitation on the applicability of this result is that the performance must be specified in terms of a bound on the closed-loop system ∞-norm. The following examples demonstrate the use of performance blocks and the SSV to evaluate robust performance.

EXAMPLE 5.13 We are given the plant with an uncertain pole location:

$$G(s) = \frac{1}{s + 1 + \delta}; \ \delta \in [-0.2, 0.2].$$

A controller for the nominal plant is obtained by using unity feedback with the gain K, as shown in Figure 5.20a. The object of the controller is to make the output follow the reference input so that the steady-state gain between the reference input and the tracking error is −20 dB or less. The reference input is assumed to be slowly varying; specifically, it is band-limited to less than 10 rad/sec. These specifications can be translated into a bound on the closed-loop transfer function from the reference input to the tracking error:

$$|H(j\omega)| \leq \begin{cases} 0.1 & \omega \leq 10 \\ \infty & \omega > 10 \end{cases}.$$

Note that the infinity in this bound indicates that the gain above 10 rad/sec is unimportant. The perturbation can be normalized as shown in Figure 5.20b.

The performance goal is normalized to 1 by using the following weighting function:

$$W(j\omega) = \begin{cases} 10 & \omega \leq 10 \\ 0 & \omega > 10 \end{cases}.$$

A low-order rational approximation of this weighting function is

$$W(j\omega) = \frac{150}{j\omega + 10}.$$

The gain of this transfer function is greater than 10 (20 dB) over the bandwidth from 0 to 10 rad/sec and has a cutoff frequency at 10 rad/sec.

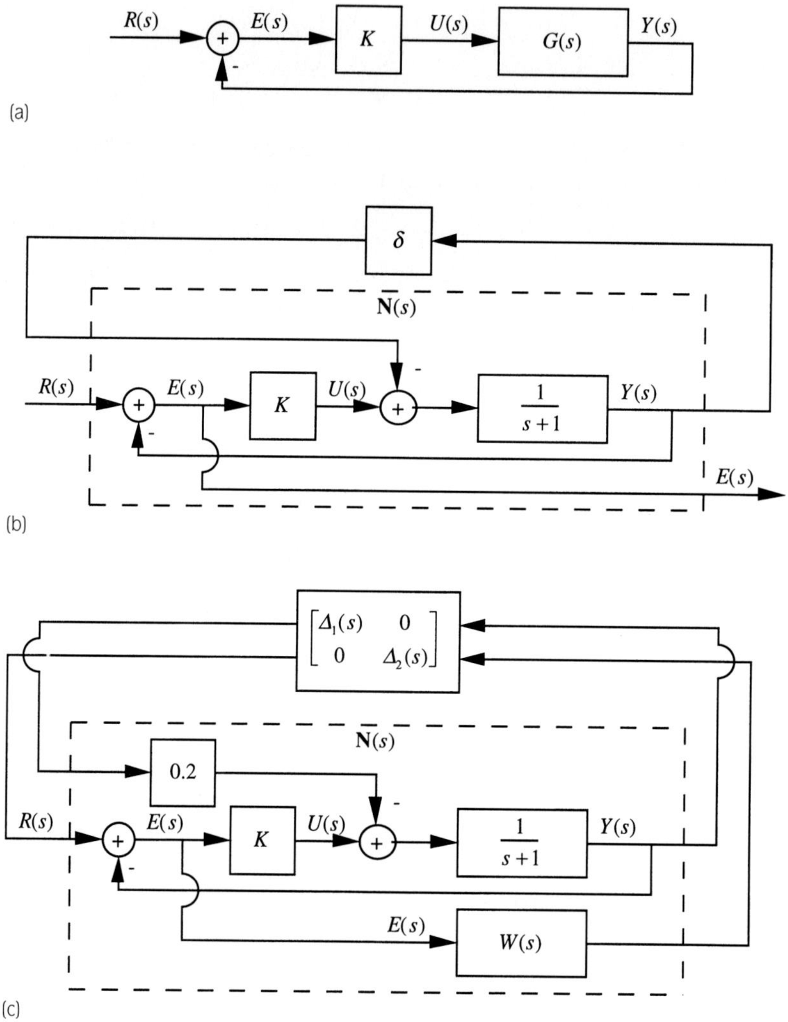

FIGURE 5.20 Performance robustness analysis of Example 5.13: (a) the closed-loop system; (b) the closed-loop system with uncertainty; (c) the closed-loop system with uncertainty and performance blocks

Robust performance can be analyzed using the SSV by appending the weighting function to the plant and adding the performance block $\mathbf{\Delta}_0$, as shown in Figure 5.20c. The transfer function of this system is

$$\mathbf{N}(s) = \begin{bmatrix} \dfrac{-0.2}{s + 1 + K} & \dfrac{K}{s + 1 + K} \\ \dfrac{30}{(s + 1 + K)(s + 10)} & \dfrac{150(s + 1)}{(s + 1 + K)(s + 10)} \end{bmatrix}.$$

Robust performance is tested by generating the SSV of $\mathbf{N}(j\omega)$. The system is stable and meets the performance specifications for all admissible perturbations if this SSV is less

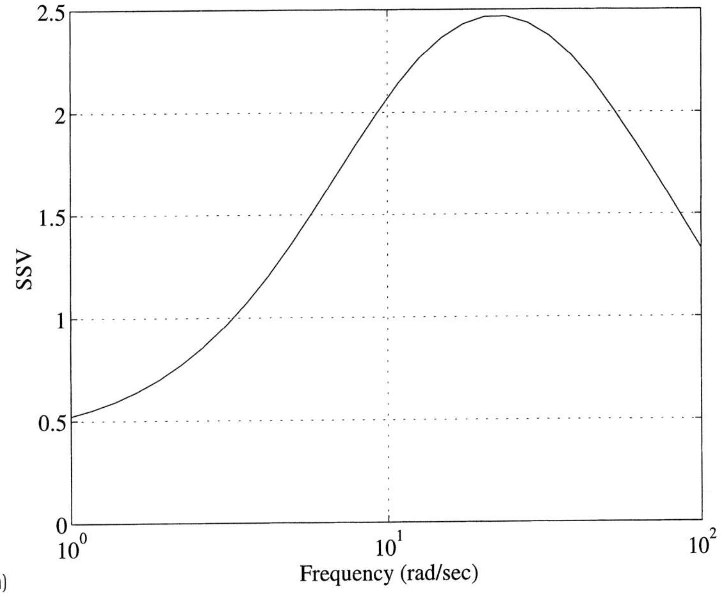

(a)

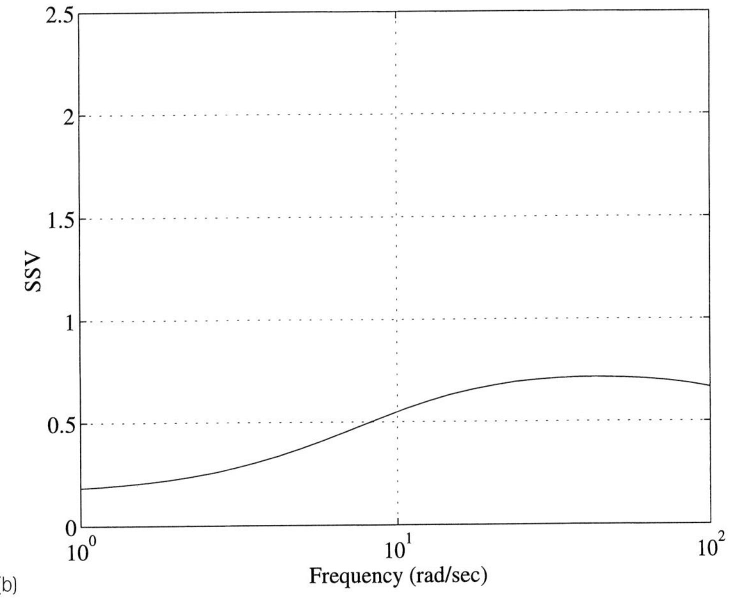

(b)

FIGURE 5.21 SSV plots for Example 5.13: (a) $K = 50$; (b) $K = 200$

than 1 for all frequencies. The computation of the SSV as a function of frequency requires a computer in even the simplest cases. Computer-generated plots of the SSV are given in Figure 5.21 for two values of K. The system does not perform robustly for $K = 50$. This is expected since the nominal system does not meet the performance specifications for this value of K. The nominal system is not required to meet the performance specifications when applying the SSV performance robustness test, but stability of the nominal system is required for application of this test. Better tracking can be obtained by increasing the feedback gain beyond 50. The system is seen to remain stable and meets the performance specifications for all admissible perturbations when $K = 200$. The design in this particular example could be accomplished using more traditional methods due to the simplicity of the system, but the example still serves to illustrate the application of the SSV to performance robustness analysis. ◆

EXAMPLE 5.14 We are given the vertical-takeoff-and-landing remotely piloted vehicle of Example 5.7. A block diagram of the pitch and yaw control system for this vehicle is given in Figure 5.13a. A reasonable performance goal is that the pitch and yaw errors be smaller than 1°, or about 5% of the largest commanded values. Note that the gain requirement is derived by comparing the allowable error to the largest commanded input expected (about 20°). The commanded values are assumed to be slowly varying; specifically, they are band-limited to less than 1 rad/sec. These specifications can be translated to a bound on the perturbed closed-loop transfer function:

$$\bar{\sigma}\{\mathbf{H}(j\omega)\} \le \begin{cases} 0.05 & \omega \le 1 \\ \infty & \omega > 1 \end{cases}.$$

Note that the infinity in this bound indicates that the performance above 1 rad/sec is unimportant. The performance goal is normalized to 1 by the use of the following weighting function:

$$W(j\omega) = \begin{cases} 20 & \omega \le 1 \\ 0 & \omega > 1 \end{cases}.$$

A rational approximation of this weighting function is

$$W(s) = \frac{20}{(s+1)^2} = \frac{20}{s^2 + 2s + 1}.$$

The magnitude of the weighting function and its rational approximation are shown in Figure 5.22a. A second-order weighting function is used to achieve a rapid roll-off at high frequencies (the specifications call for an infinitely rapid roll-off). Appending this transfer function to the system and adding the performance block yields the block diagram in Figure 5.22b. Note that the feedback gains are given in Example 5.7. This system was shown to be robustly stable in Example 5.7, but no information was generated on performance. Nominal performance is evaluated by numerically computing the maximum singular value of the nominal closed-loop transfer function:

$$\mathbf{N}_{y_d w_d}(s) = W(s)\{\mathbf{I} - \mathbf{C}_y(s\mathbf{I} - \mathbf{A} + \mathbf{B}_u\mathbf{K}\mathbf{C}_m)^{-1}\mathbf{B}_u\mathbf{K}\mathbf{K}_r\}.$$

The principle gains for this transfer function are plotted in Figure 5.23. The infinity norm of the nominal system is 0.81, which indicates that the nominal system achieves the performance specifications. The total nominal closed-loop transfer function is

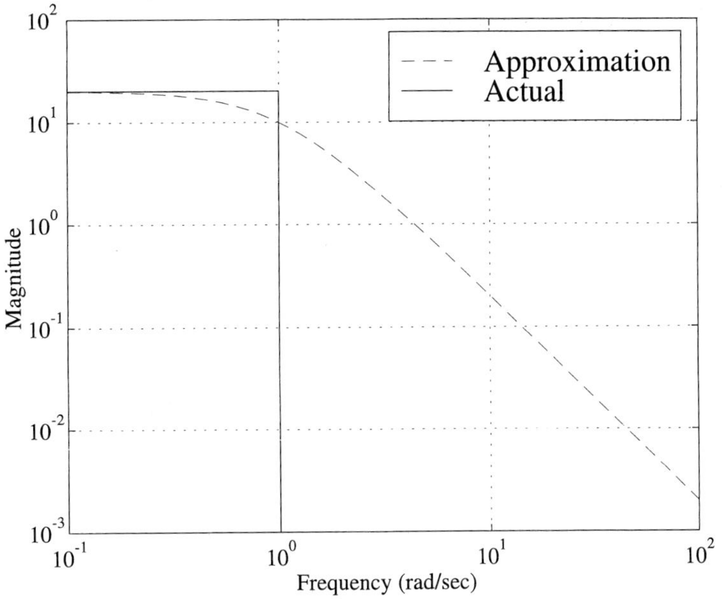

(a)

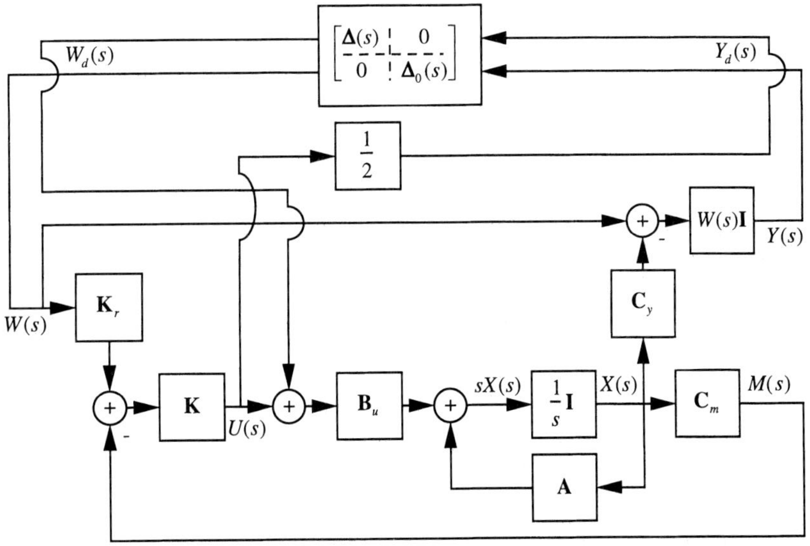

(b)

FIGURE 5.22 Robust performance analysis for Example 5.14: (a) frequency response of the weighting function; (b) block diagram of the system with weighting functions and the performance block

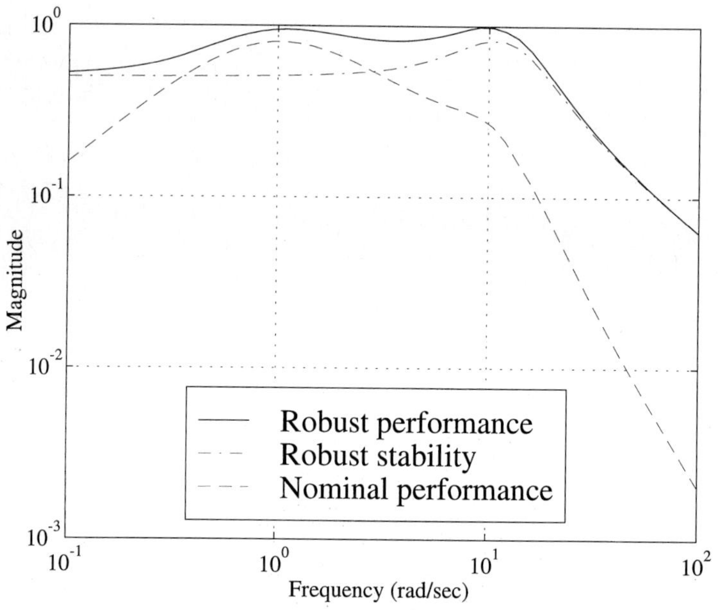

FIGURE 5.23 Robust performance analysis for Example 5.14: The structured singular value of **N**(*s*)

$$\mathbf{N}(s) =$$

$$\left[\begin{array}{c:c} -\frac{1}{2}\mathbf{KC}_m(s\mathbf{I}-\mathbf{A}+\mathbf{B}_u\mathbf{KC}_m)^{-1}\mathbf{B}_u & -\frac{1}{2}\mathbf{KC}_m(s\mathbf{I}-\mathbf{A}+\mathbf{B}_u\mathbf{KC}_m)^{-1}\mathbf{B}_u\mathbf{KK}_r+\frac{1}{2}\mathbf{KK}_r \\ \hdashline -W(s)\mathbf{C}_y(s\mathbf{I}-\mathbf{A}+\mathbf{B}_u\mathbf{KC}_m)^{-1}\mathbf{B}_u & W(s)\{\mathbf{I}-\mathbf{C}_y(s\mathbf{I}-\mathbf{A}+\mathbf{B}_u\mathbf{KC}_m)^{-1}\mathbf{B}_u\mathbf{KK}_r\} \end{array}\right]$$

Robust performance is evaluated by computing and plotting (see Figure 5.23) the structured singular value of $\mathbf{N}(s)$. The maximum structured singular value is 0.99, which indicates that the system is stable and meets the performance specifications for all admissible perturbations. For completeness, the maximum singular value of $\mathbf{N}_{y_d w_d}(s)$, which indicates robust stability, is also included in Figure 5.23. ◆

5.7 Summary

The internal stability of feedback systems is considered in this chapter. For SISO systems, the feedback connection of two systems, $G(s)$ and $K(s)$, is shown to be internally stable if all of the solutions of

$$\{1 + G(s)K(s)\} = 0$$

have negative real parts and there are no pole-zero cancellations between $G(s)$ and $K(s)$. This result then led to the Nyquist stability criterion for SISO systems. For MIMO systems, the feedback connection of two systems, $\mathbf{G}(s)$ and $\mathbf{K}(s)$, is shown to be internally stable if all of the solutions of:

$$\det\{\mathbf{I} + \mathbf{G}(s)\mathbf{K}(s)\} = 0$$

have negative real parts and both $\mathbf{G}(s)$ and $\mathbf{K}(s)$ are stable. This result is used in the analysis of robustness.

The description of uncertainty in systems is addressed. A number of uncertainty models are presented: gain perturbations for SISO plants; phase perturbations for SISO plants; single transfer function perturbations, termed unstructured uncertainty; and multiple transfer function perturbations, termed structured uncertainty. Parametric perturbations, which arise frequently in applications, are a special case of transfer function perturbations where the transfer function is a scalar. Any system with parametric uncertainty can then be modeled as either an unstructured or structured uncertainty, depending on the number of uncertain parameters.

A feedback system is termed robustly stable if the resulting system is internally stable for all admissible perturbations. The analysis of stability robustness of SISO feedback systems with gain and phase perturbations led to the classical control concepts of gain and phase margins. Results are presented for stability robustness of systems with both unstructured uncertainty and structured uncertainty. The analysis of stability robustness of systems with structured uncertainty leads to the definition of the structured singular value. The $\mathcal{D}$-scaling method for computing bounds on the SSV was presented.

A feedback system is said to possess robust performance if the resulting system is internally stable and meets certain performance objectives for all admissible perturbations. Robust performance of a system can be analyzed by considering the robust stability of an augmented system and utilizing the SSV.

The results presented on internal stability are all directly applicable to discrete-time systems provided that $\mathcal{Z}$ transfer functions are used in place of Laplace transfer functions. For discrete-time systems, stability is guaranteed when all $\mathcal{Z}$ transfer function poles have a magnitude less than 1. Discrete-time Nyquist plots are generated by mapping the unit circle through the loop transfer function. Phase and gain margins are then computed in an analogous manner to continuous-time systems. The stability robustness results and the definition of the SSV are identical in both continuous time and discrete time. For discrete-time systems, the SSV of the $\mathcal{Z}$ transfer function evaluated on the unit circle, instead of the imaginary axis, is used when determining stability and performance robustness.

REFERENCES

[1] H. Nyquist, "Regeneration theory," *Bell Systems Technical Journal,* January 1932, pp. 126–147.

[2] R. V. Churchill, J. W. Brown, and R. F. Verhey, *Complex Variables and Applications,* 3d ed., McGraw-Hill, New York, 1974.

[3] N. S. Nise, *Control Systems Engineering,* Benjamin-Cummings, Redwood City, CA, 1992.

[4] R. C. Dorf, *Modern Control Systems,* 5th ed., Addison-Wesley, Reading, MA, 1989.

[5] P. M. Young, M. P. Newlin, and J. C. Doyle, "μ analysis with real parametric uncertainty," *Proceedings of the 30th IEEE Conference on Decision and Control,* 1991, pp. 1251–1256.

[6] P. M. Young, M. P. Newlin, and J. C. Doyle, "Practical computation of the mixed μ problem," *Proceedings of the American Control Conference,* 1992, pp. 2190–2194.

[7] J. C. Doyle, "Analysis of feedback systems with structured uncertainties," *IEE Proceedings, Part D,* 1982, vol. 129, pp. 242–250.

[8] D. G. Luenberger, *Linear and Nonlinear Programming,* 2d ed., Addison-Wesley, Reading, MA, 1984.

[9] M. G. Safonov, "Stability margins of diagonally perturbed multivariable feedback systems," *IEE Proceedings, Part D,* 1982, vol. 129, pp. 251–256.

[10] J. M. Maciejowski, *Multivariable Feedback Design,* Addison-Wesley, Reading, MA, 1989.

Some additional references on robustness analysis follow:

[11] J. C. Doyle, B. A. Francis, and A. R. Tannenbaum, *Feedback Control Theory,* Macmillan, New York, 1992.

[12] J. C. Doyle, J. E. Wall, and G. Stein, "Performance and robustness analysis for structured uncertainty," *Proceedings of the IEEE Conference on Decision and Control,* 1982, pp. 629–636.

EXERCISES

5.1 Show that a SISO feedback system, which has no pole-zero cancellations between $G(s)$ and $K(s)$, is internally stable if and only if the transfer function $1 + G(s)K(s)$ has no zeros with positive (or infinite) real parts.

5.2 Is the feedback system shown in Figure 5.1 internally stable given the following?

a. $$\mathbf{G}(s) = \begin{bmatrix} \dfrac{1}{s^2+4} \\ \dfrac{s}{s^2+4} \end{bmatrix}; \quad \mathbf{K}(s) = [1 \quad 2].$$

b. $$\mathbf{G}(s) = \begin{bmatrix} \dfrac{11}{s^2+3} & \dfrac{-18}{s(s^2+3)} \\ \dfrac{22}{s(s^2+3)} & \dfrac{12}{s^2+3} \end{bmatrix}; \quad \mathbf{K}(s) = \begin{bmatrix} \dfrac{10(s+1)}{s+4} & 0 \\ 0 & \dfrac{10(s+1)}{s+4} \end{bmatrix}.$$

c. $$\mathbf{G}(s) = \begin{bmatrix} \dfrac{10}{s^2-1} & \dfrac{3}{s(s+3)} \\ \dfrac{10s}{s^2-1} & \dfrac{3}{s+3} \end{bmatrix}; \quad \mathbf{K}(s) = \begin{bmatrix} 1 & 2 \\ 4 & 6 \end{bmatrix}.$$

5.3 Compute the phase and gain margins for the following feedback systems:

a. $$G(s) = \frac{15{,}000}{s(s+50)(s+300)}; \quad K(s) = 50.$$

b. $$G(s) = \frac{15{,}000}{s(s+50)(s+300)}; \quad K(s) = \frac{20s+200}{s+4}.$$

c. $$G(s) = \frac{10}{s^2+16}; \quad K(s) = s + \frac{1}{s} + 1.$$

5.4 Generate block diagrams in standard form for:

a. the input multiplicative uncertainty in Figure 5.7b.
b. the output multiplicative uncertainty in Figure 5.7c.

c. the input feedback uncertainty in Figure 5.7d.
d. the output feedback uncertainty in Figure 5.7e.

5.5 You are given a plant whose transfer function is uncertain. The nominal plant is given by the transfer function

$$G(s) = \frac{500}{s(s+50)}.$$

The plant's transfer function is well known (the gain is known to within 0.1%) at low frequencies (frequencies below 50 rad/sec) and becomes increasingly uncertain at high frequencies until at very high frequencies (frequencies above 5000 rad/sec), the transfer function is only known to lie within 90% of the nominal gain. Model this system as a nominal plant plus an input multiplicative perturbation.

a. Generate a frequency-dependent bound for the perturbation.
b. Generate a first-order rational approximation of this bound.
c. Generate a second-order rational approximation of this bound. Is there a significant advantage to using this second-order bound?
d. Generate a block diagram in standard form for this plant and feedback controller.

5.6 Show that the transfer functions $\mathbf{G}_{e_1u_1}(s)$, $\mathbf{G}_{e_2u_1}(s)$, and $\mathbf{G}_{e_2u_2}(s)$ in (5.9) are stable if $\|\boldsymbol{\Delta}(s)\|_\infty \le 1$ and $\|\mathbf{N}_{y_dw_d}(s)\|_\infty < 1$.

5.7 You are given the nominal plant:

$$G(s) = \frac{10}{s^2+4},$$

with an unstructured input feedback uncertainty $\|\boldsymbol{\Delta}_{fi}(s)\|_\infty \le 0.5$, and the controller

$$K(s) = \frac{4(s+2)}{s+8},$$

connected via negative feedback. Determine if this system is robustly stable.

5.8 You are given the nominal plant and controller

$$\mathbf{G}(s) = \begin{bmatrix} \dfrac{10}{s^2+4} \\ \dfrac{10s}{s^2+4} \end{bmatrix}; \mathbf{K}(s) = [4 \quad 4],$$

connected via negative feedback. Assuming the unstructured perturbation is bounded by $\|\boldsymbol{\Delta}(s)\|_\infty \le 0.8$, determine the following.

a. Is this system robustly stable when the uncertainty is input multiplicative?
b. Is this system robustly stable when the uncertainty is output multiplicative?
c. Compare the results of parts *a* and *b* and comment.

5.9 You are given the plant

$$\dot{x}(t) = \begin{bmatrix} 0 & 1 \\ -g & -2 \end{bmatrix} x(t) + \begin{bmatrix} 0 \\ 1 \end{bmatrix} u(t)\,;\, y(t) = \begin{bmatrix} 1 & 0 \\ 0 & 1 \end{bmatrix} x(t),$$

with an output multiplicative uncertainty and an uncertain parameter g. These perturbations are bounded as follows:

$$g \in [1, 7]\,;\, \|\boldsymbol{\Delta}_o\|_\infty \le 0.5.$$

a. Draw a block diagram of the plant in standard form.
b. Generate a state model of the plant with the inputs u and w_d and the outputs y and y_d.

5.10 You are given a system $\mathbf{N}(j\omega)$ in standard form with an uncertainty block

$$\boldsymbol{\Delta}(j\omega) = \begin{bmatrix} \Delta_1(j\omega) & 0 & 0 \\ 0 & \Delta_2(j\omega) & 0 \\ 0 & 0 & \Delta_2(j\omega) \end{bmatrix},$$

where $\Delta_1(j\omega)$ and $\Delta_2(j\omega)$ are scalar complex perturbations and $\Delta_2(j\omega)$ is repeated. Show that the structured singular value is bounded:

$$\mu[\mathbf{N}(j\omega)] \leq \min_{d_1, \mathfrak{D}_2} \bar{\sigma}[\mathfrak{D}(j\omega)\mathbf{M}(j\omega)\mathfrak{D}^{-1}(j\omega)],$$

when

$$\mathfrak{D}(j\omega) = \begin{bmatrix} d_1 & \mathbf{0} \\ \mathbf{0} & \mathfrak{D}_2 \end{bmatrix}$$

and $\mathfrak{D}_2 \in \mathscr{C}^{2\times 2}$. Note that, in general, the bound generated above is tighter than the bound generated using diagonal scaling with all scalar blocks. This is apparent because the diagonal scaling matrices with scalar blocks are a subset of the given scaling matrices.

5.11 Given

$$\mathbf{N} = \begin{bmatrix} 1 & 2 & 3 \\ 2 & 1 & 1 \\ 2 & 2 & 2 \end{bmatrix}; \boldsymbol{\Delta} = \begin{bmatrix} \Delta_{11} & 0 & 0 \\ \Delta_{21} & 0 & 0 \\ 0 & \Delta_{32} & \Delta_{33} \end{bmatrix},$$

generate $\mathbf{N}_s$, $\boldsymbol{\Delta}_s$, and $\rho[\mathbf{N}_s]$.

5.12 Show that the structured singular value is bounded from below:

$$\mu_{\bar{\Delta}}[\mathbf{N}] \geq \rho[\mathbf{N}_s\mathbf{U}],$$

where $\mathbf{U} \in \bar{\boldsymbol{\Delta}}_s$ and $\mathbf{U}^\dagger\mathbf{U} = \mathbf{I}$.

5.13 You are given the plant and the unity feedback controller shown in Figure P5.1. The uncertain gain is bounded $g \in [15, 25]$ and the performance specification is $|H(j\omega)| \leq 0.1$ for frequencies less than 0.5 rad/sec. The closed-loop transfer function is between the reference input r and the error e.

a. Put this system in standard form and compute the SSV to demonstrate that the system meets the robust performance objectives.
b. Show by directly computing the closed-loop transfer function $E(s)/R(s)$ as a function of the perturbation that the system is robustly stable and meets the performance specifications for all perturbations.

5.14 You are given the system shown in Figure P5.2.

a. Generate the transfer function matrix $\mathbf{N}(s)$ used in computing performance robustness for this system.

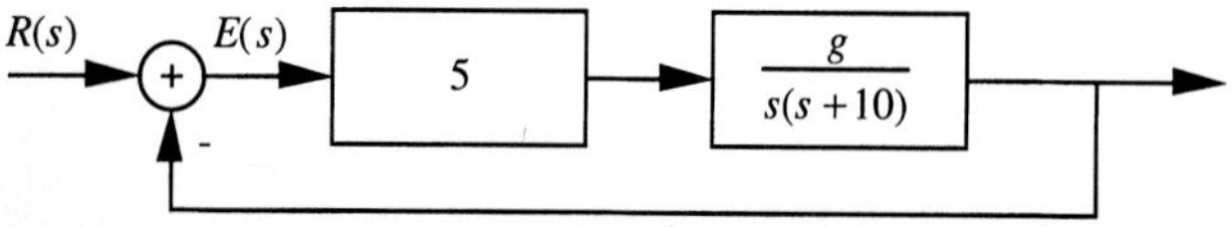

FIGURE P5.1 Unity feedback control of a servomotor

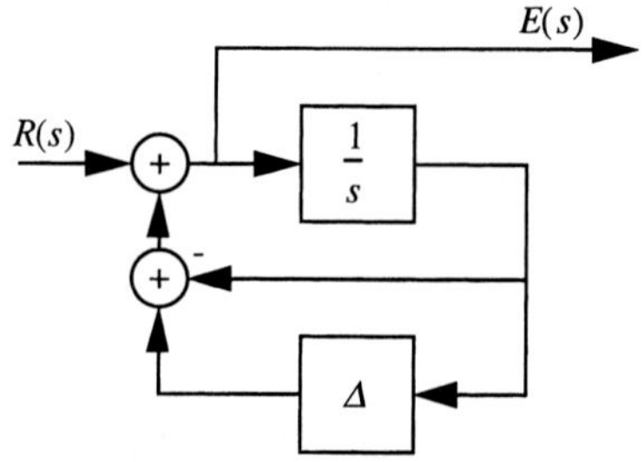

FIGURE P5.2 Control system with uncertain feedback gain.

b. Using the definition of the structured singular value, analytically compute the SSV of $\mathbf{N}(j\omega)$.
c. What is the maximum of this SSV over all frequencies?

COMPUTER EXERCISES

5.1 Rotorcraft Autopilots with Unstructured Uncertainty
The robustness of systems for controlling the pitch and yaw of a rotorcraft will be evaluated. The available sensors consist of gyros for measuring pitch and yaw, and rate gyros for measuring pitch rate and yaw rate (the entire state). The control inputs consist of vanes that supply torques in the pitch and yaw axes. Disturbances consist of torques in both pitch and yaw (measurement noise is ignored). A model of the plant follows:

$$\begin{bmatrix} \dot{\theta} \\ \ddot{\theta} \\ \dot{\Psi} \\ \ddot{\Psi} \end{bmatrix} = \begin{bmatrix} 0 & 1 & 0 & 1 \\ 0 & 0 & 0 & -7 \\ 0 & 0 & 0 & 1 \\ 0 & 7 & 0 & 0 \end{bmatrix} \begin{bmatrix} \theta \\ \dot{\theta} \\ \Psi \\ \dot{\Psi} \end{bmatrix} + \begin{bmatrix} 0 & 0 \\ -15 & 0 \\ 0 & 0 \\ 0 & -15 \end{bmatrix} \begin{bmatrix} \delta_\theta \\ \delta_\Psi \end{bmatrix} + \begin{bmatrix} 0 & 0 \\ 1 & 0 \\ 0 & 0 \\ 0 & 1 \end{bmatrix} \begin{bmatrix} w_\theta \\ w_\Psi \end{bmatrix},$$

where θ is the pitch angle, ψ is the yaw angle, δ_θ is the pitch vane deflection, δ_ψ is the yaw vane deflection, w_θ is the disturbance torque in pitch, and w_ψ is the disturbance torque in yaw. The outputs of interest are the pitch and the yaw.

a. Design a state feedback controller to drive the pitch and yaw to zero by placing the closed-loop poles at the following locations:

i. $-2 \pm 4j;\ -2 \pm 4j.$ **ii.** $-1 \pm 2j;\ -1 \pm 2j.$

b. Assume an additive unstructured uncertainty (between the control input and the states), with

$$\|\Delta_a\|_\infty \leq 0.25.$$

Are the systems in parts *i* and *ii* robustly stable? Why? For what maximum value of $\Delta_{\max}$ is each system robustly stable? For the system in part *i*, what is a destabilizing perturbation that falls within the given set?

c. Assume an input multiplicative unstructured uncertainty, with

$$\|\Delta_i\|_\infty \leq 0.25.$$

Are the systems in parts *i* and *ii* robustly stable? Why? For what maximum value of $\Delta_{\max}$ is each system robustly stable? What guarantees can be made on "gain" and "phase" margins for these systems based on these results. For the system in part *ii*, what is a destabilizing perturbation that falls within the given set? Comment on the effect of the changes in pole locations on robustness.

5.2 Blood Gas Controller for Extracorporeal Support
The robustness of a system for controlling the partial pressure of oxygen and the partial pressure of carbon dioxide in the arterial blood (the blood exiting a membrane oxygenator) is evaluated. The membrane oxygenator is used for extracorporeal support; that is, it provides external lung function during open heart surgery. The available measurements consist of the oxygen partial pressure and the carbon dioxide partial pressure of the arterial blood. The control inputs consist of the commanded flow rates (valve commands) of the sweep gases, both oxygen and air. The dynamics of the oxygenator are given as follows:

$$\begin{bmatrix} \dot{x}_1 \\ \dot{x}_2 \\ \dot{x}_3 \\ \dot{x}_4 \end{bmatrix} = \begin{bmatrix} -10 & 0 & 0 & 0 \\ 0 & -10 & 0 & 0 \\ k_1 & -k_2 & -5 & 0 \\ 0 & k_3 & 0 & -5 \end{bmatrix} \begin{bmatrix} x_1 \\ x_2 \\ x_3 \\ x_4 \end{bmatrix} + \begin{bmatrix} 10 & 0 \\ 0 & 10 \\ 0 & 0 \\ 0 & 0 \end{bmatrix} \begin{bmatrix} u_1 \\ u_2 \end{bmatrix},$$

where x_1 is the flow rate of oxygen, x_2 is the flow rate of carbon dioxide, x_3 is the arterial partial pressure of oxygen, x_4 is the arterial partial pressure of carbon dioxide, u_1 is the commanded oxygen flow rate, and u_2 is the commanded carbon dioxide flow rate. The outputs of interest are x_3 and x_4. The values of k_1, k_2, and k_3 are uncertain:

$$k_1 \in [4,\ 8]\ ;\ k_2 \in [2,\ 4]\ ;\ k_3 \in [0.1,\ 0.9].$$

This system is generated by linearization around a nominal operating point, so all state and input variables are the errors between the actual value and the value at the operating point.

A proportional-plus-integral controller for the nominal system is designed to drive the partial pressures of oxygen and carbon dioxide to the desired values $[r_1\ r_2]^T$. The equation for this controller is

$$\begin{bmatrix} u_1(t) \\ u_2(t) \end{bmatrix} = \begin{bmatrix} 2[r_1(t) - x_3(t)] + 10 \int_0^t [r_1(\tau) - x_3(\tau)]d\tau \\ 2[r_2(t) - x_4(t)] + 10 \int_0^t [r_2(\tau) - x_4(\tau)]d\tau \end{bmatrix}.$$

a. Is this system robustly stable? For what maximum value of the uncertainties, keeping the ratios of the maximum perturbations constant, is this system robustly stable?

b. Assuming the following performance requirement,

$$\bar{\sigma}[\mathbf{H}(j\omega)] \leq 0.2 \quad \text{for } |\omega| \leq 0.1 \text{ rad/sec},$$

where $\mathbf{H}(s)$ is the closed-loop transfer function from the reference inputs r to the errors

$$\begin{bmatrix} e_1 \\ e_2 \end{bmatrix} = \begin{bmatrix} r_1 \\ r_2 \end{bmatrix} - \begin{bmatrix} x_3 \\ x_4 \end{bmatrix},$$

does the nominal system meet the performance criteria?

c. Does the system meet the performance bound for all possible perturbations?

PART 2

$\mathcal{H}_2$ Control

6 The Linear Quadratic Regulator

The cost function, introduced in Chapter 4, provides a measure of the performance of a control system that is dependent on the plant, the test conditions, and the control law. The control system designer is typically charged with the task of developing a control law for a given plant.[1] A comparison of the performance of control laws can then be made after choosing the test conditions. In fact, it makes sense to search for the best control law possible, that is, the control law that yields the smallest possible cost. A control law that minimizes a cost function is termed an optimal control.

An optimal control depends on the selection of the cost function and on the test conditions. A control law may be both optimal and suboptimal with reference to distinct cost functions or distinct test conditions. The suitability of the optimal control system depends on how accurately the cost function and test conditions reflect the design requirements.

Selection of the cost function and test conditions is based on the following two criteria: (1) they should accurately reflect the designer's concept of good performance; (2) the computation of the optimal control should be tractable; that is, an optimal control can be found with a reasonable amount of effort. This second criterion is satisfied by quadratic cost functions and infinity-norm cost functions. Additionally, these cost functions have sufficient flexibility that a good match between the cost function and the specifications can typically be achieved.

This chapter addresses the problem of finding an optimal control law for quadratic cost functions with test conditions consisting of initial conditions and no external inputs. The plant is assumed to be governed by linear state equations, and the measurement consists of the entire state. This particular problem is selected because the resulting optimal control, known as the linear quadratic regulator (LQR), is readily computed and widely used in applications. In addition, the mathematics developed in solving for this optimal control are useful in solving subsequent optimal control problems.

The LQR can be applied to a wide range of control problems due to the flexibility available when selecting quadratic cost functions. The major limitations of the LQR are (1) the test conditions for which this design is optimal are limited to initial conditions; and (2) during implementation, the entire state must be measured. This first restriction

[1] The control engineer also contributes to the design of the plant in many applications. The actuators and sensors are typically chosen by the control engineer. Modifications to the plant design are also recommended in order to simplify the control system design. We assume a given plant, since this is an important subset of control problems. Modifications to sensors, actuators, and the plant are typically suggested on the basis of the results of control system design applied to a given "preliminary" plant.

is primarily moot because the LQR is also the solution of a more general optimal control problem known as the stochastic regulator (see Section 6.4). The test conditions for the stochastic regulator include random initial conditions and white noise disturbance inputs. While the second restriction is valid, the linear quadratic regulator forms a component within the linear quadratic Gaussian (LQG) controller. The LQG controller is the solution of an optimal control problem where the test conditions include random initial conditions and white noise disturbance inputs, and the measurements are noisy linear combinations of the states. Therefore, the study of the LQR is warranted both by its utility as a controller, and by its utility as a component within the more general framework of LQG controller design.

6.1 Optimization

A brief summary of the mathematics of optimization is given before addressing the derivation of the linear quadratic regulator. Differentiation is the primary tool for optimization in elementary calculus texts. Therefore, it is reasonable to assume that optimal controls can be found by taking the derivative of the cost function with respect to the control input. This approach is reasonable, but requires the differentiation of a real scalar cost function with respect to the control input, which is a function of time (this is not covered in elementary calculus). Optimization in this case can be accomplished by using a generalization of the differential called the variation.

6.1.1 Variations

The real scalar function of a scalar $J(x)$ has a local minimum at x^* if and only if

$$J(x^* + \delta x) \geq J(x^*) \tag{6.1}$$

for all δx sufficiently small; that is, the magnitude of δx is less than some positive number ε. An equivalent statement is that

$$\Delta J(x^*, \delta x) = J(x^* + \delta x) - J(x^*) \geq 0$$

for all δx sufficiently small. The term $\Delta J(x^*, \delta x)$ is called the increment of J. Expanding $J(x^* + \delta x)$ in a Taylor series around the point x^*, the optimality condition (6.1) can be written

$$\begin{aligned} \Delta J(x^*, \delta x) &= J(x^* + \delta x) - J(x^*) \\ &= \frac{dJ(x^*)}{dx}\delta x + \frac{d^2 J(x^*)}{dx^2}\delta x^2 + H.O.T. \geq 0, \end{aligned} \tag{6.2}$$

where $H.O.T.$ stands for higher-order terms. Note that the term in $\Delta J(x^*, \delta x)$ that is linear in δx is the differential of J, and the coefficient multiplying δx, in this term, is the derivative of J. When dealing with a functional (a real scalar function of functions), δx is called the variation of x, and the term in the increment that is linear in δx is called the variation of J and is denoted $\delta J(x^*, \delta x)$. The variation of J is a generalization of the differential and can be applied to the optimization of a functional.

Equation (6.2) can be used to develop necessary conditions for optimality. In the limit as δx approaches zero, the terms δx^2, δx^3, and so on become arbitrarily small compared to δx. A necessary condition for x^* to be a local minimum is then as follows: The variation of J is zero at x^* for all δx.

EXAMPLE 6.1 Consider the following function:

$$J(x) = x^2 + 6x + 8.$$

The increment of J is

$$\begin{aligned}\Delta J(x, \delta x) &= (x + \delta x)^2 + 6(x + \delta x) + 8 - (x^2 + 6x + 8)\\ &= x^2 + 2x\delta x + \delta x^2 + 6x + 6\delta x + 8 - x^2 - 6x - 8\\ &= (2x + 6)\delta x + \delta x^2.\end{aligned}$$

A necessary condition for x to be a minimum is that the variation of $J(x)$ equals zero:

$$\delta J(x, \delta x) = (2x + 6)\delta x = 0$$

for all δx. This requires that

$$2x + 6 = 0 \Rightarrow x = -3.$$

Note that $(2x + 6)$ equals the derivative of J.

The process of optimization using variations is more cumbersome than differentiation for a scalar function of a scalar, but has the advantage that it can be directly extended to the optimization of a functional.

6.1.2 Lagrange Multipliers

The optimal control problem is a constrained minimization problem; that is, minimization of the cost function is subject to constraints on the state and the control. The necessary condition for optimality, given above, is only applicable to unconstrained minimization problems. Lagrange multipliers provide a method of converting a constrained minimization problem into an unconstrained minimization problem of higher order. Optimization can then be performed using the above necessary condition.

Consider the problem of minimizing $J(x)$, where x is a vector, such that

$$\boxed{c(x) = 0.} \tag{6.3}$$

Equation (6.3) specifies a surface in the space of x. Necessary conditions for optimality of J at a point x^* are that x^* satisfies (6.3) and that the directional derivative of J at x^* equals zero in all directions along the surface. This second condition is satisfied if the gradient of J is normal to the surface at x^*. Note that the gradient of $c(x)$ is normal to the surface at all points, including x^*. Therefore, the second condition is satisfied if the gradient of J is parallel to the gradient of $c(x)$ at x^*, or equivalently,

$$\frac{\delta J(x^*)}{\delta x} + p\frac{\delta c(x^*)}{\delta x} = 0 \tag{6.4}$$

for some scalar p. Equations (6.3) and (6.4) form a set of necessary conditions for a solution of the constrained optimization problem. The necessary conditions for optimality, (6.3) and (6.4), can be generated as the solution to an unconstrained optimization problem with the following cost function:

$$\boxed{J_a(x,\, p) = J(x) + pc(x).} \tag{6.5}$$

Taking the gradient of J_a with respect to x yields (6.4), and taking the derivative of J_a with respect to p yields (6.3). The parameter p is called a Lagrange multiplier. The procedure of solving the constrained optimization problem for J by solving the unconstrained optimization problem for J_a is called the method of Lagrange multipliers. This method is also applicable to the optimal control problem, which involves the constrained minimization of a functional.

EXAMPLE 6.2 We are given the cost function and the constraint

$$J(x,\, y) = x^2 + y^2;$$

$$c(x,\, y) = 2x + y + 4 = 0.$$

The augmented cost function is

$$J_a(x,\, y,\, p) = x^2 + y^2 + p(2x + y + 4).$$

The increment of the augmented cost function is:

$$\begin{aligned}
\Delta J_a(x,\, y,\, p,\, \delta x,\, \delta y,\, \delta p) &= J_a(x + \delta x,\, y + \delta y,\, p + \delta p) - J_a(x,\, y,\, p) \\
&= (x + \delta x)^2 + (y + \delta y)^2 \\
&\quad + (p + \delta p)\{2(x + \delta x) + (y + \delta y) + 4\} \\
&\quad - x^2 - y^2 - p\{2x + y + 4\} \\
&= (2x + 2p)\delta x + (2y + p)\delta y + (2x + y + 4)\delta p \\
&\quad + \delta x^2 + \delta y^2 + 2\delta p \delta x + \delta p \delta y.
\end{aligned}$$

A necessary condition for optimality is that the variation of the augmented cost function, which consists of the linear terms in the increment, is zero for all δx, δy, and δp:

$$\delta J_a(x,\, y,\, p,\, \delta x,\, \delta y,\, \delta p) = (2x + 2p)\delta x + (2y + p)\delta y + (2x + y + 4)\delta p = 0.$$

Therefore, the coefficients that multiply δx, δy, and δp must all equal zero:

$$\frac{\delta J_a(x,\, y,\, p)}{\delta x} = 2x + 2p = 0;$$

$$\frac{\delta J_a(x,\, y,\, p)}{\delta y} = 2y + p = 0;$$

$$\frac{\delta J_a(x,\, y,\, p)}{\delta p} = 2x + y + 4 = 0.$$

Note that the coefficients of δx, δy, and δp are the partial derivatives of the augmented cost function. Solving for the values of x, y, and p that satisfy these necessary conditions yields

$$\begin{bmatrix} x \\ y \\ p \end{bmatrix} = \begin{bmatrix} -1.6 \\ -0.8 \\ 1.6 \end{bmatrix}.$$

These values of x and y correspond to the global solution of the constrained minimization problem.[2] ◆

The Lagrange multiplier method of constrained optimization is quite general. Additional constraint equations can be appended to the cost function with additional Lagrange multipliers, provided that the constraint equations are consistent; that is, there is a solution that satisfies all of the constraint equations. The Lagrange multiplier method will be applied to solving the following optimal control problem.

6.2 The Linear Quadratic Regulator

The linear quadratic regulator (LQR) is an optimal control problem where the state equation of the plant is linear, the cost function is quadratic, and the test conditions consist of initial conditions on the state and no disturbance inputs. The plant state equation is

$$\boxed{\dot{x}(t) = \mathbf{A}x(t) + \mathbf{B}_u u(t).} \tag{6.6a}$$

The reference output is[3]

$$\boxed{y(t) = \mathbf{C}_y x(t).} \tag{6.6b}$$

The cost function is

$$\boxed{J(x(t),\, u(t)) = \frac{1}{2} y^T(t_f)\mathbf{H}_y y(t_f) + \frac{1}{2}\int_0^{t_f} \{y^T(t)\mathbf{Q}_y y(t) + u^T(t)\mathbf{R}u(t)\}dt,} \tag{6.7a}$$

[2] The determination of whether or not this point is a local minimum, a local maximum, or a saddle point can be made by considering the second variation, that is, the term in the increment that involves δx^2. The fact that the minimum is global can be ascertained by considering the properties of the cost function and the constraint. In this case, the cost function is quadratic and has a single global minimum for any linear or affine constraint.

[3] A more general reference output, which includes the control, can be used in LQR plant model. This more restrictive reference output equation is used to match the majority of the literature, and to simplify the subsequent discussion of tracking systems (presented in Chapter 8).

where $\mathbf{H}_y$, $\mathbf{Q}_y$, and $\mathbf{R}$ are positive definite. The cost is evaluated subject to the initial condition

$$\boxed{x(0) = x_0,}$$

and no disturbance input. The cost function (6.7a) is frequently written directly in terms of the state and the control:

$$\boxed{J(x(t), u(t)) = \frac{1}{2}x^T(t_f)\mathbf{H}x(t_f) + \frac{1}{2}\int_0^{t_f} \{x^T(t)\mathbf{Q}x(t) + u^T(t)\mathbf{R}u(t)\}dt,} \tag{6.7b}$$

where

$$\mathbf{H} = \mathbf{C}_y^T\mathbf{H}_y\mathbf{C}_y \;;\; \mathbf{Q} = \mathbf{C}_y^T\mathbf{Q}_y\mathbf{C}_y\,.$$

Note that $\mathbf{H}$ and $\mathbf{Q}$ are positive semidefinite. This optimal control problem is a constrained optimization problem, with the cost being a functional of both $u(t)$ and $x(t)$ and the state equation providing a family of constraint equations. It is assumed that there are no other constraints on the state or control.

6.2.1 The Hamiltonian Equations

The optimal control problem, posed in (6.6) and (6.7), can be converted to an unconstrained optimization problem of higher dimension by the application of Lagrange multipliers. An augmented cost function is constructed by adding a constant times each of the constraints to the cost function. The state equation represents a family of constraint equations. These constraints can be appended to the cost function by the addition of an integral:

$$J_a(x(t), u(t), p(t)) = J(x(t), u(t)) + \int_0^{t_f} p^T(t)\{\mathbf{A}x(t) + \mathbf{B}_u u(t) - \dot{x}(t)\}dt,$$

where $p(t)$ is the family of Lagrange multipliers, one at each point in time in the interval from 0 to t_f. The augmented cost is then

$$\begin{aligned} &J_a(x(t), u(t), p(t)) \\ &= \frac{1}{2}x^T(t_f)\mathbf{H}x(t_f) \\ &+ \int_0^{t_f}\left\{\frac{1}{2}x^T(t)\mathbf{Q}x(t) + \frac{1}{2}u^T(t)\mathbf{R}u(t) + p^T(t)[\mathbf{A}x(t) + \mathbf{B}_u u(t) - \dot{x}(t)]\right\}dt. \end{aligned}$$

This augmented cost function is a function of the state $x(t)$, the control $u(t)$, and the Lagrange multiplier $p(t)$. The Lagrange multiplier is often referred to as the "costate" in optimal control applications.

The optimal control is found by forming the increment of J_a with respect to the state, the control, and the costate:

$$
\begin{aligned}
\Delta J_a(x, u, p, \delta x, \delta u, \delta p) \\
&= J_a(x + \delta x, u + \delta u, p + \delta p) - J_a(x, u, p) \\
&= \frac{1}{2}[x(t_f) + \delta x(t_f)]^T \mathbf{H}[x(t_f) + \delta x(t_f)] \\
&+ \int_0^{t_f} \left\{ \frac{1}{2}(x + \delta x)^T \mathbf{Q}(x + \delta x) + \frac{1}{2}(u + \delta u)^T \mathbf{R}(u + \delta u) \right. \\
&\qquad \left. + (p + \delta p)^T \{\mathbf{A}(x + \delta x) + \mathbf{B}_u(u + \delta u) - (\dot{x} + \delta \dot{x})\} \right\} dt \\
&- \frac{1}{2} x^T(t_f)\mathbf{H}x(t_f) - \int_0^{t_f} \left\{ \frac{1}{2} x^T \mathbf{Q} x + \frac{1}{2} u^T \mathbf{R} u + p^T(\mathbf{A}x + \mathbf{B}_u u - \dot{x}) \right\} dt.
\end{aligned}
$$

The time index has been deleted, except at the final time, to simplify the notation. Note that the variation of $\dot{x}(t)$ results from taking the derivative of the variation of $x(t)$. Expanding this expression and grouping terms yields[4]

$$
\begin{aligned}
\Delta J_a = &\frac{1}{2} \delta x^T(t_f)\mathbf{H}\delta x(t_f) \\
&+ \int_0^{t_f} \left\{ \frac{1}{2} \delta x^T \mathbf{Q} \delta x + \frac{1}{2} \delta u^T \mathbf{R} \delta u + \delta p^T(\mathbf{A}\delta x + \mathbf{B}_u \delta u - \delta \dot{x}) \right\} dt \\
&+ x^T(t_f)\mathbf{H}\delta x(t_f) + \int_0^{t_f} \{x^T \mathbf{Q}\delta x + u^T \mathbf{R} \delta u + \delta p^T(\mathbf{A}x + \mathbf{B}_u u - \dot{x}) \\
&\qquad + p^T(\mathbf{A}\delta x + \mathbf{B}_u \delta u - \delta \dot{x})\} dt.
\end{aligned}
$$

A necessary condition for the trajectory $x(t)$, $p(t)$, and $u(t)$ to be a minimum is that the variation of J_a equals zero:

$$
\begin{aligned}
\delta J_a(x, u, p, \delta x, \delta u, \delta p) = &\, x^T(t_f)\mathbf{H}\delta x(t_f) \\
&+ \int_0^{t_f} \{(x^T \mathbf{Q} + p^T \mathbf{A})\delta x + (u^T \mathbf{R} + p^T \mathbf{B}_u)\delta u \\
&\qquad + \delta p^T(\mathbf{A}x + \mathbf{B}_u u - \dot{x}) - p^T \delta \dot{x}\} dt = 0.
\end{aligned}
\tag{6.8}
$$

The last term in this integral involves $\delta \dot{x}(t)$, which is a function of $\delta x(t)$ and therefore not an independent variable. The term $\delta \dot{x}(t)$ can be eliminated from (6.8) using integration by parts:

$$
\int_0^{t_f} p^T(t)\delta \dot{x}(t) dt = p^T(t_f)\delta x(t_f) - p^T(0)\delta x(0) - \int_0^{t_f} \dot{p}^T(t)\delta x(t) dt. \tag{6.9}
$$

---

[4]Note that

$$
x^T(t)\mathbf{Q}\delta x(t) = (x^T(t)\mathbf{Q}\delta x(t))^T = \delta x^T(t)\mathbf{Q}^T x(t) = \delta x^T(t)\mathbf{Q}x(t),
$$

since $\mathbf{Q}$ is symmetric. This result also applies to the similar products involving $x(t_f)$ and $u(t)$.

The initial condition on the state is fixed, so $\delta x(0) = 0$. Using this fact, substituting (6.9) into (6.8), and grouping terms yields the necessary condition for optimality:

$$\begin{aligned}\delta J(x, u, p, \delta x, \delta u, \delta p) &= \{x^T(t_f)\mathbf{H} - p^T(t_f)\}\delta x(t_f) \\ &+ \int_0^{t_f} \{(x^T\mathbf{Q} + p^T\mathbf{A} + \dot{p}^T)\delta x + (u^T\mathbf{R} + p^T\mathbf{B}_u)\delta u \\ &+ \delta p^T(\mathbf{A}x + \mathbf{B}_u u - \dot{x})\}dt = 0.\end{aligned}$$

Since the variations $\delta x(t_f)$, δx, δu, and δp are all arbitrary, the only way this expression can equal zero is if

$$p^T(t_f) = x^T(t_f)\mathbf{H}; \tag{6.10}$$

$$\dot{p}^T(t) = -x^T(t)\mathbf{Q} - p^T(t)\mathbf{A}; \tag{6.11a}$$

$$0 = u^T(t)\mathbf{R} + p^T(t)\mathbf{B}_u; \tag{6.11b}$$

$$\dot{x}(t) = \mathbf{A}x(t) + \mathbf{B}_u u(t). \tag{6.11c}$$

Solving for the optimal control $u(t)$ in (6.11b) yields

$$u(t) = -\mathbf{R}^{-1}\mathbf{B}_u^T p(t), \tag{6.12}$$

where the inverse is guaranteed to exist since $\mathbf{R}$ is positive definite. Eliminating $u(t)$ from (6.11a) and (6.11c), and combining the resulting equations into a single state equation yields

$$\begin{bmatrix} \dot{x}(t) \\ \hline \dot{p}(t) \end{bmatrix} = \left[\begin{array}{c|c} \mathbf{A} & -\mathbf{B}_u\mathbf{R}^{-1}\mathbf{B}_u^T \\ \hline -\mathbf{Q} & -\mathbf{A}^T \end{array}\right] \begin{bmatrix} x(t) \\ \hline p(t) \end{bmatrix} = \mathfrak{X} \begin{bmatrix} x(t) \\ \hline p(t) \end{bmatrix}. \tag{6.13}$$

This system is referred to as the Hamiltonian system, and the state matrix $\mathfrak{X}$ is called the Hamiltonian. The Hamiltonian system (along with the initial and final values) represents a set of necessary conditions for the control to minimize the cost function. These equations are also sufficient; that is, the solution of the Hamiltonian system is the unique control that minimizes the cost function.[5]

The optimal control depends on the costate, which can be found by solving the homogeneous state equation (6.13) subject to the initial and final conditions:

$$x(0) = x_0; \tag{6.14a}$$

$$p(t_f) = \mathbf{H}x(t_f). \tag{6.14b}$$

The general solution of the state equation (6.13), at the final time, given initial conditions at time t is

$$\begin{bmatrix} x(t_f) \\ \hline p(t_f) \end{bmatrix} = e^{\mathfrak{X}(t_f - t)} \begin{bmatrix} x(t) \\ \hline p(t) \end{bmatrix} = \left[\begin{array}{c|c} \mathbf{\Phi}_{11}(t_f - t) & \mathbf{\Phi}_{12}(t_f - t) \\ \hline \mathbf{\Phi}_{21}(t_f - t) & \mathbf{\Phi}_{22}(t_f - t) \end{array}\right] \begin{bmatrix} x(t) \\ \hline p(t) \end{bmatrix}.$$

[5] The uniqueness of the optimal control results from the fact that the cost function is quadratic. Quadratic functions have only a single critical point, which is either a global maximum or minimum. The fact that this critical point yields a global minimum results from the positivity requirements on the weighting matrices.

Note that each submatrix in the state-transition matrix is computed from the entire Hamiltonian matrix in general. Substituting the final condition (6.14b) into this equation yields

$$\begin{bmatrix} x(t_f) \\ \hline \mathbf{H}x(t_f) \end{bmatrix} = \begin{bmatrix} \mathbf{\Phi}_{11}(t_f - t) & \mathbf{\Phi}_{12}(t_f - t) \\ \hline \mathbf{\Phi}_{21}(t_f - t) & \mathbf{\Phi}_{22}(t_f - t) \end{bmatrix} \begin{bmatrix} x(t) \\ \hline p(t) \end{bmatrix}, \tag{6.15}$$

where the unknowns are $x(t)$, $p(t)$, and $x(t_f)$. Eliminating $x(t_f)$ from these equations by substituting the first equation within (6.15) into the second, we have

$$\mathbf{H}\{\mathbf{\Phi}_{11}(t_f - t)x(t) + \mathbf{\Phi}_{12}(t_f - t)p(t)\} = \mathbf{\Phi}_{21}(t_f - t)x(t) + \mathbf{\Phi}_{22}(t_f - t)p(t).$$

The costate can then be found from the state:

$$\begin{aligned} p(t) &= \{\mathbf{\Phi}_{22}(t_f - t) - \mathbf{H}\mathbf{\Phi}_{12}(t_f - t)\}^{-1}\{\mathbf{H}\mathbf{\Phi}_{11}(t_f - t) - \mathbf{\Phi}_{21}(t_f - t)\}x(t) \\ &= \mathbf{P}(t)x(t) \end{aligned} \tag{6.16}$$

where $\mathbf{P}(t)$ is the matrix of proportionality between the costate and the state. This matrix is fully specified by the state-transition matrix of the Hamiltonian system, since the inverse in (6.16) exists at all times between the initial time and final time [1]. The optimal control is found from (6.12):

$$\boxed{u(t) = -\mathbf{R}^{-1}\mathbf{B}_u^T\mathbf{P}(t)x(t) = -\mathbf{K}(t)x(t),} \tag{6.17}$$

where $\mathbf{K}(t)$ is called the optimal feedback gain matrix. The optimal control is linear, time-varying state feedback, in general, where the optimal feedback gain matrix can be found from the state-transition matrix of the Hamiltonian. The optimal control can then be found provided that the entire state is measured perfectly. This assumption is very restrictive, but will be relaxed in subsequent chapters.

EXAMPLE 6.3 We are given the plant

$$\dot{x}(t) = x(t) + u(t)$$

and the cost function

$$J = \frac{1}{2}8x^2(10) + \frac{1}{2}\int_0^{10} \{3x^2(t) + u^2(t)\}dt.$$

The Hamiltonian system is

$$\begin{bmatrix} \dot{x}(t) \\ \dot{p}(t) \end{bmatrix} = \begin{bmatrix} 1 & -1 \\ -3 & -1 \end{bmatrix} \begin{bmatrix} x(t) \\ p(t) \end{bmatrix}.$$

The state-transition matrix of the Hamiltonian system is

$$e^{\mathfrak{X}t} = \begin{bmatrix} \frac{3}{4}e^{2t} + \frac{1}{4}e^{-2t} & -\frac{1}{4}e^{2t} + \frac{1}{4}e^{-2t} \\ -\frac{3}{4}e^{2t} + \frac{3}{4}e^{-2t} & \frac{1}{4}e^{2t} + \frac{3}{4}e^{-2t} \end{bmatrix}.$$

The optimal feedback gain matrix is

$$K(t) = \frac{8\{\frac{3}{4}e^{2(10-t)} + \frac{1}{4}e^{-2(10-t)}\} - \{-\frac{3}{4}e^{2(10-t)} + \frac{3}{4}e^{-2(10-t)}\}}{\{\frac{1}{4}e^{2(10-t)} + \frac{3}{4}e^{-2(10-t)}\} - 8\{-\frac{1}{4}e^{2(10-t)} + \frac{1}{4}e^{-2(10-t)}\}},$$

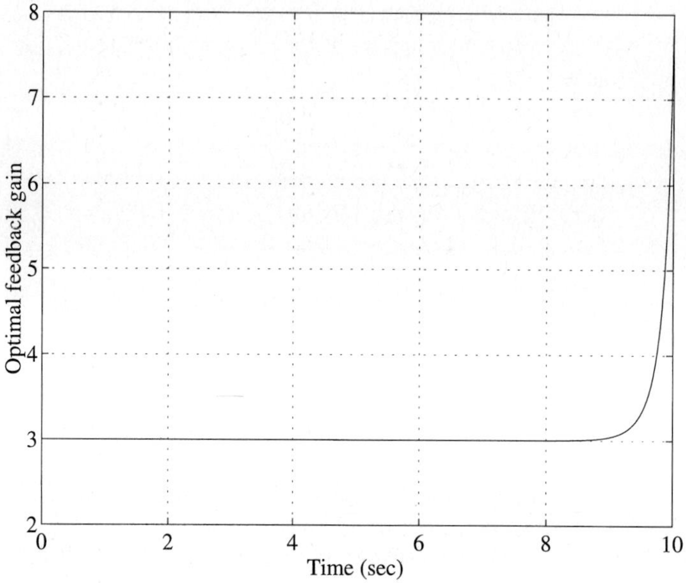

FIGURE 6.1 The time-varying optimal feedback gain for Example 6.3

which can be reduced to

$$K(t) = \frac{27e^{2(10-t)} + 5e^{-2(10-t)}}{9e^{2(10-t)} - 5e^{-2(10-t)}}.$$

A plot of the optimal feedback gain versus time is given in Figure 6.1. Note that this feedback gain experiences a transient near the final time, and maintains a steady-state value at times far from the final time. ◆

6.2.2 The Riccati Equation

The determination of the state-transition matrix for the Hamiltonian system is often a very tedious process. An alternative method of finding the optimal feedback gain matrix utilizes a nonlinear matrix differential equation, known as the Riccati equation. The Riccati equation has only final conditions and can, therefore, be solved backward in time using any numerical integration method.

A linear relationship between the costate and the state is given by (6.16):

$$p(t) = \mathbf{P}(t)x(t). \tag{6.18}$$

The solution of the optimal control problem can be reduced to finding the matrix $\mathbf{P}(t)$, since the optimal control is given:

$$\boxed{u(t) = -\mathbf{R}^{-1}\mathbf{B}_u^T\mathbf{P}(t)x(t) = -\mathbf{K}(t)x(t),} \tag{6.19}$$

and $x(t)$ is assumed to be measured perfectly.

A differential equation for $\mathbf{P}(t)$ can be generated by taking the derivative of (6.18):

$$\dot{p}(t) = \dot{\mathbf{P}}(t)x(t) + \mathbf{P}(t)\dot{x}(t).$$

Substituting for $\dot{p}(t)$ and $\dot{x}(t)$ from (6.13) yields

$$-\mathbf{Q}x(t) - \mathbf{A}^T p(t) = \dot{\mathbf{P}}(t)x(t) + \mathbf{P}(t)\{\mathbf{A}x(t) - \mathbf{B}_u\mathbf{R}^{-1}\mathbf{B}_u^T p(t)\}.$$

Substituting for $p(t)$ using (6.18) and rearranging, we have

$$\{\dot{\mathbf{P}}(t) + \mathbf{P}(t)\mathbf{A} + \mathbf{A}^T\mathbf{P}(t) + \mathbf{Q} - \mathbf{P}(t)\mathbf{B}_u\mathbf{R}^{-1}\mathbf{B}_u^T\mathbf{P}(t)\}x(t) = 0.$$

This equation is valid for an arbitrary state $x(t)$ (resulting from an arbitrary initial condition x_0), which implies that $\mathbf{P}(t)$ must satisfy

$$\boxed{\dot{\mathbf{P}}(t) = -\mathbf{P}(t)\mathbf{A} - \mathbf{A}^T\mathbf{P}(t) - \mathbf{Q} + \mathbf{P}(t)\mathbf{B}_u\mathbf{R}^{-1}\mathbf{B}_u^T\mathbf{P}(t).} \tag{6.20}$$

This differential equation is known as the Riccati equation. The value of $\mathbf{P}(t)$ corresponding to the optimal trajectory is found by solving (6.20) backward in time from the final condition, which is given by (6.14b):[6]

$$\boxed{\mathbf{P}(t_f) = \mathbf{H}.} \tag{6.21}$$

The optimal control is then found by using (6.19). The solution of the Riccati equation yields a unique optimal control that minimizes the cost function.

EXAMPLE 6.4 A servo motor can be described by the following state equation:

$$\begin{bmatrix} \dot{x}_1(t) \\ \dot{x}_2(t) \end{bmatrix} = \begin{bmatrix} 0 & 1 \\ 0 & -1 \end{bmatrix}\begin{bmatrix} x_1(t) \\ x_2(t) \end{bmatrix} + \begin{bmatrix} 0 \\ 1 \end{bmatrix} u(t),$$

where $x_1(t)$ is the angular position of the motor shaft, $x_2(t)$ is the angular velocity of the motor shaft, and the $u(t)$ is the voltage applied to the field coils of the motor. The angular position is measured by a potentiometer, and the angular velocity is measured by a tachometer, so the entire state is accessible. The control system is designed to drive the angular position to zero using the control input $u(t)$. The maximum control voltage is 5 volts. A reasonable cost function should include the angular position and the control input:

$$J(x(t), u(t)) = \frac{1}{2}x^T(t_f)\begin{bmatrix} h & 0 \\ 0 & 0 \end{bmatrix}x(t_f) + \frac{1}{2}\int_0^{t_f} x^T(t)\begin{bmatrix} q & 0 \\ 0 & 0 \end{bmatrix}x(t) + ru^2(t)dt.$$

The parameters h, q, and r specify the weighting on the final angular position, the angular position before the final time, and the control input, respectively. These parameters are varied to observe their influence on the optimal feedback gains and the optimal trajectories.

[6] The Riccati equation can be solved backward in time numerically by using Euler's approximation,

$$\mathbf{P}(t - T) = \mathbf{P}(t) = T\dot{\mathbf{P}}(t),$$

and iterating.

The feedback gains are generated for $t_f = 10$ sec and various values of h, q, and r. In addition, the system was simulated, using these feedback gains, with an initial condition of

$$x(0) = \begin{bmatrix} 1 \\ 1 \end{bmatrix}.$$

The feedback gains, motor shaft angle, and control input are given in Figure 6.2 for $h = 10, q = 0$, and $r = 0.1$; in Figure 6.3 for $h = 0, q = 1$, and $r = 0.1$; in Figure 6.4 for $h = 0, q = 1$, and $r = 1$; and in Figure 6.5 for $h = 0, q = 10$, and $r = 1$. In Figure 6.2, the feedback gains are observed to be initially very small and peak near the final time. This peak in the gains ensures that enough control is used to drive the final output to very near zero. The motor shaft angle is driven to very near zero at the final time. Note that the motor shaft is rotating at the final time. The cost function included a requirement that the motor shaft angle be near zero at the final time but no requirement on the rotation rate of the motor shaft. Adding an appropriate weighting on the final state $x_2(t_f)$ would yield a trajectory where the rotation rate at the final time is near zero. In Figure 6.3, the feedback gains are at steady-state values except near the final time, where they decay to zero. The motor shaft angle is driven and held near zero. The control is initially large and then decreases to zero. The control decays because the motor shaft angle has no tendency to move from zero after both the shaft angle and the angular velocity are near zero. The results in Figure 6.4 are for a cost function where the control

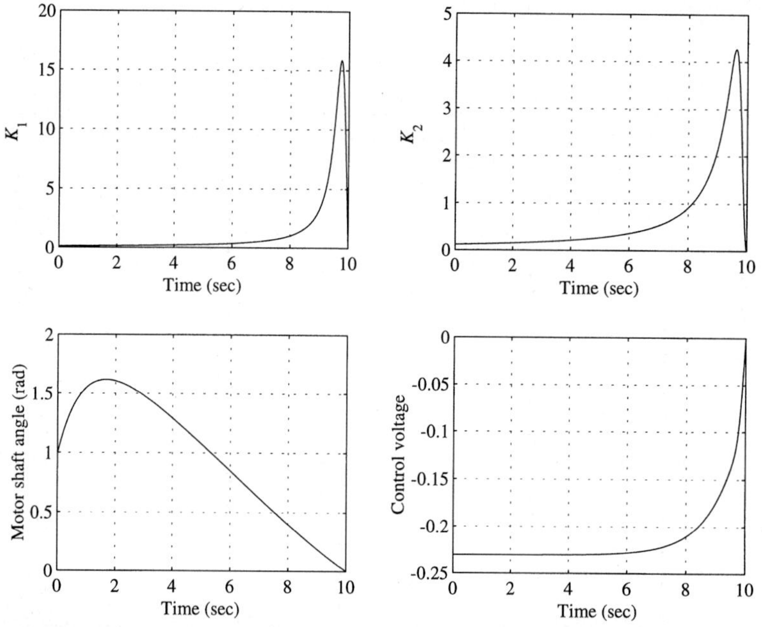

FIGURE 6.2 Results of Example 6.4 with $h = 10$, $q = 0$, and $r = 0.1$

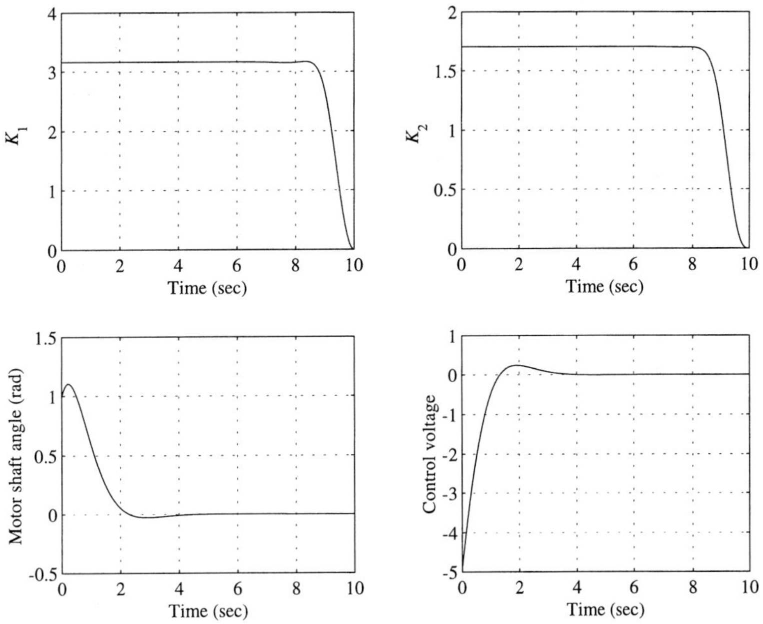

FIGURE 6.3 Results of Example 6.4 with $h = 0$, $q = 1$, and $r = 0.1$

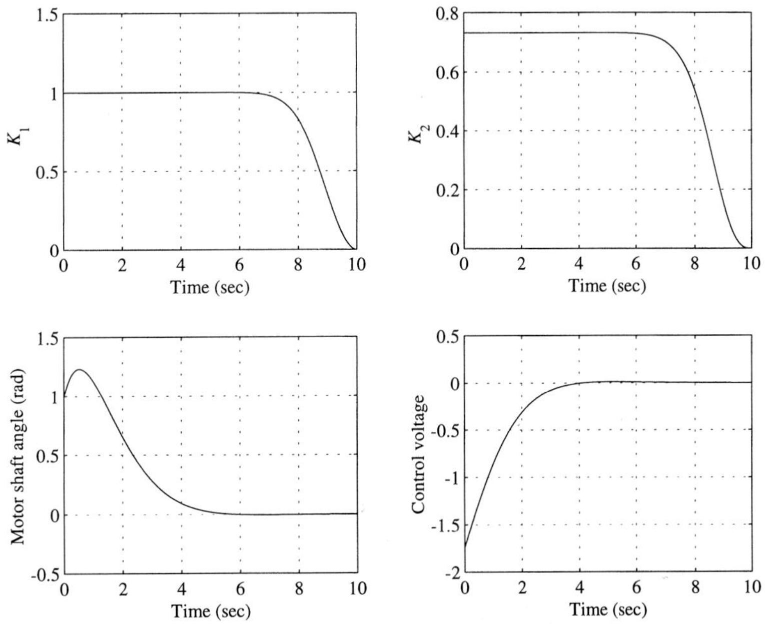

FIGURE 6.4 Results of Example 6.4 with $h = 0$, $q = 1$, and $r = 1$

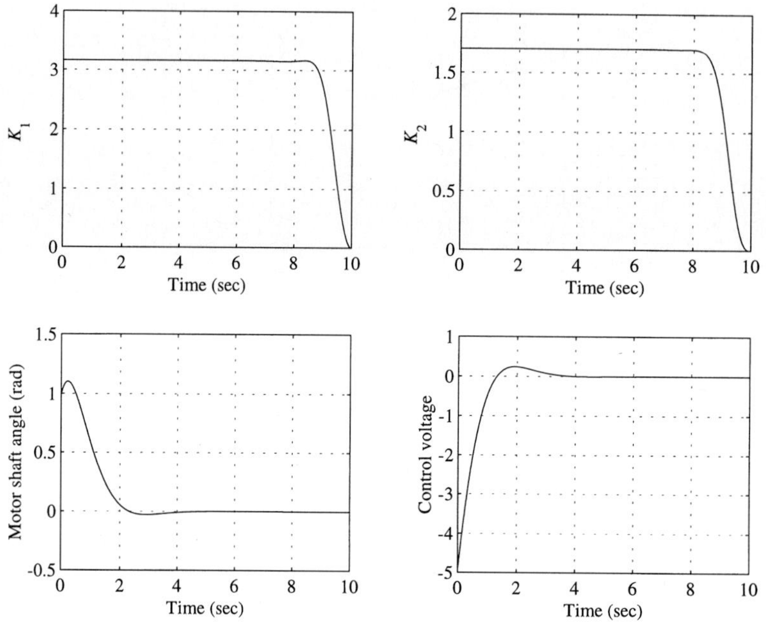

FIGURE 6.5 Results of Example 6.4 with $h = 0$, $q = 10$, and $r = 1$

weighting is increased in comparison to the cost function used to generate the results in Figure 6.3. The effect of the increased control weighting is that the feedback gains become smaller, less control is used, and the motor shaft angle approaches zero more slowly. The results in Figure 6.5 are for a cost function where the control and state weighting are increased by equal amounts in comparison to the cost function used to generate the results in Figure 6.3. The feedback gains, state trajectory, and optimal control are seen to be unchanged. This is reasonable, since multiplying the entire cost function by a positive constant changes the entire cost, but should have no effect on the control that minimizes the cost function. ◆

The Riccati solution is symmetric. This fact is demonstrated by expressing the Riccati solution as the integral of the derivative

$$\mathbf{P}(t) = \int_{t_f}^{t} \dot{\mathbf{P}}(\tau)d\tau + \mathbf{P}(t_f)$$

and considering the symmetry of the terms on the right in this equation. The final condition $\mathbf{P}(t_f) = \mathbf{H}$ is symmetric. The derivative $\dot{\mathbf{P}}(t)$ is also symmetric provided $\mathbf{P}(t)$ is symmetric:

$$\dot{\mathbf{P}}^T(t) = -\mathbf{A}^T\mathbf{P}(t) - \mathbf{P}(t)\mathbf{A} + \mathbf{Q} - \mathbf{P}(t)\mathbf{B}_u\mathbf{R}^{-1}\mathbf{B}_u^T\mathbf{P}(t) = \dot{\mathbf{P}}(t).$$

Therefore, the Riccati solution remains symmetric at all times.

6.2.3 Computation of the Optimal Cost

The Riccati equation was generated by solving for the matrix of proportionality between the state and the costate in the Hamiltonian equations. It turns out that the Riccati solution has significance other than as a tool for computing the optimal feedback gain matrix. The Riccati solution can be used to generate the cost associated with the optimal control:[7]

$$J(x(0)) = \frac{1}{2}x^T(0)\mathbf{P}(0)x(0). \tag{6.22}$$

Note that the optimal cost depends only on the initial state, since the control and state trajectories are specified as those yielding the optimal cost.

The result (6.22) is derived by noting that for the optimal trajectory, the state equation is satisfied at each point in time, since

$$\mathbf{A}x(t) + \mathbf{B}_u u(t) - \dot{x}(t) = 0.$$

The optimal cost is, therefore, unaffected by the addition of any product of terms, which includes the state constraint equation:

$$\begin{aligned} J(x(0)) &= \frac{1}{2}x^T(t_f)\mathbf{H}x(t_f) \\ &+ \int_0^{t_f}\left\{\frac{1}{2}x^T\mathbf{Q}x + \frac{1}{2}u^T\mathbf{R}u + \frac{1}{2}p^T(\mathbf{A}x + \mathbf{B}_u u - \dot{x})\right\}dt. \end{aligned} \tag{6.23}$$

From (6.11a) and (6.11b),

$$\begin{aligned} p^T(t)\mathbf{A} &= -\dot{p}^T(t) - x^T(t)\mathbf{Q}; \\ p^T(t)\mathbf{B}_u &= -u^T(t)\mathbf{R}. \end{aligned}$$

Substituting these equations into (6.23) yields

$$\begin{aligned} J(x(0)) &= \frac{1}{2}x^T(t_f)\mathbf{H}x(t_f) - \frac{1}{2}\int_0^{t_f}\{\dot{p}^Tx + p^T\dot{x}\}dt \\ &= \frac{1}{2}x^T(t_f)\mathbf{H}x(t_f) - \frac{1}{2}\int_0^{t_f}\frac{d(p^Tx)}{dt}dt; \end{aligned}$$

$$J(x(0)) = \frac{1}{2}x^T(t_f)\mathbf{H}x(t_f) - \frac{1}{2}p^T(t_f)x(t_f) + \frac{1}{2}p^T(0)x(0).$$

Using the fact that $p(t) = \mathbf{P}(t)x(t)$, we have

$$J(x(0)) = \frac{1}{2}x^T(t_f)\mathbf{H}x(t_f) - \frac{1}{2}x^T(t_f)\mathbf{P}^T(t_f)x(t_f) + \frac{1}{2}x^T(0)\mathbf{P}^T(0)x(0).$$

The expression in (6.22) is then derived by noting that $\mathbf{P}(t_f) = \mathbf{H}$ and $\mathbf{H}$ is symmetric.

---

[7] The Riccati equation can also be generated using dynamic programming [2, 3, 4]. The connection of the Riccati solution to the cost is used explicitly to derive the optimal control via dynamic programming.

EXAMPLE 6.4 CONTINUED The optimal value of the cost function in Example 6.4 is given as

$$J = x^T(0)\mathbf{P}(0)x(0) = [1 \quad 1]\mathbf{P}(0)\begin{bmatrix} 1 \\ 1 \end{bmatrix}.$$

For the four cases presented in this example, the optimal cost is

$$J = 0.05 \text{ when } h = 10 \ ; q = 0; r = 0.1;$$

$$J = 1.66 \text{ when } h = 0 \ ; q = 1; r = 0.1;$$

$$J = 4.46 \text{ when } h = 0 \ ; q = 1; r = 1;$$

$$J = 16.59 \text{ when } h = 0 \ ; q = 10; r = 1.$$

Note that increasing q and r by a factor of 10 (from case two to case four) had no effect on the optimal trajectories, but did increase the optimal cost by a factor of 10. ◆

6.2.4 Selection of the Weighting Matrices

The weighting matrices in the LQR cost function can be selected on the basis of several sets of criteria. The physics of the problem may suggest terms in the cost function. For example, assume the goal of a control system is to damp out the oscillations in an automotive suspension system. The weighting matrix for the state can be selected so that

$$\int_0^{t_f} x^T(t)\mathbf{Q}x(t)dt$$

is the average energy (both kinetic and potential) contained in the oscillation. By minimizing this term in the cost function, the LQR is minimizing the average energy in the suspension system and damping out the oscillations. As a second example, consider the field-controlled dc motor in Example 3.1. The power consumed in controlling the motor is proportional to the square of the field voltage, which is the control input. The weighting matrix for the control can be selected such that

$$\int_0^{t_f} u^T(t)\mathbf{R}u(t)dt$$

is the total energy consumed by the controller. By minimizing this term in the cost function, the LQR is minimizing the energy required to perform the control.

Trial and error can also be used to define the cost function. Even for the examples given above, the relative size of the state weighting and the control weighting matrices is not specified. In other examples, the relative sizes of terms within the weighting matrices may be unknown. Trial and error provides a means to finalize the selection of the weighting matrices in these cases.

The comparison of the performance of competing designs generated using different cost functions is made using performance criteria other than cost. Control systems are often designed to specifications that involve the settling time, the damping ratio, and the bandwidth. Control systems may also be subject to constraints on the maximum output error and the maximum control input. These specifications can typically be met using the LQR after trial-and-error selection of the weighting matrices. For example, the size of the control weighting matrix can be altered until the maximum required control input, in a worst-case scenario, is just under a bound imposed by the actuator.

The size of the weighting matrices can also be altered to yield a desired settling time or other performance criteria.

A good starting point for trial-and-error selection of the state weighting matrix is to set the various state contributions (for the states of interest) roughly equal. For example, consider a system where the state consists of a position and a velocity. The magnitude of the velocity is on the order of 1/100 times the magnitude of the position in normal operation. A reasonable state weighting matrix has the weighting on the velocity 100^2 times the weighting on the position. This results in the contribution of each state being roughly equal. Initial control weighting matrices can also be selected in this manner.

The final selection of weighting matrices often proceeds by trial and error, after initially incorporating all *a priori* information concerning weighting matrix selection. The designer first specifies which outputs are important to drive to zero and incorporates any physical insight he has concerning the relative weighting of the terms. This process typically results in a cost function that is specified except for a small number of parameters. The designer then selects values for the unspecified parameters, generates the feedback gains, and simulates or analyzes the system to evaluate the performance. The parameters selected are then adjusted to improve the performance. The performance of this new design is then evaluated, and further modifications are made until an acceptable design is obtained.

6.2.5 Perspective

The Hamiltonian equations and the Riccati equation provide two separate, albeit related, methods for finding the optimal feedback gain matrix of the linear quadratic regulator. These two methods are both presented to better prepare the reader to understand other works on optimal control, which may emphasize one of the formulations. In this book, the Riccati equation, solved using numerical integration, is typically used to find the time-varying feedback gains. The Hamiltonian formulation is typically used when solving for the steady-state feedback gain matrix, as discussed in the subsequent section.

A number of generalizations to the basic linear quadratic regulator theory exist. The LQR can be extended in a direct manner to time-varying systems (see [5], page 201). Integral control can be included in LQR designs. A thorough discussion of LQR integral controller synthesis is provided in Section 8.5.2. The LQR design method can be modified to allow frequency shaping of the controller (see [6], page 262). Solutions can be generated when cross terms between the state and the control are added to the LQR cost function (see [6], page 56). Solutions exist and can be found for some problems where the control weighting matrix $\mathbf{R}$ is not positive definite (see [7], page 246). Also, LQRs can be designed to yield a given degree of stability; that is, the state can be forced to approach zero at a rate greater than a specified function $e^{-\alpha t}$ (see [6], page 60).

6.3 The Steady-State Linear Quadratic Regulator

The feedback gains of the linear quadratic regulator typically approach steady-state values far from the final time. In the examples presented earlier, the feedback gains experienced a transient and then approached a steady-state value as time was decreased from the final time. In applications where the control system is designed to operate for time periods that are long compared to the transient time of the optimal gains, it is reasonable to ignore the transient and use the steady-state gains, exclusively. The use of the steady-state gains simplifies the controller design, since only fixed-gain amplifiers are required for analog implementation, and the need to store time-varying gains is obviated in a digital implementation.

A reasonable cost function to use when the control system is designed to operate for long time periods is

$$\boxed{J(x(t),\, u(t)) \;=\; \frac{1}{2}\int_0^{\infty} x^T(t)\mathbf{Q}x(t) \;+\; u^T(t)\mathbf{R}u(t)dt,} \tag{6.24}$$

with the test conditions consisting of the initial condition

$$x(0) \;=\; x_0 \tag{6.25}$$

and no external input. For this cost function, all finite times are infinitely far from the final time, and the gains are always in steady state. The cost function is further simplified by removing the weighting on the final state, since the final state, which occurs at infinity, is not important. Additionally, the steady-state feedback gains are independent of the weighting on the final state. Minimization of this cost function yields the steady-state linear quadratic regulator gains.

EXAMPLE 6.5 The dynamics of a three axis stabilized satellite can be separated into pitch, yaw, and roll dynamics when all angles are near the operating point. Assuming small angles and small angle rates, the dynamics for each axis are given by the state equation

$$\dot{x}(t) \;=\; \begin{bmatrix} 0 & 1 \\ 0 & 0 \end{bmatrix} x(t) \;+\; \begin{bmatrix} 0 \\ 1 \end{bmatrix} u(t),$$

where the control is normalized to equal angular acceleration. A cost function is selected that weights the angle $x_1(t)$ and the control:

$$J(x(t),\, u(t)) \;=\; \int_0^{\infty} x^T(t)\begin{bmatrix} 1 & 0 \\ 0 & 0 \end{bmatrix} x(t) \;+\; \frac{1}{10}u^2(t)dt.$$

The steady-state control that minimizes this cost is

$$u(t) \;=\; -\mathbf{K}x(t) \;=\; -[3.16 \quad 2.51]x(t).$$

The closed-loop system is simulated using both the time-varying and the steady-state feedback gains, with the initial condition

$$x(0) \;=\; \begin{bmatrix} 1 \\ 1 \end{bmatrix}$$

and various final times. The cost is computed for all cases and compared. The time-varying feedback gains, the outputs and control inputs for the simulation with the steady-state gains, and the outputs and control inputs for the simulation with the time-varying gains are given in Figure 6.6 for $t_f = 1$ second; in Figure 6.7 for $t_f = 2$ seconds; and in Figure 6.8 for $t_f = 4$ seconds. The respective costs are tabulated in Table 6.1.

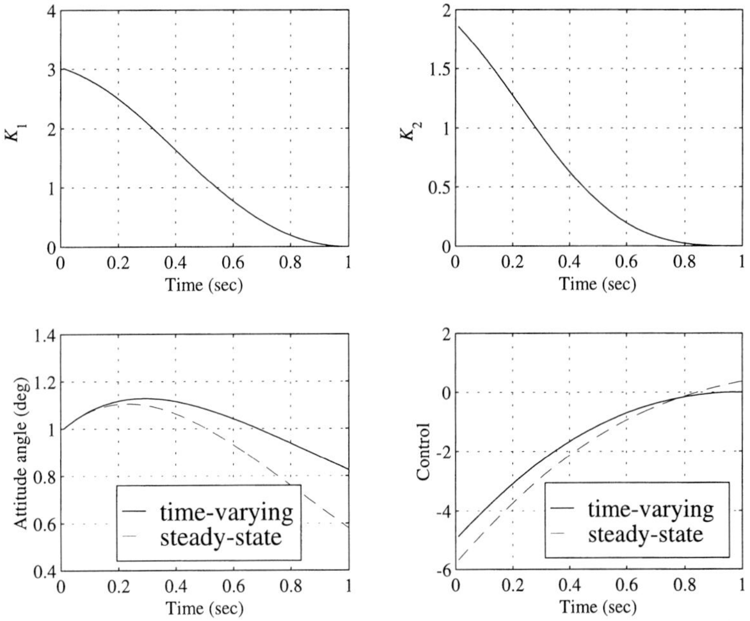

FIGURE 6.6 Results of Example 6.5 with $t_f = 1$ second

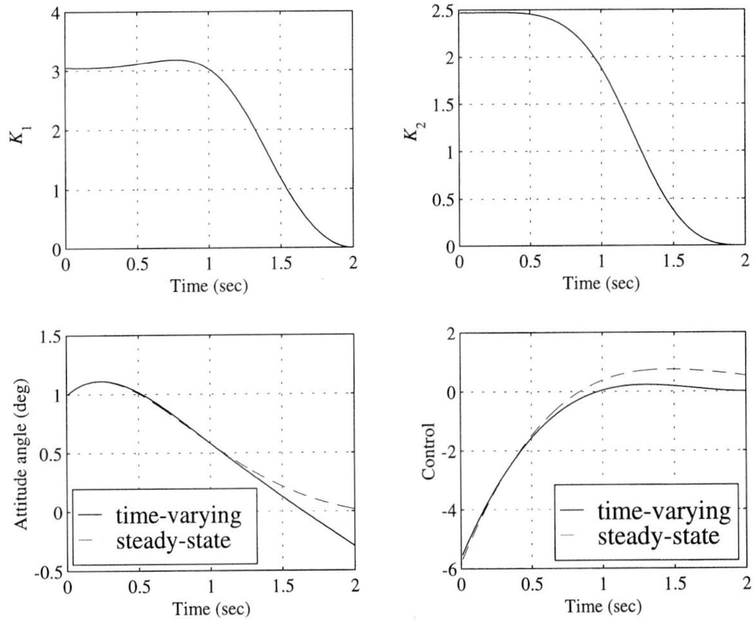

FIGURE 6.7 Results of Example 6.5 with $t_f = 2$ seconds

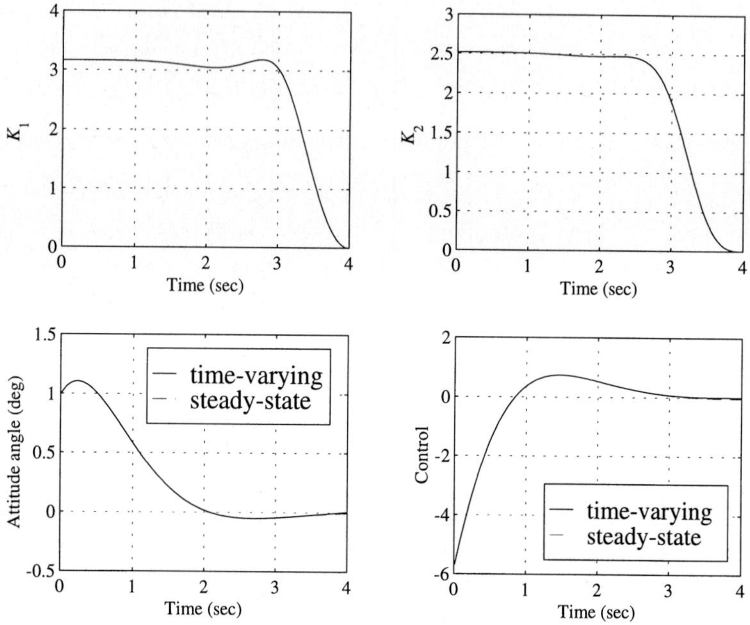

FIGURE 6.8 Results of Example 6.5 with $t_f = 4$ seconds

TABLE 6.1 Total Cost for Example 6.5

	Cost	
Final Time (seconds)	Time-Varying Gains	Steady-State Gains
1	1.51	1.56
2	1.62	1.69
4	1.68	1.70

Note that for $t_f = 4$ seconds, the trajectories resulting from the steady-state and time-varying gains are indistinguishable. Additionally, the cost using the steady-state gains is nearly identical to the optimal cost. For smaller final times, the optimal (time-varying) trajectory diverges from the trajectory obtained using the steady-state feedback gains. But even with these final times, the cost when using the steady-state gains is close to the optimal cost. ◆

EXAMPLE 6.6 The surge, sway, and yaw of a Coast Guard Iris Class ship are described by a nonlinear state model. This model can be linearized around the nominal values of zero (for station-keeping during buoy-tending operations) to yield the following state model:

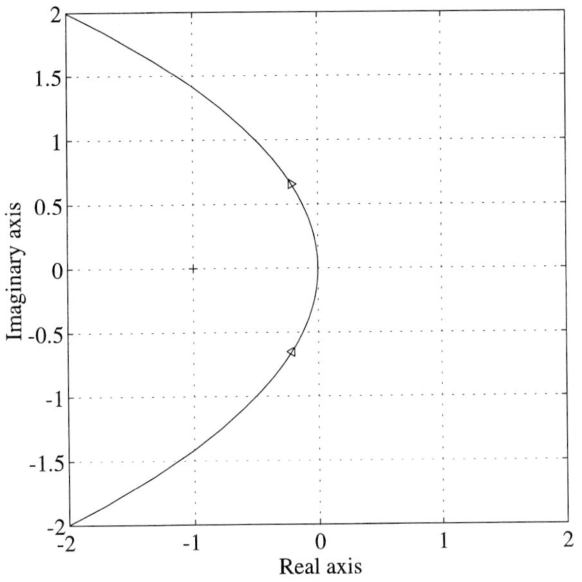

FIGURE 6.12 Nyquist plot for Example 6.7

These margins satisfy the inequalities in (6.42), with the downside gain margin being appreciably better than the bound.

Multiple-Input Systems For multiple-input systems, the inequality (6.41) yields a guaranteed bound on the smallest destabilizing input-multiplicative perturbation. The stability robustness condition (6.34) for input-multiplicative uncertainty can be written in terms of the loop transfer function. The system remains stable for all perturbations that are bounded:

$$\|\Delta\|_\infty \le \frac{1}{\|-\mathbf{G}_l(s)[\mathbf{I} + \mathbf{G}_l(s)]^{-1}\|_\infty} = \frac{1}{\|[\mathbf{I} + \mathbf{G}_l(s)]^{-1} - \mathbf{I}\|_\infty} = \frac{1}{\max\limits_{\omega} \bar{\sigma}\{[\mathbf{I} + \mathbf{G}_l(j\omega)]^{-1} - \mathbf{I}\}}. \tag{6.43}$$

The maximum singular value in this expression can be bounded using the triangle inequality:

$$\bar{\sigma}\{[\mathbf{I} + \mathbf{G}_l(j\omega)]^{-1} - \mathbf{I}\} \le 1 + \bar{\sigma}\{[\mathbf{I} + \mathbf{G}_l(j\omega)]^{-1}\}. \tag{6.44}$$

A bound on the maximum singular value of $[\mathbf{I} + \mathbf{G}_1(j\omega)]^{-1}$ can be generated using (6.41):

$$1 = \underline{\sigma}\{\mathbf{I}\} \le \underline{\sigma}\{[\mathbf{I} + \mathbf{G}_1(j\omega)]^\dagger[\mathbf{I} + \mathbf{G}_1(j\omega)]\} = \underline{\sigma}\{[\mathbf{I} + \mathbf{G}_1(j\omega)]\}^2.$$

Therefore,

$$1 \le \underline{\sigma}\{\mathbf{I} + \mathbf{G}_l(j\omega)\};$$
$$1 \ge \bar{\sigma}\{[\mathbf{I} + \mathbf{G}_l(j\omega)]^{-1}\}. \tag{6.45}$$

Substituting (6.45) into (6.44) yields

$$\bar{\sigma}\{[\mathbf{I} + \mathbf{G}_l(j\omega)]^{-1} - \mathbf{I}\} \le 2;$$

$$\max_{\omega} \bar{\sigma}\{[\mathbf{I} + \mathbf{G}_l(j\omega)]^{-1} - \mathbf{I}\} \le 2.$$

Combining this equation with (6.43), the closed-loop system is seen to remain stable for all unstructured input-multiplicative perturbations such that

$$\boxed{\|\Delta_i\|_\infty < \frac{1}{2}.} \tag{6.46}$$

This result provides a guarantee on the robustness of a linear quadratic regulator, subject to input-multiplicative perturbations, provided that the control weighting matrix is proportional to the identity matrix.

EXAMPLE 6.8 The hover mode pitch and yaw dynamics of a vertical-takeoff-and-landing remotely piloted vehicle are given by the following state equation:

$$\dot{x}(t) = \begin{bmatrix} 0 & -1 & 0 & 0 \\ 0 & 0 & 0 & -1.6 \\ 0 & 0 & 0 & -1 \\ 0 & 1.8 & 0 & 0 \end{bmatrix} x(t) + \begin{bmatrix} 0 & 0 \\ -11 & 0 \\ 0 & 0 \\ 0 & -12 \end{bmatrix} u(t).$$

A cost function is selected that weights the pitch $x_1(t)$ and yaw $x_3(t)$ equally, and weights the control input as follows:

$$J(x(t), u(t)) = \int_0^\infty x^T(t) \begin{bmatrix} 1 & 0 & 0 & 0 \\ 0 & 0 & 0 & 0 \\ 0 & 0 & 1 & 0 \\ 0 & 0 & 0 & 0 \end{bmatrix} x(t) + u^T(t) \begin{bmatrix} 1 & 0 \\ 0 & 1 \end{bmatrix} u(t) dt.$$

The control that minimizes this cost is

$$u(t) = -\mathbf{K}x(t) = -\begin{bmatrix} 0.94 & -0.41 & 0.34 & 0.00 \\ -0.34 & 0.00 & 0.94 & -0.40 \end{bmatrix} x(t).$$

The robustness of the closed-loop system to an unstructured, input-multiplicative perturbation is given by the ∞-norm of the transfer function from the perturbation input to the perturbation output:

$$\mathbf{N}_{y_d w_d}(s) = -\mathbf{K}(s\mathbf{I} - \mathbf{A} + \mathbf{B}_u\mathbf{K})^{-1}\mathbf{B}_u.$$

Note that this expression for the transfer function is generated from the state equation of the system, and is equivalent to the transfer function in (6.35). The principal gains of this transfer function are plotted in Figure 6.13. The system is therefore robustly stable for input-multiplicative uncertainties such that:

$$\|\Delta_i\|_\infty < \frac{1}{\|\mathbf{N}_{y_d w_d}(s)\|_\infty} = \frac{1}{1.25} = 0.80.$$

Note that this bound is greater than the guaranteed bound in (6.46). ◆

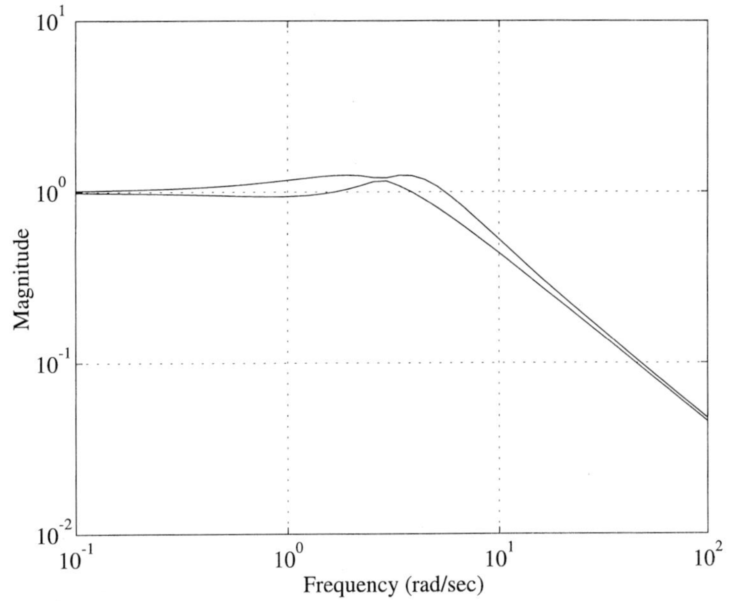

FIGURE 6.13 Principal gains for determining robustness margins in Example 6.8

Perspective A linear quadratic regulator has good stability margins and a healthy tolerance for input-multiplicative perturbations. These properties provide an indication that the system is also robust to more general perturbations. For example, the effect of a parameter variation is often to make a small change in the loop gain and/or provide a small phase shift in the loop. Additionally, good performance robustness is frequently observed in systems with large stability margins. As a rule, the LQR has both good stability and performance robustness when the system is subject to general parameter perturbations.

Unmodeled high-frequency dynamics are another source of model perturbations that are caused by a variety of sources. These dynamics often have high gains, for example, vibrational resonances, which are not effectively modeled as a small input-multiplicative uncertainty. The reason that these perturbations do not often result in instability is that the plant gain, and therefore the loop gain, falls off for increasing frequency. The loop gain is then small at the frequencies where the unmodeled dynamics are significant, and the product of the loop gain and the unmodeled dynamics is of little consequence.

The loop gain of the LQR is inversely proportional to frequency at high frequencies. This is seen by taking the limit of the loop gain:[13]

$$\lim_{\omega \to \infty} \mathbf{G}_l(j\omega) = \lim_{\omega \to \infty} \mathbf{K}(j\omega\mathbf{I} - \mathbf{A})^{-1}\mathbf{B}_u = \frac{1}{\omega}(-j)\mathbf{K}\mathbf{B}_u.$$

[13] This limit can be computed in a direct manner by writing the inverse,

$$\frac{1}{\det(j\omega\mathbf{I} - \mathbf{A})}\text{adj}(j\omega\mathbf{I} - \mathbf{A}),$$

and taking the limit of the $\text{adj}(j\omega\mathbf{I} - \mathbf{A})$ term by term.

Many designers prefer a loop gain roll-off that is proportional to the square of the frequency. This rapid roll-off of the loop gain provides assurance that the loop gain is small at frequencies where unmodeled dynamics are significant. The gist of the previous comments is that care must be taken when directly applying LQR designs to problems with significant, unmodeled, high-frequency dynamics. Methods of achieving a desired loop gain roll-off, using LQR design techniques, are discussed in Section 8.3.3 of Chapter 8.

6.3.4 The Closed-Loop Poles

The steady-state linear quadratic regulator yields an optimal control that is time-invariant state feedback and also a time-invariant closed-loop system. In this case, the closed-loop poles provide useful information concerning the transient performance of the system.

The state equation of the closed-loop system is

$$\boxed{\dot{x}(t) = \mathbf{A}x(t) + \mathbf{B}_u(-\mathbf{K}x(t)) = (\mathbf{A} - \mathbf{B}_u\mathbf{K})x(t).} \tag{6.47}$$

The poles of this system are the eigenvalues of the closed-loop state matrix $(\mathbf{A} - \mathbf{B}_u\mathbf{K})$, which are the solutions of the characteristic equation:

$$\det(s\mathbf{I} - \mathbf{A} + \mathbf{B}_u\mathbf{K}) = 0. \tag{6.48}$$

The closed-loop poles can be found by first finding the optimal feedback gain matrix, and then solving the characteristic equation (6.48). While this method of finding the closed-loop poles is satisfactory, an alternative method is developed below. This method adds insight into the meaning of the Hamiltonian system, and proves useful in developing subsequent results.

The Hamiltonian system describes both the evolution of the closed-loop system and the evolution of the costates. It is reasonable to assume that the closed-loop poles are related to the Hamiltonian poles. The characteristic equation of the Hamiltonian system is given in (6.27) and reproduced here:

$$\det(s\mathbf{I} - \mathbf{A})\det(-s\mathbf{I} - \mathbf{A}^T)\det\{\mathbf{R} + \mathbf{B}_u^T(-s\mathbf{I} - \mathbf{A}^T)^{-1}\mathbf{Q}(s\mathbf{I} - \mathbf{A})^{-1}\mathbf{B}_u\} = 0. \tag{6.49}$$

The term within the third determinant in this expression is given by (6.38), which is reproduced here:

$$\begin{aligned}&[\mathbf{I} + \mathbf{B}_u^T(-s\mathbf{I} - \mathbf{A}^T)^{-1}\mathbf{K}^T]\mathbf{R}[\mathbf{I} + \mathbf{K}(s\mathbf{I} - \mathbf{A})^{-1}\mathbf{B}_u]\\ &= \mathbf{R} + \mathbf{B}_u^T(-s\mathbf{I} - \mathbf{A}^T)^{-1}\mathbf{Q}(s\mathbf{I} - \mathbf{A})^{-1}\mathbf{B}_u.\end{aligned}$$

Substituting this equation into (6.49) yields

$$\begin{aligned}&\det(s\mathbf{I} - \mathbf{A})\det(-s\mathbf{I} - \mathbf{A}^T)\det\{[\mathbf{I} + \mathbf{B}_u^T(-s\mathbf{I} - \mathbf{A}^T)^{-1}\mathbf{K}^T]\mathbf{R}[\mathbf{I} + \mathbf{K}(s\mathbf{I} - \mathbf{A})^{-1}\mathbf{B}_u]\}\\ &= 0.\end{aligned}$$

Expanding the determinant as a product of determinants, noting that $G_1(s)$ is square, and recognizing that $\det(\mathbf{R}) \neq 0$ since $\mathbf{R}$ is positive definite gives

$$\begin{aligned}&\det(-s\mathbf{I} - \mathbf{A}^T)\det\{\mathbf{I} + \mathbf{B}_u^T(-s\mathbf{I} - \mathbf{A}^T)^{-1}\mathbf{K}^T\}\det(s\mathbf{I} - \mathbf{A})\det\{\mathbf{I} + \mathbf{K}(s\mathbf{I} - \mathbf{A})^{-1}\mathbf{B}_u\}\\ &= 0.\end{aligned}$$

Applying Appendix equation (A8.1) results in

$$\det(-s\mathbf{I} - \mathbf{A}^T)\det\{\mathbf{I} + (-s\mathbf{I} - \mathbf{A}^T)^{-1}\mathbf{K}^T\mathbf{B}_u^T\}\det(s\mathbf{I} - \mathbf{A})\det\{\mathbf{I} + (s\mathbf{I} - \mathbf{A})^{-1}\mathbf{B}_u\mathbf{K}\}$$
$$= \det(-s\mathbf{I} - \mathbf{A}^T + \mathbf{K}^T\mathbf{B}_u^T)\det(s\mathbf{I} - \mathbf{A} + \mathbf{B}_u\mathbf{K})$$
$$= \det(-s\mathbf{I} - \mathbf{A} + \mathbf{B}_u\mathbf{K})\det(s\mathbf{I} - \mathbf{A} + \mathbf{B}_u\mathbf{K}) = 0.$$

The poles of the Hamiltonian system consist of both the closed-loop poles and the negative of the closed-loop poles. The closed-loop poles are, therefore, the left half-plane poles of the Hamiltonian, since the closed-loop system is stable.

The closed-loop system poles provide insight into the performance of the controller. Understanding how the closed-loop poles vary as the weighting matrices are changed will assist the designer in selecting weighting matrices, and also in understanding the consequences of a particular choice of weighting matrices. The pole positions for the limiting cases of large and small control weighting are evaluated below. Note that the optimal control only depends on the relative sizes of the state and the control weighting matrices. The pole positions for the limiting cases of small and large state weighting are therefore also obtained.

The control weighting matrix is assumed to be of the following form:

$$\mathbf{R} = \rho\mathbf{R}_0$$

where $\mathbf{R}_0$ is a positive definite weighting matrix, and ρ is a positive scalar parameter called the relative control weighting. The closed-loop pole locations can be found from the characteristic equation of the Hamiltonian (6.27):

$$\det(s\mathbf{I} - \mathbf{A})\det(-s\mathbf{I} - \mathbf{A}^T)\det\{\rho\mathbf{R}_0 + \mathbf{B}_u^T(-s\mathbf{I} - \mathbf{A}^T)^{-1}\mathbf{Q}(s\mathbf{I} - \mathbf{A})^{-1}\mathbf{B}_u\} = 0.$$

The pole positions for the limiting cases where $\rho \to \infty$ and $\rho \to 0$ are subsequently computed.

Large Control Weighting The characteristic equation of the Hamiltonian system is approximated as follows:

$$\det(s\mathbf{I} - \mathbf{A})\det(-s\mathbf{I} - \mathbf{A}^T)\det(\rho\mathbf{R}_0) = 0;$$

$$\det(s\mathbf{I} - \mathbf{A})\det(-s\mathbf{I} - \mathbf{A}^T) = 0,$$

for large control weighting. In this case, the Hamiltonian poles consist of both the open-loop poles of the plant and the reflections of these open-loop poles about the imaginary axis. The closed-loop poles of the linear quadratic regulator consist of the stable poles of the Hamiltonian. Therefore, as the control weighting is increased, the closed-loop poles approach the stable plant poles and reflections about the imaginary axis of the unstable plant poles.

The fact that unstable poles are moved to their reflections about the imaginary axis is surprising at first glance. The control gains required to move the poles to these points are greater than the control gains required to move the poles just within the stable region. The problem with moving the poles to just within the left half-plane is that the control is applied for a longer period of time. In this case, most of the control is applied simply to counter the plant's tendency toward instability. Only a small portion of the control is then available to drive the states toward zero. Placing the unstable poles at their reflections provides a balance between the plant's tendency to blow up and the control system's effort to drive the state to zero.

Small Control Weighting The characteristic equation of the Hamiltonian system is approximated as follows:

$$\det(s\mathbf{I} - \mathbf{A})\det(-s\mathbf{I} - \mathbf{A}^T)\det\{\mathbf{B}_u^T(-s\mathbf{I} - \mathbf{A}^T)^{-1}\mathbf{Q}(s\mathbf{I} - \mathbf{A})^{-1}\mathbf{B}_u\} = 0, \tag{6.50}$$

for small control weighting. The poles of the Hamiltonian are the solutions of (6.50), which may occur when any of the individual determinants are equal to zero. A more detailed analysis shows that the Hamiltonian poles only occur when the last determinant in (6.50) is equal to zero. This can be seen by expanding the characteristic equation (6.50):

$$\det(s\mathbf{I} - \mathbf{A})\det(-s\mathbf{I} - \mathbf{A}^T)\det\left\{\mathbf{B}_u^T\frac{\mathrm{adj}(-s\mathbf{I} - \mathbf{A}^T)}{\det(-s\mathbf{I} - \mathbf{A}^T)}\mathbf{Q}\frac{\mathrm{adj}(s\mathbf{I} - \mathbf{A})}{\det(s\mathbf{I} - \mathbf{A})}\mathbf{B}_u\right\}$$

$$= \frac{1}{\det(s\mathbf{I} - \mathbf{A})^{n_u-1}}\frac{1}{\det(-s\mathbf{I} - \mathbf{A}^T)^{n_u-1}}\det\{\mathbf{B}_u^T\mathrm{adj}(-s\mathbf{I} - \mathbf{A}^T)\mathbf{Q}\mathrm{adj}(s\mathbf{I} - \mathbf{A})\mathbf{B}_u\} = 0, \tag{6.51}$$

where n_u is the number of control inputs. The first two determinants in (6.50) are canceled by terms from the last determinant, and do not yield poles when they equal zero. The limiting poles of the Hamiltonian system are therefore the solutions of the equation:

$$\det\{\mathbf{B}_u^T\mathrm{adj}(-s\mathbf{I} - \mathbf{A}^T)\mathbf{Q}\mathrm{adj}(s\mathbf{I} - \mathbf{A})\mathbf{B}_u\} = 0, \tag{6.52}$$

or infinity.

Equation (6.52) defines the zeros of the transfer function:

$$\mathbf{B}_u^T(-s\mathbf{I} - \mathbf{A}^T)^{-1}\mathbf{Q}(s\mathbf{I} - \mathbf{A})^{-1}\mathbf{B}_u. \tag{6.53}$$

The closed-loop poles of the linear quadratic regulator consist of the stable poles of the Hamiltonian. Therefore, as the control weighting is decreased, the closed-loop poles approach the left half-plane zeros of (6.53) or approach infinity in the left half-plane.

The zeros of (6.53) can be directly related to the zeros of the plant with the output

$$y(t) = \mathbf{C}_y x(t)$$

provided this plant is square; that is, $n_y = n_u$. Using the expression for the state weighting matrix,

$$\mathbf{Q} = \mathbf{C}_y^T\mathbf{Q}_y\mathbf{C}_y,$$

(6.53) can be written in terms of the plant transfer function $\mathbf{G}(s)$:

$$\mathbf{G}(-s)\mathbf{Q}_y\mathbf{G}(s) = \mathbf{B}_u^T(-s\mathbf{I} - \mathbf{A}^T)^{-1}\mathbf{C}_y^T\mathbf{Q}_y\mathbf{C}_y(s\mathbf{I} - \mathbf{A})^{-1}\mathbf{B}_u.$$

When the plant is square, the zeros of (6.53) can be written as follows:

$$\det\{\mathbf{G}(-s)\mathbf{Q}_y\mathbf{G}(s)\} = \det\{\mathbf{G}(-s)\}\det\{\mathbf{Q}_y\}\det\{\mathbf{G}(s)\} = 0.$$

Since $\mathbf{Q}_y$ is positive definite, the $\det\{\mathbf{Q}_y\} \neq 0$, and these zeros are simply the plant zeros and the reflections of the plant zeros about the imaginary axis. Therefore, for square plants, the closed-loop poles approach the left half-plane plant zeros, reflections about the imaginary axis of the right half-plane plant zeros, and infinity in the left half-plane.

The poles that go to infinity, as the control weighting is decreased, approach asymptotes radiating from the origin. For single-input plants, these asymptotes are in a Butterworth filter pattern; that is, they are at angles:

$$\pm\frac{l\pi}{n_x - N_z},\ l = 0, 1, \ldots, \frac{n_x - N_z - 1}{2},\ \text{when } n_x - N_z \text{ odd;}$$

$$\pm\frac{(l + \frac{1}{2})\pi}{n_x - N_z},\ l = 0, 1, \ldots, \frac{n_x - N_z}{2} - 1,\ \text{when } n_x - N_z \text{ even,}$$

from the negative real axis. The parameter N_z is the number of zeros of the plant, and $(n_x - N_z)$ is the number of poles that approach infinity. For multiple-input systems, the asymptotes can form multiple Butterworth patterns; that is, the poles that go to infinity can form several groups, each of which has asymptotes in a Butterworth pattern.

The root locus of the linear quadratic regulator is similar to the root loci observed in classical control. As the LQR gains are increased (the control weighting decreased), the poles start at the plant poles (or the left half-plane reflections of these poles). For square plants, these poles then move to the plant zeros (or the left half-plane reflections of these poles). For nonsquare plants, the poles terminate instead at the stable zeros of (6.53). Similar behavior is observed in classical control systems, except that the actual poles and zeros, not their left half-plane reflections, are used as initial and terminal points of the root locus. The LQR poles that have no zeros to approach go to infinity in equally spaced directions in the left half-plane. In classical control, these poles go to infinity in equally spaced directions in the entire complex plane. The major difference between the LQR root locus and root loci in classical control is that the LQR root locus is restricted to lie entirely within the left half-plane.

EXAMPLE 6.9 We are given the plant described by the following state equation:

$$\dot{x}(t) = \begin{bmatrix} 1 & -1 \\ 3 & -4 \end{bmatrix} x(t) + \begin{bmatrix} 1 \\ 0 \end{bmatrix} u(t).$$

The reference output is

$$y(t) = [1 \quad 0]x(t).$$

The transfer function of this plant with the output $\mathbf{C}_y x(t)$ is

$$G(s) = [1 \quad 0]\begin{bmatrix} s - 1 & 1 \\ -3 & s + 4 \end{bmatrix}^{-1}\begin{bmatrix} 1 \\ 0 \end{bmatrix} = \frac{s + 4}{(s + 3.3)(s - 0.3)}.$$

A control is desired that minimizes the cost function:

$$J(x(t), u(t)) = \int_0^\infty y^2(t) + Ru^2(t)dt = \int_0^\infty x^T(t)\begin{bmatrix} 1 & 0 \\ 0 & 0 \end{bmatrix}x(t) + Ru^2(t)dt,$$

where R is a variable. The root locus of the pole locations as a function of R is given in Figure 6.14. The poles approach the open-loop pole location at -3.3 and the negative

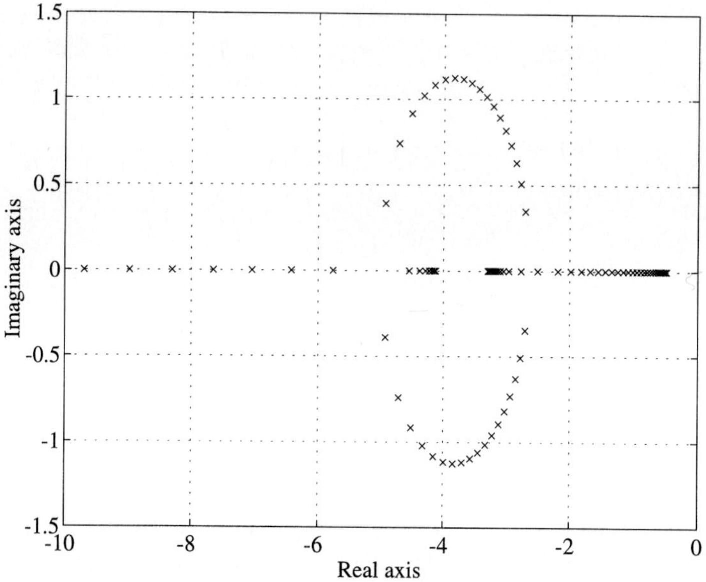

FIGURE 6.14 Linear quadratic regulator root locus for Example 6.9

of the open-loop pole location at 0.3 as the control weighting becomes large. One pole approaches the plant zero location at -4, and the other pole approaches $-\infty$ as the control weighting becomes small. ◆

EXAMPLE 6.10 We are given the system described by the following state equation:

$$\dot{x}(t) = \begin{bmatrix} 0 & 1 & 0 \\ 0 & 0 & 1 \\ -170 & -37 & -12 \end{bmatrix} x(t) + \begin{bmatrix} 0 & 0 \\ 1 & 0 \\ 0 & 1 \end{bmatrix} u(t).$$

The reference output is

$$y(t) = \begin{bmatrix} 7 & 2 & 4 \\ 5 & 1 & 0 \end{bmatrix} x(t).$$

The plant with the output $\mathbf{C}_y x(t)$ is square, with the following transfer function matrix:

$$\mathbf{G}(s) = \begin{bmatrix} \dfrac{2s^2 - 117s - 596}{s^3 + 12s^2 + 37s + 170} & \dfrac{4s^2 + 2s + 7}{s^3 + 12s^2 + 37s + 170} \\ \dfrac{s^2 + 17s + 60}{s^3 + 12s^2 + 37s + 170} & \dfrac{s + 5}{s^3 + 12s^2 + 37s + 170} \end{bmatrix}.$$

The poles and zeros of the plant are

$$\text{poles} = \{-1 + 4j,\ -1 - 4j,\ -10\}\ ;\ \text{zeros} = \{-5\}.$$

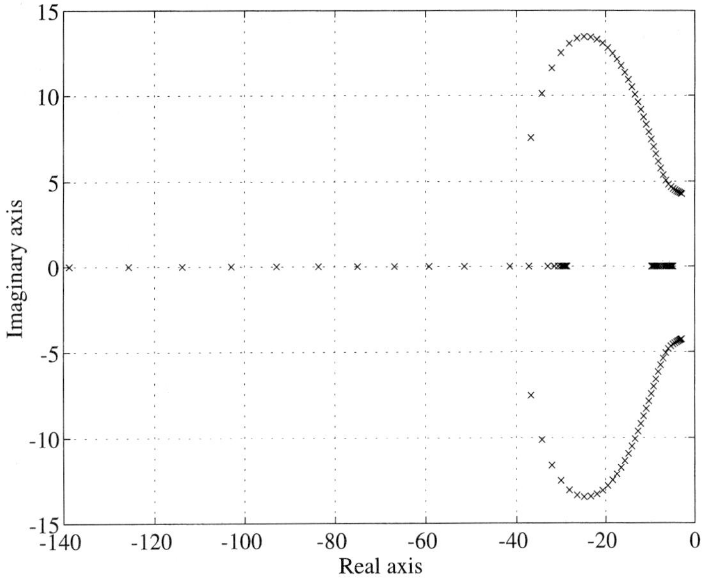

FIGURE 6.15 Linear quadratic regulator root locus for Example 6.10

The control is required to minimize the cost function

$$J(x(t),\, u(t)) = \int_0^\infty y^T(t)\begin{bmatrix} 100 & 0 \\ 0 & 1 \end{bmatrix} y(t) + u^T(t)\begin{bmatrix} \rho & 0 \\ 0 & \rho \end{bmatrix} u(t)dt$$

$$= \int_0^\infty x^T(t)\begin{bmatrix} 101 & 203 & 100 \\ 203 & 409 & 200 \\ 100 & 200 & 100 \end{bmatrix} x(t) + u^T(t)\begin{bmatrix} \rho & 0 \\ 0 & \rho \end{bmatrix} u(t)dt,$$

where ρ is a variable. The root locus of the pole locations as a function of ρ is given in Figure 6.15. The poles approach the plant pole locations as ρ becomes large and approach the plant zero locations as ρ becomes small. Note that the limiting pole locations are not affected by the output weighting matrix. The two poles that approach infinity as the control weighting is decreased both approach minus infinity along the real axis; that is, they form two Butterworth filter patterns. Another difference is observed between this root locus and the root loci from classical control: The poles change directions when moving along the real axis. Both of the poles that go to infinity change directions in this example. ◆

The previous examples illustrate the behavior of the poles of the closed-loop linear quadratic regulator as the control weighting is varied. These poles determine, at least in part, the performance of the system in terms of settling time, damping ratio, and the closed-loop frequency response. The range of these performance parameters that are achieved by the LQR may be limited by the open-loop poles and zeros of the system. While the open-loop poles are typically given, the designer can influence the open-loop zeros, since these depend on the state weighting matrix. Adding derivatives of the output of interest to the cost function is one way of manipulating the open-loop zeros. When the LQR is yielding unacceptable performance in terms of settling time, damping ratio, or frequency response for all reasonable control weightings, moving the open-loop zeros by qualitatively changing the state weighting matrix often yields improved performance. Computing the open-loop zeros that result from various types of state weighting matrices provides additional information that can be used in selecting cost functions.

6.4 The Stochastic Regulator

The linear quadratic regulator minimizes a quadratic cost function when the test conditions consist of initial conditions and no disturbance inputs. This is a fairly restrictive problem statement, since real-world control systems are typically subject to disturbance inputs. A more general optimal control problem known as the stochastic regulator is posed in this section. The stochastic regulator requires the minimization of a quadratic cost function where the test conditions consist of both initial conditions and disturbance inputs. The solution of this more general optimal control problem is shown to be identical to the LQR solution.

The stochastic regulator is an optimal control problem where the plant state equation is linear, the cost function is quadratic, and the test conditions consist of random initial conditions and random disturbance inputs. The linear plant state equation is

$$\dot{x}(t) = \mathbf{A}x(t) + \mathbf{B}_u u(t) + \mathbf{B}_w w(t), \tag{6.54}$$

and the initial condition is a zero-mean random vector with the following correlation matrix:

$$E[x(0)x^T(0)] = \mathbf{\Sigma}_x(0).$$

The disturbance input $w(t)$ is a white noise random signal with spectral density $\mathbf{S}_w$, and is assumed to be uncorrelated with the initial condition. The cost function is

$$J_{SR} = \frac{1}{2}E\left[x^T(t_f)\mathbf{H}x(t_f) + \int_0^{t_f} \{x^T(t)\mathbf{Q}x(t) + u^T(t)\mathbf{R}u(t)\}dt\right], \tag{6.55}$$

where $\mathbf{H}$ and $\mathbf{Q}$ are positive semidefinite, and $\mathbf{R}$ is positive definite. The expected value in this cost function is required since the state and control input are random. The stochastic regulator problem is to find the linear feedback controller that minimizes this cost function subject to the given test conditions. The stochastic regulator requires that the controller be a feedback system since the state is not known *a priori,* and therefore the control cannot be determined *a priori*. Linearity of the feedback controller is assumed to simplify optimization and the analysis of the closed-loop system. Note that linear feedback control is optimal over the set of all feedback controllers for the special case when the initial conditions and disturbance inputs are Gaussian.

The state at a given time contains all of the information about the past inputs and states that contribute to the future behavior of a system. In addition, white noise inputs are uncorrelated with past inputs and cannot be predicted from past inputs or states.

Therefore, the current state contains all the available information on the future behavior of the plant, and state feedback,

$$u(t) = -\mathbf{K}(t)x(t),$$

provides the most general form of linear feedback control. The stochastic regulator problem is equivalent to the problem of finding the linear state feedback gain matrix (possibly time-varying) that minimizes the cost function (6.55).

For state feedback, the closed-loop system is

$$\dot{x}(t) = [\mathbf{A} - \mathbf{B}_u\mathbf{K}(t)]x(t) + \mathbf{B}_w w(t), \tag{6.56}$$

and the cost function (6.55) can be written in terms of only the state:

$$J_{SR}(\mathbf{K}) = \frac{1}{2}E\left[x^T(t_f)\mathbf{H}x(t_f) + \int_0^{t_f} x^T(t)\mathbf{Q}_R(t)x(t)dt\right],$$

where

$$\mathbf{Q}_R(t) = \mathbf{Q} + \mathbf{K}^T(t)\mathbf{R}\mathbf{K}(t).$$

The notation $J_{SR}(\mathbf{K})$ is used to emphasize that this cost depends on the selection of the state feedback gain. Taking the trace of this expression and using the fact that the trace is invariant under cyclic permutations [see Appendix equation (A2.3)], the cost can be written in terms of the state correlation matrix as follows:

$$J_{SR}(\mathbf{K}) = \frac{1}{2}\text{tr}\left\{\mathbf{H}\mathbf{\Sigma}_x(t_f) + \int_0^{t_f} \mathbf{Q}_R(t)\mathbf{\Sigma}_x(t)dt\right\}. \tag{6.57}$$

The stochastic regulator cost can be written as a weighted sum of LQR costs. This relationship is developed by expanding the state covariance matrix using the methods in Section 3.3.1 of Chapter 3:

$$\mathbf{\Sigma}_x(t) = \mathbf{\Phi}(t, 0)\mathbf{\Sigma}_x(0)\mathbf{\Phi}^T(t, 0) + \int_0^t \mathbf{\Phi}(t, \tau)\mathbf{B}_w\mathbf{S}_w\mathbf{B}_w^T\mathbf{\Phi}^T(t, \tau)d\tau. \tag{6.58}$$

Substituting this solution into the stochastic regulator cost function yields

$$\begin{aligned} J_{SR}(\mathbf{K}) = \frac{1}{2}\text{tr}\Bigg\{&\mathbf{H}\mathbf{\Phi}(t_f, 0)\mathbf{\Sigma}_x(0)\mathbf{\Phi}^T(t_f, 0) + \int_0^{t_f} \mathbf{H}\mathbf{\Phi}(t_f, \tau)\mathbf{B}_w\mathbf{S}_w\mathbf{B}_w^T\mathbf{\Phi}^T(t_f, \tau)d\tau \\ &+ \int_0^{t_f} \mathbf{Q}_R(t)\mathbf{\Phi}(t, 0)\mathbf{\Sigma}_x(0)\mathbf{\Phi}^T(t, 0)d\tau \\ &+ \int_0^{t_f}\int_0^t \mathbf{Q}_R(t)\mathbf{\Phi}(t, \tau)\mathbf{B}_w\mathbf{S}_w\mathbf{B}_w^T\mathbf{\Phi}^T(t, \tau)d\tau\Bigg\}. \end{aligned}$$

Rearranging terms and interchanging the order of integration in the final term yields

$$\begin{aligned} J_{SR}(\mathbf{K}) = &\frac{1}{2}\text{tr}\left\{\mathbf{H}\mathbf{\Phi}(t_f, 0)\mathbf{\Sigma}_x(0)\mathbf{\Phi}^T(t_f, 0) + \int_0^{t_f} \mathbf{Q}_R(t)\mathbf{\Phi}(t, 0)\mathbf{\Sigma}_x(0)\mathbf{\Phi}^T(t, 0)d\tau\right\} \\ &+ \int_0^{t_f}\frac{1}{2}\text{tr}\Bigg\{\mathbf{H}\mathbf{\Phi}(t_f, \tau)\mathbf{B}_w\mathbf{S}_w\mathbf{B}_w^T\mathbf{\Phi}^T(t_f, \tau) \\ &\qquad + \int_\tau^{t_t} \mathbf{Q}_R(t)\mathbf{\Phi}(t, \tau)\mathbf{B}_w\mathbf{S}_w\mathbf{B}_w^T\mathbf{\Phi}^T(t, \tau)dt\Bigg\}d\tau. \end{aligned} \tag{6.59}$$

The cost function of the linear quadratic regulator with an initial condition at time t_0 can be expanded using the solution of (6.56) and the fact that the trace is invariant under cyclic permutations [Appendix equation (A2.3)]:

$$J_{LQR}(x_0, t_0, \mathbf{K}) = \frac{1}{2}\mathrm{tr}\left\{\mathbf{H\Phi}(t_f, t_0)x(t_0)x^T(t_0)\mathbf{\Phi}^T(t_f, t_0) + \int_{t_0}^{t_f} \mathbf{Q}_R(t)\mathbf{\Phi}(t, t_0)x(t_0)x^T(t_0)\mathbf{\Phi}^T(t, t_0)dt\right\}.$$

The relationship between this expression and the stochastic regulator cost can be formalized by expanding $\mathbf{\Sigma}_x(0)$ and $\mathbf{S}_w$ in (6.59) using their singular value decompositions:

$$\begin{aligned}J_{SR}(\mathbf{K}) = \sum_{k=1}^{n_x} \sigma_k(\mathbf{\Sigma})\frac{1}{2}\mathrm{tr}&\left\{\mathbf{H\Phi}(t_f, 0)U_k(\mathbf{\Sigma})U_k^T(\mathbf{\Sigma})\mathbf{\Phi}^T(t_f, 0) + \int_0^{t_f} \mathbf{Q}_R(t)\mathbf{\Phi}(t, 0)U_k(\mathbf{\Sigma})U_k^T(\mathbf{\Sigma})\mathbf{\Phi}^T(t, 0)d\tau\right\}\\ + \int_0^{t_f}\sum_{k=1}^{n_u} \sigma_k(\mathbf{S})\frac{1}{2}\mathrm{tr}&\left\{\mathbf{H\Phi}(t_f, \tau)\mathbf{B}_w U_k(\mathbf{S})U_k^T(\mathbf{S})\mathbf{B}_w^T\mathbf{\Phi}^T(t_f, \tau) + \int_\tau^{t_t} \mathbf{Q}_R(t)\mathbf{\Phi}(t, \tau)\mathbf{B}_w U_k(\mathbf{S})U_k^T(\mathbf{S})\mathbf{B}_w^T\mathbf{\Phi}^T(t, \tau)dt\right\}d\tau,\end{aligned}$$

where $\{\sigma_k(\mathbf{\Sigma})\}$ are the singular values of $\mathbf{\Sigma}_x(0)$, $\{U_k(\mathbf{\Sigma})\}$ are the singular vectors of $\mathbf{\Sigma}_x(0)$, $\{\sigma_k(\mathbf{S})\}$ are the singular values of $\mathbf{S}_w$, and $\{U_k(\mathbf{S})\}$ are the singular vectors of $\mathbf{S}_w$. Note that the right and left singular vectors are equal for symmetric matrices. The stochastic regulator cost is then a weighted sum of LQR cost functions:

$$\begin{aligned}J_{SR}(\mathbf{K}) &= \frac{1}{2}\sum_{k=1}^{n_x} \sigma_k(\mathbf{\Sigma})J_{LQR}(U_k(\mathbf{\Sigma}), 0, \mathbf{K})\\ &+ \frac{1}{2}\int_0^{t_f}\sum_{k=1}^{n_u} \sigma_k(\mathbf{S})J_{LQR}(\mathbf{B}_w U_k(\mathbf{S}), t, \mathbf{K})dt,\end{aligned} \tag{6.60}$$

where all the weights are non-negative. The same feedback gain minimizes the LQR cost for any initial condition at any time before the final time. Therefore, the optimal feedback gain for the LQR also minimizes the stochastic regulator cost function. Note that the optimal feedback gain for the stochastic regulator is independent of the initial condition covariance matrix and the input spectral density matrix.

The derivation of the equivalence between the optimal solutions of the stochastic regulator and the linear quadratic regulator is somewhat convoluted. This result can be intuitively understood by noting that the LQR feedback gain is optimal for all initial conditions, both deterministic and random. In addition, the white disturbance input to the stochastic regulator can be thought of as generating new initial conditions at each point in time. The LQR feedback gain is optimal for each of these initial conditions and therefore also optimal for controlling the effects of the white disturbance input.

A time-invariant solution to the stochastic regulator is obtained by letting $\mathbf{H} = \mathbf{0}$ and $t_f = \infty$, which results in the following cost function:

$$J_{SR} = \frac{1}{2}E\left[\int_0^\infty x^T(t)\mathbf{Q}x(t) + u^T(t)\mathbf{R}u(t)dt\right].$$

This cost function is infinite, in general, since the state and control are asymptotically stationary for a stable closed-loop system, and the integrand never approaches zero. This problem can be obviated by normalizing the cost function:

$$\begin{aligned} J_{SR} &= \frac{1}{2}E\left[\lim_{t_f\to\infty}\frac{2}{t_f}\int_0^{t_f} x^T(t)\mathbf{Q}x(t) + u^T(t)\mathbf{R}u(t)dt\right] \\ &= E[x^T(\infty)\mathbf{Q}x(\infty) + u^T(\infty)\mathbf{R}u(\infty)], \end{aligned} \tag{6.61}$$

where the notation (∞) in the final expectation denotes the steady-state value. The feedback gain that minimizes this cost function for all finite final times is the LQR solution. Therefore, the limit must equal the limiting LQR gain; that is, the steady-state stochastic regulator gain equals the steady-state LQR gain.

6.4.1 Cost Computation

The minimum cost can be computed from (6.60) by substituting for the optimum LQR cost using (6.22):

$$J_{SR} = \frac{1}{2}\sum_{k=1}^{n}\sigma_k(\mathbf{\Sigma})U_k^T(\mathbf{\Sigma})\mathbf{P}(0)U_k(\mathbf{\Sigma}) + \frac{1}{2}\int_0^{t_f}\sum_{k=1}^{m}\sigma_k(\mathbf{S})U_k^T(\mathbf{S})\mathbf{B}_w^T\mathbf{P}(t)\mathbf{B}_wU_k(\mathbf{S})dt,$$

where $\mathbf{P}(\bullet)$ is the solution of the Riccati equation (6.20). This expression can be simplified using Appendix equation (A2.3) and the definition of the singular value decomposition:

$$\boxed{J_{SR} = \frac{1}{2}\mathrm{tr}\left\{\mathbf{P}(0)\mathbf{\Sigma}_x(0) + \int_0^{t_f}\mathbf{P}(\tau)\mathbf{B}_w\mathbf{S}_w\mathbf{B}_w^T d\tau\right\}.} \tag{6.62}$$

The cost for the steady-state stochastic regulator can be generated in a similar manner:

$$\boxed{J_{SR} = \mathrm{tr}\{\mathbf{P}(\infty)\mathbf{B}_w\mathbf{S}_w\mathbf{B}_w^T\},} \tag{6.63}$$

where $\mathbf{P}(\infty)$ is the steady-state solution of the Riccati equation.

The cost can also be obtained using (6.57) after directly solving for the closed-loop state correlation matrix. The closed-loop state correlation matrix can be found either by using (6.58) or by solving the following matrix differential equation:

$$\dot{\mathbf{\Sigma}}_x(t) = \{\mathbf{A} - \mathbf{BK}(t)\}\mathbf{\Sigma}_x(t) + \mathbf{\Sigma}_x(t)\{\mathbf{A} - \mathbf{BK}(t)\}^T + \mathbf{B}_w\mathbf{S}_w\mathbf{B}_w^T.$$

Note that the cost can be computed in this manner for both optimal and suboptimal gain matrices, allowing the performance of feedback controllers with suboptimal gains to be compared. This is particularly useful when deciding if the complexity of using an optimal, time-varying state feedback is warranted, or if the time-invariant, steady-state feedback gain produces acceptable performance.

The state correlation matrix can be found by solving the Lyapunov equation:

$$\{\mathbf{A} - \mathbf{BK}\}\boldsymbol{\Sigma}_x(\infty) + \boldsymbol{\Sigma}_x(\infty)\{\mathbf{A} - \mathbf{BK}\}^T + \mathbf{B}_w\mathbf{S}_w\mathbf{B}_w^T = \mathbf{0}, \tag{6.64}$$

in the steady-state case where the feedback gain, and therefore the closed-loop system, is time-invariant. The cost (6.61) is then computed:

$$J_{SR} = \mathrm{tr}\{\mathbf{Q}_R\boldsymbol{\Sigma}_x(\infty)\}.$$

EXAMPLE 6.11 A satellite tracking antenna, subject to random wind torques, can be modeled as follows:

$$\begin{bmatrix} \dot{\theta}(t) \\ \ddot{\theta}(t) \end{bmatrix} = \begin{bmatrix} 0 & 1 \\ 0 & -0.1 \end{bmatrix}\begin{bmatrix} \theta(t) \\ \dot{\theta}(t) \end{bmatrix} + \begin{bmatrix} 0 \\ 0.001 \end{bmatrix} u(t) + \begin{bmatrix} 0 \\ 0.001 \end{bmatrix} w(t),$$

where $\theta(t)$ is the pointing error of the antenna in degrees, $u(t)$ is the control torque in N-m, and $w(t)$ is the wind torque (disturbance input) in N-m. The wind torque is assumed to be white noise with a spectral density $S_w = 5000$ N^2-m^2/Hz. A cost function for this system is chosen:

$$J = E\left[[\theta(\infty) \quad \dot{\theta}(\infty)]\begin{bmatrix} 180 & 0 \\ 0 & 0 \end{bmatrix}\begin{bmatrix} \theta(\infty) \\ \dot{\theta}(\infty) \end{bmatrix} + u^2(\infty)\right].$$

The optimal feedback controller is

$$u(t) = -[13.4 \quad 91.9]\begin{bmatrix} \theta(t) \\ \dot{\theta}(t) \end{bmatrix},$$

which yields the optimal cost of $J = 1.98$.

The state weighting used in this example was approximately the minimum value that yielded a pointing error variance less than or equal to one degree. This state weighting minimizes the required control torque subject to this specification. The state covariance matrix of the closed-loop system is found by solving the Lyapunov equation (6.64):

$$\boldsymbol{\Sigma}_x(\infty) = \begin{bmatrix} 0.971 \text{ deg}^2 & 0 \\ 0 & 0.013\dfrac{\text{deg}^2}{\text{sec}^2} \end{bmatrix},$$

and the standard deviation of the pointing error is the square root of the (1, 1) element of this matrix: $\sigma_\theta = 0.985°$. ◆

6.4.2 $\mathcal{H}_2$ Optimal Control

The performance of a feedback system can be quantified in terms of the closed-loop gain from the disturbance inputs to the reference outputs. The system 2-norm represents an average gain and can be used as a performance function for an optimal control problem. The optimal control for the LQR is also optimal in terms of minimizing the 2-norm of the closed-loop system when the plant is appropriately defined. Formulating the LQR problem as a system 2-norm optimization problem adds insight into the LQR, and yields a formulation that can be easily generalized to include frequency-domain performance specifications. This generalization is discussed in Chapter 8.

The $\mathcal{H}_2$ optimal control problem is to find the linear, time-invariant controller for the plant

$$\dot{x}(t) = \mathbf{A}x(t) + [\mathbf{B}_u \mid \mathbf{I}]\begin{bmatrix} u(t) \\ \hdashline w(t) \end{bmatrix}; \tag{6.69a}$$

$$\begin{bmatrix} m(t) \\ \hdashline y_1(t) \\ u_1(t) \end{bmatrix} = \begin{bmatrix} \mathbf{I} \\ \hdashline \mathbf{Q}^{1/2} \\ \mathbf{0} \end{bmatrix} x(t) + \begin{bmatrix} \mathbf{0} & \mathbf{0} \\ \hdashline \mathbf{0} & \mathbf{0} \\ \mathbf{R}^{1/2} & \mathbf{0} \end{bmatrix}\begin{bmatrix} u(t) \\ \hdashline w(t) \end{bmatrix}, \tag{6.69b}$$

that stabilizes the closed-loop system and minimizes the system 2-norm,

$$J_2 = \left[\int_0^\infty \operatorname{tr}\{\mathbf{g}_{cl}^T(t)\mathbf{g}_{cl}(t)\}dt\right]^{1/2} = \|\mathbf{G}_{cl}\|_2, \tag{6.70}$$

where $\mathbf{g}_{cl}(t)$ is the impulse response matrix of the closed-loop system from the disturbance input to the reference output. A block diagram of the system of (6.69) is given in Figure 6.16. The notation $\mathcal{H}_2$ derives from the Hardy space ($\mathcal{H}$) of all stable, linear, time-invariant systems, with the subscript 2 denoting the applicable system norm.

The $\mathcal{H}_2$ optimal control problem is equivalent to the steady-state stochastic regulator with $\mathbf{S}_w = \mathbf{I}$. In this case, the optimal feedback gain is time-invariant and stabilizes the system. The equivalence of the $\mathcal{H}_2$ optimal control problem and the stochastic regulator can be seen by noting that the steady-state mean square value of the reference output,

$$E\left\{[y_1^T(\infty) \mid u_1^T(\infty)]\begin{bmatrix} y_1(\infty) \\ \hdashline u_1(\infty) \end{bmatrix}\right\} = E[x^T(\infty)\mathbf{Q}x(\infty) + u^T(\infty)\mathbf{R}u(\infty)],$$

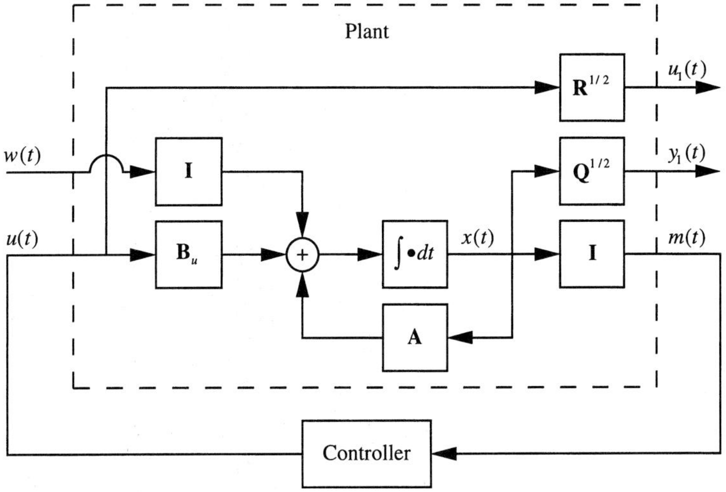

FIGURE 6.16 $\mathcal{H}_2$ optimal control block diagram

is equal to the steady-state stochastic regulator cost function. This mean square value of the output can also be given in terms of the closed-loop system 2-norm:

$$E\left\{[y_1^T(\infty) \vdots u_1^T(\infty)]\begin{bmatrix} y_1(\infty) \\ \hline u_1(\infty) \end{bmatrix}\right\} = \|\mathbf{G}_{cl}\|_2^2.$$

Equating these two expressions, the stochastic regulator cost function is seen to equal the square of the system 2-norm:

$$J_{SR} = J_2^2$$

provided the spectral density matrix is the identity. Since the squaring operation is monotonic, the control that minimizes J_{SR} also minimizes J_2. The $\mathcal{H}_2$ optimal feedback then equals state feedback, where the feedback gain is the steady-state stochastic regulator gain or, equivalently, the steady-state LQR gain.

The linear quadratic regulator optimal control is more general than originally presented. The LQR was originally presented as an open-loop optimization problem; that is, find a control that minimizes a quadratic cost function with no restrictions placed on the control. The test conditions for this problem consist of initial conditions and no disturbance inputs. The optimal solution for the LQR was then found to be state feedback. The same state feedback has now been shown to be the solution of the stochastic regulator problem, where both random initial conditions and random disturbance inputs are present. In addition, state feedback with the LQR feedback gains also minimizes the $\mathcal{H}_2$ cost. These additional results make the LQR solution useful in a wide range of control applications. The one restriction on the use of the LQR is that the entire state must be available for feedback. The need to measure all of the states adds complexity and real economic cost to the control system. This restriction is removed in Chapter 8 after first presenting material on state estimation using incomplete measurements.

6.5 Summary

An optimal control is defined as the control that minimizes a cost function subject to the test conditions. Note that a particular control law can be both optimal and suboptimal with respect to distinct cost functions and test conditions. The optimal control problem is, in general, a constrained minimization problem, since the cost function is subject to the constraints imposed by the state equation. The solution of constrained optimization problems is facilitated by application of the Lagrange multiplier method, which converts a constrained minimization into an unconstrained minimization over a higher-dimensional parameter space. The variational approach to optimization can then be applied to find the optimal control.

The linear quadratic regulator is an optimal control problem where the state equation is linear, the cost function is quadratic, the entire state is measured, there are no constraints on the control, and the test conditions consist of initial conditions on the state with no external inputs. The quadratic cost function can be used in a wide range of applications by appropriately selecting the weights on the state and the control input.

The general solution of the linear quadratic regulator can be derived using variations and the Lagrange multiplier method. The determination of the optimal gains requires the solution of a linear system, of dimension equal to $2n_x$, called the Hamiltonian system. The boundary conditions for this system consist of mixed initial and final conditions. The solution of a system with these boundary conditions, called a two-point boundary-value problem, is not trivial in general.

A linear relationship between the state and the costate can be defined and used in solving the Hamiltonian system. Applying this definition yields a nonlinear matrix differential equation, of dimension n_x^2, with boundary conditions consisting of only final conditions. This equation, known as the Riccati equation, can be solved numerically by working backward from the final condition. Note that the solution of either the Hamiltonian system or the Riccati equation yields the optimal control. The selection of which equation to use is a matter of convenience.

The solution of the linear quadratic regulator problem is time-varying state feedback. The feedback gains vary through a transient, at the final time, and then settle to a steady-state value, far from the final time. In many applications, the control system is designed to operate for extended periods of time, and the steady-state optimal control is sufficient. The steady-state optimal control can be found from the eigensolution of the Hamiltonian system and yields a stable, closed-loop system, provided that the plant is both observable and controllable.

The steady-state LQR, for single-input systems, has good gain and phase margins. For multiple-input systems, the LQR is found to have good stability robustness when subject to input-multiplicative perturbations. This last result requires that the control weighting matrix be proportional to the identity matrix. This is a mild requirement, which can typically be achieved in practice.

The steady-state linear quadratic regulator yields time-invariant state feedback and a time-invariant, closed-loop system. The poles of a time-invariant system provide useful information on closed-loop transient performance. The closed-loop poles are the left half-plane poles of the Hamiltonian. These poles describe a root locus as the control weighting is varied from large to small. This root locus starts at the plant poles (or the left half-plane reflections of these poles) and terminates at the plant zeros (or the left half-plane reflections of these zeros).

The stochastic regulator is an optimal control problem where the state is linear, the cost function quadratic, and the test conditions include random initial conditions and random disturbance inputs. The linear quadratic regulator solution is also the optimal linear solution of the stochastic regulator. This result greatly extends the number of applications that LQR theory can address.

The linear quadratic regulator also yields the solution of an $\mathcal{H}_2$ optimization problem. This $\mathcal{H}_2$ optimization problem is formulated with the closed-loop input matrix equal to the identity matrix, and the output equal to the combination of $\mathbf{Q}^{1/2}$ times the state and $\mathbf{R}^{1/2}$ times the control. Thinking of the LQR in terms of $\mathcal{H}_2$ optimization is useful in generalizing the LQR to include frequency-domain performance specifications.

Optimal control and the linear quadratic regulator can be applied directly to the discrete-time state model of the system (see [9], page 422). Since most optimal controllers are implemented using a digital computer, the discrete-time state model is often the most appropriate model to use for LQR design. The discrete-time LQR equations are very similar to the continuous-time LQR equations, as are the related properties. The exception is in the robustness of the steady-state LQR. The act of sampling can be modeled as the addition of a one-half sample time delay into the feedback loop. This

TABLE 6.2 Summary of Formulas for Continuous-Time and Discrete-Time Systems

Formula	Continuous-Time	Discrete-Time
The linear quadratic regulator: Problem definition	$\dot{x}(t) = \mathbf{A}x(t) + \mathbf{B}_u u(t)$ $J = \frac{1}{2}x^T(t_f)\mathbf{H}x(t_f) + \frac{1}{2}\int_0^{t_f} x^T(t)\mathbf{Q}x(t) + u^T(t)\mathbf{R}u(t)dt$	$x(k+1) = \mathbf{\Phi}x(k) + \mathbf{\Gamma}u(k)$ $J = \frac{1}{2}x^T(k_f)\mathbf{H}_d x(k_f) + \frac{1}{2}\sum_{k=0}^{k_f-1} x^T(k)\mathbf{Q}_d x(k) + u^T(k)\mathbf{R}_d u(k)$
The Hamiltonian system	$\begin{bmatrix} \dot{x}(t) \\ \dot{p}(t) \end{bmatrix} = \begin{bmatrix} \mathbf{A} & -\mathbf{B}_u\mathbf{R}^{-1}\mathbf{B}_u^T \\ -\mathbf{Q} & -\mathbf{A}^T \end{bmatrix}\begin{bmatrix} x(t) \\ p(t) \end{bmatrix};$ $x(0) = x_0;\ p(t_f) = \mathbf{H}x(t_f)$	$\begin{bmatrix} x(k+1) \\ p(k+1) \end{bmatrix} = \begin{bmatrix} \mathbf{\Phi} + \mathbf{\Gamma}\mathbf{R}_d^{-1}\mathbf{\Gamma}^T\mathbf{\Phi}^{-T}\mathbf{Q}_d & -\mathbf{\Gamma}\mathbf{R}_d^{-1}\mathbf{\Gamma}^T\mathbf{\Phi}^{-T} \\ -\mathbf{\Phi}^{-T}\mathbf{Q}_d & \mathbf{\Phi}^{-T} \end{bmatrix}\begin{bmatrix} x(k) \\ p(k) \end{bmatrix}$ $x(0) = x_0;\ p(k_f) = \mathbf{H}_d x(k_f)$
The optimal control	$u(t) = -\mathbf{R}^{-1}\mathbf{B}_u^T p(t) = -\mathbf{K}(t)x(t)$	$u(k) = -\mathbf{R}_d^{-1}\mathbf{\Gamma}^T p(k+1) = -\mathbf{K}(k)x(k)$
The Riccati equation	$\dot{\mathbf{P}}(t) = -\mathbf{P}(t)\mathbf{A} - \mathbf{A}^T\mathbf{P}(t) - \mathbf{Q} + \mathbf{P}(t)\mathbf{B}_u\mathbf{R}^{-1}\mathbf{B}_u^T\mathbf{P}(t);$ $\mathbf{P}(t_f) = \mathbf{H}$	$\mathbf{P}(k) = \mathbf{\Phi}^T\mathbf{P}(k+1)\mathbf{\Phi} + \mathbf{Q}_d - \mathbf{\Phi}^T\mathbf{P}(k+1)\mathbf{\Gamma} \cdot [\mathbf{R}_d + \mathbf{\Gamma}^T\mathbf{P}(k+1)\mathbf{\Gamma}]^{-1}\mathbf{\Gamma}^T\mathbf{P}(k+1)\mathbf{\Phi};$ $\mathbf{P}(k_f) = \mathbf{H}_d$
The optimal feedback gains	$\mathbf{K}(t) = \mathbf{R}^{-1}\mathbf{B}_u^T\mathbf{P}(t)$	$\mathbf{K}(k) = [\mathbf{R}_d + \mathbf{\Gamma}^T\mathbf{P}(k+1)\mathbf{\Gamma}]^{-1}\mathbf{\Gamma}^T\mathbf{P}(k+1)\mathbf{\Phi}$
The optimal cost	$J = \frac{1}{2}x^T(0)\mathbf{P}(0)x(0)$	$J = \frac{1}{2}x^T(0)\mathbf{P}(0)x(0)$
The weighting matrices	$\mathbf{H};\ \mathbf{Q};\ \mathbf{R}$	$\mathbf{H}_d = \mathbf{H};\ \mathbf{Q}_d = \mathbf{Q}T;\ \mathbf{R}_d = \mathbf{R}T$

delay results in phase shifts that destroy the guaranteed robustness margins of the continuous-time LQR. Note that these robustness margins are restored when the sampling time approaches zero. The fundamental equations presented in this chapter are summarized for both continuous-time and discrete-time systems in Table 6.2.

REFERENCES

[1] R. E. Kalman, "Contributions to the theory of optimal control," *Bol. Soc. Mat. Mexicana,* 5 (1960): 102–19.

[2] D. E. Kirk, *Optimal Control Theory: An Introduction,* Prentice-Hall, Englewood Cliffs, NJ, 1970.

[3] R. E. Bellman, *Dynamic Programming,* Princeton University Press, Princeton, NJ, 1957.

[4] M. Athans and P. L. Falb, *Optimal Control: An Introduction to the Theory and Its Applications,* McGraw-Hill, New York, 1966.

[5] H. Kwakernaak and R. Sivan, *Linear Optimal Control Systems,* Wiley-Interscience, New York, 1972.

[6] B. D. O. Anderson and J. B. Moore, *Optimal Control: Linear Quadratic Methods,* Prentice-Hall, Englewood Cliffs, NJ, 1990.

[7] A. E. Bryson, Jr. and Y.-C. Ho, *Applied Optimal Control: Optimization, Estimation, and Control,* Hemisphere Publishing Co., Washington, DC, 1975.

[8] W. F. Arnold, III and A. J. Laub, "Generalized eigenproblem algorithms and software for algebraic Riccati equations," *Proceedings of IEEE,* 72, no. 12 (1984): 1746–54.

[9] G. E. Franklin, J. D. Powell, and M. L. Workman, *Digital Control of Dynamic Systems,* 2d ed. Addison-Wesley, Reading, MA, 1990.

Some additional references on optimal control and the linear quadratic regulator follow:

[10] B. Friedland, *Control System Design: An Introduction to State-Space Methods,* McGraw-Hill, NY, 1986.

[11] F. L. Lewis, *Optimal Control,* John Wiley & Sons, New York, 1986.

[12] A. P. Sage and C. C. White, III, *Optimum Systems Control,* 2d ed., Prentice-Hall, Englewood Cliffs, NJ, 1977.

EXERCISES

6.1 Use variations to derive a set of necessary conditions for the minimum of

$$J(x) = x^T\mathbf{Q}x + v^Tx,$$

where $\mathbf{Q}$ is positive definite, and v is fixed. Define the gradient of $J(x)$ with respect to x, using the variation.

6.2 Given the cost function and the constraint,

$$J(x, y) = 5x^2 - 10x + 5 + y^2;\ y = x^2,$$

use the Lagrange multiplier method to find the values of x and y that minimize the cost.

6.3 The precise positioning of an end effector on a robot arm is required to perform a welding operation. The end effector must be in position at a specified time. Generate a reasonable state equation (specify units) and cost function to use for designing an optimal control for this system.

6.4 The location of a geosynchronous satellite must be maintained at a given location in orbit (satellites tend to drift over southeast Asia if left untended). Assume that the state vector of the satellite is $x(t) = [\theta(t)\quad \dot{\theta}(t)\quad r(t)\quad \dot{r}(t)\quad \phi(t)\quad \dot{\phi}(t)]^T$, where $\theta(t)$ is the angle of the satellite (from the vernal equinox) in the equatorial plane of the Earth, $r(t)$ is the radius of the orbit, and $\phi(t)$ is the tilt of the orbit relative to Earth's equatorial plane. Generate a reasonable cost function for this control task.

6.5 Given the system:

$$\dot{x}(t) = 2x(t) + u(t),$$

and the cost function:

$$J(x(t), u(t)) = 5x^2(10) + \int_0^{10} 2x^2(t) + u^2(t)dt,$$

do the following.

a. Generate the Hamiltonian for this control problem.
b. Generate the Riccati equation for this control problem.
c. Solve for the optimal feedback gain (this can be done numerically).
d. Solve for the steady-state solution of the optimal gain, using the eigenvector decomposition method.
e. What is the pole of the steady-state, closed-loop system?
f. How does this pole change if the state weighting is increased by a factor of 10?
g. If $x(0) = 100$, what is the optimal cost?

6.6 You are given the time-varying plant

$$\dot{x}(t) = \mathbf{A}(t)x(t) + \mathbf{B}(t)u(t),$$

and the cost function

$$J(x(t), u(t)) = \frac{1}{2}x^T(t_f)\mathbf{H}x(t_f) + \frac{1}{2}\int_0^{t_f} x^T(t)\mathbf{Q}(t)x(t) + u^T(t)\mathbf{R}(t)u(t)dt.$$

Note that the state and control weighting matrices in the cost function are time-varying. Use variations and the Lagrange multiplier method to derive the optimal control for this system.

6.7 Given the discrete-time plant

$$x(k+1) = \mathbf{\Phi}x(k) + \mathbf{\Gamma}u(k),$$

and the cost function

$$J(x(k), u(k)) = \frac{1}{2}x^T(k_f)\mathbf{H}_d x(k_f) + \frac{1}{2}\sum_{k=0}^{k_f-1} x^T(k)\mathbf{Q}_d x(k) + u^T(k)\mathbf{R}_d u(k),$$

do the following.

a. Use the Lagrange multiplier method and differentiation to derive a set of necessary conditions for optimality for this control problem.
b. Assume $p(k) = \mathbf{P}(k)x(k)$, and generate an expression for the optimal control in terms of $\mathbf{P}(k)$ and $x(k)$. This expression should match that in Table 6.2.
c. Generate a recursive expression for updating $\mathbf{P}(k)$ by substituting for $p(k)$ in the necessary conditions. *Hint:* Put this expression into a form where a function of $\mathbf{P}(k)$ is all multiplied by $x(k)$. This function of $\mathbf{P}(k)$ must then equal zero, since $x(k)$ is arbitrary. The resulting recursive expression for $\mathbf{P}(k)$ should match that in Table 6.2.
d. Generate a final condition for $\mathbf{P}(k)$ by differentiating J_a with respect to $x(k_f)$.

6.8 Show that the function

$$J(x) = x^T\mathbf{Q}x + v^Tx$$

has a unique minimum if $\mathbf{Q}$ is positive definite and v is given.

6.9 Given the plant

$$\dot{x}(t) = \mathbf{A}x(t) + \mathbf{B}u(t),$$

and the cost function

$$J(x(t), u(t)) = \frac{1}{2}x^T(t_f)\mathbf{H}x(t_f) + \frac{1}{2}\int_0^{t_f} [x(t) - r(t)]^T\mathbf{Q}[x(t) - r(t)] + u^T(t)\mathbf{R}u(t)dt,$$

use variations and the Lagrange multiplier method to derive the optimal control for this system. *Hint:* Assume that the costate is related to the state by the equation

$$p(t) = \mathbf{P}(t)x(t) + s(t),$$

and generate differential equations and final conditions for both $\mathbf{P}(t)$ and $s(t)$. The optimal control combines feedback and feedforward control. Is this a desirable control structure? Why?

COMPUTER EXERCISES

6.1 Optimal Control of a DC Motor
Design a system for controlling the shaft angle of a dc motor. The entire state is measured and can be used to generate the motor control voltage. The desired shaft angle is zero degrees. Note that this can yield any desired shaft angle by the appropriate definition of reference direction for shaft angle.

A block diagram for the dc motor is given in Figure P6.1, where $\theta(t)$ is the motor shaft angle, and $u(t)$ is the dc voltage applied to the motor. Design a control system to minimize the cost function

$$J(x(t), u(t)) = \frac{1}{2}\int_0^{0.4} [\theta(t) \quad \dot{\theta}(t)]\begin{bmatrix} 1 & 0 \\ 0 & 0 \end{bmatrix}\begin{bmatrix} \theta(t) \\ \dot{\theta}(t) \end{bmatrix} + 10^{-8}u^2(t)dt.$$

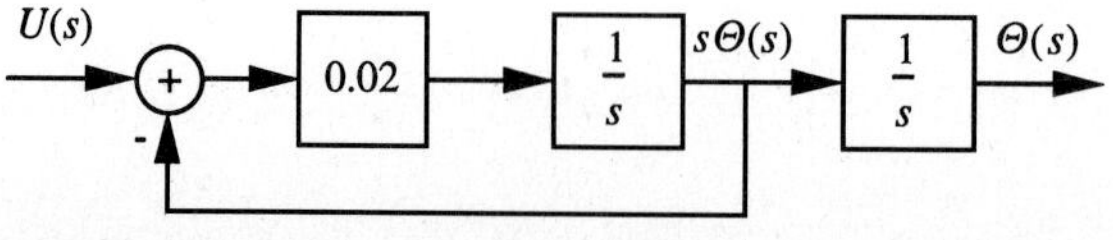

FIGURE P6.1 Block diagram for a dc motor

Simulate the closed-loop system with an initial shaft angle of 10 degrees and an initial shaft velocity of zero. Using the steady-state gains, simulate the closed-loop system with the same initial conditions. Plot the plant state and the control input for these two simulations and compare the results. Compute the cost for each of these two controllers and compare.

a. Repeat the above when the final time in the cost function is changed to 0.2. Comment on the results.
b. Repeat the above when the final time in the cost function is changed to 0.1. Comment on the results.

6.2 Optimal Control of the Altitude of an Unmanned Surveillance Vehicle

Design a system for controlling the altitude of an unmanned surveillance vehicle. The entire state is measured and can be used to generate the throttle input. A constant altitude of zero feet is desired. Note that this can yield any desired height by appropriate definition of the zero altitude reference plane.

A block diagram for the vertical motion of the vehicle is given in Figure P6.2, where $h(t)$ is the height of the vehicle, $u(t)$ is the throttle input, and $\tau = 2$ is a time constant indicating that acceleration does not change instantaneously. Design a control system to minimize the cost:

$$J(x(t), u(t)) = \frac{1}{2}\int_0^\infty [h(t) \quad \dot{h}(t) \quad \ddot{h}(t)]\begin{bmatrix} 1 & 0 & 0 \\ 0 & 0 & 0 \\ 0 & 0 & 0 \end{bmatrix}\begin{bmatrix} h(t) \\ \dot{h}(t) \\ \ddot{h}(t) \end{bmatrix} + 2u^2(t)dt.$$

Simulate the closed-loop system with the initial height equal to 10 feet and the other states equal to zero, and plot the plant state and the control input. Generate a Nyquist plot for the system and find the stability margins. Compare these to the guaranteed margins given in (6.42).

Design additional controllers, simulate, evaluate robustness, and plot, as above, for the following:

a. $$\mathbf{Q} = \begin{bmatrix} 1 & 0 & 0 \\ 0 & 0 & 0 \\ 0 & 0 & 0 \end{bmatrix}; R = 2000;$$

b. $$\mathbf{Q} = \begin{bmatrix} 10 & 0 & 0 \\ 0 & 0 & 0 \\ 0 & 0 & 0 \end{bmatrix}; R = 2;$$

c. $$\mathbf{Q} = \begin{bmatrix} 1 & 0 & 0 \\ 0 & 100 & 0 \\ 0 & 0 & 0 \end{bmatrix}; R = 2.$$

Comment on the effects of changing the cost function on the states, the control inputs, the optimal feedback gains, and the robustness.

Generate a root locus for the closed-loop poles as R is varied and $\mathbf{Q}$ is as given for the original design. Compare the initial and terminating points of this locus to the limiting pole locations given in Section 6.3.4.

6.3 Optimal Control of Blood Gases During Extracorporeal Support

Design a system for controlling the partial pressure of oxygen and the partial pressure of carbon dioxide during extracorporeal support. The entire state is measured and can be fed back to control the oxygenator.

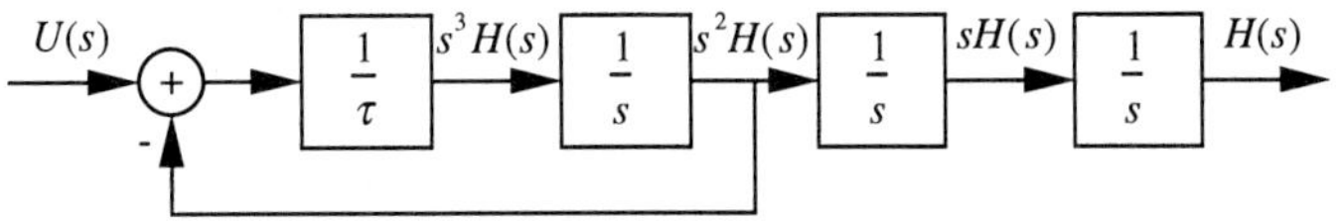

FIGURE P6.2 Block diagram for the vertical motion of an unmanned surveillance vehicle

The state model of the oxygenator is

$$\begin{bmatrix} \dot{x}_1(t) \\ \dot{x}_2(t) \\ \dot{x}_3(t) \\ \dot{x}_4(t) \end{bmatrix} = \begin{bmatrix} -10 & 0 & 0 & 0 \\ 0 & -10 & 0 & 0 \\ 6 & -3 & -5 & 0 \\ 0 & 0.5 & 0 & -5 \end{bmatrix} \begin{bmatrix} x_1(t) \\ x_2(t) \\ x_3(t) \\ x_4(t) \end{bmatrix} + \begin{bmatrix} 10 & 0 \\ 0 & 10 \\ 0 & 0 \\ 0 & 0 \end{bmatrix} \begin{bmatrix} u_1(t) \\ u_2(t) \end{bmatrix},$$

where x_1 is the flow rate of oxygen, x_2 is the flow rate of carbon dioxide, x_3 is the arterial partial pressure of oxygen, x_4 is the arterial partial pressure of carbon dioxide, u_1 is the commanded oxygen flow rate, and u_2 is the commanded carbon dioxide flow rate. The outputs of interest are x_3 and x_4. Design a control system to minimize the cost function

$$J(x(t), u(t)) = \frac{1}{2} \int_0^\infty x_3^2(t) + x_4^2(t) + \rho u^T(t) u(t) dt,$$

where $\rho = 2$. Simulate the closed-loop system with an initial state

$$x(0) = [0 \quad 0 \quad 50 \quad 20]^T,$$

and plot the outputs and the control input. Compute the norm of the smallest destabilizing input-multiplicative perturbation. Compare this to the guaranteed value given in (6.46).

Design additional controllers, simulate, evaluate robustness, and plot, as above, for the following:

a. $\rho = 0.02$;
b. $\rho = 200$.

Comment on the effect of changing the cost function on the states, the control inputs, the optimal feedback gains, and the robustness.

Generate a root locus for the closed-loop poles as ρ is varied. Compare the initial and terminating points of this locus to the limiting pole locations given in Section 6.3.4.

7 The Kalman Filter

The linear quadratic regulator provides a powerful controller design methodology that is applicable to both single- and multiple-input plants. A fundamental limitation of the LQR is imposed by the need to measure the entire state. In many applications, some states are unmeasurable; that is, there are no currently available sensors capable of measuring these states. In many other applications, the cost of including sensors for measuring the entire state is prohibitive or undesirable. Therefore, a methodology is needed for designing controllers when only partial state measurements are available.

A controller can be designed by using partial state measurements to estimate the entire state, and then using these estimates in place of the actual states in the LQR. A design generated in this way should perform similarly to the LQR provided the state estimates are close to the actual states. In fact, the resulting design is optimal in terms of minimizing a quadratic cost function if the state estimates are optimal. The controller formed by combining the LQR with an optimal state estimator is known as a linear quadratic Gaussian controller (see Chapter 8).

The Kalman filter is an optimal estimator of the state, where optimal is defined in terms of minimizing the mean square estimation error. The Kalman filter has been utilized in an extremely wide range of applications both as a signal processing tool and as an integral component in the linear quadratic Gaussian controller. Signal processing is the extraction of information from sensor signals. For example, the estimation of an aircraft's range and azimuth from a surveillance radar is a signal processing problem that is typically solved using Kalman filters. At the same time, a Kalman filter may also be an integral part of the autopilot for the airplane being tracked by the radar. Due to this dichotomy of function, the Kalman filter is probably used more often than any other technique presented in this book.

The Kalman filter is developed in the remainder of this chapter after first presenting some results on linear minimal mean square estimation.

7.1 Linear Minimum Mean Square Estimation

The minimal mean square estimate $\hat{x}(m)$ of a vector x, called the state (since state estimation is of interest), given a set of measurements (or data) $\{m(\tau), 0 \leq \tau \leq t\}$ is defined as the estimate that minimizes the mean square error:

$$J = E[e^T e] = E[\|e\|^2] = \sum_{k=1}^{n} E[\{x_i - \hat{x}_i(m)\}^2], \tag{7.1a}$$

where

$$e = x - \hat{x}(m) \tag{7.1b}$$

is the estimation error. The estimate that minimizes the mean square error is the conditional mean of the state, conditioned on the measurements (see [1], page 313):

$$\hat{x}(m) = E[x \mid m(\tau),\ 0 \le \tau \le t]. \tag{7.2}$$

Computing the conditional mean is not tractable in many applications. A special case of great interest is when the state and the measurements are jointly Gaussian and zero mean.[1] In this case, the optimal estimate is a linear function of the measurements:

$$\hat{x}(m) = \int_0^t \mathbf{F}(t,\ \tau)m(\tau)d\tau. \tag{7.3}$$

The determination of the optimal estimator then only requires finding the parameters in $\mathbf{F}(t, \tau)$.

A linear estimator can also be used in applications where the state and measurements are non-Gaussian. The optimal linear estimator is defined as the linear estimator that minimizes the mean square error. Note that the estimates obtained in this case are not necessarily optimal over the set of all possible estimators. However, using linear estimators typically simplifies the computation required for implementing the estimator, as well as the computation required for designing the estimator. In addition, optimal linear estimators often yield excellent estimates in practical applications.

The remainder of this section presents some results on the determination of the optimal linear estimator. These results are derived for the case when the measurements form a single vector, and the linear estimator (7.3) simplifies to

$$\hat{x}(m) = \mathbf{F}m. \tag{7.4}$$

This restriction on the data is made to simplify the derivations, but the results presented are quite general and can be applied when an infinite number of measurements exist. The derivations presented can be directly extended to the more general case by replacing (7.4) with (7.3).

7.1.1 The Orthogonality Principle

The optimal linear estimate can be obtained by differentiating the mean square error with respect to each of the estimator parameters, and setting the results to zero. This procedure yields the correct estimator, but is a bit tedious. The optimal linear estimator can also be obtained by setting the differential (or variation) of the mean square error, with respect to the estimator parameters, to zero. This approach has the advantage that it is matrix-based, and can be directly extended to the case when the number of measurements is infinite.

---

[1] The restriction to zero mean can be removed, in general, by subtracting the mean from the data and using the result to estimate the difference between the state and its mean. The estimate of the state is then obtained by adding the estimated difference to the known mean value of the state.

The differential can be found by rewriting the mean square error:

$$J = E[\|e\|^2] = E[\mathrm{tr}\{ee^T\}] = E[\mathrm{tr}\{(x - \mathbf{F}m)(x - \mathbf{F}m)^T\}].$$

The increment of J with respect to the estimator gain $\mathbf{F}$ is then

$$\Delta J(\mathbf{F}, \Delta_\mathbf{F}) = J(\mathbf{F} + \Delta_\mathbf{F}) - J(\mathbf{F}) = E[\mathrm{tr}\{-em^T\Delta_\mathbf{F}^T - \Delta_\mathbf{F} me^T + \Delta_\mathbf{F} mm^T\Delta_\mathbf{F}^T\}].$$

The differential with respect to the estimator gain matrix is the portion of the increment that is linear in $\Delta_\mathbf{F}$:

$$\delta J(\mathbf{F}, \Delta_\mathbf{F}) = E[\mathrm{tr}\{-em^T\Delta_\mathbf{F}^T - \Delta_\mathbf{F} me^T\}] = -2E[\mathrm{tr}\{em^T\Delta_\mathbf{F}^T\}].$$

This final expression is obtained by noting that the trace is unaffected by the transpose. For the optimal estimator, this differential must equal zero for all perturbations $\Delta_\mathbf{F}$, which implies that

$$E[em^T] = E[(x - \mathbf{F}m)m^T] = \mathbf{0}. \tag{7.5}$$

Equation (7.5) is a necessary condition for an optimal linear estimator. An important geometric interpretation of (7.5) follows:

> A necessary condition for a linear estimate to be optimal is that the estimation error be orthogonal (in the probabilistic sense) to all of the data:

$$\boxed{E[em_i] = 0 \text{ for all } i.} \tag{7.6}$$

This result is known as the orthogonality principle.

EXAMPLE 7.1 A sensor is used to measure the torque on a radio telescope due to wind. This information is used to place the telescope in a safe position when wind torques become excessive. The measurements are digitally sampled every second and have significant errors. These errors can be reduced by averaging multiple measurements, but caution is required since the wind velocity changes with time, causing old data to be invalid. Consider using two measurements, the current measurement m_0 and the previous measurement m_{-1}, to estimate the true torque. Two measurements are used to simplify the algebra. A practical system would probably utilize more measurements. The measured torques are modeled by

$$m_k = T_k + v_k,$$

where T_k is the true torque, and v_k is the measurement noise. The measurement noises v_{-1} and v_0 are assumed to be jointly independent and to be zero mean with a standard deviation of $\sigma_v = 100$ N-m. The true torque is assumed to be zero mean, have a standard deviation of 100 N-m, and to have a correlation time of 5 seconds. A correlation function for the torque that satisfies these conditions is

$$R_T(\tau) = 10{,}000e^{-|\tau|/5} \text{ N}^2\text{-m}^2.$$

While this is not the only possible correlation function that matches the data, it will be assumed to accurately describe the torques. The noise is also assumed to be independent of the true torque.

The general linear estimate is of the following form:

$$\hat{T}_0 = [f_1 \quad f_2]\begin{bmatrix} m_{-1} \\ m_0 \end{bmatrix}.$$

The coefficients in this estimator can be found by using (7.5):

$$E\left[\left(T_0 - [f_1 \quad f_2]\begin{bmatrix} m_{-1} \\ m_0 \end{bmatrix}\right)[m_{-1} \quad m_0]\right] = \mathbf{0}.$$

Solving for $\mathbf{F}$, we have

$$[f_1 \quad f_2] = E\{T_0[m_{-1} \quad m_0]\}E\left\{\begin{bmatrix} m_{-1} \\ m_0 \end{bmatrix}[m_{-1} \quad m_0]\right\}^{-1},$$

which yields the optimal estimator:

$$\hat{x}(m) = E\{T_0[m_{-1} \quad m_0]\}E\left\{\begin{bmatrix} m_{-1} \\ m_0 \end{bmatrix}[m_{-1} \quad m_0]\right\}^{-1}\begin{bmatrix} m_{-1} \\ m_0 \end{bmatrix}.$$

The first expected value in this equation can be expanded:

$$E\{T_0[m_{-1} \quad m_0]\} = E\{T_0[T_{-1} + v_{-1} \quad T_0 + v_0]\} = [R_T(1) \quad R_T(0)],$$

where $E[T_k v_j] = 0$ due to the independence of the torque and the noise, and the fact that the noise is zero mean. The correlation matrix of the measurement noise can also be expanded:

$$\begin{aligned} E\left\{\begin{bmatrix} m_{-1} \\ m_0 \end{bmatrix}[m_{-1} \quad m_0]\right\} &= E\left\{\begin{bmatrix} T_{-1} + v_{-1} \\ T_0 + v_0 \end{bmatrix}[T_{-1} + v_{-1} \quad T_0 + v_0]\right\} \\ &= \begin{bmatrix} R_T(0) + \sigma_v^2 & R_T(1) \\ R_T(1) & R_T(0) + \sigma_v^2 \end{bmatrix}, \end{aligned}$$

where $E[v_{-1}v_0] = 0$ due to the independence of the measurement noises and the fact that the noise is zero mean. Combining these results yields the estimator:

$$\hat{T}_0 = [0.25 \quad 0.40]\begin{bmatrix} m_{-1} \\ m_0 \end{bmatrix}.$$

Note that the most current measurement is weighted more heavily than the previous measurement. This is reasonable since the wind velocity, and therefore the torque, is likely to have changed since the previous measurement was made. ◆

7.1.2 The Optimal Estimation Error

The estimation error covariance matrix and the mean square error provide quantitative measures of the performance of the optimal estimator. Both of these performance indicators can be computed from the statistical information used in generating the optimal estimator.

The estimation error covariance matrix is defined as follows:

$$\mathbf{\Sigma}_e = E[ee^T] = E[\{x - \hat{x}(m)\}\{x - \hat{x}(m)\}^T]. \tag{7.7}$$

For the optimal estimator (7.4), this covariance matrix can be expanded:

$$\begin{aligned}\mathbf{\Sigma}_e = E[ee^T] &= E[\{x - \hat{x}(m)\}x^T] - E[\{x - \hat{x}(m)\}m^T]\mathbf{F}^T \\ &= E[\{x - \hat{x}(m)\}x^T] = E[ex^T],\end{aligned} \tag{7.8}$$

where the orthogonality of the optimal estimation error and the data has been used to simplify this expression.

The mean square error is the trace of the estimation error covariance matrix:

$$J = \text{tr}\{E[ex^T]\} = E[e^Tx]. \tag{7.9}$$

The final expression in (7.9) is generated by noting that the trace is invariant under cyclic permutations; see Appendix equation (A2.3).

EXAMPLE 7.1 CONTINUED The mean square error equals the covariance matrix of the estimation error in Example 7.1, since the quantity being estimated is a scalar. The mean square error is

$$J = E[\{T_0 - \hat{T}_0\}T_0] = E[\{T_0 - f_1(T_{-1} + v_{-1}) - f_2(T_0 + v_0)\}T_0].$$

Using the fact that the torque is independent of the measurement noise yields

$$J = R_T(0) - f_1R_T(1) - f_2R_T(0) = 4000 \text{ N}^2\text{-m}^2.$$

The standard deviation of the estimation error is then 63 N-m. This is less than the standard deviation of a single measurement, which is 100 N-m.

7.1.3 Updating an Estimate Given New Data

The data arrive sequentially in most real-time estimation problems, and an estimate is generated as each measurement arrives. The number of operations required to compute the optimal gain matrix and to implement the estimator (7.4) increases as the size of the data set increases. For long data sequences, the amount of computation eventually grows to the point where the estimate can no longer be generated in real time using (7.4). An alternative form of the optimal estimator can be developed that updates the old estimate as new data arrives. This form of the estimator requires a fixed amount of computation for each update, and can be applied to long data sequences in real time.

Consider the problem of estimating x given first the measurement vector m_1, and then updating this estimate given a second measurement vector m_2. The optimal estimate can be obtained by first estimating the vector x from m_1. A second estimate of the vector x is then obtained from the portion of m_2 that is orthogonal to m_1. These two estimates are summed to yield the final optimal estimate. Formally, the linear minimum mean square estimator of x, given m_1 and m_2 is

$$\hat{x}(m_1, m_2) = \hat{x}(m_1) + \hat{x}(m_2 - \hat{m}_2(m_1)) \tag{7.10}$$

where $\hat{x}(m_1, m_2)$ is the optimal linear estimate of x given both m_1 and m_2, $\hat{x}(m_1)$ is the optimal linear estimate of x given only m_1, $\hat{m}_2(m_1)$ is the optimal linear estimate of m_2 given m_1, and $\hat{x}(m_2 - \hat{m}_2(m_1))$ is the optimal linear estimate of x given only $(m_2 - \hat{m}_2(m_1))$.

The optimality of (7.10) can be verified using the orthogonality principle. The optimal linear estimate of x satisfies the orthogonality conditions:

$$E[\{x - \hat{x}(m_1, m_2)\}m_1^T] = \mathbf{0}; \tag{7.11a}$$

$$E[\{x - \hat{x}(m_1, m_2)\}m_2^T] = \mathbf{0}. \tag{7.11b}$$

Substituting the estimate (7.10) into the expected value in (7.11a) and regrouping yields

$$E[\{x - \hat{x}(m_1, m_2)\}m_1^T] = E[\{x - \hat{x}(m_1)\}m_1^T] - E[\hat{x}(m_2 - \hat{m}_2(m_1))m_1^T].$$

Noting that the linear estimate in the second term is simply a gain matrix times the data, this expression becomes

$$E[\{x - \hat{x}(m_1, m_2)\}m_1^T] = E[\{x - \hat{x}(m_1)\}m_1^T] - \mathbf{F}_1 E[\{m_2 - \hat{m}_2(m_1)\}m_1^T].$$

Both of the terms in this equation are zero since the estimates in these terms are assumed to be optimal. Therefore, the estimate (7.10) satisfies the first of the conditions for optimality (7.11a). Substituting the estimate (7.10) into the expected value in (7.11b) and adding and subtracting $\hat{m}_2^T(m_1)$, we have

$$\begin{aligned} &E[\{x - \hat{x}(m_1, m_2)\}m_2^T] \\ &= E[\{x - \hat{x}(m_1) - \hat{x}(m_2 - \hat{m}_2(m_1))\}\{m_2^T - \hat{m}_2^T(m_1) + \hat{m}_2^T(m_1)\}]. \end{aligned}$$

Rearranging yields

$$\begin{aligned} E[\{x - \hat{x}(m_1, m_2)\}m_2^T] = {} & E[\{x - \hat{x}(m_2 - \hat{m}_2(m_1))\}\{m_2^T - \hat{m}_2^T(m_1)\}] \\ & - E[\hat{x}(m_1)\{m_2^T - \hat{m}_2^T(m_1)\}] + E[\{x - \hat{x}(m_1)\}\hat{m}_2^T(m_1)] \\ & - E[\hat{x}(m_2 - \hat{m}_2(m_1))\hat{m}_2^T(m_1)]. \end{aligned}$$

Noting that a linear estimate is simply a gain matrix times the data, this expression becomes

$$\begin{aligned} E[\{x - \hat{x}(m_1, m_2)\}m_2^T] = {} & E[\{x - \hat{x}(m_2 - \hat{m}_2(m_1))\}\{m_2^T - \hat{m}_2^T(m_1)\}] \\ & - \mathbf{F}_2 E[m_1\{m_2^T - \hat{m}_2^T(m_1)\}] + E[\{x - \hat{x}(m_1)\}m_1^T]\mathbf{F}_3^T \\ & - \mathbf{F}_1 E[\{m_2 - \hat{m}_2(m_1)\}m_1^T]\mathbf{F}_3^T. \end{aligned} \tag{7.12}$$

All of the terms on the right in (7.12) are zero since the estimates in these terms are assumed to be optimal. Therefore, the estimate (7.10) satisfies (7.11b) and is the optimal estimate.

EXAMPLE 7.1 CONTINUED The optimal linear estimator in Example 7.1 can also be found using (7.10):

$$\hat{T}_0(m_{-1}, m_0) = \hat{T}_0(m_{-1}) + \hat{T}_0(m_0 - \hat{m}_0(m_{-1})).$$

The estimate of the current torque, given the previous measurement is

$$\hat{T}_0(m_{-1}) = \frac{R_T(1)}{R_T(1) + \sigma_v^2} m_{-1} = 0.41 m_{-1}.$$

The estimate of the current torque measurement, given the previous torque measurement is

$$\hat{m}_0(m_1) = \frac{R_T(1)}{R_T(1) + \sigma_v^2} m_{-1} = 0.41 m_{-1}.$$

The estimate of the torque, given the orthogonal portion of the current measurement is

$$\begin{aligned}\hat{T}_0[m_0 - \hat{m}_0(m_1)] &= \frac{[R_T(0) - 0.41R_T(1)][m_0 - \hat{m}_0(m_1)]}{R_T(0) + \sigma_v^2 + 0.41^2 R_T(0) + 0.41^2\sigma_v^2 - 0.82R_T(1)} \\ &= 0.40[m_0 - \hat{m}_0(m_1)]\end{aligned}$$

Combining these results yields the optimal linear estimate of the current torque:

$$\hat{T}_0(m_{-1}, m_0) = 0.41m_{-1} + 0.40(m_0 - 0.41m_{-1}) = 0.25m_{-1} + 0.40m_0.$$

Note that this result is identical to the optimal estimate obtained previously. ◆

The derivation of the estimator in this example was more complex than the direct derivation presented in Subsection 7.1.1. The advantage of updating an existing estimate is that it does not require recomputing all of the estimator gains for the data used in the existing estimate. This ability to sequentially process data is a significant practical advantage when processing long data strings in real time.

To summarize, the linear optimal estimate of the state, given two data vectors, can be computed sequentially. The state is initially estimated using the first data vector. The state is then estimated using the portion of the second data vector that is not predictable from the first data vector. The final state estimate is obtained by summing these two estimates. This procedure is reasonably intuitive when thinking of the unpredictable portion of the second data vector as new information. The summing of the two estimates can then be thought of as correcting the original estimate based on this new information.

7.2 The Kalman Filter

The Kalman filter estimates the state of a plant given a set of known inputs and a set of measurements. The plant is described by the state model:[2]

$$\dot{x}(t) = \mathbf{A}x(t) + \mathbf{B}_u u(t) + \mathbf{B}_w w(t); \tag{7.13a}$$

$$m(t) = \mathbf{C}_m x(t) + v(t), \tag{7.13b}$$

which is driven by both a known, deterministic input $u(t)$, and an unknown random input $w(t)$ called the plant noise. The measurements from the plant $m(t)$ are corrupted by a random measurement noise $v(t)$. The plant and measurement noises are assumed to be white noise vectors with the spectral densities $\mathbf{S}_w$ and $\mathbf{S}_v$, respectively:

$$E[w(t)w^T(t + \tau)] = \mathbf{S}_w \delta(\tau); \tag{7.14a}$$

$$E[v(t)v^T(t + \tau)] = \mathbf{S}_v \delta(\tau). \tag{7.14b}$$

[2] The plant and measurement noises can be combined into a single disturbance input. Separating these inputs as in (7.13) matches the majority of the literature, and simplifies the discussion of the noise properties and the resulting filter performance.

The requirement that the plant noise and the measurement noise be white is subsequently removed in Section 7.4.

The plant noise and the measurement noise are assumed to be uncorrelated since they are typically generated by differing phenomena.[3] Since the mean values of these noises are zero, they are also orthogonal:

$$\boxed{E[v(t)w^T(t + \tau)] = \mathbf{0}.} \tag{7.15a}$$

The plant noise and the initial state are assumed to be uncorrelated and therefore orthogonal:

$$\boxed{E[x(0)w^T(t)] = \mathbf{0} \text{ for all } t \geq 0.} \tag{7.15b}$$

This assumption follows from the fact that the plant noise can only affect future states, and white noise is uncorrelated from one time instant to the next. The state and the measurement noise are also assumed to be uncorrelated and orthogonal:

$$\boxed{E[x(t)v^T(t + \tau)] = \mathbf{0} \text{ for all } \tau,} \tag{7.15c}$$

due to their differing origins.

The Kalman filter generates the linear estimate of the plant state (7.13) that minimizes the mean square estimation error:

$$\boxed{J = E[\{x(t) - \hat{x}(t)\}^T\{x(t) - \hat{x}(t)\}] = \sum_{k=1}^{n} E[\{x_i(t) - \hat{x}_i(t)\}^2].} \tag{7.16}$$

The data used in generating this estimate are the measurements from time zero to the present:

$$\boxed{\text{data} = m(\tau) \mid \tau \in [0, t].} \tag{7.17}$$

Additionally, the known input at the current and past times is assumed to be available for use by the Kalman filter.

7.2.1 The Kalman Filter Equation

The optimal linear state estimator is a dynamic system, since the measurements and known inputs are functions of time. A convenient model for such a system is the state model:

$$\dot{\hat{x}}(t) = \mathbf{F}(t)\hat{x}(t) + \mathbf{G}(t)m(t) + \mathbf{H}(t)u(t),$$

[3] This assumption is valid in most applications, with the notable exception of when the measurement noise is nonwhite. In this case, modified plant equations are obtained with correlated plant and measurement noises. The optimal estimator for nonwhite measurement noise is given in Subsection 7.4.2. The optimal estimator for the case of correlated plant and measurement noises is also provided in that subsection.

where all of the matrices are assumed to be time-varying, in general. The state in this model is chosen to be the state estimate for the plant, a choice that fixes the order of the estimator. This selection of estimator order is justified below by showing optimality of the resulting estimates.

The matrices in the optimal estimator are found by considering the estimation of the state derivative. The data is segmented into "old data," measurements from time 0 through time $(t - \varepsilon)$ where ε is a small positive number, and "new data," the measurement at time t.[4] Equation (7.10) can then be used to generate the optimal estimate. Applying this equation, and noting that a linear estimate is just a gain matrix times the data, yields

$$\hat{\dot{x}}(t) = \hat{\dot{x}}_\varepsilon(t) + \mathbf{G}(t)[m(t) - \hat{m}_\varepsilon(t)],$$

where the subscript ε denotes that the estimate is based on data through time $(t - \varepsilon)$. The order of differentiation and estimation can be interchanged when both operations exist:[5]

$$\dot{\hat{x}}(t) = \hat{\dot{x}}_\varepsilon(t) + \mathbf{G}(t)[m(t) - \hat{m}_\varepsilon(t)]. \tag{7.18}$$

The derivative of the state estimate is then fully specified by $\hat{\dot{x}}_\varepsilon(t)$, $\hat{m}_\varepsilon(t)$, and the Kalman gain $\mathbf{G}(t)$.

The estimate $\hat{\dot{x}}_\varepsilon(t)$ can be computed from the optimal state estimate. Given the optimal estimate $\hat{x}_\varepsilon(t)$ and no new information, it is reasonable to estimate the state derivative using the state equation:

$$\hat{\dot{x}}_\varepsilon(t) = \mathbf{A}\hat{x}_\varepsilon(t) + \mathbf{B}_u u(t). \tag{7.19}$$

The known input is included in this state equation, and the plant noise is set equal to its expected value of zero, since this noise is totally unpredictable. The estimate (7.19) is indeed optimal, as shown by applying the orthogonality principle:

$$\begin{aligned}
&E[\{\dot{x}(t) - \hat{\dot{x}}_\varepsilon(t)\}m^T(\tau)] \\
&= E[\{\mathbf{A}x(t) + \mathbf{B}_u u(t) + \mathbf{B}_w w(t) - \mathbf{A}\hat{x}_\varepsilon(t) - \mathbf{B}_u u(t)\}m^T(\tau)] \\
&= \mathbf{A}E[\{x(t) - \hat{x}_\varepsilon(t)\}m^T(\tau)] + \mathbf{B}_w E[w(t)m^T(\tau)] = \mathbf{0},
\end{aligned} \tag{7.20}$$

where $\tau \in [0, t - \varepsilon]$. The first term in the last line of (7.20) equals zero due to the orthogonality principle, which is applicable since $\hat{x}_\varepsilon(t)$ is the optimal linear state estimate, given the measurements. The second term in the last line of (7.20) equals zero, since the plant noise at time t is uncorrelated with the measurement at an earlier time τ, and the expected value of the plant noise equals zero. Note that for a causal system, plant inputs only influence future measurements. In conclusion, (7.19) yields the optimal estimate of the state derivative.

---

[4] Partitioning the data this way leaves a gap in time between $(t - \varepsilon)$ and t. This data is subsequently included in the estimate when taking the limit as ε approaches zero.

[5] This is an application of Fubini's theorem. Fubini's theorem can be found in most advanced calculus books, for example, [2], page 302.

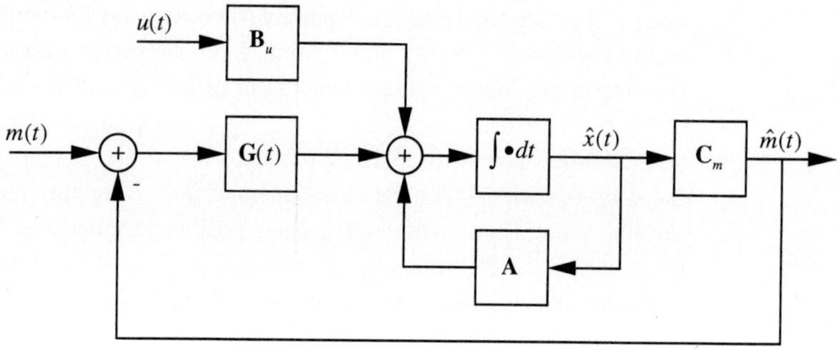

FIGURE 7.1 The Kalman filter in predictor/corrector form

The predicted measurement $\hat{m}_\varepsilon(t)$ is the estimate of $m(t)$ given data through time $(t-\varepsilon)$. This estimate can be written in terms of $\hat{x}_\varepsilon(t)$. It is reasonable to estimate $m(t)$ by using the estimate $\hat{x}_\varepsilon(t)$ in the measurement equation:

$$\hat{m}_\varepsilon(t) = \mathbf{C}_m \hat{x}_\varepsilon(t), \tag{7.21}$$

where the unknown measurement noise is assumed equal to its expected value of zero. This is the optimal estimate as shown by applying the orthogonality principle:

$$E[\{m(t) - \hat{m}_\varepsilon(t)\}m^T(\tau)] = \mathbf{C}_m E[\{x(t) - \hat{x}_\varepsilon(t)\}m^T(\tau)] + E[v(t)m^T(\tau)] = \mathbf{0}$$

where $\tau \in [0, t-\varepsilon]$. The first term in the second expression above equals zero due to the orthogonality principle, which is again applicable since $\hat{x}_\varepsilon(t)$ is the optimal linear estimate of $x(t)$. The second term in this expression equals zero, since the measurement noise at time t is uncorrelated with the measurements at an earlier time τ, and the expected value of the measurement noise equals zero. In conclusion, (7.21) yields the optimal estimate of the measurement.

The final version of the Kalman filter equation is obtained by combining (7.18), (7.19), (7.21), and taking the limit as ε approaches 0:

$$\boxed{\dot{\hat{x}}(t) = \mathbf{A}\hat{x}(t) + \mathbf{B}_u u(t) + \mathbf{G}(t)[m(t) - \mathbf{C}_m \hat{x}(t)].} \tag{7.22}$$

Note that the subscript ε has been dropped, since all estimates are now based on all of the data, and this notation is no longer needed for clarity. This form of the Kalman filter equation is referred to as the predictor/corrector form, since the derivative of the state is "predicted" from old data and then corrected using current data.[6] A block diagram of the predictor/corrector form of the Kalman filter is given in Figure 7.1. An alternative form of the Kalman filter equation is

$$\begin{aligned} \dot{\hat{x}}(t) &= [\mathbf{A} - \mathbf{G}(t)\mathbf{C}_m]\hat{x}(t) + \mathbf{B}_u u(t) + \mathbf{G}(t)m(t) \\ &= \mathbf{F}(t)\hat{x}(t) + \mathbf{B}_u u(t) + \mathbf{G}(t)m(t), \end{aligned} \tag{7.23}$$

[6] This terminology was originally developed for discrete-time Kalman filters, where the state is predicted and then corrected.

where $\mathbf{F}(t) = \mathbf{A} - \mathbf{G}(t)\mathbf{C}_m$. This form of the Kalman filter equation is referred to as the observer form, since it is identical to the equation for a time-varying Luenberger observer.

7.2.2 The Kalman Gain Equations

The Kalman gain is found using the orthogonality principle. The resulting Kalman filter is the linear estimator that minimizes the mean square estimation error.

The orthogonality principle states that the estimation error is orthogonal to the data:

$$E[\{x(t) - \hat{x}(t)\}m^T(\tau)] = \mathbf{0},$$

where $\tau \le t$. Expanding this equation using (7.13) yields

$$E[\{x(t) - \hat{x}(t)\}x^T(\tau)]\mathbf{C}_m^T + E[x(t)v^T(\tau)] - E[\hat{x}(t)v^T(\tau)] = \mathbf{0}.$$

Using the fact that the state is orthogonal to the measurement noise (7.15c), this equation can be simplified:

$$E[\{x(t) - \hat{x}(t)\}x^T(\tau)]\mathbf{C}_m^T = E[\hat{x}(t)v^T(\tau)]. \tag{7.24}$$

The orthogonality condition (7.24) implicitly includes the Kalman gain, since the state estimate depends on this gain. In order to solve for the gain, the state estimate must be explicitly written in terms of the gain. This can be accomplished by noting that the Kalman filter (7.23) is a time-varying state equation with the following solution:

$$\hat{x}(t) = \mathbf{\Phi}(t, 0)\hat{x}(0) + \int_0^t \mathbf{\Phi}(t, \gamma)\mathbf{B}_u u(\gamma)d\gamma + \int_0^t \mathbf{\Phi}(t, \gamma)\mathbf{G}(\gamma)m(\gamma)d\gamma,$$

where $\mathbf{\Phi}(t, \gamma)$ is the state-transition matrix of the Kalman filter equation. Substituting this expression for the state estimate into the right side of (7.24), we have

$$E[\{x(t) - \hat{x}(t)\}x^T(\tau)]\mathbf{C}_m^T$$
$$= E\left[\left\{\mathbf{\Phi}(t, 0)\hat{x}(0) + \int_0^t \mathbf{\Phi}(t, \gamma)\mathbf{B}_u u(\gamma)d\gamma + \int_0^t \mathbf{\Phi}(t, \gamma)\mathbf{G}(\gamma)m(\gamma)d\gamma\right\}v^T(\tau)\right].$$

Using the facts that $u(t)$ is deterministic and $\hat{x}(0)$ is deterministic (since $\hat{x}(0)$ does not depend on the measurements), this equation can be written

$$E[\{x(t) - \hat{x}(t)\}x^T(\tau)]\mathbf{C}_m^T = \mathbf{\Phi}(t, 0)\hat{x}(0)E[v^T(\tau)]$$
$$+ \int_0^t \mathbf{\Phi}(t, \gamma)\mathbf{B}_u u(\gamma)d\gamma E[v^T(\tau)] + \int_0^t \mathbf{\Phi}(t, \gamma)\mathbf{G}(\gamma)E[m(\gamma)v^T(\tau)]d\gamma.$$

The expected value of the white measurement noise is zero:

$$E[\{x(t) - \hat{x}(t)\}x^T(\tau)]\mathbf{C}_m^T$$
$$= \int_0^t \mathbf{\Phi}(t, \gamma)\mathbf{G}(\gamma)E[m(\gamma)v^T(\tau)]d\gamma$$
$$= \int_0^t \mathbf{\Phi}(t, \gamma)\mathbf{G}(\gamma)\mathbf{C}_m E[x(\gamma)v^T(\tau)]d\gamma + \int_0^t \mathbf{\Phi}(t, \gamma)\mathbf{G}(\gamma)E[v(\gamma)v^T(\tau)]d\gamma.$$

Using the fact that the measurement noise is orthogonal to the state and recognizing that the final expected value is the correlation function of the measurement noise yields

$$E[\{x(t) - \hat{x}(t)\}x^T(\tau)]\mathbf{C}_m^T = \int_0^t \mathbf{\Phi}(t, \gamma)\mathbf{G}(\gamma)\mathbf{S}_v\delta(\gamma - \tau)d\gamma. \tag{7.25}$$

For the optimal estimate, (7.25) is satisfied for all $\tau \le t$ and in particular for $\tau = t - \varepsilon$ (where ε is a small positive number):[7]

$$\begin{aligned} E[\{x(t) - \hat{x}(t)\}x^T(t - \varepsilon)]\mathbf{C}_m^T &= \int_0^t \mathbf{\Phi}(t, \gamma)\mathbf{G}(\gamma)\mathbf{S}_v\delta(\gamma - t + \varepsilon)d\gamma \\ &= \mathbf{G}(t - \varepsilon)\mathbf{S}_v. \end{aligned}$$

Taking the limit as ε approaches 0 then yields[8]

$$\mathbf{\Sigma}_e(t)\mathbf{C}_m^T = \mathbf{G}(t)\mathbf{S}_v, \tag{7.26}$$

where

$$\mathbf{\Sigma}_e(t) = E[\{x(t) - \hat{x}(t)\}x^T(t)] = E[\{x(t) - \hat{x}(t)\}\{x(t) - \hat{x}(t)\}^T] \tag{7.27}$$

is the covariance matrix of the estimation error. Using (7.26), the Kalman gain can be written in terms of the estimation error covariance matrix:

$$\boxed{\mathbf{G}(t) = \mathbf{\Sigma}_e(t)\mathbf{C}_m^T\mathbf{S}_v^{-1},} \tag{7.28}$$

where the inverse is guaranteed to exist since $\mathbf{S}_v$ is positive definite.

The Error Model The computation of the Kalman gain requires the covariance matrix of the estimation error. Before generating this covariance matrix, a state equation for the estimation error is developed. The derivative of the estimation error is

$$\dot{e}(t) = \dot{x}(t) - \dot{\hat{x}}(t).$$

Substituting (7.13a) and (7.22) into this equation, we have

$$\begin{aligned} \dot{e}(t) = {} & \mathbf{A}x(t) + \mathbf{B}_u u(t) + \mathbf{B}_w w(t) - \mathbf{A}\hat{x}(t) - \mathbf{B}_u u(t) \\ & - \mathbf{G}(t)[\mathbf{C}_m x(t) + v(t) - \mathbf{C}_m\hat{x}(t)]. \end{aligned}$$

Collecting terms and canceling yields

$$\boxed{\dot{e}(t) = [\mathbf{A} - \mathbf{G}(t)\mathbf{C}_m]e(t) + [\mathbf{B}_w \;\vdots\; -\mathbf{G}(t)]\begin{bmatrix} w(t) \\ \hdashline v(t) \end{bmatrix}.} \tag{7.29}$$

[7] The evaluation of this integral at $(t - \varepsilon)$ is a trick that avoids the mathematical difficulties associated with white noise and the Dirac delta function. A more rigorous derivation of the gain equation can be obtained by using the stochastic Ito calculus, or by taking the limit of the discrete-time Kalman filter, as in [3].

[8] The existence of these limits requires continuity of the estimation error correlation function [the expected value in (7.25)] and continuity of the Kalman gain. Continuity of these quantities is guaranteed since they are generated by a differential equation with finite inputs, as is shown subsequently.

This state equation for the estimation error of the Kalman filter is referred to as the error model.

The error model can be used in generating the estimation error covariance matrix and analyzing the performance of the Kalman filter. This model also provides insight into the selection of the Kalman gain. For example, the estimation error resulting from the measurement noise increases as the gain is increased, since the Kalman gain is the input coupling matrix for this noise. Note that the negative sign on this gain does not affect the resulting estimation error variance. The estimation error resulting from the plant noise decreases as the gain is increased, since the increased Kalman gain results in a larger state matrix. This larger state matrix (i.e., more feedback) produces a smaller closed-loop gain. The optimal Kalman gain can be thought of as a compromise between these two effects.

The Riccati Equation The estimation error covariance matrix is required to specify the Kalman gain. This covariance matrix can be generated from the error model. The input to the error model is white noise, with the correlation function

$$E[w^T(t) \mid v^T(t)]\begin{bmatrix} w(t+\tau) \\ \hline v(t+\tau) \end{bmatrix} = \begin{bmatrix} \mathbf{S}_w & \mathbf{0} \\ \mathbf{0} & \mathbf{S}_v \end{bmatrix}\delta(\tau).$$

A differential equation that describes the evolution of the state covariance matrix of a system with a white noise input is generated in Section 3.3.1. Applying this result to the error model, we have

$$\begin{aligned} \dot{\mathbf{\Sigma}}_e(t) &= \mathbf{\Sigma}_e(t)\{\mathbf{A} - \mathbf{G}(t)\mathbf{C}_m\}^T + \{\mathbf{A} - \mathbf{G}(t)\mathbf{C}_m\}\mathbf{\Sigma}_e(t) \\ &+ [\mathbf{B}_w \mid -\mathbf{G}(t)]\begin{bmatrix} \mathbf{S}_w & \mathbf{0} \\ \mathbf{0} & \mathbf{S}_v \end{bmatrix}\begin{bmatrix} \mathbf{B}_w^T \\ \hline -\mathbf{G}^T(t) \end{bmatrix}. \end{aligned}$$

Expanding this equation gives

$$\begin{aligned} \dot{\mathbf{\Sigma}}_e(t) &= \mathbf{\Sigma}_e(t)\mathbf{A}^T - \mathbf{\Sigma}_e(t)\mathbf{C}_m^T\mathbf{G}^T(t) + \mathbf{A}\mathbf{\Sigma}_e(t) \\ &- \mathbf{G}(t)\mathbf{C}_m\mathbf{\Sigma}_e(t) + \mathbf{B}_w\mathbf{S}_w\mathbf{B}_w^T + \mathbf{G}(t)\mathbf{S}_v\mathbf{G}^T(t). \end{aligned}$$

Substituting for the Kalman gain using (7.28) and reducing yields

$$\boxed{\dot{\mathbf{\Sigma}}_e(t) = \mathbf{\Sigma}_e(t)\mathbf{A}^T + \mathbf{A}\mathbf{\Sigma}_e(t) + \mathbf{B}_w\mathbf{S}_w\mathbf{B}_w^T - \mathbf{\Sigma}_e(t)\mathbf{C}_m^T\mathbf{S}_v^{-1}\mathbf{C}_m\mathbf{\Sigma}_e(t).} \quad (7.30)$$

This differential equation, called the Riccati equation, can be used to solve for the estimation error covariance matrix.

The Kalman filter is defined by the Riccati equation (7.30), the Kalman gain equation (7.28), the Kalman filter equation (7.22) or (7.23), and the initial conditions on the two differential equations. The equations for the Kalman filter are summarized as follows:

$$\dot{\hat{x}}(t) = \mathbf{A}\hat{x}(t) + \mathbf{B}_u u(t) + \mathbf{G}(t)[m(t) - \mathbf{C}_m\hat{x}(t)]; \quad (7.22)$$

$$\dot{\mathbf{\Sigma}}_e(t) = \mathbf{\Sigma}_e(t)\mathbf{A}^T + \mathbf{A}\mathbf{\Sigma}_e(t) + \mathbf{B}_w\mathbf{S}_w\mathbf{B}_w^T - \mathbf{\Sigma}_e(t)\mathbf{C}_m^T\mathbf{S}_v^{-1}\mathbf{C}_m\mathbf{\Sigma}_e(t); \quad (7.30)$$

$$\mathbf{G}(t) = \mathbf{\Sigma}_e(t)\mathbf{C}_m^T\mathbf{S}_v^{-1}. \quad (7.28)$$

The equations for the gain (7.30) and (7.28) are independent of the measurements. The Kalman gain can therefore be found in advance of filtering the data, which greatly reduces the real-time computational burden of the Kalman filter. Additionally, the time-varying estimation error covariance matrix can be found in advance of filtering the data. This covariance matrix is a measure of the accuracy of each state estimate. The time variation of this matrix can then be used to provide information on how long the filter must operate before reaching a given level of accuracy.

EXAMPLE 7.2 A Kalman filter is used to estimate the range and radial velocity of an aircraft from noisy radar range measurements. The application of the Kalman filter requires a plant state model, a plant noise spectral density matrix, a measurement noise spectral density matrix, an initial condition on the state estimate, and an initial estimation error covariance matrix. Assuming a random white noise acceleration for the aircraft, a state equation for the range and radial velocity is

$$\begin{bmatrix} \dot{r}(t) \\ \ddot{r}(t) \end{bmatrix} = \begin{bmatrix} 0 & 1 \\ 0 & 0 \end{bmatrix} \begin{bmatrix} r(t) \\ \dot{r}(t) \end{bmatrix} + \begin{bmatrix} 0 \\ 1 \end{bmatrix} w(t),$$

where $r(t)$ is the actual range of the aircraft. The range measurements are given:

$$m(t) = [1 \quad 0] \begin{bmatrix} r(t) \\ \dot{r}(t) \end{bmatrix} + v(t).$$

The Kalman gains and the estimation error variances are generated for this plant with various values for the plant noise spectral density matrix, the measurement noise spectral density matrix, and the initial estimation error covariance matrix. In addition, the plant is simulated, and the simulated measurements are put into the Kalman filter to yield estimated states, which are compared with the actual states. In all cases, the simulations utilize the assumptions made in generating the Kalman filter. The initial condition on the state in all of the simulations is

$$\begin{bmatrix} r(0) \\ \dot{r}(0) \end{bmatrix} = \begin{bmatrix} 10{,}000 \text{ m} \\ -150 \text{ m/s} \end{bmatrix}.$$

The Kalman gains, estimation error variances, and estimated states are given in Figures 7.2 through 7.5 for several combinations of plant noise spectral density, measurement noise spectral density, and initialization. Figure 7.2 shows the results for the baseline case defined by the following parameters:

$$S_w = 4\frac{\text{m}^2}{\text{sec}^4 \cdot \text{Hz}}; \; S_v = 10^4\frac{\text{m}^2}{\text{Hz}}; \; \mathbf{\Sigma}_e(0) = \begin{bmatrix} 10^6 \text{ m}^2 & 0 \\ 0 & 4 \times 10^5 \text{ m}^2/\text{sec}^2 \end{bmatrix};$$

$$\begin{bmatrix} \hat{r}(0) \\ \hat{\dot{r}}(0) \end{bmatrix} = \begin{bmatrix} 9000 \text{ m} \\ 0 \text{ m/sec} \end{bmatrix}.$$

Figure 7.3 shows the results obtained when the plant noise spectral density is increased by a factor of 100:

$$S_w = 400\frac{\text{m}^2}{\text{sec}^4 \cdot \text{Hz}}.$$

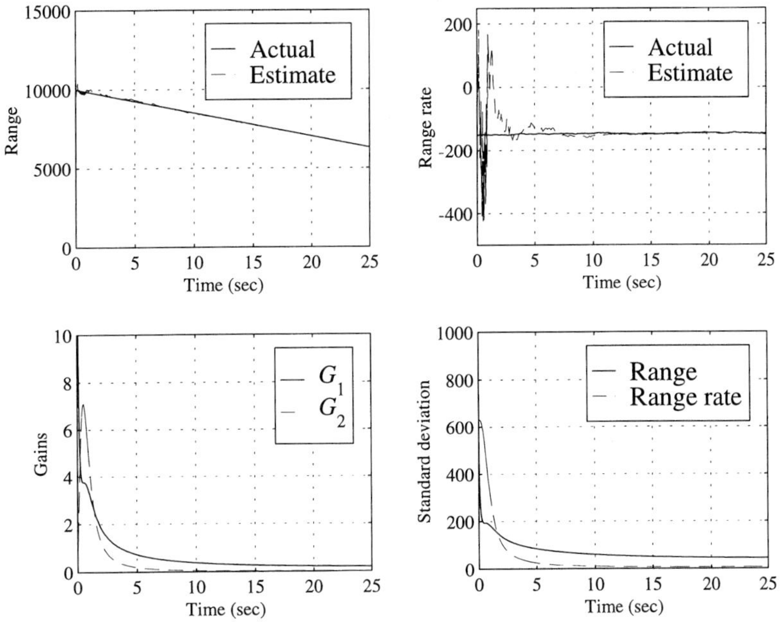

FIGURE 7.2 Baseline results for Example 7.2

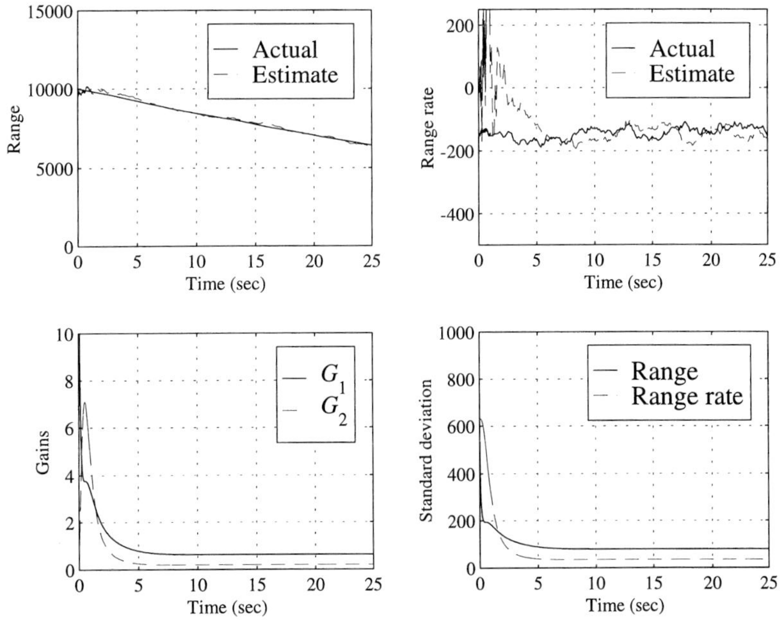

FIGURE 7.3 Results for Example 7.2 with increased plant noise

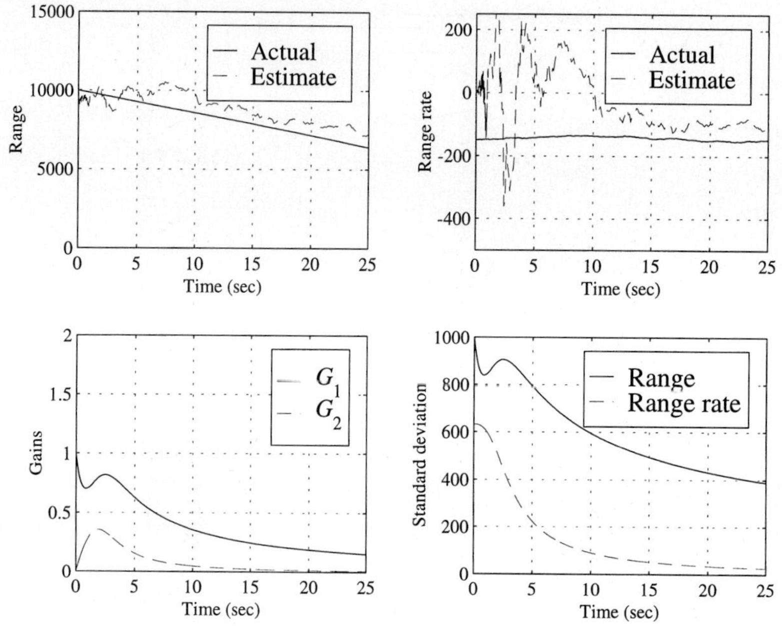

FIGURE 7.4 Results for Example 7.2 with increased measurement noise

All other parameters are kept at the baseline values. Figure 7.4 shows the results obtained when the measurement noise spectral density is increased by a factor of 100:

$$S_v = 10^6 \frac{\text{m}^2}{\text{Hz}}.$$

All other parameters are kept at the baseline values. Figure 7.5 shows the results obtained when the initialization is perfect; that is, exact *a priori* information on the range and radial velocity of the aircraft is available:

$$\mathbf{\Sigma}_e(0) = \begin{bmatrix} 0 & 0 \\ 0 & 0 \end{bmatrix}; \begin{bmatrix} \hat{r}(0) \\ \hat{\dot{r}}(0) \end{bmatrix} = \begin{bmatrix} 10{,}000 \text{ m} \\ -150 \text{ m/sec} \end{bmatrix}.$$

All other parameters are kept at the baseline values.

The estimates approach the actual states in all of the cases presented. In the baseline case, the Kalman gains experience a transient and then decrease to a small steady-state value, while the standard deviations of the estimation errors decrease to steady-state values. The steady-state values are achieved after a settling time of approximately 25 sec.

The plant states depart from a predictable pattern more rapidly when the plant noise spectral density is increased. As a consequence, the steady-state estimation errors increase, since less averaging can be performed on the measurements. This fact is reflected in the larger steady-state Kalman gains; that is, a greater emphasis is placed on the current measurement in generating the estimates. Additionally, the settling time of the filter is decreased, since old data becomes obsolete more rapidly.

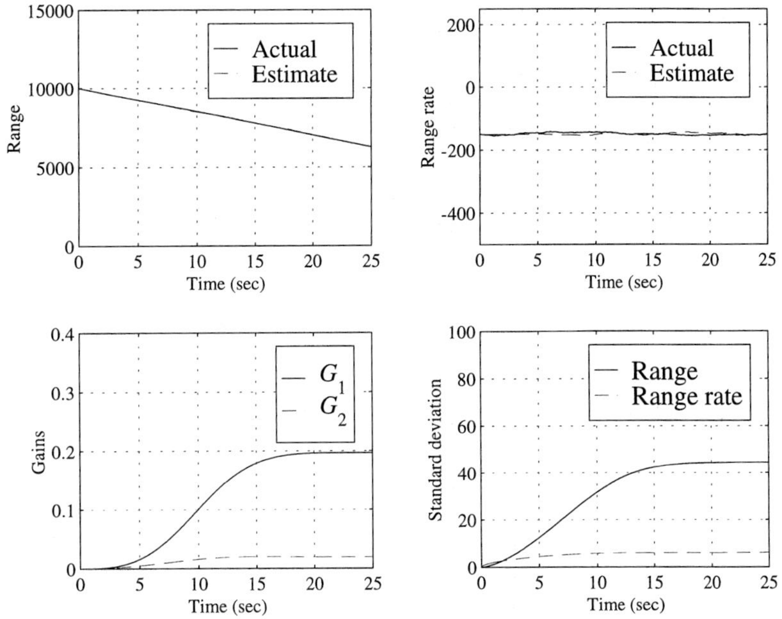

FIGURE 7.5 Results for Example 7.2 with perfect initial estimates

The measurements become less reliable when the measurement noise spectral density is increased. The estimation errors also increase due to this decrease in measurement quality. The Kalman gains decrease in an attempt to put less emphasis on the current measurement. The settling time of the filter is also increased, since more measurements must be averaged in order to beat down the measurement noise.

In the last case the initial estimated state equals the actual state. In this case the large transient errors that occurred in the baseline case are not present, since the filter can largely ignore the initial measurements and rely instead on the perfect *a priori* information. As a consequence, the Kalman gains and the estimation errors are initially zero and then increase as the *a priori* information becomes outdated. The steady-state gains, the steady-state estimation error standard deviations, and the settling time required to reach steady state are all the same as obtained in the baseline case. Note that this fact may not be apparent due to the different scales used in the plots.

In summary, the Kalman filter is able to estimate the plant state in all of the above cases. The estimation error variances and the Kalman gains were found to approach steady-state values as time increases. Further, these gains and variances are modified in a logical manner when the parameters that define the Kalman filter are changed. ◆

The Hamiltonian Equation The Riccati equation used to find the estimation error covariance matrix has a form similar to the Riccati equation for the linear quadratic regulator. Indeed, the Kalman Riccati equation is identical to the LQR Riccati equation with the proper substitutions. This correspondence between the estimation and control equations is called duality and is subsequently discussed in detail. The Riccati equation for the linear quadratic regulator resulted from defining a relationship between the state and the costate in the Hamiltonian equations. A question immediately comes to mind: Is there a Hamiltonian system associated with the Kalman Riccati equation? The answer is yes. The Hamiltonian system for the Kalman filter is derived in the remainder of this subsection. The relationship of this system to the Kalman gain is also given.

The solution of the Riccati equation for the linear quadratic regulator can be factored as in (6.16):[9]

$$\mathbf{\Sigma}_e(t) = \mathbf{T}_2(t)\mathbf{T}_1^{-1}(t).$$

Making this substitution into the Riccati equation (7.30) yields

$$\begin{aligned}\frac{d[\mathbf{T}_2(t)\mathbf{T}_1^{-1}(t)]}{dt} &= \frac{d\mathbf{T}_2(t)}{dt}\mathbf{T}_1^{-1}(t) + \mathbf{T}_2(t)\frac{d[\mathbf{T}_1^{-1}(t)]}{dt}\\ &= \mathbf{T}_2(t)\mathbf{T}_1^{-1}(t)\mathbf{A}^T + \mathbf{A}\mathbf{T}_2(t)\mathbf{T}_1^{-1}(t) + \mathbf{B}_w\mathbf{S}_w\mathbf{B}_w^T\\ &\quad - \mathbf{T}_2(t)\mathbf{T}_1^{-1}(t)\mathbf{C}_m^T\mathbf{S}_v^{-1}\mathbf{C}_m\mathbf{T}_2(t)\mathbf{T}_1^{-1}(t).\end{aligned}$$

Using the formula for the derivative of the inverse of a matrix in Appendix equation (A1.5):

$$\frac{d[\mathbf{T}_1^{-1}(t)]}{dt} = -\mathbf{T}_1^{-1}(t)\dot{\mathbf{T}}_1(t)\mathbf{T}_1^{-1}(t),$$

the Riccati equation becomes

$$\begin{aligned}&\dot{\mathbf{T}}_2(t)\mathbf{T}_1^{-1}(t) - \mathbf{T}_2(t)\mathbf{T}_1^{-1}(t)\dot{\mathbf{T}}_1(t)\mathbf{T}_1^{-1}(t)\\ &= \mathbf{T}_2(t)\mathbf{T}_1^{-1}(t)\mathbf{A}^T + \mathbf{A}\mathbf{T}_2(t)\mathbf{T}_1^{-1}(t) + \mathbf{B}_w\mathbf{S}_w\mathbf{B}_w^T\\ &\quad - \mathbf{T}_2(t)\mathbf{T}_1^{-1}(t)\mathbf{C}_m^T\mathbf{S}_v^{-1}\mathbf{C}_m\mathbf{T}_2(t)\mathbf{T}_1^{-1}(t).\end{aligned}$$

Right-multiplying by $\mathbf{T}_1(t)$ and regrouping yields

$$\begin{aligned}&\dot{\mathbf{T}}_2(t) - \mathbf{T}_2(t)\mathbf{T}_1^{-1}(t)[\dot{\mathbf{T}}_1(t)]\\ &= \mathbf{B}_w\mathbf{S}_w\mathbf{B}_w^T\mathbf{T}_1(t) + \mathbf{A}\mathbf{T}_2(t) - \mathbf{T}_2(t)\mathbf{T}_1^{-1}(t)[-\mathbf{A}^T\mathbf{T}_1(t) + \mathbf{C}_m^T\mathbf{S}_v^{-1}\mathbf{C}_m\mathbf{T}_2(t)].\end{aligned}$$

Comparing the two sides of this equation, a solution is obtained when $\mathbf{T}_1(t)$ and $\mathbf{T}_2(t)$ are solutions of the following linear differential equations:

$$\begin{aligned}\dot{\mathbf{T}}_1(t) &= -\mathbf{A}^T\mathbf{T}_1(t) + \mathbf{C}_m^T\mathbf{S}_v^{-1}\mathbf{C}_m\mathbf{T}_2(t);\\ \dot{\mathbf{T}}_2(t) &= \mathbf{B}_w\mathbf{S}_w\mathbf{B}_w^T\mathbf{T}_1(t) + \mathbf{A}\mathbf{T}_2(t).\end{aligned}$$

Combining these equations yields the Hamiltonian system for the Kalman filter:

$$\begin{bmatrix}\dot{\mathbf{T}}_1(t)\\ \dot{\mathbf{T}}_2(t)\end{bmatrix} = \begin{bmatrix}-\mathbf{A}^T & \mathbf{C}_m^T\mathbf{S}_v^{-1}\mathbf{C}_m\\ \mathbf{B}_w\mathbf{S}_w\mathbf{B}_w^T & \mathbf{A}\end{bmatrix}\begin{bmatrix}\mathbf{T}_1(t)\\ \mathbf{T}_2(t)\end{bmatrix} = \mathcal{Y}\begin{bmatrix}\mathbf{T}_1(t)\\ \mathbf{T}_2(t)\end{bmatrix}. \tag{7.31}$$

The matrix $\mathcal{Y}$ is termed the Hamiltonian of the Kalman filter.

The initial conditions for the Hamiltonian system must satisfy

$$\mathbf{\Sigma}_e(0) = \mathbf{T}_2(0)\mathbf{T}_1^{-1}(0). \tag{7.32}$$

[9]The representation of a matrix as the product of a matrix and the inverse of a matrix is termed a matrix fraction decomposition in the control literature.

The initial conditions $\mathbf{T}_1(0)$ and $\mathbf{T}_2(0)$ are not uniquely specified by (7.32), but any initial conditions satisfying (7.32) yield a solution of the initial-value problem specified by the Riccati equation and $\mathbf{\Sigma}_e(0)$. When the Riccati equation has a unique solution, any initial conditions $\mathbf{T}_1(0)$ and $\mathbf{T}_2(0)$ that satisfy (7.32) result in the same covariance matrix. Selecting the initial conditions

$$\boxed{\begin{bmatrix} \mathbf{T}_1(0) \\ \hdashline \mathbf{T}_2(0) \end{bmatrix} = \begin{bmatrix} \mathbf{I} \\ \hdashline \mathbf{\Sigma}_e(0) \end{bmatrix},} \tag{7.33}$$

the Hamiltonian system is solved as follows:

$$\begin{bmatrix} \mathbf{T}_1(t) \\ \hdashline \mathbf{T}_2(t) \end{bmatrix} = \mathbf{\Phi}(t) \begin{bmatrix} \mathbf{I} \\ \hdashline \mathbf{\Sigma}_e(0) \end{bmatrix} = \begin{bmatrix} \mathbf{\Phi}_{11}(t) & \mathbf{\Phi}_{12}(t) \\ \hdashline \mathbf{\Phi}_{21}(t) & \mathbf{\Phi}_{22}(t) \end{bmatrix} \begin{bmatrix} \mathbf{I} \\ \hdashline \mathbf{\Sigma}_e(0) \end{bmatrix}$$

where $\mathbf{\Phi}(t)$ is the state-transition matrix of the Hamiltonian system. Solving for the Riccati solution yields

$$\boxed{\mathbf{\Sigma}_e(t) = \{\mathbf{\Phi}_{21}(t) + \mathbf{\Phi}_{22}(t)\mathbf{\Sigma}_e(0)\}\{\mathbf{\Phi}_{11}(t) + \mathbf{\Phi}_{12}(t)\mathbf{\Sigma}_e(0)\}^{-1}.} \tag{7.34}$$

Writing the Kalman gain in terms of the state transition matrix of the Hamiltonian system, we have

$$\boxed{\mathbf{G}(t) = \{\mathbf{\Phi}_{21}(t) + \mathbf{\Phi}_{22}(t)\mathbf{\Sigma}_e(0)\}\{\mathbf{\Phi}_{11}(t) + \mathbf{\Phi}_{12}(t)\mathbf{\Sigma}_e(0)\}^{-1}\mathbf{C}_m^T\mathbf{S}_v^{-1}.} \tag{7.35}$$

EXAMPLE 7.3 An ac motor is described by the following state model:

$$\dot{x}(t) = -x(t) + u(t) + w(t),$$

where $x(t)$ is the rotational velocity of the shaft, $u(t)$ is the applied voltage (10 volts rms), and $w(t)$ is noise on the applied voltage. A Kalman filter is used to estimate the true rotational velocity from the noisy tachometer measurements:

$$m(t) = x(t) + v(t),$$

where $v(t)$ is the measurement noise. The noises $w(t)$ and $v(t)$ are both assumed to be white with the spectral densities:

$$S_w = 3\frac{\text{Volts}^2}{\text{Hz}}; \ S_v = 1\frac{\text{rad}^2}{\text{sec}^2 \cdot \text{Hz}},$$

respectively. The initial state estimate and initial covariance matrix are

$$\hat{x}(0) = 10\frac{\text{rad}}{\text{sec}}; \ \Sigma_e(0) = 1.5\frac{\text{rad}^2}{\text{sec}^2}.$$

The Hamiltonian is

$$\mathcal{Y} = \begin{bmatrix} 1 & 1 \\ 3 & -1 \end{bmatrix},$$

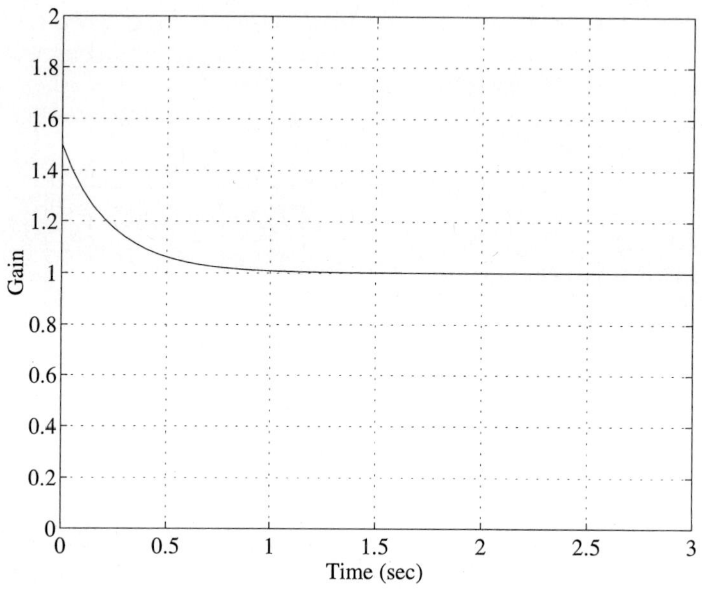

FIGURE 7.6 Kalman gain for Example 7.3

and the state-transition matrix of the Hamiltonian is

$$e^{\mathcal{H}t} = \begin{bmatrix} \frac{1}{4}e^{-2t} + \frac{3}{4}e^{2t} & -\frac{1}{4}e^{-2t} + \frac{1}{4}e^{2t} \\ -\frac{3}{4}e^{-2t} + \frac{3}{4}e^{2t} & \frac{3}{4}e^{-2t} + \frac{1}{4}e^{2t} \end{bmatrix}.$$

The Kalman gain is

$$G(t) = \frac{\left(-\frac{3}{4}e^{-2t} + \frac{3}{4}e^{2t}\right) + 1.5\left(\frac{3}{4}e^{-2t} + \frac{1}{4}e^{2t}\right)}{\left(\frac{1}{4}e^{-2t} + \frac{3}{4}e^{2t}\right) + 1.5\left(-\frac{1}{4}e^{-2t} + \frac{1}{4}e^{2t}\right)} = \frac{3e^{-2t} + 9e^{2t}}{-e^{-2t} + 9e^{2t}}.$$

A plot of the Kalman gain verses time is given in Figure 7.6. Note that this gain experiences a transient at the initial time and then decreases to a steady-state value. ◆

7.2.3 Application Notes

The application of the Kalman filter requires the Kalman filter equations, the Kalman gain equations, a state model of the plant, an initial state estimate, an initial estimation error covariance matrix, a measurement noise spectral density matrix, and a plant noise spectral density matrix. The filter and gain equations are specified in Subsections 7.2.1 and 7.2.2, respectively. The generation of the remainder of these quantities is discussed below along with some comments on computation.

The Plant Model The state model of the plant can be generated using physics (or another appropriate science) or generated empirically by using system identification. Mathematical modeling via physics is addressed in a wide range of material. This material tends to be application-specific and is beyond the scope of this book. System identification is also a large field of study [4, 5] and is beyond the scope of this book. This subsection includes some general properties of mathematical models that make them suitable for use in Kalman filtering.

The plant model must be driven by noise and should include as states, or linear combinations of the states, all of the quantities that the filter is required to estimate. Additionally, the measurements, which are the outputs of this model, should be linear combinations of the states. As in all engineering applications, the model should be as simple as possible while still describing the relevant plant behavior. The modeling of the plant may be one of the most difficult tasks facing the Kalman filter designer.

The Measurement Noise Spectral Density Matrix The measurement noise spectral density matrix may be given by the sensor manufacturer, determined theoretically, or determined empirically. These methods of generating spectral density matrices are illustrated in the following examples.

EXAMPLE 7.4 An accelerometer is being used to estimate the vibration of a rotating machine. The manufacturer of the accelerometer provides information on the error bounds of the sensor,

$$\text{Error} \leq 0.001 \frac{\text{m}}{\text{sec}^2},$$

and the sensor bandwidth:

$$\text{Bandwidth} = 100 \text{ Hz}.$$

Assuming a uniform distribution of the error within the bounds (a reasonable assumption), the error variance is found to be

$$\sigma^2 = \frac{0.001^2}{3} \frac{\text{m}^2}{\text{sec}^4}.$$

The measurement noise can be approximated by white noise provided the bandwidth of this noise is large compared to the plant bandwidth. In this case, the noise can be assumed to be white provided that all of the significant vibrational frequencies of the rotating machine are much less than 100 rad/sec. The measurement noise spectral density then equals the variance divided by twice the bandwidth:

$$S_v = \frac{0.001^2}{600} = 1.67 \times 10^{-9} \frac{\text{m}^2}{\text{sec}^2 - \text{Hz}}.$$

EXAMPLE 7.5 A Kalman filter is used to estimate the range and the radial velocity of an aircraft given noisy range measurements provided by a fixed radar. The measurement noise is caused primarily by thermal electrons at the front end of the receiver. Careful modeling and a detailed theoretical analysis of this thermal noise yields the spectral density of a radar range estimate:

$$S_v = \frac{c^2}{16B^3(S/N)},$$

where c is the speed of light, B is the bandwidth of the radar, and (S/N) is the signal-to-noise ratio at the front end of the receiver. ◆

This example only presents the results of a theoretical analysis of noise to generate a measurement spectral density. The modeling of noise sources is an extensive subject and application-specific. Many results can be obtained from the literature, which should be consulted when trying to specify noise spectral densities.

EXAMPLE 7.6 A Kalman filter is used to estimate the angular position and angular velocity of a motor shaft. The angular position sensor is a resistance pot attached to the motor shaft. The error in this measurement is due to a combination of thermal noise in the pot and noise on the voltage source. A theoretical analysis of the noise may not be warranted due to the complexity of the noise source. The spectral density of the measurement can be obtained empirically by positioning the motor shaft at a known location and inputting the measurement error into a spectrum analyzer. The measurement noise spectral density is the magnitude of the spectrum analyzer output at low frequencies. The spectrum is assumed to be relatively flat over the bandwidth of the motor. If this assumption is violated, the measurement noise is not white, and the techniques of Section 7.4 can be applied to yield an optimal estimator.

Generating noise spectrums for a number of motor shaft positions and averaging the results improves the estimate of the spectral density and reduces the possibility of a systematic error associated with a given motor shaft position. A measurement error dependent on the motor shaft angle is called state-dependent measurement noise, and violates the assumptions used in deriving the Kalman filter. In practice, the Kalman filter performs well in the presence of state-dependent measurement noise when the state is rapidly changing, or when the state-dependent component is small compared to the white component of the measurement noise. Note that the Kalman filter is no longer optimal when state dependent noise is present. ◆

A number of techniques exist for the empirical estimation of the spectral density. These include (1) putting the estimates into a bank of filters and measuring the power out of each filter; (2) sampling the signal and applying the fast Fourier transform (FFT), which should be applied to a number of segments of the data and the spectrums averaged for best results; and (3) estimating parameters in a model whose input is unit spectral density white noise and whose output matches the observed signal; and so forth. Spectral estimation is a complex subject that cannot be treated in a single paragraph. For a thorough treatment of this subject, see [6, 7].

The Plant Noise Spectral Density Matrix The plant noise spectral density matrix can be determined theoretically or empirically. The theoretical determination of the plant noise spectral density can be made from a detailed analysis of the input. This process is similar to that followed in the theoretical determination of the measurement noise spectral density, where an in-depth understanding of the application and the related literature is usually required.

The empirical determination of this spectral density is complicated by the fact that the plant input is not always directly measurable. In this case, the plant noise spectral density can be obtained from the spectral density matrix of the plant output, as illustrated in the following example.

EXAMPLE 7.7 A Kalman filter is used to estimate the radial velocity of a motor shaft from measurements provided by an inexpensive and noisy, tachometer. Note that the cost of the extra circuitry required by the Kalman filter is more than compensated for by savings on the tachometer. The input to the motor is a noisy field voltage. This noise is created, in part, by induction within the motor and is not directly measurable. The mathematical model of the plant is

$$\dot{x}(t) = -0.1x(t) + 3u(t) + 3w(t);$$
$$m(t) = x(t) + v(t).$$

An expensive tachometer, which provides very accurate measurements, is used in the laboratory to provide data on the motor's angular velocity during operation. Using this tachometer, the variance of the angular velocity is found to be 0.4 rad^2/sec^2. This variance is assumed to be caused solely by the plant noise, since this tachometer is very accurate. The variance of the angular velocity due to plant noise can be computed from the plant noise spectral density:

$$\sigma_m^2 = S_w \int_0^\infty h^2(t)dt = S_w \int_0^\infty (3e^{-0.1t})^2 dt = 45S_w.$$

Setting this result equal to the measured variance, we find the spectral density of the plant noise:

$$S_w = \frac{0.4}{45} \frac{\text{Volts}^2}{\text{Hz}}.$$

Initialization The estimation of the state using a Kalman filter requires the initialization of the differential equations for the filter. Specifically, an initial state estimate and an initial value of the estimation error covariance matrix must be specified.

The initial state estimate is the designer's best guess of the state at the initial time. This best guess is the conditional mean:

$$\hat{x}(0) = E[x(0) \mid \text{data}],$$

where the data consists of measurements at times before zero. In many applications, there is no data available before time zero, and this conditional expectation reduces to

$$\hat{x}(0) = E[x(0)]. \tag{7.36}$$

The expected value of the plant state in (7.13) is

$$\begin{aligned} E[x(0)] &= \int_{-\infty}^{0} e^{-\mathbf{A}\tau}\mathbf{B}_u u(\tau)d\tau + \int_{-\infty}^{0} e^{-\mathbf{A}\tau}\mathbf{B}_w E[w(\tau)]d\tau \\ &= \int_{-\infty}^{0} e^{-\mathbf{A}\tau}\mathbf{B}_u u(\tau)d\tau. \end{aligned} \tag{7.37}$$

The use of this initial estimate with the Kalman filter yields the optimal estimate at all times.

◆**EXAMPLE 7.8**   Assume that before time equals zero, the deterministic input to the motor in Example 7.7 is a constant $u(t) = 15$ volts. The Kalman filter for estimating the motor shaft angular velocity is initialized before any measurements are available:

$$\hat{x}(0) = E[x(0)] = \int_{-\infty}^{0} e^{0.1t} \cdot 3 \cdot 15dt = 150 \text{ rad/sec.} \qquad ◆$$

The expected value in (7.37) may be difficult to evaluate or may not exist. This expected value does not exist when the plant (7.13) is unstable. Unstable plants are encountered in a number of practical applications. An alternative method of initializing the Kalman filter is therefore required in these applications.

The initial state can be estimated using an algorithm other than the Kalman filter. For example, maximum likelihood estimation can be applied when the second-moment description of the parameter is unavailable. In particular, consider the initial detection of an aircraft by a radar. Before initializing the Kalman filter, the presence of a target must be verified. The detection algorithm that determines the presence of the target must take measurements of the target for a period of time. These measurements may be used by a maximum likelihood estimator (or some nonoptimal algorithm that localizes the target) to estimate the initial state.

A priori information is often available at the initial time. For example, consider initializing a Kalman filter for tracking the position of a rocket after launch. The initial state (consisting of position and velocity) is known *a priori,* since the rocket is launched from a known location and has zero velocity at launch. In this case, the initial state estimate is not derived from the modeled measurements. The following example further illustrates the use of *a priori* information in initializing the Kalman filter.

◆**EXAMPLE 7.9**   A radar has detected a target, and range tracking must be initialized, but no range measurements are available. A reasonable guess of the initial range can be obtained by noting that most targets are first observed at a range between 50 and 150 km. Assuming the range is uniformly distributed between these values, the expected value of 100 km is a reasonable initial range estimate. The initial radial velocity is assumed to lie between 0 and -200 m/sec, since the target must be moving toward the radar to move within the detection range. Again, assuming a uniform distribution, the initial radial velocity estimate is -100 m/sec. Note that the initial estimation error covariance matrix should reflect the large uncertainty of these initial states.   ◆

The initial covariance matrix of the estimation error is

$$\mathbf{\Sigma}_e(0) = E[\{x(0) - \hat{x}(0)\}\{x(0) - \hat{x}(0)\}^T]. \qquad (7.38)$$

Assuming the initial estimate is the expected value of the state in (7.36), this covariance matrix can be computed as follows:

$$\mathbf{\Sigma}_e(0) = \int_0^{\infty} \mathbf{g}_{xw}(t)\mathbf{S}_w\mathbf{g}_{xw}^T(t)dt = \int_0^{\infty} e^{\mathbf{A}t}\mathbf{B}_w\mathbf{S}_w\mathbf{B}_w^T e^{\mathbf{A}^T t}dt, \qquad (7.39)$$

where $\mathbf{g}_{xw}(t)$ is the impulse response matrix from the plant noise to the state.

EXAMPLE 7.8 CONTINUED The initial estimation error covariance matrix for the motor in Example 7.7 is

$$\Sigma_e(0) = \int_0^{\infty} e^{-0.1t} 3 \frac{0.4}{45} 3 e^{-0.1t} dt = 0.4 \frac{\text{rad}^2}{\text{sec}^2}.$$

Note that this value equals the variance of the angular velocity of the motor shaft angle generated by the plant noise. ◆

The initial estimation error covariance matrix must be computed directly from (7.38) when the initial estimate is not equal to the expected value of the initial state, as is the case when (1) the expected value of the initial state does not exist; (2) an algorithm other than the Kalman filter is used to generate estimates prior to Kalman filter initialization; or (3) *a priori* information makes it advantageous to use another initialization.

EXAMPLE 7.9 CONTINUED The initial estimation error covariance matrix for the radar system is

$$\mathbf{\Sigma}_e(0) = \begin{bmatrix} E[\{r(0) - \hat{r}(0)\}^2] & E[\{r(0) - \hat{r}(0)\}\{\dot{r}(0) - \hat{\dot{r}}(0)\}] \\ E[\{r(0) - \hat{r}(0)\}\{\dot{r}(0) - \hat{\dot{r}}(0)\}] & E[\{\dot{r}(0) - \hat{\dot{r}}(0)\}^2] \end{bmatrix}.$$

The initial range is assumed to be uniformly distributed between 50 and 150 km, and the initial radial velocity is assumed to be uniformly distributed between 0 and −200 m/s. Further, assuming the range and the radial velocity at detection are uncorrelated (a reasonable assumption), the initial estimation error covariance matrix becomes

$$\mathbf{\Sigma}_e(0) = \begin{bmatrix} 833 \text{ km}^2 & 0 \\ 0 & 3330 \text{ m}^2/\text{sec}^2 \end{bmatrix}.$$

◆

The estimation error covariance matrix is infinite when the plant is unstable and has been operating for a long time. The optimal estimate can then be obtained by using an arbitrary initial state estimate and setting the estimation error covariance matrix to infinity. In practice, a best guess for initializing the state estimates, along with a large estimation error covariance matrix, yields nearly optimal estimates for an unstable plant.

Computation Roundoff errors and/or time quantization may cause the estimation error covariance matrix to become indefinite (losing positive semidefiniteness) when numerically solving the Riccati equation. An indefinite Riccati solution is physically impossible (since this solution is a covariance matrix) and may result in instability of the Kalman filter. These difficulties occur more frequently when the steady-state error variance is small for one or more of the states (or linear combinations of the states).

The occurrence of an indefinite estimation error covariance matrix can often be avoided by using the discrete-time Riccati equation, as opposed to discretizing the continuous-time Riccati equation (see Table 7.1). If a problem still occurs, increasing the numerical precision used in the calculations decreases the roundoff errors. Decreased roundoff error can also be obtained by using a "square root" Riccati solution algorithm. Square root Riccati solution algorithms operate on a factorization of the covariance matrix, thereby reducing the dynamic range (the span between the largest and smallest terms) of the equations used to find the Riccati solution. A good survey of square root

Riccati solution algorithms can be found starting on page 234 of [8]. As a last resort, the Riccati solution can be tested for definiteness at each iteration. If a nonpositive, semidefinite Riccati solution results, it can be modified. This adjustment is performed by replacing the negative eigenvalues by zero (or by a small positive number) in the eigenvector decomposition of the Riccati solution:

$$\mathbf{\Sigma}_e(t) = \sum_{k=1}^{n_x} \lambda_k \boldsymbol{\phi}_k \boldsymbol{\phi}_k^T.$$

The resulting matrix is positive semidefinite, and can be used both to compute the Kalman gain and for further iteration of the discrete-time Riccati equation.

7.3 The Steady-State Kalman Filter

The Kalman gain matrix experiences a transient and then approaches steady state as time increases from the initial time in the examples presented. In applications where the estimator is designed to operate for time periods that are long compared to the transient time of the Kalman gains, it is reasonable to ignore the transient and exclusively use the steady-state gains. The resulting filter is called the steady-state Kalman filter. This filter is suboptimal, in general, but still yields excellent estimates in many applications. Additionally, the steady-state Kalman filter is nearly optimal when estimating the state far from the initial time.

The use of the steady-state gains simplifies the estimator design, since only fixed-gain amplifiers are required for an analog implementation, and the need to store time-varying gains is obviated for a digital implementation. Additionally, the steady-state Kalman gain matrix is independent of the initial estimation error covariance matrix, making computation of this covariance matrix unnecessary. Note that the steady-state Kalman filter must still be initialized with a state estimate. This initial state estimate affects the early performance of the estimator, but has little affect on performance after a sufficient period of time has elapsed.

The estimation error matrix tends to the steady state as time increases. This steady-state covariance matrix provides information on the accuracy of the optimal estimates and provides a fundamental limit on estimator accuracy.[10] The estimates can only be improved beyond this limit by increasing the accuracy of the sensor, or by adding additional sensors. Therefore, the steady-state estimation error covariance matrix can be used as an aid in sensor selection.

7.3.1 $\mathcal{H}_2$ Optimal Estimation

The performance of an estimator can be quantified in terms of the gain from the plant noise and measurement noise to the estimation error. The system 2-norm and the system ∞-norm were introduced in Chapter 4 as two measures of this gain. The steady-

[10] The steady-state covariance matrix provides a fundamental limit on the accuracy of estimates provided the random inputs are Gaussian and no additional information (other than the measurements) is available. For non-Gaussian noise, a nonlinear estimator may provide estimates superior to those of the Kalman filter. But even in this case, the estimates provided by a nonlinear estimator are typically only slightly better than those provided by the Kalman filter.

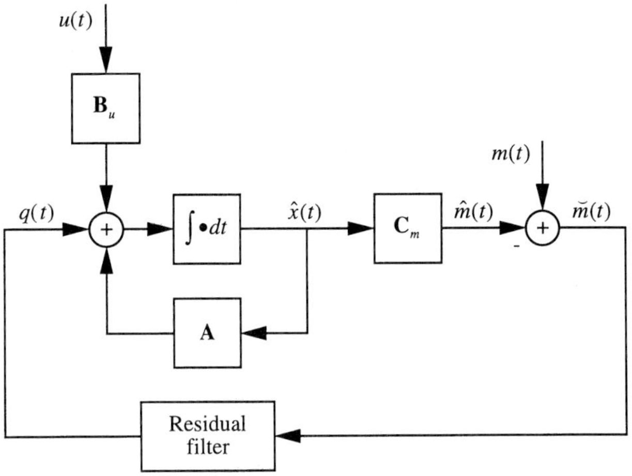

FIGURE 7.7 A general, linear, time-invariant, unbiased estimator

state Kalman filter is optimal in terms of minimizing the system 2-norm when the noise inputs are appropriately normalized. Optimization of the estimator with respect to the system ∞-norm is discussed in a subsequent chapter. Formulating the Kalman filtering problem as a 2-norm optimization problem adds insight into the Kalman filter, and provides a link between the Kalman filter and the linear quadratic regulator.

The steady-state Kalman estimator is a linear, time-invariant filter, whose inputs are the known plant inputs and the measured plant outputs. The plant is subject to the following random inputs: plant noise and measurement noise, with the properties given in Section 7.2. The steady-state Kalman filter generates state estimates that minimize the mean square estimation error. This filter can be written in predictor/corrector form:

$$\dot{\hat{x}}(t) = \mathbf{A}\hat{x}(t) + \mathbf{B}_u u(t) + \mathbf{G}[m(t) - \mathbf{C}_m \hat{x}(t)].$$

All linear, time-invariant, unbiased estimators can be written in predictor/corrector form provided the gain matrix **G** is replaced by a linear, time-invariant filter.[11] A general, linear, time-invariant, unbiased estimator is shown in Figure 7.7, where

$$\breve{m}(t) = m(t) - \mathbf{C}_m \hat{x}(t)$$

is called the residual or innovations, and $q(t)$ is the correction to the estimate of the state derivative, or simply the correction. The residual is the portion of the measurement that is not predictable from the current state estimate, and the residual filter is a linear, time-invariant system that generates the correction from the residual. Note that the selection of the residual filter is a feedback control problem, where the filter is selected to both stabilize the estimator and minimize the mean square error. The relationship of this problem to the linear quadratic regulator is discussed further in the following subsection.

[11] An estimator is unbiased if the expected value of the estimate equals the quantity being estimated.

A state model for the estimation error of the Kalman filter was generated in Subsection 7.2.2. Following the same procedure, a state model for the estimation error of a general, linear, time-invariant, unbiased estimator can be generated:

$$\dot{e}(t) = \mathbf{A}e(t) + [-\mathbf{I} \mid \mathbf{B}_w\mathbf{S}_w^{1/2} \quad \mathbf{0}]\begin{bmatrix} q(t) \\ \hline w_1(t) \\ v_1(t) \end{bmatrix}; \tag{7.40a}$$

$$\begin{bmatrix} \breve{m}(t) \\ \hline e(t) \end{bmatrix} = \begin{bmatrix} \mathbf{C}_m \\ \hline \mathbf{I} \end{bmatrix} e(t) + \begin{bmatrix} \mathbf{0} & \mathbf{0} & \mathbf{S}_v^{1/2} \\ \hline \mathbf{0} & \mathbf{0} & \mathbf{0} \end{bmatrix}\begin{bmatrix} q(t) \\ \hline w_1(t) \\ v_1(t) \end{bmatrix}. \tag{7.40b}$$

Note that the correction and the residual are included in this model as an input and an output, respectively, and the model is in the standard form for the plant in a feedback control system. Additionally, the noise inputs are replaced by the normalized noise inputs defined by the following transformation:

$$\begin{bmatrix} w(t) \\ v(t) \end{bmatrix} = \begin{bmatrix} \mathbf{S}_w^{1/2} & \mathbf{0} \\ \mathbf{0} & \mathbf{S}_v^{1/2} \end{bmatrix}\begin{bmatrix} w_1(t) \\ v_1(t) \end{bmatrix}.$$

The correlation function of the normalized noise is

$$E\left\{[w_1^T(t) \quad v_1^T(t)]\begin{bmatrix} w_1(t+\tau) \\ v_1(t+\tau) \end{bmatrix}\right\} = \begin{bmatrix} \mathbf{I}\delta(\tau) & \mathbf{0} \\ \mathbf{0} & \mathbf{I}\delta(\tau) \end{bmatrix} = \mathbf{I}\delta(\tau).$$

The $\mathcal{H}_2$ optimal estimation problem is to find the linear, time-invariant residual filter that stabilizes the closed-loop error model (7.40) and minimizes the mean square estimation error:

$$J = \mathrm{tr}\{E[e(t)e^T(t)]\}.$$

A block diagram of the closed-loop error model is given in Figure 7.8. The mean square estimation error can be written in terms of the system impulse response matrix $\mathbf{g}(t)$ formed by closing the loop between the plant (7.40) and the residual filter (see Subsection 4.5.4):

$$J = \int_0^\infty \mathrm{tr}\{\mathbf{g}(t)\mathbf{g}^T(t)\}dt = \|\mathbf{g}(t)\|_2^2. \tag{7.41}$$

The steady-state Kalman filter is the optimal linear state estimator, is unbiased, uses a time-invariant residual filter, and yields a stable closed-loop error model, given the easily satisfied conditions presented in Subsection 7.3.4. The steady-state Kalman filter is therefore the optimal solution of the $\mathcal{H}_2$ optimal estimation problem. Intuitively, this is reasonable, since minimizing the 2-norm of the closed-loop error model is equivalent to minimizing the mean square estimation error caused by the normalized noise.

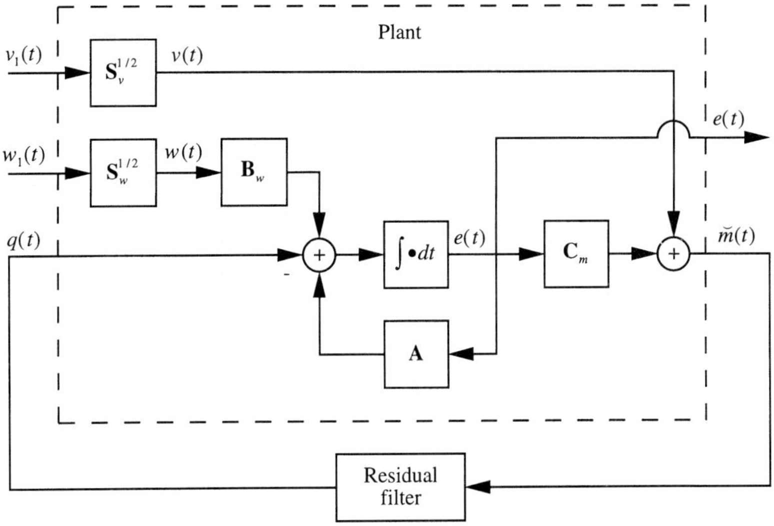

FIGURE 7.8 The error model with normalized inputs

The normalization of the noise inputs provides a weighting on the various gains so that the noise inputs with the largest spectral densities have the largest attenuation.

7.3.2 Duality

The $\mathcal{H}_2$ optimal estimation and the $\mathcal{H}_2$ optimal control problems are very similar. In both cases, a linear, time-invariant feedback that minimizes the 2-norm of the closed-loop system is desired. It is therefore not surprising that the equations for computing the Kalman gain bear a striking resemblance to the equations for computing the linear quadratic regulator gain. In fact these two problems are identical when the appropriate correspondences are made between the various terms. This fact, along with these correspondences, are collectively referred to as duality. The duality relations between the steady-state Kalman filter and the steady-state LQR are derived below.

The $\mathcal{H}_2$ optimal estimator is the steady-state Kalman filter, and the residual filter for this optimal estimator is a gain matrix. The Kalman filter is therefore also the solution of the constrained $\mathcal{H}_2$ optimization problem, which is summarized as follows: Find the gain matrix **G** that, when used as a residual filter, minimizes the 2-norm of the closed-loop error model:

$$\dot{e}(t) = [\mathbf{A} - \mathbf{G}\mathbf{C}_m]e(t) + [\mathbf{B}_w\mathbf{S}_w^{1/2} \;\vdots\; -\mathbf{G}\mathbf{S}_v^{1/2}]\begin{bmatrix} w_1(t) \\ \hdashline v_1(t) \end{bmatrix};$$

$$e(t) = \mathbf{I}e(t).$$

The impulse response matrix of this system is

$$\mathbf{g}(t) = e^{(\mathbf{A}-\mathbf{G}\mathbf{C}_m)t}[\mathbf{B}_w\mathbf{S}_w^{1/2} \;\vdots\; -\mathbf{G}\mathbf{S}_v^{1/2}],$$

and the 2-norm of this system is

$$J = \mathrm{tr}\left\{\int_0^\infty e^{(\mathbf{A}-\mathbf{G}\mathbf{C}_m)t}[\mathbf{B}_w\mathbf{S}_w^{1/2} \mid -\mathbf{G}\mathbf{S}_v^{1/2}]\left[\begin{array}{c}\mathbf{S}_w^{1/2}\mathbf{B}_w^T \\ \hline -\mathbf{S}_v^{1/2}\mathbf{G}^T\end{array}\right]e^{(\mathbf{A}^T-\mathbf{C}_m^T\mathbf{G}^T)t}dt\right\}$$
$$= \mathrm{tr}\left\{\int_0^\infty e^{(\mathbf{A}-\mathbf{G}\mathbf{C}_m)t}\{\mathbf{B}_w\mathbf{S}_w\mathbf{B}_w^T + \mathbf{G}\mathbf{S}_v\mathbf{G}^T\}e^{(\mathbf{A}^T-\mathbf{C}_m^T\mathbf{G}^T)t}dt\right\}. \quad (7.42)$$

The $\mathcal{H}_2$ optimal control is the steady-state linear quadratic regulator. The optimal controller for the LQR is state feedback. The LQR is therefore also the solution of the constrained $\mathcal{H}_2$ optimization problem, which is summarized as follows: Find the state feedback gain matrix that, when used as a controller, minimizes the 2-norm of the closed-loop system:

$$\dot{x}(t) = [\mathbf{A} - \mathbf{B}\mathbf{K}]x(t) + \mathbf{I}w(t);$$

$$y(t) = \left[\begin{array}{c}\mathbf{Q}^{1/2} \\ \hline \mathbf{R}^{1/2}\mathbf{K}\end{array}\right]x(t).$$

The impulse response matrix of this closed-loop system is

$$\mathbf{g}(t) = \left[\begin{array}{c}\mathbf{Q}^{1/2} \\ \hline \mathbf{R}^{1/2}\mathbf{K}\end{array}\right]e^{(\mathbf{A}-\mathbf{B}\mathbf{K})t},$$

and the 2-norm of this system is

$$J = \mathrm{tr}\left\{\int_0^\infty e^{(\mathbf{A}^T-\mathbf{K}^T\mathbf{B}^T)t}[\mathbf{Q}^{1/2} \mid \mathbf{K}^T\mathbf{R}^{1/2}]\left[\begin{array}{c}\mathbf{Q}^{1/2} \\ \hline \mathbf{R}^{1/2}\mathbf{K}\end{array}\right]e^{(\mathbf{A}-\mathbf{B}\mathbf{K})t}dt\right\}$$
$$= \mathrm{tr}\left\{\int_0^\infty e^{(\mathbf{A}^T-\mathbf{K}^T\mathbf{B}^T)t}\{\mathbf{Q} + \mathbf{K}^T\mathbf{R}\mathbf{K}\}e^{(\mathbf{A}-\mathbf{B}\mathbf{K})t}dt\right\}. \quad (7.43)$$

The optimal Kalman gain and the optimal LQR gain are the gains that minimize (7.42) and (7.43), respectively. These cost functions are identical, given the following correspondences:

$$\mathbf{A} \Leftrightarrow \mathbf{A}^T; \quad (7.44a)$$
$$\mathbf{B}_u \Leftrightarrow \mathbf{C}_m^T; \quad (7.44b)$$
$$\mathbf{K} \Leftrightarrow \mathbf{G}^T; \quad (7.44c)$$
$$\mathbf{Q} \Leftrightarrow \mathbf{B}_w\mathbf{S}_w\mathbf{B}_w^T; \quad (7.44d)$$
$$\mathbf{R} \Leftrightarrow \mathbf{S}_v. \quad (7.44e)$$

Note that all of the expressions on the right refer to the Kalman filter, and all of the expressions on the left are for the linear quadratic regulator. These results indicate that the linear quadratic regulator problem is mathematically equivalent to the Kalman filtering problem. Therefore, software developed for solution and analysis of the linear quadratic regulator can be applied to the solution and analysis of the Kalman filter with the appropriate substitutions. In addition, considerable time can be saved in the derivation of properties and results for the Kalman filter, since any property of the linear quadratic regulator must have an equivalent result for the Kalman filter. The remainder of this section presents properties of the steady-state Kalman filter. The derivations of these results are analogous to the derivations of similar results for the LQR. Note that while duality is demonstrated for the steady-state Kalman and the steady-state LQR, this result is quite general, and similar correspondences to those in (7.44) can be obtained in the time-varying case (see Section A11 of the Appendix or [9], page 364).

7.3.3 Computation of the Kalman Gain

The steady-state Kalman gain matrix can be computed by integrating the Riccati equation until steady state is obtained. The steady-state Riccati solution is then substituted into (7.28) to compute the steady-state Kalman gain. While this algorithm works, a numerically better-conditioned method of finding the Riccati solution utilizes the Hamiltonian equation. This solution algorithm, called the eigenvector solution algorithm, is analogous to that presented for finding the steady-state linear quadratic regulator gain matrix, and can be derived using the duality relations.

The eigenvector solution algorithm begins by finding the eigensolution of the Hamiltonian matrix. Note that the eigenvalues of the Hamiltonian matrix are symmetric around both the real and the imaginary axes; that is, if λ is an eigenvalue then $-\lambda$ is an eigenvalue. The eigenequation of the Hamiltonian is

$$\left[\begin{array}{c:c} -\mathbf{A}^T & \mathbf{C}_m^T\mathbf{S}_v^{-1}\mathbf{C}_m \\ \hdashline \mathbf{B}_w\mathbf{S}_w\mathbf{B}_w^T & \mathbf{A} \end{array}\right]\boldsymbol{\Psi} = \boldsymbol{\Psi}\boldsymbol{\Lambda},$$

where $\boldsymbol{\Lambda}$ is a diagonal matrix containing the eigenvalues of the Hamiltonian (the first n_x of which are in the left half-plane), and $\boldsymbol{\Psi}$ is the matrix of eigenvectors. The eigenvector matrix is partitioned:

$$\boldsymbol{\Psi} = \left[\begin{array}{c:c} \boldsymbol{\Psi}_{11} & \boldsymbol{\Psi}_{12} \\ \hdashline \boldsymbol{\Psi}_{21} & \boldsymbol{\Psi}_{22} \end{array}\right].$$

The steady-state Riccati solution is computed from the eigenvector matrix:

$$\boxed{\boldsymbol{\Sigma}_e = \boldsymbol{\Psi}_{22}(\boldsymbol{\Psi}^{-1})_{12},} \tag{7.45}$$

and the steady-state feedback gain matrix is then

$$\boxed{\mathbf{G} = \boldsymbol{\Psi}_{22}(\boldsymbol{\Psi}^{-1})_{12}\mathbf{C}_m^T\mathbf{S}_v^{-1}.} \tag{7.46}$$

The steady-state solution of the Riccati equation and subsequent computation of the Kalman gain matrix only require the determination of the eigenspace associated with the unstable poles of the Hamiltonian. This eigenspace can be found without generating the complete solution to the eigenequation. Alternative methods of solving the Riccati equation that determine a basis for this eigenspace, but do not determine the complete eigensolution, exist. In particular, a Riccati equation solution algorithm, which utilizes the Schur decomposition, is numerically more stable than the eigenvector solution algorithm. The details of this Schur algorithm can be found in [10].

7.3.4 Existence and Uniqueness

The steady-state Kalman gain matrix is computed under the assumption that the mean square estimation error exists (i.e., is finite). A finite mean square error is a desirable quality for an estimator. It is also desirable that the error model of (7.40) be stable since this forces estimation errors due to initial conditions or unmodeled transient inputs to approach zero as time goes to infinity. Additionally, the steady-state Kalman filter is suboptimal, in general, and a stable estimator drives the errors due to the initially suboptimal gains to zero with time. Note that the poles (and therefore the stability) of the error model are the same as the poles of the steady-state Kalman filter, so the stability of these systems can be discussed interchangeably. A question immediately comes to mind: Under what conditions does the Kalman filter yield a stable error model with a finite mean square error?

A stabilizing steady-state optimal estimator exists if either the plant is stable or the plant is both observable and controllable.[12] The observability and controllability of the plant is defined with respect to the state model given in (7.13). Note that controllability is defined with reference to the plant noise input. The conditions for the existence of the steady-state optimal estimator are intuitively justified below. A formal derivation of these conditions, based on duality, can be found in [9], page 365.

An estimator gain matrix exists that yields a finite mean square estimation error and a stable error model (7.40) when the plant (7.13) is stable. This fact is demonstrated by setting the estimator gain to zero. In this case, the error model is stable, since its state matrix is identical to the state matrix of the plant. In addition, the mean square estimation error is finite. This can be seen by direct evaluation of the mean square error when the inputs are white noise. The root mean square error equals the 2-norm of the impulse response matrix. This 2-norm is finite, since each term in the impulse response matrix decays exponentially when the system is stable. Note that while the estimator with zero gain may not be optimal, the optimal estimator must have a smaller, and therefore finite, mean square error.

An estimator gain matrix exists that yields a finite mean square estimation error and a stable error model when the plant is observable and controllable. The existence of a constant estimator gain that stabilizes the error model is guaranteed by the observability of the plant. The steady-state Kalman filter is a Luenberger observer, and the poles of an observer can be moved to any desired location using a constant gain whenever the plant is observable. Moving all of the poles into the left half-plane results in a stable estimator. Why then is the plant required to be controllable? If the plant is not controllable, there are states (or linear combinations of states) that are not excited by the plant noise. These states can be perfectly estimated, given an infinite history of data. Therefore, the Kalman gain goes to zero, in the steady-state, for these states, since new

[12] In general, a stabilizing steady-state Kalman gain exists if the plant is both stabilizable and detectable.

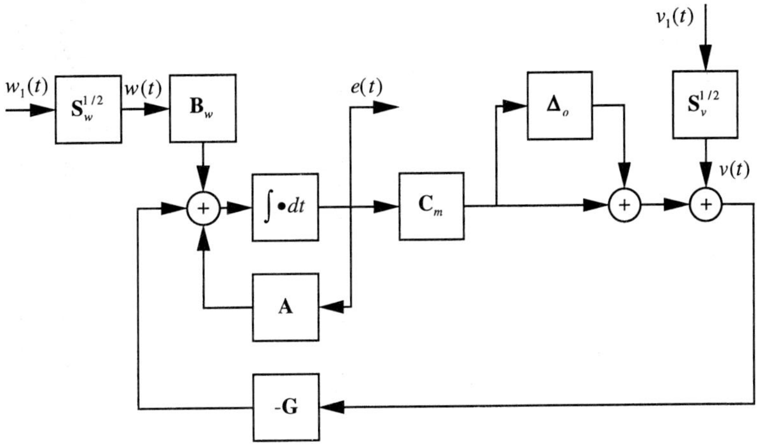

FIGURE 7.9 The error model with an output-multiplicative perturbation

data is not required. Assuming that an uncontrollable state is unstable, an initial estimation error will not decay to zero. Note that the time-varying Kalman filter drives these initial errors to zero before shutting off the gain. The steady-state Kalman filter has the gain shut off from the initial time and therefore does not estimate these initial conditions. The presence of plant noise insures that the steady-state gains are nonzero.

The Kalman filter is unique when the plant model (7.13) is observable and controllable.[13] This fact is true for both the steady-state Kalman filter and the general time-varying Kalman filter. The uniqueness of the Kalman filter is demonstrated in [9], page 365.

7.3.5 Robustness

The robustness results for the linear quadratic regulator are applicable to the Kalman filter when modified by the duality relations. These results state that the error model remains stable in the presence of output-multiplicative perturbations:

$$\|\mathbf{\Delta}_o\|_\infty \leq \frac{1}{2}.$$

A block diagram of the error model with an output multiplicative perturbation is shown in Figure 7.9. While this result is of theoretical value, it does not directly address the robustness of the Kalman filter. Perturbations to the error model are composed of perturbations to the plant and perturbations to the Kalman filter. These two perturbations are not the same, since perturbations to the plant are caused by modeling errors, and perturbations to the filter are caused by component quantization. Note that perturbations to the filter are often small enough to be ignored. Combining these terms into an error model of order n_x is not possible, in general. Therefore, modeling uncertainty in the error model as an output-multiplicative perturbation is contrived, and caution must

[13] In general, stabilizability and detectability can be used in place of controllability and observability.

be used in interpreting and applying this robustness result. Still, stability robustness in the presence of output-multiplicative perturbations implies that the system is relatively insensitive to state-dependent noise injected at the perturbation. This provides an indication that the Kalman filter is insensitive to modeling errors that generate state-dependent noise. Indeed, the Kalman filter has proven to be extremely robust in large numbers of applications where the plant is poorly modeled.

7.3.6 The Kalman Filter Poles

The error model poles provide information on the transient performance of the steady-state Kalman filter. For example, the time required for an initial estimation error to decay to zero and the amount of oscillation observed in this decay can both be estimated from these poles. In addition, insight can be gained by considering the motion of the poles as the magnitude of the measurement noise is changed relative to the plant noise.

The poles of the error model are the solutions of

$$\det(s\mathbf{I} - \mathbf{A} + \mathbf{G}\mathbf{C}_m) = 0.$$

These poles are the stable eigenvalues of the Kalman Hamiltonian:

$$\mathcal{Y} = \left[\begin{array}{c:c} -\mathbf{A}^T & \mathbf{C}_m^T \mathbf{S}_v^{-1} \mathbf{C}_m \\ \hdashline \mathbf{B}_w \mathbf{S}_w \mathbf{B}_w^T & \mathbf{A} \end{array}\right].$$

The poles of the error model depend on the steady-state Kalman gain, which in turn depends on the relative size of the spectral density of the measurement and plant noises. Limiting pole locations can be derived when the measurement noise is large and small. These results are based on the duality of the Kalman filter and the linear quadratic regulator, and follow directly from the results in Subsection 6.3.4.

The error model poles approach the left half-plane reflections of the plant poles (7.13) when the measurement noise spectral density approaches infinity. These limiting pole locations are quite intuitive in the case of a stable plant. In this case, the measurements are nearly useless, and a Kalman gain of near zero is optimal. This near-zero gain causes the filter to mostly ignore the poor measurements and instead rely heavily on the previous estimate. The state matrix of the error model is then approximately the state matrix of the plant. For an unstable plant, the fact that the poles of the error model approach the left half-plane reflections of the plant poles is surprising at first glance. These pole locations require a nonzero Kalman gain, which in turn implies the use of the poor measurements. The use of the measurements is required since the estimation error resulting from an unstable error model grows without bound if this gain is zero. Placing the error model poles at the left half-plane reflections of the unstable plant poles represents a compromise between the desire to rely on the previous estimate when the measurements are poor, and the need to drive existing estimation errors to zero.

The poles of the error model approach either the left half-plane reflections of the plant zeros or infinity within the left half-plane when the measurement noise spectral density approaches zero. The poles that approach infinity do so in a Butterworth filter pattern in the single-measurement case. When multiple measurements are used, these poles may approach infinity in either a single Butterworth filter pattern or in multiple Butterworth filter patterns.

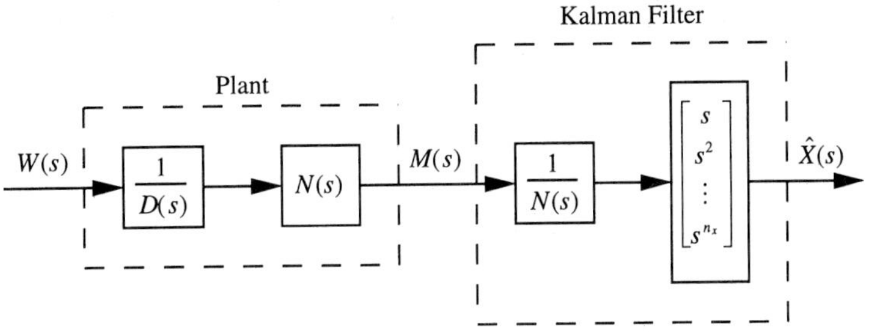

FIGURE 7.10 The asymptotic Kalman filter inverts the zeros

The limiting pole locations are quite intuitive when the plant has no zeros. In this case, the measurements are nearly perfect, and the Kalman filter relies heavily on the current measurement while ignoring past estimates. The optimal filter is therefore very fast, as indicated by the real part of the poles going to negative infinity. In this case, the Kalman filter reconstructs the state by differentiating the output, since the Kalman filter becomes an all-zero transfer function when the poles go to infinity. Note that differentiating a noisy signal is not desirable in practice. Therefore, the Kalman filter designer should always assume a reasonable amount of measurement noise to avoid this differentiation.

The plant zeros limit the speed of the Kalman filter. This can be understood by noting that to reconstruct the state in the absence of noise, the output is sent through a filter that inverts the plant zeros, and the result is differentiated to obtain the state estimates (see Figure 7.10). The speed of the optimal filter is therefore limited by the inverse filter.

7.4 Nonwhite Noise Inputs

The requirements that the plant noise and measurement noise be white is overly restrictive in some applications. White noise is a useful idealization of the actual noise whenever the correlation time of the noise is small compared to the time constants of the plant whose states are being estimated. Nonwhite noise should be assumed when the correlation time of the noise is comparable to or larger than the time constants of the plant. Optimal estimators can be obtained when nonwhite noises are present by making minor modifications of the basic Kalman filter equations. These modifications are given below.

7.4.1 Nonwhite Plant Noise

The assumption of white plant noise is violated frequently in applications. For example, consider the case of estimating the range of an aircraft from an airport control tower, given noisy radar range measurements. The plant in this case is the range dynamics of the airplane. A reasonable assumption for this model is that the radial acceleration is a random input. This random input is not, in general, white noise. A standard-rate turn for aircraft is 3 degrees per second. An aircraft turning 90 degrees, as in turning to the final approach heading in a normal traffic pattern, is accelerating in a consistent fashion for 30 seconds. The radial acceleration can then be expected to be correlated for 30 seconds, which is a significant time and violates the assumption of a white plant noise.

A time-correlated plant noise can be characterized by its frequency-dependent spectral density matrix $\mathbf{S}_w(\omega)$. A random signal with this spectral density can be generated as the output of a shaping filter with a unit spectral density white noise input. A state model for this filter is given:

$$\dot{x}_f(t) = \mathbf{A}_f x_f(t) + \mathbf{B}_f w_1(t); \tag{7.47a}$$

$$w(t) = \mathbf{C}_f x_f(t). \tag{7.47b}$$

To specify this filter, note that the plant noise spectral density is

$$\mathbf{S}_w(\omega) = \mathbf{G}_f(j\omega)\mathbf{I}\mathbf{G}_f^{\dagger}(j\omega) = \mathbf{C}_f(j\omega\mathbf{I} - \mathbf{A}_f)^{-1}\mathbf{B}_f\mathbf{B}_f^T(-j\omega\mathbf{I} - \mathbf{A}_f^T)^{-1}\mathbf{C}_f^T,$$

where $\mathbf{G}_f(j\omega)$ is the transfer function matrix from w_1 to w. The synthesis of a filter that yields a given spectral density is addressed in Section 3.4.

The information available on the spectral density of the plant noise is not very extensive in many applications. For example, in the previous discussion of target acceleration, only the correlation time τ_c is specified. In these cases, very simple filters can be designed to generate the plant noise. As always, the simplest possible model (7.47) that corresponds to the available information should be used.

An augmented plant model can be developed by combining the plant model (7.13) with the model that generates the plant noise (7.47):

$$\begin{bmatrix} \dot{x}(t) \\ \dot{x}_f(t) \end{bmatrix} = \begin{bmatrix} \mathbf{A} & \mathbf{B}_w\mathbf{C}_f \\ \mathbf{0} & \mathbf{A}_f \end{bmatrix}\begin{bmatrix} x(t) \\ x_f(t) \end{bmatrix} + \begin{bmatrix} \mathbf{B}_u \\ \mathbf{0} \end{bmatrix} u(t) + \begin{bmatrix} \mathbf{0} \\ \mathbf{B}_f \end{bmatrix} w_1(t); \tag{7.48a}$$

$$m(t) = [\mathbf{C}_m \;\vdots\; \mathbf{0}]\begin{bmatrix} x(t) \\ x_f(t) \end{bmatrix} + v(t). \tag{7.48b}$$

This augmented plant is driven by white noise, and the augmented state can be estimated using the Kalman filter. The optimal estimates of the original state are generated as part of the augmented state.

EXAMPLE 7.10 A Kalman filter is being used to estimate the range and radial velocity of a aircraft from noisy range measurements. Assuming a random acceleration, the plant model is

$$\begin{bmatrix} \dot{R}(t) \\ \ddot{R}(t) \end{bmatrix} = \begin{bmatrix} 0 & 1 \\ 0 & 0 \end{bmatrix}\begin{bmatrix} R(t) \\ \dot{R}(t) \end{bmatrix} + \begin{bmatrix} 0 \\ 1 \end{bmatrix} w(t);$$

$$m(t) = [1 \quad 0]\begin{bmatrix} R(t) \\ \dot{R}(t) \end{bmatrix} + v(t).$$

Note that there are no known inputs to this plant. The spectral density of the measurement noise is

$$S_v = 10{,}000 \text{ m}^2/\text{Hz}.$$

The plant noise is correlated for 30 seconds (this correlation time is discussed above) and has a variance of 100 m²/sec⁴. A simple state model for the noise, which allows the correlation time and gain to be specified, is

$$\dot{x}_f(t) = -a_f x_f(t) + b_f w_1(t);$$
$$w(t) = x_f(t),$$

where $w_1(t)$ is white noise with a unit spectral density. The filter parameter and the spectral density of the white noise $w_1(t)$ are

$$a_f = \frac{1}{\tau_c} = 0.033;$$
$$b_f = \sqrt{\frac{2\sigma_w^2}{\tau_c}} = 2.6.$$

The augmented state model is then

$$\begin{bmatrix} \dot{R}(t) \\ \ddot{R}(t) \\ \dot{x}_f(t) \end{bmatrix} = \begin{bmatrix} 0 & 1 & 0 \\ 0 & 0 & 1 \\ 0 & 0 & -0.033 \end{bmatrix} \begin{bmatrix} R(t) \\ \dot{R}(t) \\ x_f(t) \end{bmatrix} + \begin{bmatrix} 0 \\ 0 \\ 2.6 \end{bmatrix} w_1(t);$$

$$m(t) = [1 \quad 0 \quad 0] \begin{bmatrix} R(t) \\ \dot{R}(t) \\ w(t) \end{bmatrix} + v(t).$$

◆

The disturbance input is only roughly approximated by specifying the variance and the correlation time. Fortunately, the Kalman filter typically provides very good estimates, even when the plant and noise models contain errors.

7.4.2 Nonwhite Measurement Noise

The plant augmentation procedure used for nonwhite plant noise is not applicable to the case of nonwhite measurement noise. This fact is demonstrated by forming the augmented state model and observing that this model has no measurement noise.

The plant whose states are being estimated is given by (7.13). The time-correlated measurement noise is characterized by its frequency-dependent spectral density matrix $\mathbf{S}_v(\omega)$. A random signal with this spectral density can be generated as the output of a shaping filter with a unit spectral density white noise input. A state model for this filter is given:

$$\dot{x}_f(t) = \mathbf{A}_f x_f(t) + \mathbf{B}_f v_1(t); \tag{7.49a}$$
$$v(t) = \mathbf{C}_f x_f(t). \tag{7.49b}$$

An augmented plant model can be generated by combining (7.49) with the plant dynamics (7.13):

$$\begin{bmatrix} \dot{x}(t) \\ \hline \dot{x}_f(t) \end{bmatrix} = \left[\begin{array}{c:c} \mathbf{A} & \mathbf{0} \\ \hline \mathbf{0} & \mathbf{A}_f \end{array}\right] \begin{bmatrix} x(t) \\ \hline x_f(t) \end{bmatrix} + \begin{bmatrix} \mathbf{B}_u \\ \hline \mathbf{0} \end{bmatrix} u(t) + \left[\begin{array}{c:c} \mathbf{B}_w & \mathbf{0} \\ \hline \mathbf{0} & \mathbf{B}_f \end{array}\right] \begin{bmatrix} w(t) \\ \hline v_1(t) \end{bmatrix};$$

$$m(t) = [\mathbf{C}_m \;\vdots\; \mathbf{C}_f] \begin{bmatrix} x(t) \\ \hline v(t) \end{bmatrix}.$$

Note that measurement noise is not present in the augmented plant model, since the measurement noise in the plant model is included within the state. Therefore, the spectral density matrix of the measurement noise in this model is not positive definite, the inverse in the Kalman gain equation (7.28) does not exist, the Kalman gain is no longer given by (7.28), and plant augmentation does not directly lead to the optimal estimator.

The optimal estimator for the case of nonwhite measurement noise can be obtained by a judicious redefinition of the measurements and a minor extension of the Kalman theory. A simplified noise model is used in this derivation to keep the notation from getting too cumbersome:

$$\dot{v}(t) = \mathbf{A}_f v(t) + v_1(t),$$

where the $v_1(t)$ is white noise with spectral density $\mathbf{S}_{v_1}$. This noise model limits the order of the filter approximation to the number of measurements. This limitation is consistent with previous examples.

A new measurement can be constructed:

$$z(t) = \dot{m}(t) - \mathbf{A}_f m(t) - \mathbf{C}_m \mathbf{B}_u u(t).$$

This new measurement, called the constructed measurement, can be formed from the known plant measurement and the plant input.[14] The constructed measurement can be written in terms of the plant state by substituting for $m(t)$ and $\dot{m}(t)$ using (7.13b) and its derivative:

$$z(t) = \{\mathbf{C}_m\mathbf{A} - \mathbf{A}_f\mathbf{C}_m\}x(t) + [\mathbf{C}_m\mathbf{B}_w \mid \mathbf{I}]\begin{bmatrix} w(t) \\ \hdashline v_1(t) \end{bmatrix} = \mathbf{C}_z x(t) + v_2(t), \tag{7.50a}$$

where the terms $\mathbf{C}_z$ and the constructed measurement noise $v_2(t)$ are defined by this equation. Note that the constructed measurement noise is white. The Kalman filter can be applied to the plant formed by using the original plant state equation,

$$\dot{x}(t) = \mathbf{A}x(t) + \mathbf{B}_u u(t) + \mathbf{B}_w w(t), \tag{7.50b}$$

and the constructed measurement equation. The application of the Kalman filter requires the spectral density of the constructed measurement noise. This spectral density is

$$\mathbf{S}_{v_2} = \mathbf{C}_m\mathbf{B}_w\mathbf{S}_w\mathbf{B}_w^T\mathbf{C}_m^T + \mathbf{S}_{v_1}, \tag{7.51}$$

assuming the plant noise and the shaping filter input are uncorrelated. This is a reasonable assumption, since these two noises are typically formed from very different physical processes. There is, however, a correlation between the plant noise and the constructed measurement noise, since the plant noise is included within the constructed measure-

[14] The enterprising student is probably concerned at this point, because taking the derivative of a measurement, as is done in generating the constructed measurement, is not a desirable operation due to the extreme noise sensitivity of the differentiation operation. Be advised that after developing the optimal estimator, this concern is addressed.

ment noise. The cross-covariance function of the plant noise and constructed measurement noise is

$$\mathbf{R}_{wv_2}(\tau) = E[w(t)v_2^T(t + \tau)] = \mathbf{S}_w\mathbf{B}_w^T\mathbf{C}_m^T\delta(\tau) = \mathbf{S}_{wv_2}\delta(\tau), \tag{7.52}$$

where $\mathbf{S}_{wv_2}$ is the cross–spectral density of these two processes. Note that it is again assumed that $w(t)$ and $v_1(t)$ are uncorrelated in deriving this equation.

The fact that the plant noise and the constructed measurement noise are correlated requires a modification to the Kalman filter. The basic Kalman filter equation (7.22) remains unchanged when this correlation is present. For the plant in (7.50), the optimal estimator is

$$\begin{aligned}\dot{\hat{x}}(t) &= \mathbf{A}\hat{x}(t) + \mathbf{B}u(t) + \mathbf{G}(t)[z(t) - \mathbf{C}_z\hat{x}(t)] \\ &= \mathbf{A}\hat{x}(t) + \mathbf{B}u(t) + \mathbf{G}(t)[\dot{m}(t) - \mathbf{A}_f m(t) - \mathbf{C}_m\mathbf{B}_u u(t) - \{\mathbf{C}_m\mathbf{A} - \mathbf{A}_f\mathbf{C}_m\}\hat{x}(t)].\end{aligned} \tag{7.53}$$

But the Kalman gain equations must be modified when the plant noise and the measurement noise are correlated:

$$\dot{\mathbf{\Sigma}}_e(t) = \mathbf{\Sigma}_e(t)\mathbf{A}^T + \mathbf{A}\mathbf{\Sigma}_e(t) + \mathbf{B}_w\mathbf{S}_w\mathbf{B}_w^T - \mathbf{\Sigma}_e(t)\mathbf{C}_z^T\mathbf{S}_{v_2}^{-1}\mathbf{C}_z\mathbf{\Sigma}_e(t); \tag{7.54a}$$

$$\mathbf{G}(t) = [\mathbf{\Sigma}_e(t)\mathbf{C}_z^T + \mathbf{B}_w\mathbf{S}_{wv_2}]\mathbf{S}_{v_2}^{-1}. \tag{7.54b}$$

The details of the derivation of the Kalman filter equations for correlated plant and measurement noises can be found in [11], page 124 (see also Exercise 7.12).

The derivative of the measurement appears in the estimator equation (7.53). The differentiation of a measured quantity is undesirable since this operation is very noise- and error-sensitive. This differentiation can be avoided by noting that

$$\mathbf{G}(t)\dot{m}(t) = \frac{d\{\mathbf{G}(t)m(t)\}}{dt} - \dot{\mathbf{G}}(t)m(t).$$

Substituting this expression into (7.53) yields the following differential equation:

$$\begin{aligned}&\frac{d[\hat{x}(t) - \mathbf{G}(t)m(t)]}{dt} \\ &= \mathbf{A}\hat{x}(t) + \mathbf{B}_u u(t) - \dot{\mathbf{G}}(t)m(t) - \mathbf{G}(t)[\mathbf{A}_f m(t) + \mathbf{C}_m\mathbf{B}_u u(t) + \mathbf{C}_z\hat{x}(t)].\end{aligned} \tag{7.55}$$

The state estimate is generated by integrating this equation and then subtracting $\mathbf{G}(t)m(t)$ from the result. A block diagram of this optimal estimator is shown in Figure 7.11.

Initial conditions on the state estimate and the estimation error covariance matrix are required for implementing the optimal estimator. The generation of these initial conditions for the filter specified above is complicated by the fact that discontinuities occur in the state estimate and the estimation error covariance matrix at the initial time. These discontinuities occur because the initial measurement, when the measurement noise is nonwhite, has a finite variance and can be used to provide an instantaneous

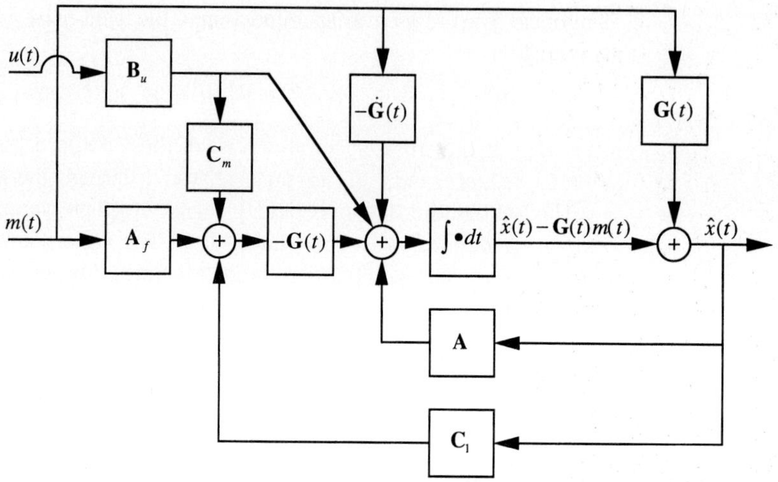

FIGURE 7.11 The optimal estimator when the measurement noise is nonwhite

estimate of the state. This is in contrast to the case of white measurement noise, where the measurements have an infinite variance and can only generate useful estimates after integration. For nonwhite measurement noise, the new information represented by each subsequent measurement is infinitesimal, since new measurements are strongly correlated with the past measurements. Therefore, these measurements must be filtered to obtain additional information, and the differential estimator equation applies.

The initial condition for the estimator (7.55) is defined an instant after receiving the initial measurement. This initial condition is $[\hat{x}(0^+) - \mathbf{G}(0)m(0)]$, where

$$\hat{x}(0^+) = \hat{x}(0) + \mathbf{\Sigma}_e(0)\mathbf{C}_m^T[\mathbf{C}_m\mathbf{\Sigma}_e(0)\mathbf{C}_m^T + \sigma_v^2]^{-1}[m(0) - \mathbf{C}_m\hat{x}(0)]. \tag{7.56a}$$

Note that this initialization requires σ_v^2, which is the variance of the measurement noise, and the initial measurement $m(0)$.

The estimation error covariance matrix experiences an instantaneous reduction at time zero due to the instantaneous improvement in the estimate when the initial measurement arrives. This covariance matrix is

$$\mathbf{\Sigma}_e(0^+) = \mathbf{\Sigma}_e(0) - \mathbf{\Sigma}_e(0)\mathbf{C}_m^T[\mathbf{C}_m\mathbf{\Sigma}_e(0)\mathbf{C}_m^T + \sigma_v^2]^{-1}\mathbf{C}_m\mathbf{\Sigma}_e(0) \tag{7.56b}$$

an instant after time zero. Note that this estimation error covariance matrix is independent of the measurements and can therefore be computed in advance of performing the filtering. The initial conditions (7.56) are formally derived in [12].

The initial conditions (7.56), the estimator equation (7.55), the gain equations (7.54), the plant model (7.50), the plant noise spectral density, the constructed measurement noise spectral density (7.51), and the cross–spectral density of the plant noise and the constructed measurement noise (7.52) fully specify the optimal estimator when the measurement noise is nonwhite.

EXAMPLE 7.11 The states of a controlled dc motor are being estimated in order to detect faults in the motor or controller. The state model of the motor is

$$\begin{bmatrix} \dot{\theta}(t) \\ \ddot{\theta}(t) \end{bmatrix} = \begin{bmatrix} 0 & 1 \\ -1 & -1 \end{bmatrix} \begin{bmatrix} \theta(t) \\ \dot{\theta}(t) \end{bmatrix} + \begin{bmatrix} 0 \\ 1 \end{bmatrix} w(t);$$

$$m(t) = [1 \quad 0] \begin{bmatrix} \theta(t) \\ \dot{\theta}(t) \end{bmatrix} + v(t),$$

where $\theta(t)$ is the motor shaft angle. Note that the control input to the motor in this example is inaccessible and cannot be used as an input to the Kalman filter. The plant noise is white, with the following spectral density: $S_w = 1$ rad^2/(sec^4-Hz). The measurement noise is nonwhite, with a variance and correlation time of $\sigma_v^2 = 1$ rad^2 and $\tau_c =$ 3 sec, respectively. A measurement noise with this variance and time constant can be generated by putting white noise into the filter:

$$\dot{v}(t) = -\frac{1}{3} v(t) + v_1(t),$$

where the spectral density of $v_1(t)$ is $S_{v_1} = 0.67$ rad^2/(sec^2-Hz). The constructed measurement is

$$z(t) = [-\tfrac{1}{3} \quad 1] \begin{bmatrix} \theta(t) \\ \dot{\theta}(t) \end{bmatrix} + [0 \quad 1] \begin{bmatrix} w(t) \\ v_1(t) \end{bmatrix} = [-\tfrac{1}{3} \quad 1] \begin{bmatrix} \theta(t) \\ \dot{\theta}(t) \end{bmatrix} + v_1(t).$$

In this example, the plant noise does not appear in the constructed measurement equation and can be eliminated. This simplification results from the fact that the product $\mathbf{C}_m \mathbf{B}_w$ is zero, or physically, that the derivative of the measurement does not include the plant noise. In this case, the measurement noise is not correlated with the plant noise, and the usual Kalman gain equations can be employed in the estimator (7.55).

The initial conditions for the optimal estimator are computed by first generating the initial state estimate and the initial covariance matrix just before the arrival of the first measurement:

$$\begin{bmatrix} \hat{\theta}(0^-) \\ \hat{\dot{\theta}}(0^-) \end{bmatrix} = \begin{bmatrix} 0 \text{ rad} \\ 0 \text{ rad/sec} \end{bmatrix}, \quad \mathbf{\Sigma}_e(0^-) = \begin{bmatrix} \frac{1}{2} \text{ rad}^2 & 0 \\ 0 & \frac{1}{2} \text{ rad}^2/\text{sec}^2 \end{bmatrix}.$$

Note that these values are generated assuming that the motor has been operating for a long time. The initial conditions for the estimator and the Riccati equation are then

$$\begin{bmatrix} \hat{\theta}(0^+) \\ \hat{\dot{\theta}}(0^+) \end{bmatrix} = \begin{bmatrix} \frac{1}{3}\text{m}(0) \text{ rad} \\ 0 \text{ rad/sec} \end{bmatrix}, \quad \mathbf{\Sigma}_e(0^+) = \begin{bmatrix} \frac{1}{3} \text{ rad}^2 & 0 \\ 0 & \frac{1}{2} \text{ rad}^2/\text{sec}^2 \end{bmatrix}.$$

The estimates, the Kalman gains, and the estimation error covariance matrix for the motor are given Figure 7.12. Note that the system is operated long enough that the

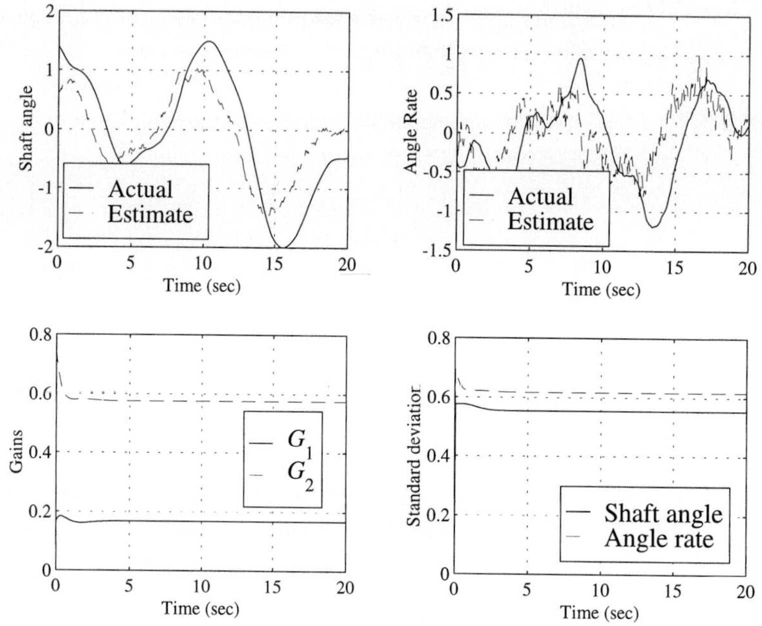

FIGURE 7.12 Results for Example 7.11

states and the measurement noise have reached steady state before state estimation commences. The graphs show that the Kalman filter provides reasonable estimates. The improvement in the estimation error due to measurements after the initial measurement is modest. This is reasonable, since the measurements have a long correlation time compared to the correlation time of the plant, and extensive averaging of the measurements is not warranted. ◆

Augmenting the plant with a measurement noise dynamic model, as is done for nonwhite plant noise, yields measurements with a noise spectral density that is not positive definite. It is therefore reasonable to assume that the results of this subsection can also be applied when the spectral density of the measurement noise is not positive definite. In this case, a reduced-order Kalman filter is the optimal estimator. A derivation of this filter is given in [13], page 427.

7.5 Summary

The estimation of the plant state from the measurements and the inputs is an important problem in signal processing and control. The optimal state estimate is defined as the estimate that minimizes the mean square estimation error. An optimal linear state estimator, known as the Kalman filter, is developed in this chapter. Optimal linear estimates are defined by the orthogonality principle: For the optimal linear estimate, the estimation error is orthogonal, in the probabilistic sense, to the data.

The precise mathematical formulation of the Kalman filtering problem is presented, and the Kalman filter equations developed. These equations require that a gain, known as the Kalman gain, be computed. The equation for the Kalman gain is also developed. This gain depends on the estimation error covariance matrix which can be computed using a nonlinear matrix differential equation known as the Riccati equation. As in the case of the linear quadratic regulator, this Riccati equation is equivalent to a system of linear differential equations known as the Hamiltonian system. The Riccati equation and the Hamiltonian system provide two distinct methods of generating the estimation error covariance matrix, which is used in computing the Kalman gain.

The application of the Kalman filter requires a plant model, a measurement noise spectral density matrix, a plant noise spectral density matrix, an initial state estimate, and an initial estimation error covariance matrix. The generation of these quantities is discussed in detail and illustrated by a number of examples.

The Kalman filter is a time-varying Luenberger observer. The Kalman gain varies through an initial transient, and then settles to a steady-state value far from the initial time. In many applications, the Kalman filter is designed to operate for extended periods of time, and the steady-state Kalman gain can be used exclusively. A stable observer results when using the steady-state gain provided that the plant is both observable and controllable. The steady-state Kalman gain can be found from the eigensolution of the Hamiltonian matrix.

The steady-state Kalman filter problem is shown to be equivalent to an $\mathcal{H}_2$ optimal estimation problem. In fact, this optimization problem is equivalent to the linear quadratic regulator problem when the appropriate substitutions are made. This correspondence is termed duality. Duality of these two mathematical problems can be used to generate properties of the Kalman filter, which correspond to properties of the LQR. In particular, duality is used to generate results on existence, uniqueness, and robustness of the Kalman filter. Additionally, duality is used to find the limiting pole locations of the error model as the size of the measurement noise spectral density matrix is varied.

The Kalman filter is derived assuming that the plant noise and the measurement noise are both white. When one or both of these noises are nonwhite, an optimal estimator can be generated using the basic Kalman filter equations. The optimal estimator in the case of nonwhite plant noise is a Kalman filter applied to an augmented system consisting of the original plant, and a shaping filter that generates the plant noise from a white noise input. The optimal estimator for the case of nonwhite measurement noise is obtained by defining a constructed measurement that has a white measurement noise and applying the Kalman filter. In general, the measurement noise on this constructed measurement is correlated with the plant noise. The Kalman gain equations are modified to yield the optimal estimator in this case. The initialization of the optimal estimator must be altered when the measurement noise is nonwhite. The appropriate initialization equations are presented.

A Kalman filter can be derived for the discrete-time state model of the plant (see [14], page 444). The discrete-time state model is often the most appropriate model to use for Kalman filter design, since this filter is typically implemented with a digital computer. The fundamental equations presented in this chapter are summarized for both continuous-time and discrete-time systems in Table 7.1, where T is the sampling time.

TABLE 7.1 Summary of Formulas for Continuous-Time and Discrete-Time Systems

Formula	Continuous-Time	Discrete-Time
The plant model	$\dot{x}(t) = \mathbf{A}x(t) + \mathbf{B}_u u(t) + \mathbf{B}_w w(t)$ $m(t) = \mathbf{C}_m x(t) + v(t)$	$x(k+1) = \mathbf{\Phi} x(k) + \mathbf{\Gamma}_u u(k) + \mathbf{\Gamma}_w w(k)$ $m(k) = \mathbf{C}_m x(k) + v(k)$
The noise model	$E[w(t)w^T(t+\tau)] = \mathbf{S}_w \delta(\tau)$	$E[w(k)w^T(k+p)] = \mathbf{\Sigma}_w \delta(p) = \frac{S_w}{T}\delta(p)$
	$E[v(t)v^T(t+\tau)] = \mathbf{S}_v \delta(\tau)$	$E[v(k)v^T(k+p)] = \mathbf{\Sigma}_v \delta(p) = \frac{\mathbf{S}_v}{T}\delta(p)$
	$E[v(t)w^T(t+\tau)] = \mathbf{0}$	$E[v(k)w^T(k+p)] = \mathbf{0}$
The Kalman filter equation[1]	$\dot{\hat{x}}(t) = \mathbf{A}\hat{x}(t) + \mathbf{B}_u u(t)$ $+ \mathbf{G}(t)[m(t) - \mathbf{C}_m \hat{x}(t)]$	$\hat{x}(k+1) = \mathbf{\Phi}\hat{x}(k) + \mathbf{\Gamma}_u u(k) + \mathbf{G}_d(k)$ $\cdot [m(k) - \mathbf{C}_m \hat{x}(k)]$
The Kalman gain	$\mathbf{G}(t) = \mathbf{\Sigma}_e(t)\mathbf{C}_m^T \mathbf{S}_v^{-1}$	$\mathbf{G}_d(k) = \mathbf{\Phi}\mathbf{\Sigma}_e(k)\mathbf{C}_m^T[\mathbf{C}_m \mathbf{\Sigma}_e(k)\mathbf{C}_m^T + \mathbf{\Sigma}_v]^{-1}$
The Riccati equation	$\dot{\mathbf{\Sigma}}_e(t) = \mathbf{A}\mathbf{\Sigma}_e(t) + \mathbf{\Sigma}_e(t)\mathbf{A}^T + \mathbf{B}_w \mathbf{S}_w \mathbf{B}_w^T$ $- \mathbf{\Sigma}_e(t)\mathbf{C}_m^T \mathbf{S}_v^{-1}\mathbf{C}_m \mathbf{\Sigma}_e(t)$	$\mathbf{\Sigma}_e(k+1) = \mathbf{\Phi}\mathbf{\Sigma}_e(k)\mathbf{\Phi}^T + \mathbf{\Gamma}_w \mathbf{\Sigma}_w \mathbf{\Gamma}_w'$ $- \mathbf{\Phi}\mathbf{\Sigma}_e(k)\mathbf{C}_m^T[\mathbf{C}_m \mathbf{\Sigma}_e(k)\mathbf{C}_m^T + \mathbf{\Sigma}_v]^{-1}\mathbf{C}_m \mathbf{\Sigma}_e(k)\mathbf{\Phi}^T$
The Hamiltonian	$\mathcal{Y} = \begin{bmatrix} -\mathbf{A}^T & \mathbf{C}_m^T \mathbf{S}_v^{-1}\mathbf{C}_m \\ \mathbf{B}_w \mathbf{S}_w \mathbf{B}_w^T & \mathbf{A} \end{bmatrix}$	$\mathcal{Y}_d = \begin{bmatrix} \mathbf{\Phi}^T + \mathbf{C}_m^T \mathbf{\Sigma}_v^{-1}\mathbf{C}_m \mathbf{\Phi}^{-1}\mathbf{\Gamma}_w \mathbf{\Sigma}_w \mathbf{\Gamma}_w^T & \mathbf{C}_m^T \mathbf{\Sigma}_v^{-1}\mathbf{C}_m \mathbf{\Phi}^{-1} \\ -\mathbf{\Phi}^{-1}\mathbf{\Gamma}_w \mathbf{\Sigma}_w \mathbf{\Gamma}_w^T & \mathbf{\Phi}^{-1} \end{bmatrix}$

[1] The Kalman filter equations are for the one-step-ahead prediction of the state; that is, the estimate of $x(k)$ is based on the data $m(j)$ where $j = 0, 1, \ldots, k-1$. For an estimator of $x(k)$, given data that includes $m(k)$, see [14], page 444.

REFERENCES

[1] C. W. Therrien, *Discrete Random Signals and Statistical Signal Processing,* Prentice-Hall, Englewood Cliffs, NJ, 1992.

[2] J. E. Marsden, *Elementary Classical Analysis,* W. H. Freeman, New York, 1974.

[3] R. E. Kalman and R. S. Bucy, "New results in linear filtering and prediction theory," *ASME Journal of Basic Engineering,* series D, 83 (1961): 95–108.

[4] L. Ljung, *System Identification: Theory for the User,* Prentice-Hall, Englewood Cliffs, NJ, 1987.

[5] T. Soderstrom and P. Stoica, *System Identification,* Prentice-Hall, Englewood Cliffs, NJ, 1989.

[6] L. Marple, Jr., *Digital Spectral Analysis with Applications,* Prentice-Hall, Englewood Cliffs, NJ, 1987.

[7] S. Kay, *Modern Spectral Estimation: Theory and Applications,* Prentice-Hall, Englewood Cliffs, NJ, 1988.

[8] M. S. Grewal and A. P. Andrews, *Kalman Filtering: Theory and Practice,* Prentice-Hall, Englewood Cliffs, NJ, 1993.

[9] H. Kwakernaak and R. Sivan, *Linear Optimal Control Systems,* Wiley-Interscience, New York, 1972.

[10] W. F. Arnold, III and A. J. Laub, "Generalized eigenproblem algorithms and software for algebraic Riccati equations," *Proc. IEEE,* 72, no. 12 (1984): 1746–54.

[11] A. Gelb, *Applied Optimal Estimation,* MIT Press, Cambridge, MA, 1986.

[12] A. E. Bryson, Jr. and D. E. Johansen, "Linear filtering for time-varying systems using measurements containing colored noise," *IEEE Transactions on Automatic Control,* AC-10, no. 1 (1965): 4–10.

[13] B. Friedland, *Control System Design: An Introduction to State-Space Methods,* McGraw-Hill, New York, 1986.

[14] G. F. Franklin, J. D. Powell, and M. L. Workman, *Digital Control of Dynamic Systems,* 2d ed., Addison-Wesley, Reading, MA, 1990.

Some additional references on optimal estimation and Kalman filtering follow:

[15] B. D. O. Anderson and J. B. Moore, *Optimal Filtering,* Prentice-Hall, Englewood Cliffs, NJ, 1979.

[16] A. E. Bryson, Jr., and Y. C. Ho, *Applied Optimal Control,* Blaisdell, Waltham, MA, 1969.

[17] A. P. Sage and C. C. White, III, *Optimum Systems Control,* 2d ed., Prentice-Hall, Englewood Cliffs, NJ, 1977.

EXERCISES

7.1 Use the orthogonality principle to generate the optimal estimator of s, where the measurements are

$$m(k) = s + v(k),\ k \in \text{Integers}[1,\ 10].$$

The measurement noise has the following property:

$$E[v(k)v(l)] = \begin{cases} 4 & k = l \\ 0 & k \neq l \end{cases}.$$

a. The signal s is an unknown parameter (assume $E[s^2] = \infty$ to indicate a total lack of *a priori* information concerning s).
b. The signal s is a random variable (constant in time) with a mean of zero and a variance of 1.
c. What is the difference between the estimator gains in parts a and b? Why?
d. Compute the mean square error when s is given as in a and b. What is the difference between the two optimal estimation errors? Why?

7.2 Given the estimate of s as specified in Exercise 7.1b, generate the optimal update of this estimate, when an additional measurement is obtained.

7.3 Given the cost function

$$J = E[\{x - \hat{x}(m)\}^T \mathbf{W} \{x - \hat{x}(m)\}],$$

where $\mathbf{W}$ is a symmetric, positive definite matrix, show that the linear estimate that minimizes this cost function is identical to the linear estimate that minimizes the mean square error. Assume that the state and measurement are both vectors. Note that this result implies that the Kalman filter is also optimal in terms of minimizing this cost function.

7.4 Solve for the estimator gains in Example 7.1 using the discrete-time Kalman filter equations. Verify that these gains equal the gains derived in the example.

7.5 You are given the following plant:

$$\dot{x}(t) = 2x(t) + 2u(t) + w(t);$$
$$m(t) = 3x(t) + v(t).$$

The correlation function of the plant and measurement noises are

$$E[w(t)w(t + \tau)] = \delta(\tau);$$
$$E[v(t)v(t + \tau)] = 5\delta(\tau),$$

respectively, and the initial condition on the state is known *a priori*: $x(0) = 0$.

a. Generate the Riccati equation for this system. Solve this equation analytically and generate an analytic expression for the Kalman gain.
b. Generate the Hamiltonian of this system. Solve this system analytically and generate an analytic expression for the Kalman gain.
c. Solve for the steady-state solution of the Kalman gain using the eigenvector decomposition method. Verify that this matches the results of part a and b.
d. What is the pole of the steady-state Kalman filter?

e. How does this pole change if the measurement noise spectral density is increased by a factor of 10?

7.6 You are given the time-varying plant

$$\dot{x}(t) = \mathbf{A}(t)x(t) + \mathbf{B}_u(t)u(t) + \mathbf{B}_w(t)w(t);$$
$$m(t) = \mathbf{C}_m(t)x(t) + v(t).$$

The plant and measurement noises are assumed to be white, with the following correlation functions:

$$E[w(t_1)w(t_2)] = \mathbf{S}_w(t_1)\delta(t_1 - t_2);$$
$$E[v(t_1)v(t_2)] = \mathbf{S}_v(t_1)\delta(t_1 - t_2),$$

respectively. Note that the spectral density matrices are time-varying. Show that the optimal state estimate is given by the Kalman filter:

$$\dot{\hat{x}}(t) = \mathbf{A}(t)\hat{x}(t) + \mathbf{B}_u(t)u(t) + \mathbf{G}(t)[m(t) - \mathbf{C}_m(t)\hat{x}(t)].$$

Also show that the Kalman gain is

$$\mathbf{G}(t) = \mathbf{\Sigma}_e(t)\mathbf{C}_m^T(t)\mathbf{S}_v^{-1}(t),$$

and the estimation error covariance matrix is given by the time-varying Riccati equation:

$$\dot{\mathbf{\Sigma}}_e(t) = \mathbf{\Sigma}_e(t)\mathbf{A}^T(t) + \mathbf{A}(t)\mathbf{\Sigma}_e(t) + \mathbf{B}_w(t)\mathbf{S}_w(t)\mathbf{B}_w^T(t) - \mathbf{\Sigma}_e(t)\mathbf{C}_m^T(t)\mathbf{S}_v^{-1}(t)\mathbf{C}_m(t)\mathbf{\Sigma}_e(t).$$

Hint: The development of the Kalman filter for a time-varying plant is analogous to the development in the time-invariant case.

7.7 Verify that closing the loop in the error model (7.40) results in the state equation for the error (7.29).

7.8 You are given the dc motor described by the following state model:

$$\begin{bmatrix}\dot{x}_1(t)\\ \dot{x}_2(t)\end{bmatrix} = \begin{bmatrix}0 & 1\\ 0 & -1\end{bmatrix}\begin{bmatrix}x_1(t)\\ x_2(t)\end{bmatrix} + \begin{bmatrix}0\\ 1\end{bmatrix}u(t) + \begin{bmatrix}0\\ 1\end{bmatrix}w(t);$$
$$m(t) = [1 \quad 0]\begin{bmatrix}x_1(t)\\ x_2(t)\end{bmatrix} + v(t).$$

Generate an augmented plant model under the following conditions.

a. The plant noise has a variance of 1 and a correlation time of 2 seconds. For correlation times less than what value (approximately) could the plant noise be assumed to be white? Why?

b. The plant noise has a spectral density of

$$S_w(\omega) = \frac{1}{\omega^4 - 1.99\omega^2 + 1}.$$

7.9 You are given the plant described by the following state model:

$$\begin{bmatrix}\dot{x}_1(t)\\ \dot{x}_2(t)\end{bmatrix} = \begin{bmatrix}0 & 1\\ -4 & -1\end{bmatrix}\begin{bmatrix}x_1(t)\\ x_2(t)\end{bmatrix} + \begin{bmatrix}0\\ 1\end{bmatrix}w(t);$$
$$m(t) = [1 \quad 2]\begin{bmatrix}x_1(t)\\ x_2(t)\end{bmatrix} + v(t).$$

The plant and measurement noises are assumed to have the following correlation functions:

$$E[w(t)w(t + \tau)] = \delta(\tau);$$
$$E[v(t)v(t + \tau)] = e^{-|\tau|/2},$$

respectively. Generate the optimal estimator for this plant. Your estimator design should include the filter equations, the gain equations, and the initialization for both the filter and the Riccati equation.

7.10 Use the orthogonality principle to verify that the discrete-time Kalman filter equation,

$$\hat{x}(k+1) = \mathbf{\Phi}\hat{x}(k) + \mathbf{\Gamma}_u u(k) + \mathbf{G}(k)[m(k) - \mathbf{C}_m\hat{x}(k)],$$

yields the optimal estimate of the state at time $(k + 1)$, given measurements through time k, where the Kalman gain and the plant model are given in Table 7.1.

7.11 Derive the following discrete-time Riccati equation:

$$\mathbf{\Sigma}_e(k+1) = \mathbf{\Phi}\mathbf{\Sigma}_e(k)\mathbf{\Phi}^T + \mathbf{\Gamma}_w\mathbf{\Sigma}_w\mathbf{\Gamma}_w^T - \mathbf{\Phi}\mathbf{\Sigma}_e(k)\mathbf{C}_m^T[\mathbf{C}_m\mathbf{\Sigma}_e(k)\mathbf{C}_m^T + \mathbf{\Sigma}_v]^{-1}\mathbf{C}_m\mathbf{\Sigma}_e(k)\mathbf{\Phi}^T,$$

where the Kalman filter equation, Kalman gain, and the plant model are given in Table 7.1. Note the estimate at time $(k + 1)$ is based on measurements through time k. *Hint:* Let

$$\mathbf{\Sigma}_e(k+1) = E[\{x(k+1) - \hat{x}(k+1)\}\{x(k+1) - \hat{x}(k+1)\}^T]$$

and substitute for $x(k + 1)$ and $\hat{x}(k + 1)$ using the discrete-time state equation and the discrete-time Kalman filter equation, respectively. Then substitute for the Kalman gain.

7.12 You are given the plant (7.13). The plant and measurement noises are assumed to be white, with the correlation functions given by (7.14). Further, these noises are correlated with the following cross-correlation function:

$$E[w(t)v(t+\tau)] = \mathbf{S}_{wv}\delta(\tau).$$

Assume that the Kalman filter equation is given by (7.22), and derive an expression for the Kalman gain in terms of the estimation error covariance matrix, the plant parameters, the spectral densities of the noises, and $\mathbf{S}_{wv}$.

7.13 Derive the initial condition on the state, an instant after the first nonwhite measurement becomes available:

$$\hat{x}(0^+) = \hat{x}(0) + \mathbf{\Sigma}_e(0)\mathbf{C}_m^T[\mathbf{C}_m\mathbf{\Sigma}_e(0)\mathbf{C}_m^T + \sigma_v^2]^{-1}[m(0) - \mathbf{C}_m\hat{x}(0)].$$

Assume that the variance of the measurement noise is σ_v^2. Also assume that the state estimate $\hat{x}(0)$ and the covariance matrix of the estimation error $\mathbf{\Sigma}_e(0)$ are known before the first measurement becomes available.

COMPUTER EXERCISES

7.1 Optimal Tracking of an Airborne Target

An air traffic control radar system furnishes measurements of azimuth to an airborne target. These measurements are noisy due to thermal noise in the receiver. The accuracy of the azimuth measurements can be improved by filtering the measurements, thereby providing an improved "estimate" of the target azimuth.

A mathematical model for the airborne target is

$$\dot{\theta}(t) = w(t),$$

where the azimuth rate is assumed to equal white noise with the following spectral density:

$$S_w = 1\frac{\text{deg}^2}{\text{sec}^2 \cdot \text{Hz}}.$$

This is a very simplistic model, but it is reasonable in some applications where the airborne target is maneuvering rapidly. The radar provides a noisy measurement of the azimuth:

$$m(t) = \theta(t) + v(t),$$

where the measurement noise $v(t)$ has the following spectral density:

$$S_v = 100\,\frac{\text{deg}^2}{\text{Hz}}.$$

A filter for estimating the azimuth angle is

$$\dot{\hat{\theta}}(t) = G[m(t) - \hat{\theta}(t)].$$

Note that this filter is the Kalman filter when using the optimal gain. Generate a state model for the estimation error of this system. The steady-state variance of $e(t)$ for various positive values of G is

$$\sigma_e^2 = S_w \int_0^\infty g_w^2(t)dt + S_v \int_0^\infty g_v^2(t)dt,$$

where $g_w(t)$ is the impulse response of $e(t)$ due to the input $w(t)$ and $g_v(t)$ is the impulse response of $e(t)$ due to the input $v(t)$. Note that the two terms in this expression are the estimation error variance due to $w(t)$ and $v(t)$, respectively. Generate these variances by simulating the system and numerically integrating the square of the impulse responses. Plot the estimation error variances due to $w(t)$, the estimation error variance due to $v(t)$, and the total estimation error variance as a function of G, for various positive values of G. Find the value of G that minimizes the estimation error variance. Compare this value of G to the steady-state Kalman gain. Also, compare the minimum variance of $e(t)$ on your plot with the estimation error variance obtained when solving for the steady-state Kalman gain. Comment on the effects of changing G on the estimation error variances due to the individual noises.

7.2 Estimating the States of a DC Motor

The state of a dc motor, consisting of the shaft angle and the angular velocity of the shaft, is estimated from noisy measurements of the shaft angle. The input to this motor is the applied voltage $u(t)$ plus noise on this voltage signal $w(t)$. The output $m(t)$ is the measured motor shaft angle. A state model for this motor is

$$\begin{bmatrix} \dot\theta(t) \\ \ddot\theta(t) \end{bmatrix} = \begin{bmatrix} 0 & 1 \\ 0 & -0.02 \end{bmatrix} \begin{bmatrix} \theta(t) \\ \dot\theta(t) \end{bmatrix} + \begin{bmatrix} 0 \\ 1 \end{bmatrix} u(t) + \begin{bmatrix} 0 \\ 1 \end{bmatrix} w(t);$$

$$m(t) = [1 \quad 0] \begin{bmatrix} \theta(t) \\ \dot\theta(t) \end{bmatrix} + v(t).$$

Estimates of both states are required to monitor the status of the motor. For this plant, design a Kalman filter for each of the following cases

a. $S_w = 2\dfrac{\text{deg}^2}{\text{sec}^4 \cdot \text{Hz}}; S_v = 40\dfrac{\text{deg}^2}{\text{Hz}}; \begin{bmatrix} \hat\theta(0) \\ \hat{\dot\theta}(0) \end{bmatrix} = \begin{bmatrix} 0 \text{ deg} \\ 0\dfrac{\text{deg}}{\text{sec}} \end{bmatrix}; \boldsymbol{\Sigma}_e(0) = \begin{bmatrix} 0 \text{ deg}^2 & 0\dfrac{\text{deg}^2}{\text{sec}} \\ 0\dfrac{\text{deg}^2}{\text{sec}} & 0\dfrac{\text{deg}^2}{\text{sec}^2} \end{bmatrix}.$

b. The same as case a, except: $S_w = 20\dfrac{\text{deg}^2}{\text{sec}^4 \cdot \text{Hz}}$.

c. The same as case a, except: $S_v = 400\dfrac{\text{deg}^2}{\text{Hz}}$.

d. The same as case a, except: $\begin{bmatrix} \hat\theta(0) \\ \hat{\dot\theta}(0) \end{bmatrix} = \begin{bmatrix} 5 \text{ deg} \\ 1\dfrac{\text{deg}}{\text{sec}} \end{bmatrix}; \boldsymbol{\Sigma}_e(0) = \begin{bmatrix} 100 \text{ deg}^2 & 0\dfrac{\text{deg}^2}{\text{sec}} \\ 0\dfrac{\text{deg}^2}{\text{sec}} & 10\dfrac{\text{deg}^2}{\text{sec}^2} \end{bmatrix}.$

e. The same as case a, except use the steady-state Kalman gain.

For each case, generate the Kalman gain and the estimation error covariance matrix, simulate the system, and implement the Kalman filter. For case e, the estimation error covariance matrix should be computed assuming the steady-state Kalman gain is used at all times. The initial condition for all of the simulations is:

$$\begin{bmatrix} \theta(0) \\ \dot\theta(0) \end{bmatrix} = \begin{bmatrix} 0 \text{ deg} \\ 0\dfrac{\text{deg}}{\text{sec}} \end{bmatrix}.$$

Assume that the applied voltage $u(t)$ is zero.

Plot the Kalman gains and the standard deviations of each of the estimation errors as a function of time. Plot the state and the estimated state (together on one graph) as a function of time, and plot the estimation error as a function of time. Compare each result to those obtained in case *a* and comment on how the changes made in the other cases affect the filter performance, the rate of convergence of the gains, and the standard deviations of the estimates.

7.3 Radar Range Tracking Utilizing Doppler Shift

Design a Kalman filter to estimate the range, radial velocity, and radial acceleration of an aircraft from noisy radar measurements. The radial acceleration is a random process with a correlation time of 30 seconds and a variance of 100 m^2/s^4. The augmented state equation for this plant is given (as in Example 7.10):

$$\begin{bmatrix} \dot{R}(t) \\ \ddot{R}(t) \\ \dot{w}(t) \end{bmatrix} = \begin{bmatrix} 0 & 1 & 0 \\ 0 & 0 & 1 \\ 0 & 0 & 0.033 \end{bmatrix} \begin{bmatrix} R(t) \\ \dot{R}(t) \\ w(t) \end{bmatrix} + \begin{bmatrix} 0 \\ 0 \\ 1 \end{bmatrix} w_1(t),$$

where $S_{w_1} = 6.7\ m^2/(sec^6\text{-Hz})$. The measurements consist of range and Doppler frequency. The Doppler frequency is the frequency shift on the reflected radar signal that results from radial motion, and is related to the range rate

$$f_d = \frac{2f}{c}\dot{R},$$

where $f = 10$ GHz is the radar frequency, and c is the speed of light. Both the range and Doppler frequency measurements are corrupted by white measurement noise:

$$\begin{bmatrix} m_R(t) \\ m_f(t) \end{bmatrix} = \begin{bmatrix} 1 & 0 & 0 \\ 0 & 66.7 & 0 \end{bmatrix} \begin{bmatrix} R(t) \\ \dot{R}(t) \\ w(t) \end{bmatrix} + \begin{bmatrix} v_R(t) \\ v_f(t) \end{bmatrix}.$$

Design a steady-state Kalman filter for this plant with each of the following measurement noise spectral density matrices.

a. $$\mathbf{S}_v = \begin{bmatrix} 20 & 0 \\ 0 & 20 \end{bmatrix};$$

b. $$\mathbf{S}_v = \begin{bmatrix} 2000 & 0 \\ 0 & 20 \end{bmatrix};$$

c. $$\mathbf{S}_v = \begin{bmatrix} 20 & 0 \\ 0 & 2000 \end{bmatrix}.$$

For each case, give the Kalman gain, give the estimation error covariance matrix, and simulate the system with the following initial conditions:

$$\begin{bmatrix} R(0) \\ \dot{R}(0) \\ w(0) \end{bmatrix} = \begin{bmatrix} 10{,}000 \\ 0 \\ 0 \end{bmatrix}.$$

Use perfect initialization for the Kalman filter, and use the steady-state gains. Plot the actual and estimated states, and relate the results to the estimation error covariance matrix. Compare each result to those obtained in case *a*. Comment on how changes in the measurement noise spectral density affects the filter performance and the standard deviations of the estimates.

8 Linear Quadratic Gaussian Control

The linear quadratic regulator (LQR) is an optimal control methodology that can be employed in a wide range of applications. The quadratic cost function provides the designer with lots of flexibility to perform trade-offs among various performance criteria. The relationship between cost function weights and performance criteria hold even for high-order and multiple-input systems, where classical control becomes cumbersome. Therefore, the LQR methodology can improve low-order designs via optimization and also enables the systematic design of controllers for high-order and multiple-input systems.

A major limitation of the LQR is that the entire state must be measured exactly when generating the control. This limitation becomes increasingly troublesome for high-order systems, where measuring all the states can be very expensive. In addition, no measurement is ever exact. Therefore, an optimal design methodology that results in controllers that utilize noisy, partial state information is desirable. The linear quadratic Gaussian (LQG) methodology provides a means of designing such controllers.[1]

The LQG controller overcomes the need to measure the entire state by estimating the state, using a Kalman filter. The estimated state is then used in the LQR. The LQG controller is, therefore, a combination of an LQR and a Kalman filter, a fact known as the separation principle. This separation of the LQG controller into components allows the reader to apply knowledge and insight gained with both the LQR and the Kalman filter to LQG design.

A number of modifications can be made to the linear quadratic Gaussian methodology to tailor the resulting controller to specific applications. Loop transfer recovery is a procedure for increasing the robustness of the basic LQG controller. This procedure results in a family of controllers that trade off robustness and performance.

The basic LQG controller is a regulator, but can be modified to allow tracking of reference inputs. Feedforward control, error feedback, and integral control are all means of tracking reference inputs that can be implemented within the LQG framework. Note that these methods of obtaining tracking systems can also be applied within the LQR setting, but are presented with reference to LQG control so that issues associated with estimator design can also be discussed.

The basic LQG is designed to reject white noise disturbance inputs. Modifications to this design can be made that enhance the rejection of nonwhite noise disturbance inputs.

[1] LQG control is also known as $\mathcal{H}_2$ control. The $\mathcal{H}$ stands for the Hardy space of all stabilizable controllers, and the 2 denotes the system 2-norm (which is equivalent to a quadratic cost function).

The linear quadratic Gaussian optimal control methodology is presented in this chapter along with the basic equations governing performance evaluation. Methods of modifying this controller to allow tracking, enhance disturbance rejection, and increase robustness are also presented. These modifications greatly increase the utility of the LQG controller.

8.1 Combined Estimation and Control: LQG Control

Linear quadratic Gaussian[2] (LQG) control refers to an optimal control problem where the plant model is linear, the cost function is quadratic, and the test conditions consist of random initial conditions, a white noise disturbance input, and a white measurement noise. The plant is described by the following linear state equation:

$$\boxed{\dot{x}(t) = \mathbf{A}x(t) + \mathbf{B}_u u(t) + \mathbf{B}_w w(t),} \tag{8.1}$$

where $u(t)$ is the control input and $w(t)$ is a random disturbance input known as plant noise. The measurements available for feedback are

$$\boxed{m(t) = \mathbf{C}_m x(t) + v(t),} \tag{8.2}$$

where $v(t)$ is a random signal known as measurement noise. The plant noise and measurement noise are both white and uncorrelated:

$$E\left[\begin{bmatrix} w(t) \\ \hline v(t) \end{bmatrix} \left[w^T(t+\tau) \mid v^T(t+\tau)\right]\right] = \begin{bmatrix} \mathbf{S}_w & \mathbf{0} \\ \mathbf{0} & \mathbf{S}_v \end{bmatrix} \delta(\tau).$$

Note that the description of the plant and measurement noises is equivalent to that used in the Kalman filter problem description.

The plant output to be controlled is

$$y(t) = \mathbf{C}_y x(t). \tag{8.3}$$

A quadratic cost function including both this output and the control input is

$$\boxed{J(x(t), u(t)) = E\left[\frac{1}{2} y^T(t_f)\mathbf{H}_y y(t_f) + \frac{1}{2}\int_0^{t_f} \{y^T(t)\mathbf{Q}_y y(t) + u^T(t)\mathbf{R}u(t)\}dt\right],} \tag{8.4a}$$

where $\mathbf{H}_y$, $\mathbf{Q}_y$, and $\mathbf{R}$ are positive definite. This cost function is frequently written directly in terms of the state:

[2] The term *Gaussian* is used because the resulting controller is optimal over all feedback controllers (both linear and nonlinear) when the disturbance inputs and measurement noises are Gaussian. The LQG controller is developed here without specifying the distribution of these signals, but with the implicit constraint that the control law is linear.

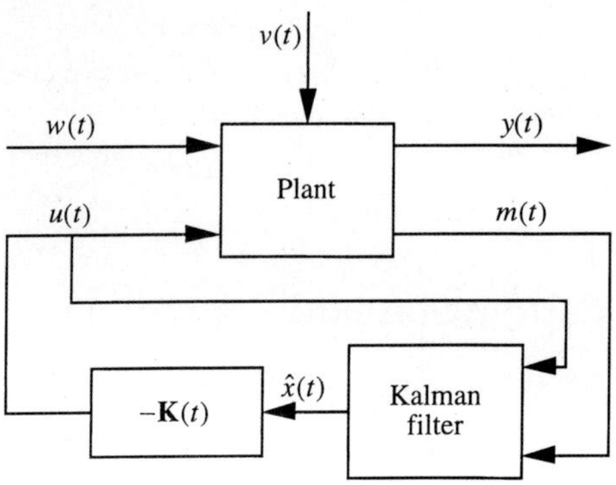

FIGURE 8.1 A linear quadratic Gaussian optimal control system

$$J(x(t), u(t)) = E\left[\frac{1}{2}x^T(t_f)\mathbf{H}x(t_f) + \frac{1}{2}\int_0^{t_f}\{x^T(t)\mathbf{Q}x(t) + u^T(t)\mathbf{R}u(t)\}dt\right], \tag{8.4b}$$

where

$$\mathbf{H} = \mathbf{C}_y^T\mathbf{H}_y\mathbf{C}_y;\ \mathbf{Q} = \mathbf{C}_y^T\mathbf{Q}_y\mathbf{C}_y.$$

The matrices **H** and **Q** are positive semidefinite, in general. It is assumed that there are no constraints on the state or the control. Note that these cost functions are equal to the cost functions specified for the stochastic regulator.

A reasonable controller design for the LQG control problem can be obtained by using the linear quadratic regulator feedback gain matrix operating on the state estimate generated by the Kalman filter:

$$u(t) = -\mathbf{K}(t)\hat{x}(t). \tag{8.5}$$

A block diagram of this feedback control system is given in Figure 8.1. This control law is, in fact, the optimal solution to the LQG optimal control problem.

The optimality of the control law (8.5) can be demonstrated by rewriting the expected value:[3]

$$\begin{aligned} E[x^T(t)\mathbf{Q}x(t)] &= E[\{x(t) - \hat{x}(t) + \hat{x}(t)\}^T\mathbf{Q}\{x(t) - \hat{x}(t) + \hat{x}(t)\}] \\ &= E[\{x(t) - \hat{x}(t)\}^T\mathbf{Q}\{x(t) - \hat{x}(t)\}] \\ &\quad + 2E[\{x(t) - \hat{x}(t)\}^T\mathbf{Q}\hat{x}(t)] + E[\hat{x}^T(t)\mathbf{Q}\hat{x}(t)]. \end{aligned}$$

Each term in this expression is a scalar that is unchanged by the trace operator:

$$E[x^T\mathbf{Q}x] = \mathrm{tr}\{E[\{x - \hat{x}\}^T\mathbf{Q}\{x - \hat{x}\}]\} + 2\mathrm{tr}\{E[\{x - \hat{x}\}^T\mathbf{Q}\hat{x}]\} + E[\hat{x}^T\mathbf{Q}\hat{x}],$$

[3] This demonstration of optimality follows the proof of optimality presented by Kwakernaak and Sivan [1], p. 400.

where the time argument has been dropped to simplify the notation. Using the fact that the trace is invariant under cyclic perturbations [see Appendix equation (A2.3)]:

$$E[x^T\mathbf{Q}x] = \mathrm{tr}\{E[(x - \hat{x})(x - \hat{x})^T]\mathbf{Q}\} + 2\mathrm{tr}\{E[\hat{x}(x - \hat{x})^T]\mathbf{Q}\} + E[\hat{x}^T\mathbf{Q}\hat{x}].$$

The second term on the right in this equation is zero, since the state estimate is a linear combination of the measurements, and the estimation error is orthogonal to the measurements. In addition, the expected value in the first term is the covariance matrix of the state estimation error. This expression then reduces to

$$E[x^T\mathbf{Q}x] = \mathrm{tr}\{\Sigma_e\mathbf{Q}\} + E[\hat{x}^T\mathbf{Q}\hat{x}]. \tag{8.6}$$

Using (8.6), the cost function (8.4b) can be written as follows:

$$\begin{aligned} J(x(t), u(t)) = E&\left[\frac{1}{2}\hat{x}^T(t_f)\mathbf{H}\hat{x}(t_f) + \frac{1}{2}\int_0^{t_f}\{\hat{x}^T(t)\mathbf{Q}\hat{x}(t) + u^T(t)\mathbf{R}u(t)\}dt\right] \\ &+ \frac{1}{2}\mathrm{tr}\left\{\Sigma_e(t_f)\mathbf{H} + \int_0^{t_f}\Sigma_e(t)\mathbf{Q}dt\right\}, \end{aligned} \tag{8.7}$$

where (8.6) has also been applied to the term involving the state at the final time. The terms within the trace operator in (8.7) are independent of the control input, and $J(x(t), u(t))$ is minimized whenever

$$J(\hat{x}(t), u(t)) = E\left[\frac{1}{2}\hat{x}^T(t_f)\mathbf{H}\hat{x}(t_f) + \frac{1}{2}\int_0^{t_f}\{\hat{x}^T(t)\mathbf{Q}\hat{x}(t) + u^T(t)\mathbf{R}u(t)\}dt\right] \tag{8.8}$$

is minimized. The state equation for the Kalman filter estimate is given in (7.22):

$$\dot{\hat{x}}(t) = \mathbf{A}\hat{x}(t) + \mathbf{B}_u u(t) + \mathbf{G}(t)[m(t) - \mathbf{C}_m\hat{x}(t)].$$

The innovations process $[m(t) - \mathbf{C}_m\hat{x}(t)]$ is white noise, as shown in Section A12 of the Appendix. Therefore, this state equation and the cost function (8.8) form a stochastic regulator with the optimal solution given by (8.5). This optimal solution also minimizes the cost (8.4), and is the solution of the LQG optimal control problem.

The state model for the optimal controller is

$$\dot{\hat{x}}(t) = [\mathbf{A} - \mathbf{G}(t)\mathbf{C}_m - \mathbf{B}_u\mathbf{K}(t)]\hat{x}(t) + \mathbf{G}(t)m(t); \tag{8.9a}$$

$$u(t) = -\mathbf{K}(t)\hat{x}(t), \tag{8.9b}$$

where (8.9b) has been used to eliminate the control input in the Kalman filter equation (8.9a). The gain matrices in this controller are found by solving a pair of Riccati equations. The state feedback gain $\mathbf{K}(t)$ is found by solving the following:

$$\dot{\mathbf{P}}(t) = -\mathbf{P}(t)\mathbf{A} - \mathbf{A}^T\mathbf{P}(t) - \mathbf{Q} + \mathbf{P}(t)\mathbf{B}_u\mathbf{R}^{-1}\mathbf{B}_u^T\mathbf{P}(t);$$

$$\mathbf{K}(t) = \mathbf{R}^{-1}\mathbf{B}_u^T\mathbf{P}(t),$$

subject to the final condition

$$\mathbf{P}(t_f) = \mathbf{H}.$$

The Kalman gain $\mathbf{G}(t)$ is found by solving the following:

$$\dot{\mathbf{\Sigma}}_e(t) = \mathbf{\Sigma}_e(t)\mathbf{A}^T + \mathbf{A}\mathbf{\Sigma}_e(t) + \mathbf{B}_w\mathbf{S}_w\mathbf{B}_w^T - \mathbf{\Sigma}_e(t)\mathbf{C}_m^T\mathbf{S}_v^{-1}\mathbf{C}_m\mathbf{\Sigma}_e(t);$$

$$\mathbf{G}(t) = \mathbf{\Sigma}_e(t)\mathbf{C}_m^T\mathbf{S}_v^{-1},$$

subject to the initial condition

$$\mathbf{\Sigma}_e(0).$$

Note that these Riccati equations are independent of the measured data, and can be solved when designing the controller. The resulting gains can then be stored for use during controller operation. Therefore, only the filter state model (8.9) needs to be implemented in real time.

In summary, the solution of the linear quadratic Gaussian optimal control problem can be broken into two parts: (1) Find the linear quadratic regulator feedback gains that minimize the cost, assuming perfect state information; (2) generate a Kalman filter to estimate the state. This is a remarkable result known as the *stochastic separation principle*. This result greatly simplifies controller design and testing, since the linear quadratic regulator and the Kalman filter can both be designed and tested separately to validate performance. In addition, intuitive insight gained with the LQR concerning the affects of cost function parameters on gain matrices, transient performance, and required control magnitude can be directly applied to linear quadratic Gaussian control.

EXAMPLE 8.1 A satellite tracking antenna, subject to random wind torques, can be modeled (as in Example 6.11):

$$\begin{bmatrix} \dot{\theta}(t) \\ \ddot{\theta}(t) \end{bmatrix} = \begin{bmatrix} 0 & 1 \\ 0 & -0.1 \end{bmatrix}\begin{bmatrix} \theta(t) \\ \dot{\theta}(t) \end{bmatrix} + \begin{bmatrix} 0 \\ 0.001 \end{bmatrix}u(t) + \begin{bmatrix} 0 \\ 0.001 \end{bmatrix}w(t)$$

where $\theta(t)$ is the pointing error of the antenna in degrees, $u(t)$ is the control torque in N-m, and $w(t)$ is the wind torque (disturbance input) in N-m. The wind torque is assumed to be white noise with a spectral density $S_w = 5000$ N^2-m^2/Hz. The pointing error is measured as

$$m(t) = [1 \quad 0]\begin{bmatrix} \theta(t) \\ \dot{\theta}(t) \end{bmatrix} + v(t),$$

where the measurement noise is white with a spectral density $S_v = 1$ deg^2/Hz. A cost function for this system is chosen:

$$J = E\left[\int_0^{1000} q\theta^2(t) + ru^2(t)dt\right],$$

where $q = 180$ and $r = 1$. Note that this cost function was chosen in Example 6.11 to yield a pointing error with a standard deviation of approximately 1 while using minimal control. The solution of this optimal control problem requires the specification of the initial state estimate and the initial estimation error covariance matrix. The initial state is assumed to be known and is identically equal to zero, which yields the following Kalman filter initialization:

$$\hat{x}(0) = \begin{bmatrix} 0 \\ 0 \end{bmatrix}; \ \mathbf{\Sigma}_e(0) = \begin{bmatrix} 0 & 0 \\ 0 & 0 \end{bmatrix}.$$

The optimal state feedback gain and optimal Kalman gain are given in Figure 8.2. The resulting closed-loop system is simulated with random inputs using the optimal gains. The trajectories of θ, $\hat{\theta}$, and the control input are also given in Figure 8.2. Note that the optimal state feedback gain approaches steady-state far from the final time, and that the Kalman gain approaches steady-state far from the initial time.

The feedback gains, θ, $\hat{\theta}$, and the control input are given in Figure 8.3 for q = 18,000, $r = 1$, $S_w = 5000$, and $S_v = 1$; in Figure 8.4 for $q = 180$, $r = 100$, $S_w = 5000$, and $S_v = 1$; in Figure 8.5 for $q = 180$, $r = 1$, S_w = 500,000, and $S_v = 1$; and in Figure 8.6 for $q = 180$, $r = 1$, $S_w = 5000$, and $S_v = 100$. Figures 8.3 through 8.6 each represent a change to the baseline system presented in Figure 8.2.

The state weighting in the cost function has been increased in Figure 8.3. In this case, the state feedback gains are larger and approach steady state more rapidly, while the Kalman gains remain the same. The larger state feedback gains generate larger control inputs that yield better control performance, as evidenced by a decrease in the angle error.

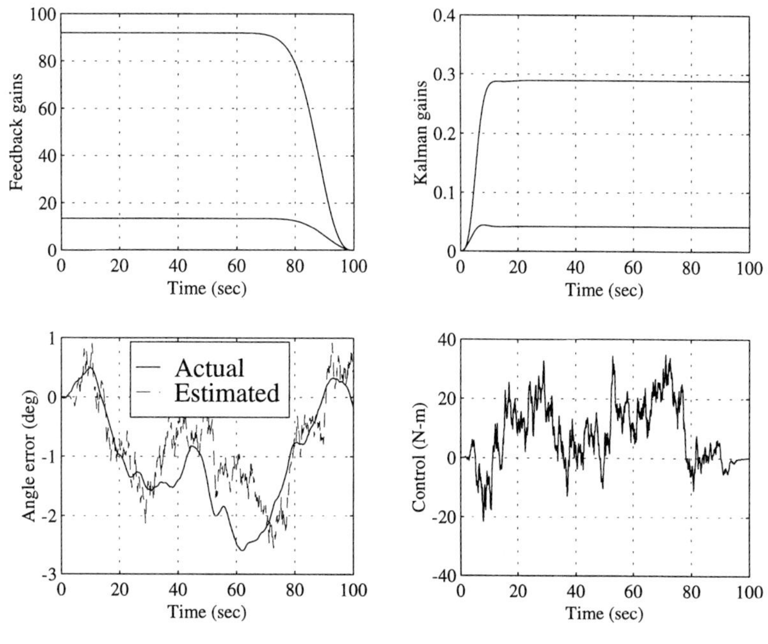

FIGURE 8.2 Results of Example 8.1 with $q = 180$, $r = 1$, $S_w = 5000$, and $S_v = 1$

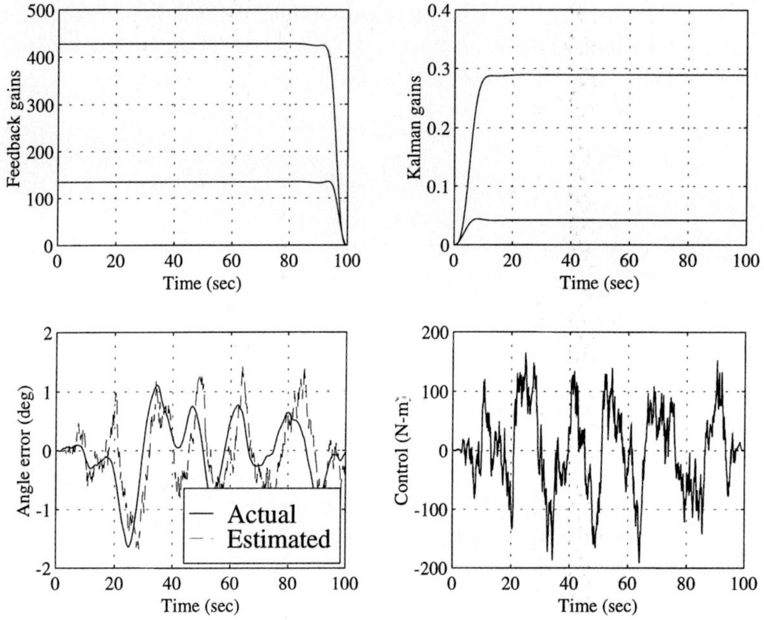

FIGURE 8.3 Results of Example 8.1 with $q = 18{,}000$, $r = 1$, $S_w = 5000$, and $S_v = 1$

The control weighting in the cost function has been increased in Figure 8.4. In this case, the state feedback gains are smaller and approach steady state more slowly, while the Kalman gains remain the same. The smaller state feedback gains generate smaller control inputs that yield increased angle errors.

The spectral density of the plant noise has been increased in Figure 8.5. In this case, the Kalman gains are larger and approach steady state more rapidly, while the state feedback gains remain the same. The larger Kalman gains make the estimator faster, but the larger plant noise means that the estimates are still less accurate. The control input also increases due to the larger gains and the increased plant noise. Note that the angle error is increased by roughly a factor of 10, which is the square root of the increase in the spectral density.

The spectral density of the measurement noise has been increased in Figure 8.6. In this case, the Kalman gains are smaller and approach steady state more slowly, while the state feedback gains remain the same. The smaller Kalman gains make the estimator slower, and the larger measurement noise makes the estimates less accurate. The angle error becomes larger due to the increased measurement noise, and this should translate into an increase in the control input. Note that the increase in angle error and control input are not readily apparent in the figure. This is because the time interval is sufficiently short that steady state is not achieved (at least for long). Sufficient data is, therefore, not available to get reasonable sample variance estimates. It is apparent from the

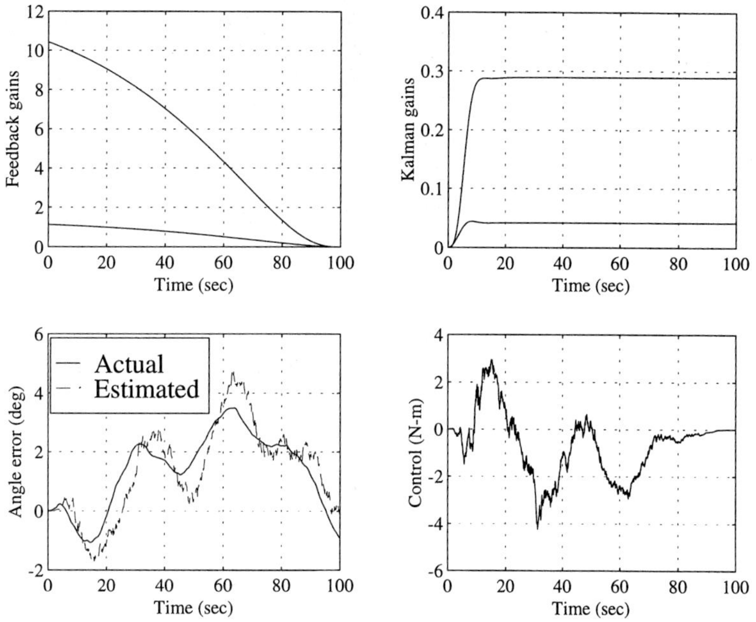

FIGURE 8.4 Results of Example 8.1 with $q = 180$, $r = 100$, $S_w = 5000$, and $S_v = 1$

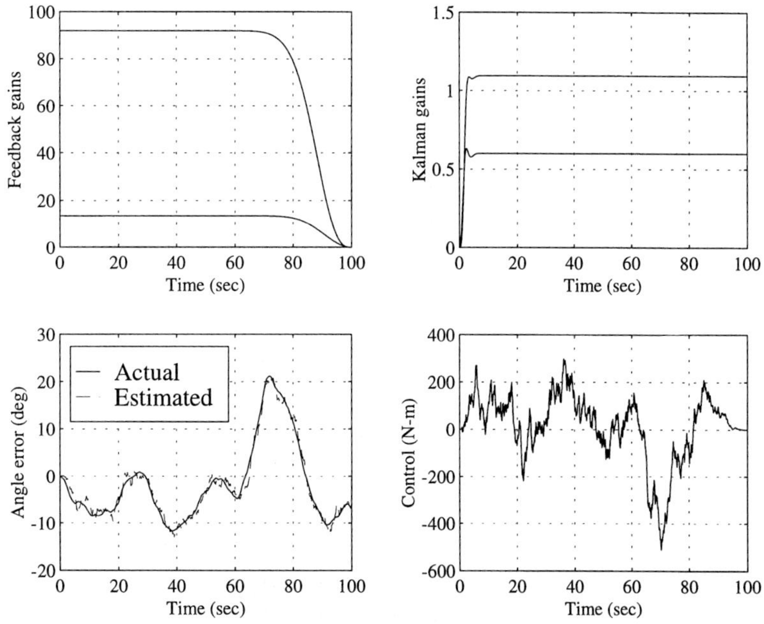

FIGURE 8.5 Results of Example 8.1 with $q = 180$, $r = 1$, $S_w = 500{,}000$, and $S_v = 1$

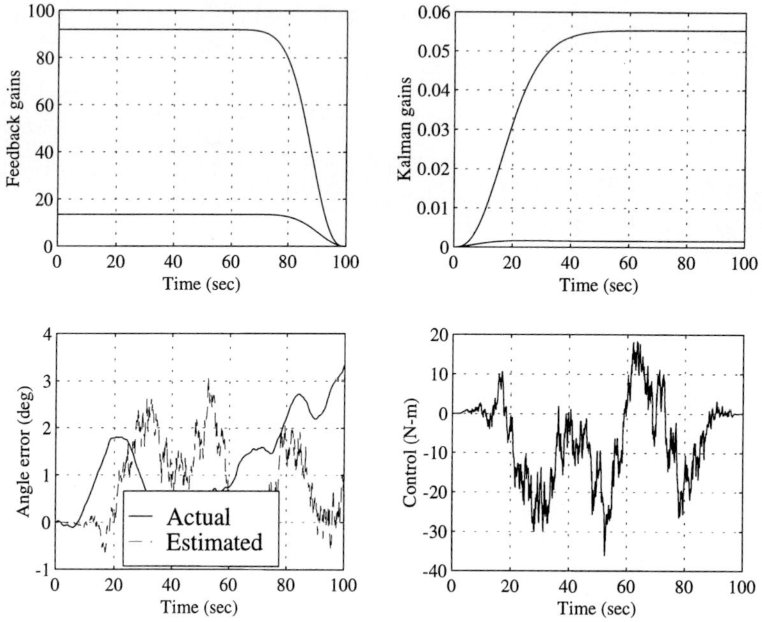

FIGURE 8.6 Results of Example 8.1 with $q = 180$, $r = 1$, $S_w = 5000$, and $S_v = 100$

figure that the estimation error is greatly increased due to the increase in measurement noise. ◆

8.1.1 Performance

The performance of a linear quadratic Gaussian optimal control system, like any control system, is composed of transient performance, tracking performance, and disturbance rejection. Transient performance can be evaluated via simulation of the time-varying LQG system. The LQG system is a regulator that drives the states to zero; that is, there is no reference input to track. Modifications to the basic LQG control system that allow the tracking of reference inputs are presented in Section 8.4. A discussion of tracking performance is, therefore, deferred until that section. The LQG controller is designed to optimize disturbance rejection for white noise disturbances, but can also be applied when other types of disturbance inputs are present. The rejection of nonwhite noise disturbances can be evaluated via simulation, and is discussed in more detail in Section 8.5. For white noise inputs, the disturbance rejection of the LQG control system is quantified, in part, by the cost function. The covariance matrices of the state, the estimation error, the reference output, and the control input all contain additional information to aid in evaluating disturbance rejection. The computation of the cost function and these covariance matrices is addressed in the remainder of this section after first presenting a state model for the optimal LQG control system.

The plant state and the estimation error form a state for the closed-loop LQG optimal control system. This choice of state is used, since the covariance matrices of these quantities are of direct interest in evaluating performance. The closed-loop state model is generated by appending the Kalman filter error equation (7.29) to the plant state equation (8.1):

$$\begin{bmatrix} \dot{x}(t) \\ \hline \dot{e}(t) \end{bmatrix} = \left[\begin{array}{c|c} \mathbf{A} & \mathbf{0} \\ \hline \mathbf{0} & \mathbf{A} - \mathbf{G}(t)\mathbf{C}_m \end{array}\right] \begin{bmatrix} x(t) \\ \hline e(t) \end{bmatrix} + \begin{bmatrix} \mathbf{B}_u \\ \hline \mathbf{0} \end{bmatrix} u(t) + \begin{bmatrix} \mathbf{B}_w \\ \hline \mathbf{B}_w \end{bmatrix} w(t) + \begin{bmatrix} \mathbf{0} \\ \hline -\mathbf{G}(t) \end{bmatrix} v(t).$$

Substituting for the control input using (8.5) and noting at $\hat{x}(t) = x(t) - e(t)$ yields the closed-loop state model yields

$$\begin{bmatrix} \dot{x}(t) \\ \hline \dot{e}(t) \end{bmatrix} = \left[\begin{array}{c|c} \mathbf{A} - \mathbf{B}_u\mathbf{K}(t) & \mathbf{B}_u\mathbf{K}(t) \\ \hline \mathbf{0} & \mathbf{A} - \mathbf{G}(t)\mathbf{C}_m \end{array}\right] \begin{bmatrix} x(t) \\ \hline e(t) \end{bmatrix} + \left[\begin{array}{c|c} \mathbf{B}_w & \mathbf{0} \\ \hline \mathbf{B}_w & -\mathbf{G}(t) \end{array}\right] \begin{bmatrix} w(t) \\ \hline v(t) \end{bmatrix}$$
$$= \mathbf{A}_{cl}(t) \begin{bmatrix} x(t) \\ \hline e(t) \end{bmatrix} + \mathbf{B}_{cl}(t) \begin{bmatrix} w(t) \\ \hline v(t) \end{bmatrix}. \tag{8.10a}$$

The reference output (8.3) and the control input can both be given in terms of this state:

$$y(t) = [\mathbf{C}_y \;\vdots\; \mathbf{0}] \begin{bmatrix} x(t) \\ \hline e(t) \end{bmatrix}; \tag{8.10b}$$

$$u(t) = -\mathbf{K}(t)\hat{x}(t) = [-\mathbf{K}(t) \;\vdots\; \mathbf{K}(t)] \begin{bmatrix} x(t) \\ \hline e(t) \end{bmatrix}. \tag{8.10c}$$

The covariance matrices of the plant state, the estimation error, the reference output, and the control input, along with the cost, can be simply computed from the covariance matrix of the closed-loop state. The covariance matrices of the plant state $\mathbf{\Sigma}_x(t)$ and the estimation error $\mathbf{\Sigma}_e(t)$ are both part of the closed-loop state covariance matrix:

$$\mathbf{\Sigma}_{\left[\frac{x}{e}\right]}(t) = E\left[\begin{bmatrix} x(t) \\ \hline e(t) \end{bmatrix} [x^T(t) \;\vdots\; e^T(t)]\right]$$
$$= \left[\begin{array}{c|c} E[x(t)x^T(t)] & E[x(t)e^T(t)] \\ \hline E[e(t)x^T(t)] & E[e(t)e^T(t)] \end{array}\right] = \left[\begin{array}{c|c} \mathbf{\Sigma}_x(t) & \mathbf{\Sigma}_{xe}(t) \\ \hline \mathbf{\Sigma}_{ex}(t) & \mathbf{\Sigma}_e(t) \end{array}\right].$$

The covariance matrices of the output of interest and the control input are

$$\boldsymbol{\Sigma}_y(t) = [\mathbf{C}_y \,\vdots\, \mathbf{0}]\boldsymbol{\Sigma}_{\left[\frac{x}{e}\right]}(t)\begin{bmatrix}\mathbf{C}_y^T \\ \hdashline \mathbf{0}\end{bmatrix} = \mathbf{C}_y\boldsymbol{\Sigma}_x(t)\mathbf{C}_y^T; \tag{8.11a}$$

$$\boldsymbol{\Sigma}_u(t) = [-\mathbf{K}(t) \,\vdots\, \mathbf{K}(t)]\boldsymbol{\Sigma}_{\left[\frac{x}{e}\right]}(t)\begin{bmatrix}-\mathbf{K}^T(t) \\ \hdashline \mathbf{K}^T(t)\end{bmatrix}. \tag{8.11b}$$

The cost function can be given in terms of the closed-loop state:

$$J = E\left[\frac{1}{2}[x^T(t_f) \,\vdots\, e^T(t_f)]\begin{bmatrix}\mathbf{H} & \mathbf{0} \\ \mathbf{0} & \mathbf{0}\end{bmatrix}\begin{bmatrix}x(t_f) \\ \hdashline e(t_f)\end{bmatrix}\right.$$
$$\left. + \frac{1}{2}\int_0^{t_f}[x^T(t) \,\vdots\, e^T(t)]\begin{bmatrix}\mathbf{Q} + \mathbf{K}^T(t)\mathbf{R}\mathbf{K}(t) & -\mathbf{K}^T(t)\mathbf{R}\mathbf{K}(t) \\ -\mathbf{K}^T(t)\mathbf{R}\mathbf{K}(t) & \mathbf{K}^T(t)\mathbf{R}\mathbf{K}(t)\end{bmatrix}\begin{bmatrix}x(t) \\ \hdashline e(t)\end{bmatrix}dt\right].$$

This cost function can then be written in terms of the closed-loop state covariance matrix:

$$J = \frac{1}{2}\mathrm{tr}\left\{\begin{bmatrix}\mathbf{H} & \mathbf{0} \\ \mathbf{0} & \mathbf{0}\end{bmatrix}\boldsymbol{\Sigma}_{\left[\frac{x}{e}\right]}(t_f) + \int_0^{t_f}\begin{bmatrix}\mathbf{Q} + \mathbf{K}^T(t)\mathbf{R}\mathbf{K}(t) & -\mathbf{K}^T(t)\mathbf{R}\mathbf{K}(t) \\ -\mathbf{K}^T(t)\mathbf{R}\mathbf{K}(t) & \mathbf{K}^T(t)\mathbf{R}\mathbf{K}(t)\end{bmatrix}\boldsymbol{\Sigma}_{\left[\frac{x}{e}\right]}(t)dt\right\}. \tag{8.12}$$

The closed-loop state covariance matrix can be computed using the methods of Chapter 3:

$$\boldsymbol{\Sigma}_{\left[\frac{x}{e}\right]}(t) = \boldsymbol{\Phi}(t, 0)\boldsymbol{\Sigma}_{\left[\frac{x}{e}\right]}(0)\boldsymbol{\Phi}^T(t, 0) + \int_0^t \boldsymbol{\Phi}(t, \tau)\mathbf{B}_{cl}(\tau)\begin{bmatrix}\mathbf{S}_w & \mathbf{0} \\ \mathbf{0} & \mathbf{S}_v\end{bmatrix}\mathbf{B}_{cl}^T(\tau)\boldsymbol{\Phi}^T(t, \tau)d\tau \tag{8.13}$$

where $\boldsymbol{\Phi}(t, \tau)$ is the state-transition matrix of closed-loop system (8.10). Alternatively, the state covariance matrix can be found by solving the following differential equation:

$$\dot{\boldsymbol{\Sigma}}_{\left[\frac{x}{e}\right]}(t) = \mathbf{A}_{cl}(t)\boldsymbol{\Sigma}_{\left[\frac{x}{e}\right]}(t) + \boldsymbol{\Sigma}_{\left[\frac{x}{e}\right]}(t)\mathbf{A}_{cl}^T(t) + \mathbf{B}_{cl}(t)\begin{bmatrix}\mathbf{S}_w & \mathbf{0} \\ \mathbf{0} & \mathbf{S}_v\end{bmatrix}\mathbf{B}_{cl}^T(t). \tag{8.14}$$

After computing the closed-loop state covariance matrix, the cost, and the covariance matrices of the plant state, the estimation error, the output of interest, and the control input can all be found using the formulas given on previous page. Note that these expressions are also valid for all control systems with the structure of the LQG controller, whether or not the gains are optimal. This allows the direct comparison of competing control system designs. In particular, the performance degradation resulting from use of constant state feedback and Kalman gains can be investigated using these results. A similar analysis methodology, if not the given formulas, can also be applied for suboptimal control systems with structures different than the LQG controller. This allows the comparison of classical control systems and reduced-order controllers with the optimal LQG controller.

EXAMPLE 8.1 CONTINUED The disturbance rejection properties of the baseline control system in Example 8.1 are investigated. The first step in this analysis is the computation of the closed-loop state covariance matrix. The initial value of this matrix is

$$\mathbf{\Sigma}_{\left[\frac{x}{e}\right]}(0) = \begin{bmatrix} 0 & 0 \\ 0 & 0 \end{bmatrix},$$

since the initial state and estimate are both zero. The closed-loop state covariance matrix is then found by numerically solving (8.14). A time-invariant control system can be obtained by using the steady-state LQR feedback gains and the steady-state Kalman gains. The closed-loop state covariance matrix of this system is also generated assuming the same initial conditions.

The costs for these two systems are $J_{TV} = 31{,}700$ and $J_{SS} = 33{,}100$, where the subscript TV refers to the optimal, time-varying controller, and the subscript SS refers to the suboptimal, time-invariant, steady-state controller. As expected, the optimal time-varying control system yields a smaller cost. The standard deviations of the angle error θ, the estimation error for this angle, and the control input are plotted in Figure 8.7. These quantities are obtained from the covariance matrices of the output, the estimation error, and the control, respectively. Note that the optimal time-varying state feedback gains and Kalman gains for this system can be found in Figure 8.2.

The standard deviation of the angle error is comparable for both controllers during the first three-quarters of the time interval, and smaller for the steady-state controller in the latter half of the time interval. Note also that the control used during the second half of the time interval is significantly larger for the steady-state controller. These facts are reasonable, since the time-varying state feedback gains decrease to zero near the final time. This allows the use of significantly less control with only a moderate increase in the final state variances. The estimation error for the steady-state system is larger over the initial transient due to the use of the non-optimal Kalman gain. Notice also that there is an early increase in control input for the steady-state controller. This is due to the fact that the steady-state Kalman filter relies less on the *a priori* initial state information (the initial state is zero) and therefore yields larger estimates. While the steady-state controller may look better due to the decrease in angle error near the final time, the cost is increased for this controller, indicating worse overall performance (provided the cost function truly reflects the designers goals). Note that better control of the final state can be obtained for the optimal system by adding final state weighting to the cost function.

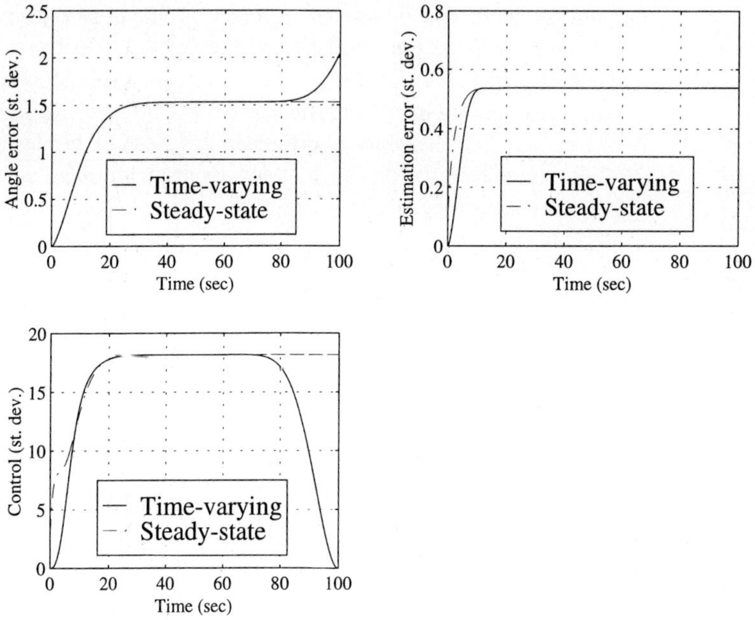

FIGURE 8.7 Performance analysis for Example 8.1

8.2 Steady-State LQG Control

The linear quadratic regulator feedback gain approaches a constant value far from the final time, and the Kalman filter gain approaches a constant value far from the initial time. For long time intervals, both of these gains are approximately constant over most of the interval. The use of these constant state feedback and Kalman gains simplifies the implementation of the controller, and allows the closed-loop system to be analyzed as a time-invariant system. Besides being convenient, the steady-state gains produce the optimal control when the cost function is

$$J = E[x^T(\infty)\mathbf{Q}x(\infty) + u^T(\infty)\mathbf{R}u(\infty)], \tag{8.15}$$

where the infinity symbol is used as an argument to indicate that the given quantity has reached steady state. In summary, the cost function (8.15) is minimized by a controller that consists of using the steady-state linear quadratic regulator gains in (8.5) and state estimates obtained using the steady-state Kalman filter.

8.2.1 Performance

The transient performance and disturbance rejection of the steady-state LQG control system should be evaluated prior to implementation. The transient performance is quantified, in part, by the closed-loop poles, since the system is time-invariant. Disturbance rejection is quantified by the cost function and the covariance matrices of the plant state, the estimation error, the output of interest, and the control input.

The Closed-Loop Poles The characteristic equation of the closed-loop LQG control system (8.10) is computed using Appendix equation (A2.5):

$$\det\left[\begin{array}{c|c} s\mathbf{I} - \mathbf{A} + \mathbf{B}_u\mathbf{K} & -\mathbf{B}_u\mathbf{K} \\ \hline \mathbf{0} & s\mathbf{I} - \mathbf{A} + \mathbf{G}\mathbf{C}_m \end{array}\right] = \det(s\mathbf{I} - \mathbf{A} + \mathbf{B}_u\mathbf{K})\det(s\mathbf{I} - \mathbf{A} + \mathbf{G}\mathbf{C}_m) = 0.$$

The closed-loop poles are then found by setting the factors in this characteristic equation to zero:

$$\det(s\mathbf{I} - \mathbf{A} + \mathbf{B}_u\mathbf{K}) = 0; \tag{8.16a}$$

$$\det(s\mathbf{I} - \mathbf{A} + \mathbf{G}\mathbf{C}_m) = 0. \tag{8.16b}$$

Note that (8.16a) equals the characteristic equation of the closed-loop LQR system that results when all the states are available for feedback. Also, (8.16b) is the characteristic equation of the Kalman filter. Therefore, the poles of the closed-loop LQG optimal control system are equal to the poles of the state feedback system, which utilizes the LQR feedback gain plus the poles of the Kalman filter.[4] This fact implies that the closed-loop system is stable, since both the LQR and the Kalman filter are stable.

Disturbance Rejection The cost function and the covariance matrices of the plant state, the estimation error, the output of interest, and the control input can all be computed from the state covariance matrix of the closed-loop system. The steady-state, closed-loop system is given by the state model (8.10), where the feedback and Kalman gains are time-invariant. The closed-loop state covariance matrix can then be computed:

$$\boldsymbol{\Sigma}_{\left[\frac{x}{e}\right]}(t) = \int_0^{\infty} \boldsymbol{\Phi}(t,\tau)\mathbf{B}_{cl}\left[\begin{array}{c|c} \mathbf{S}_w & \mathbf{0} \\ \hline \mathbf{0} & \mathbf{S}_v \end{array}\right]\mathbf{B}_{cl}^T\boldsymbol{\Phi}^T(t,\tau)\,d\tau,$$

or it can be found by solving the following Lyapunov equation:

$$\mathbf{A}_{cl}\boldsymbol{\Sigma}_{\left[\frac{x}{e}\right]}(t) + \boldsymbol{\Sigma}_{\left[\frac{x}{e}\right]}(t)\mathbf{A}_{cl}^T + \mathbf{B}_{cl}\left[\begin{array}{c|c} \mathbf{S}_w & \mathbf{0} \\ \hline \mathbf{0} & \mathbf{S}_v \end{array}\right]\mathbf{B}_{cl}^T = \mathbf{0}.$$

The covariance matrices of the plant state and the estimation error are overtly contained in this covariance matrix. The covariance matrices of the reference output and the control input can be computed using (8.11). The cost function (8.15) is given in terms of the closed-loop state covariance matrix:

$$J = \mathrm{tr}\left\{\begin{bmatrix} \mathbf{Q} + \mathbf{K}^T\mathbf{R}\mathbf{K} & -\mathbf{K}^T\mathbf{R}\mathbf{K} \\ -\mathbf{K}^T\mathbf{R}\mathbf{K} & \mathbf{K}^T\mathbf{R}\mathbf{K} \end{bmatrix}\boldsymbol{\Sigma}_{\left[\frac{x}{e}\right]}(\infty)\right\}.$$

[4] The poles of an observer feedback system (of which the LQG optimal controller is a special case) are, in general, the poles of the state feedback system plus the poles of the observer. This result is frequently referred to as the deterministic separation principle.

EXAMPLE 8.2 The hover mode pitch and yaw dynamics of a vertical-takeoff-and-landing remotely piloted vehicle are given by the following state model:

$$\dot{x}(t) = \begin{bmatrix} 0 & -1 & 0 & 0 \\ 0 & 0 & 0 & -1.6 \\ 0 & 0 & 0 & -1 \\ 0 & 1.8 & 0 & 0 \end{bmatrix} x(t) + \begin{bmatrix} 0 & 0 \\ -11 & 0 \\ 0 & 0 \\ 0 & -12 \end{bmatrix} u(t) + \begin{bmatrix} 0 & 0 \\ -11 & 0 \\ 0 & 0 \\ 0 & -12 \end{bmatrix} w(t);$$

$$m(t) = \begin{bmatrix} 1 & 0 & 0 & 0 \\ 0 & 0 & 1 & 0 \end{bmatrix} x(t) + v(t),$$

where $w(t)$ is actuator noise ($\mathbf{S}_w = 10\mathbf{I}$), and $v(t)$ is measurement noise ($\mathbf{S}_v = 10^{-4}\mathbf{I}$). The cost function to minimize is

$$J = E\left[x^T(\infty) \begin{bmatrix} 1 & 0 & 0 & 0 \\ 0 & 0 & 0 & 0 \\ 0 & 0 & 1 & 0 \\ 0 & 0 & 0 & 0 \end{bmatrix} x(\infty) + u^T(\infty) \begin{bmatrix} 0.02 & 0 \\ 0 & 0.02 \end{bmatrix} u(\infty) \right].$$

The resulting control is

$$u(t) = -\mathbf{K}\hat{x}(t) = -\begin{bmatrix} 7.01 & -1.13 & 0.94 & 0 \\ -0.94 & 0 & 7.01 & -1.08 \end{bmatrix} \hat{x}(t),$$

where $\hat{x}(t)$ is the estimate of the plant state generated by the steady-state Kalman filter with the following gain:

$$\mathbf{G} = \begin{bmatrix} 83 & 0 \\ -3478 & 70 \\ 0 & 87 \\ -75 & -3794 \end{bmatrix}.$$

The poles of the closed-loop system are then

$$-6.3 \pm 5.6j \; ; \; -6.4 \pm 7.3j \; ; \; -41.9 \pm 41.4j \; ; \; -43.4 \pm 43.8j,$$

indicating a settling time of approximately 0.6 seconds and good damping. The cost is

$$J = 6.83,$$

and the covariance matrices of the reference output and the control input are

$$\mathbf{\Sigma}_y = \begin{bmatrix} 1.09 & 0 \\ 0 & 1.14 \end{bmatrix}; \; \mathbf{\Sigma}_u = \begin{bmatrix} 113 & 0 \\ 0 & 117 \end{bmatrix}.$$

These matrices show that the control system yields effective disturbance rejection (angle errors on the order of 1 degree) while using control torques of reasonable magnitude (on the order of 10 N-m).

8.2.2 Robustness

Stability and performance robustness are important considerations in the design of any control system. The steady-state LQG controller is designed to optimize disturbance

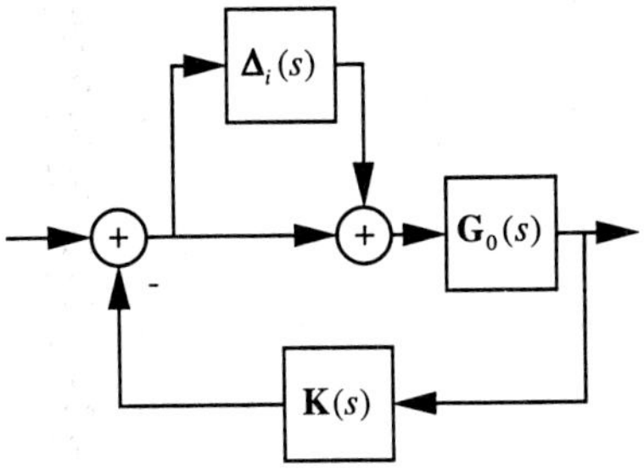

FIGURE 8.8 Input-multiplicative uncertainty

rejection, but should be analyzed to evaluate robustness before application. The robustness of this control system can be evaluated using the methods of Chapter 5, which include stability margins, the small-gain theorem, and the structured singular value robustness test.

The linear quadratic regulator and the Kalman filter were both shown to have a guaranteed minimum robustness to input- and output-multiplicative perturbations, respectively. In particular, the LQR is guaranteed to remain stable for all input-multiplicative perturbations (as shown in Figure 8.8) such that

$$\|\Delta_i\|_\infty < \frac{1}{2},$$

provided that the control weighting matrix in the cost function is of the form $\mathbf{R} = \rho\mathbf{I}$. This result translates into the following guaranteed stability margins:

$$GM^+ = \infty = \infty \text{ dB};$$

$$GM^- \leq \frac{1}{2} = -6 \text{ dB};$$

$$PM \geq 60°,$$

for SISO systems. Note that dual results hold for the Kalman filter subject to output-multiplicative perturbations. It is then reasonable to expect that the linear quadratic Gaussian optimal control, which is designed as an LQR operating on the output of a Kalman filter, would exhibit similar guaranteed robustness properties. Unfortunately, while the LQG is frequently very robust, it is not guaranteed to exhibit good robustness. The lack of guaranteed robustness of the LQG optimal controller is demonstrated by a SISO example.

FIGURE 8.9 Robot arm in Example 8.3

EXAMPLE 8.3 A robot arm (single link) is being held vertically to allow performance of a task located overhead (see Figure 8.9). Newton's law yields the mathematical model of the arm:

$$J\ddot{\theta}(t) = mg\,\sin[\theta(t)] + u(t),$$

where m is the mass of the arm, J is the moment of inertial of the arm, and $u(t)$ is an applied torque. Linearizing this model about the vertical, and inserting numbers yields

$$\begin{bmatrix} \dot{\theta}(t) \\ \ddot{\theta}(t) \end{bmatrix} = \begin{bmatrix} 0 & 1 \\ 0.5 & 0 \end{bmatrix} \begin{bmatrix} \theta(t) \\ \dot{\theta}(t) \end{bmatrix} + \begin{bmatrix} 0 \\ 0.1 \end{bmatrix} u(t) + \begin{bmatrix} 0 \\ 0.1 \end{bmatrix} w(t).$$

The disturbance input consists of a random torque caused by noise in the torquer circuit and has the spectral density $S_w = 1$. The angular position of the arm is measured as

$$m(t) = [1 \quad 0] \begin{bmatrix} \theta(t) \\ \dot{\theta}(t) \end{bmatrix} + v(t),$$

where the measurement noise spectral density is $S_v = 10$. A steady-state LQG controller is synthesized to minimize the cost function:

$$J = E[\theta^2(t) + 16u^2(t)].$$

Note that this controller is fully specified by the model, the cost function, and the parameters given above. The Nyquist plot for this LQG control system is given in Figure 8.10b, and the stability margins are

$$GM^+ = 6.0 \text{ dB} \; ; \; GM^- = -1.2 \text{ dB} \; ; \; PM = 11 \text{ deg}.$$

These stability margins are considerably less than those guaranteed for both the LQR and the Kalman filter. In fact, the downside gain margin and phase margin are unacceptably small in terms of classical controller design practice. For comparison, the Nyquist loci for the LQR and the Kalman filter are shown in Figure 8.10a. These Nyquist loci are indistinguishable in this plot, but may be considerably different in general. ◆

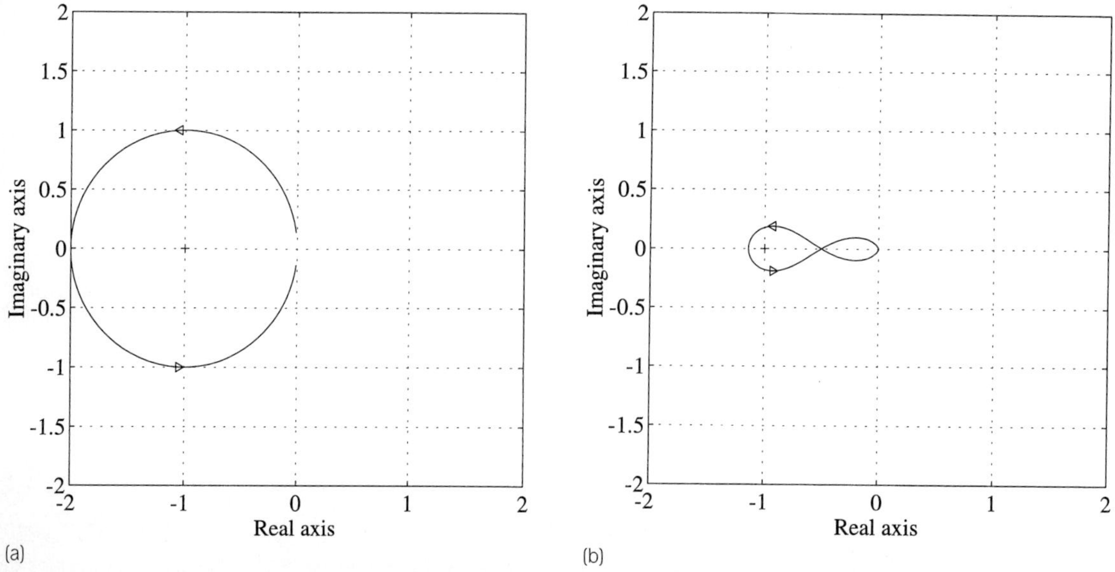

FIGURE 8.10 Nyquist plots for Example 8.3. (Parts *c–f* are referred to in continuations of Example 8.3.)

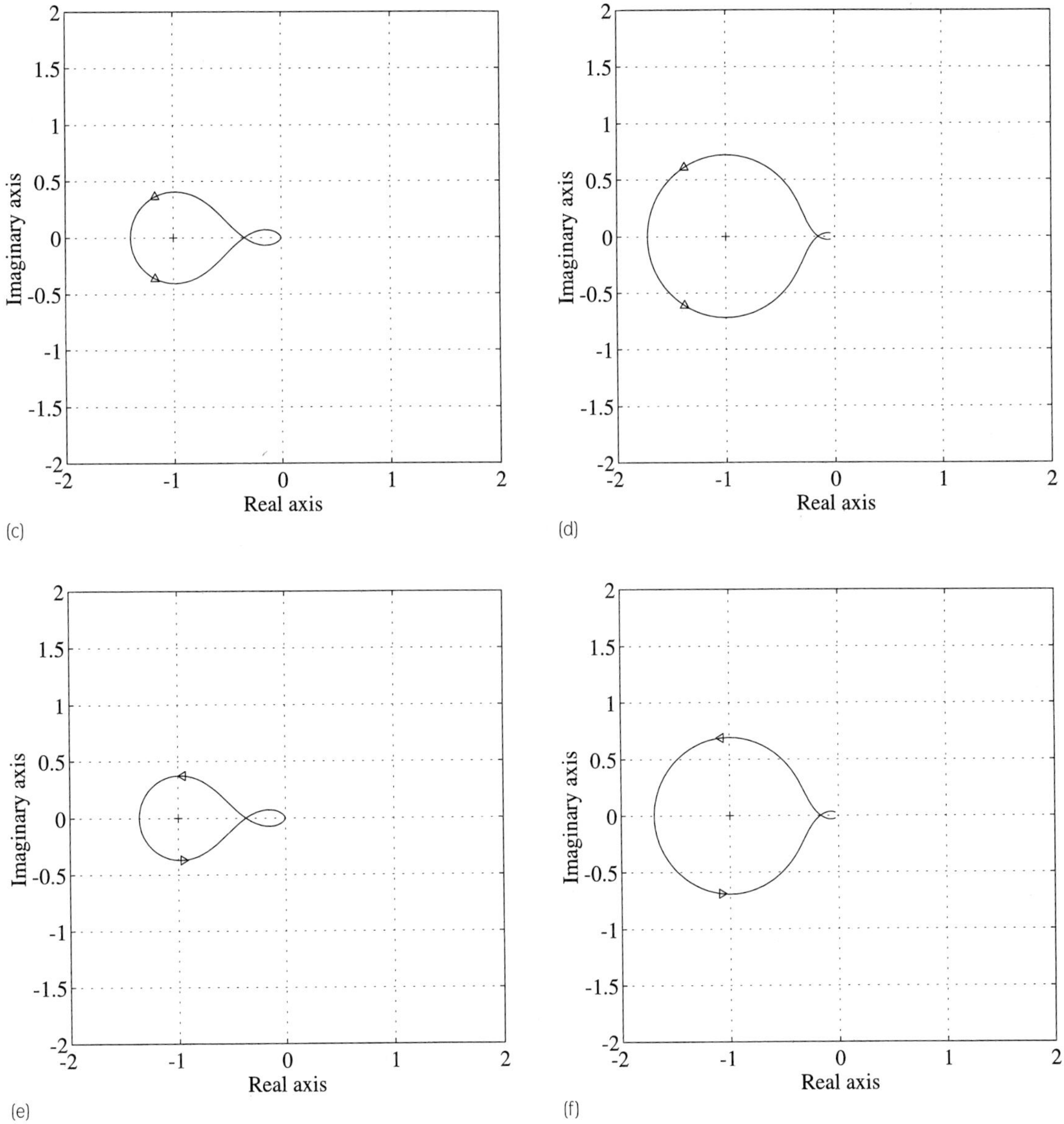

FIGURE 8.10 *Continued*

8.3 Loop Transfer Recovery

Loop transfer recovery (LTR) is a method of modifying the LQG controller to recover the robustness to input-multiplicative perturbations (and stability margins) of the LQR. LTR generates suboptimal controllers with reference to the LQG cost function, since any modification of the LQG controller increases the cost. A useful feature of the LTR design procedure is that it provides access to a family of controllers with a range of robustness properties. The final LTR controller design is selected as a compromise between robustness and performance.

Loop transfer recovery can be intuitively understood by considering the block diagram of the LQG controller in Figure 8.1. This control system can be made to look like the LQR by cutting the feedback path from the control input to the Kalman filter, and making the Kalman filter sufficiently fast that its dynamics can be ignored, as shown in Figure 8.11. The Kalman filter can be modified to accomplish both of these goals by adding fictitious noise to the control input during Kalman filter design. This noise has the effect of reducing the filter's reliance on the control input, and also making the Kalman filter faster. In general, as the fictitious noise becomes large, the poles of the Kalman filter go both to infinity and to the left half-plane reflections of the plant zeros. The slow modes associated with these finite Kalman poles are acceptable, provided they are in the left half-plane, since they are canceled by plant zeros. A more detailed discussion of the asymptotic properties of the LTR controller is given in the next subsection.

Fictitious control noise $w_f(t)$ can be added to the plant state equation:

$$\begin{aligned} \dot{x}(t) &= \mathbf{A}x(t) + \mathbf{B}_u u(t) + \mathbf{B}_w w(t) + \mathbf{B}_u w_f(t) \\ &= \mathbf{A}x(t) + \mathbf{B}_u u(t) + [\mathbf{B}_w \mid \mathbf{B}_u]\begin{bmatrix} w(t) \\ \hline w_f(t) \end{bmatrix}. \end{aligned} \tag{8.17}$$

The spectral density of the augmented white noise disturbance is then

$$\mathbf{S}_{\left[\frac{w}{w_f}\right]} = \begin{bmatrix} \mathbf{S}_w & \mathbf{0} \\ \hline \mathbf{0} & S_{w_f}\mathbf{I} \end{bmatrix}. \tag{8.18}$$

Loop transfer recovery is accomplished by increasing the spectral density of the fictitious noise until acceptable robustness properties are obtained. When the original plant noise is actuator noise (i.e., it enters the plant with the control), then this noise can be increased to yield LTR without the addition of a separate plant noise. Note that the

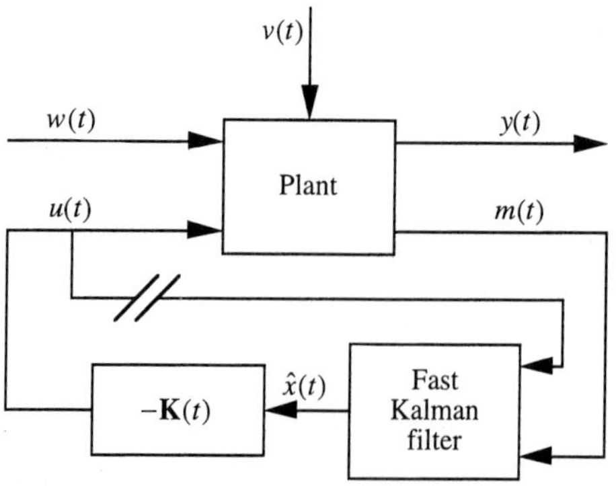

FIGURE 8.11 Loop transfer recovery modifies the LQG controller

fictitious noise should not be physically added to the control input, because this adversely affects controller performance. Instead, fictitious noise is only added to the control input during design of the Kalman filter. The addition of fictitious noise in the Kalman filter design yields a suboptimal controller for the original LQG problem, with the cost increasing as the fictitious noise is increased. A compromise is then required between the increase in robustness (a good thing) and the increase in cost (a bad thing).

The loop transfer recovery design procedure was developed intuitively. In the following subsection, this procedure is shown to yield controllers with loop transfer functions, and therefore robustness properties, that are arbitrarily close to those obtained with the LQR provided that

1. The number of measurements is greater than or equal to the number of control inputs;
2. The plant transfer function

$$\mathbf{C}_m(s\mathbf{I} - \mathbf{A})^{-1}\mathbf{B}_u$$

is minimum phase; that is, all the zeros have negative real parts.

An alternative LTR design procedure is presented in Subsection 8.3.2 for the case when the number of measurements is less than the number of control inputs.

EXAMPLE 8.3 CONTINUED Loop transfer recovery is applied to increase the gain and phase margins of the LQG controller for this example. Fictitious noise with a spectral density of 10^4 was used in the Kalman filter design, resulting in an LQG/LTR design. A second design was generated with a fictitious noise spectral density of 10^6. The Nyquist plots for these two designs are given in Figure 8.10c and 8.10d. Performance criteria and stability margins for these two LQG/LTR control systems (along with the performance criteria and margins for the LQG, and the margins for the LQR and the Kalman filter) are given in Table 8.1.

TABLE 8.1 Performance and Robustness for Example 8.3

	J	σ_θ	σ_u	GM^+	GM^-	PM
LQR	—	—	—	∞	0.50	60°
Kalman filter	—	—	—	∞	0.50	60°
LQG	9.1×10^4	9.2	75	2.0	0.87	11°
LTR ($S_{w_f} = 10^4$)	2.2×10^5	5.6	120	2.9	0.72	24°
LTR ($S_{w_f} = 10^6$)	2.9×10^6	4.3	430	6.7	0.58	42°
LTR ($\rho = 10^4$)	1.8×10^5	5.8	110	2.7	0.73	22°
LTR ($\rho = 10^6$)	2.1×10^6	4.4	370	6.1	0.59	41°

Note that increasing the fictitious noise results in an increase in the stability margins of the system and also results in the Nyquist plot more closely approximating that of the LQR (shown in Figure 8.10a). Further, increasing the fictitious noise results in a decrease in the standard deviation of θ, that is, better regulation. This improvement in regulation is obtained by greatly increasing the control standard deviation and also increasing the cost.

8.3.1 Asymptotic Properties

The validity of the loop transfer recovery design procedure is formally demonstrated after first deriving an asymptotic expression for the Kalman gain. The plant is assumed to be SISO in the following development to simplify the mathematics. Comments addressing the more general MIMO case are included, and the results obtained are valid in the more general MIMO case.

The limiting Kalman gain can be found by considering the algebraic Riccati equation (7.30) of the Kalman filter for the plant in (8.17) with spectral density (8.18):

$$\mathbf{\Sigma}_e\mathbf{A}^T + \mathbf{A}\mathbf{\Sigma}_e + \mathbf{B}_w S_w \mathbf{B}_w^T + S_{w_f}\mathbf{B}_u\mathbf{B}_u^T - \mathbf{\Sigma}_e\mathbf{C}_m^T S_v^{-1}\mathbf{C}_m\mathbf{\Sigma}_e = \mathbf{0}.$$

Dividing by the spectral density of the fictitious noise yields

$$\frac{\mathbf{\Sigma}_e}{S_{w_f}}\mathbf{A}^T + \mathbf{A}\frac{\mathbf{\Sigma}_e}{S_{w_f}} + \mathbf{B}_w\frac{S_w}{S_{w_f}}\mathbf{B}_w^T + \mathbf{B}_u\mathbf{B}_u^T - \frac{\mathbf{\Sigma}_e}{S_{w_f}}\mathbf{C}_m^T\left(\frac{S_v}{S_{w_f}}\right)^{-1}\mathbf{C}_m\frac{\mathbf{\Sigma}_o}{S_{w_f}} = \mathbf{0}.$$

For large fictitious noise spectral densities, this equation reduces to

$$\left(\frac{\mathbf{\Sigma}_e}{S_{w_f}}\right)\mathbf{A}^T + \mathbf{A}\left(\frac{\mathbf{\Sigma}_e}{S_{w_f}}\right) + \mathbf{B}_u\mathbf{B}_u^T - \left(\frac{\mathbf{\Sigma}_e}{S_{w_f}}\right)\mathbf{C}_m^T\left(\frac{S_v}{S_{w_f}}\right)^{-1}\mathbf{C}_m\left(\frac{\mathbf{\Sigma}_e}{S_{w_f}}\right) = \mathbf{0}, \tag{8.19}$$

which is the Kalman Riccati equation for the plant with unity plant noise spectral density (entering through $\mathbf{B}_u$) and measurement noise spectral density S_v/S_{w_f}. Note that this measurement noise approaches zero when the fictitious noise spectral density approaches infinity. In this limit, the Kalman filter yields perfect state estimation provided the plant is minimum phase and the number of measurements exceeds the number of control inputs.[5] Therefore,

$$\lim_{S_{w_f}\to\infty}\frac{\mathbf{\Sigma}_e}{S_{w_f}} = \mathbf{0},$$

and (8.19) can be further simplified for large fictitious noise spectral density:

$$\mathbf{B}_u\mathbf{B}_u^T - \left(\frac{\mathbf{\Sigma}_e}{S_{w_f}}\right)\mathbf{C}_m^T\left(\frac{S_v}{S_{w_f}}\right)^{-1}\mathbf{C}_m\left(\frac{\mathbf{\Sigma}_e}{S_{w_f}}\right) = \mathbf{0}. \tag{8.20}$$

Recognizing that the Kalman gain (7.28) can be written as a function of the fictitious noise spectral density,

$$\mathbf{G}(S_{w_f}) = \left(\frac{\mathbf{\Sigma}_e}{S_{w_f}}\right)\mathbf{C}_m^T\left(\frac{S_v}{S_{w_f}}\right)^{-1}$$

(8.20) becomes

$$\mathbf{B}_u\mathbf{B}_u^T - \mathbf{G}(S_{w_f})\left(\frac{S_v}{S_{w_f}}\right)\mathbf{G}^T(S_{w_f}) = \mathbf{0}. \tag{8.21}$$

[5] This result is presented in Theorem 3.14, parts *d* and *e*, on page 307 of [1]. Note that the number of controls equals the number of disturbance inputs, since the disturbances enter through $\mathbf{B}_u$. The proof of this result can be found in [2].

The limiting Kalman gain matrix is then

$$\mathbf{G}(S_{wf}) = \sqrt{\frac{S_{wf}}{S_v}}\mathbf{B}_u. \tag{8.22}$$

The limiting Kalman gain matrix can be substituted into the controller transfer function. Before making this substitution, the transfer function is rewritten:

$$\begin{aligned}\mathbf{G}_c(s) &= -\mathbf{K}(s\mathbf{I} - \mathbf{A} + \mathbf{B}_u\mathbf{K} + \mathbf{G}\mathbf{C}_m)^{-1}\mathbf{G}\\ &= -\mathbf{K}\{(s\mathbf{I} - \mathbf{A} + \mathbf{B}_u\mathbf{K})[\mathbf{I} + (s\mathbf{I} - \mathbf{A} + \mathbf{B}_u\mathbf{K})^{-1}\mathbf{G}\mathbf{C}_m]\}^{-1}\mathbf{G}\\ &= -\mathbf{K}[\mathbf{I} + (s\mathbf{I} - \mathbf{A} + \mathbf{B}_u\mathbf{K})^{-1}\mathbf{G}\mathbf{C}_m]^{-1}(s\mathbf{I} - \mathbf{A} + \mathbf{B}_u\mathbf{K})^{-1}\mathbf{G}.\end{aligned}$$

Applying the push through theorem of Appendix equation (A8.2) to this expression yields

$$\mathbf{G}_c(s) = -\mathbf{K}(s\mathbf{I} - \mathbf{A} + \mathbf{B}_u\mathbf{K})^{-1}\mathbf{G}[1 + \mathbf{C}_m(s\mathbf{I} - \mathbf{A} + \mathbf{B}_u\mathbf{K})^{-1}\mathbf{G}]^{-1}. \tag{8.23}$$

Substituting (8.22) into (8.23) yields

$$\mathbf{G}_c(s) = -\mathbf{K}(s\mathbf{I} - \mathbf{A} + \mathbf{B}_u\mathbf{K})^{-1}\sqrt{\frac{S_{wf}}{S_v}}\mathbf{B}_u\left[1 + \mathbf{C}_m(s\mathbf{I} - \mathbf{A} + \mathbf{B}_u\mathbf{K})^{-1}\sqrt{\frac{S_{wf}}{S_v}}\mathbf{B}_u\right]^{-1}.$$

In the limit as $S_{wf} \to \infty$, the 1 in the bracketed inverse is insignificant and can be deleted:

$$\begin{aligned}\mathbf{G}_c(s) &= -\mathbf{K}(s\mathbf{I} - \mathbf{A} + \mathbf{B}_u\mathbf{K})^{-1}\sqrt{\frac{S_{wf}}{S_v}}\mathbf{B}_u\left[\mathbf{C}_m(s\mathbf{I} - \mathbf{A} + \mathbf{B}_u\mathbf{K})^{-1}\sqrt{\frac{S_{wf}}{S_v}}\mathbf{B}_u\right]^{-1}\\ &= -\mathbf{K}(s\mathbf{I} - \mathbf{A} + \mathbf{B}_u\mathbf{K})^{-1}\mathbf{B}_u[\mathbf{C}_m(s\mathbf{I} - \mathbf{A} + \mathbf{B}_u\mathbf{K})^{-1}\mathbf{B}_u]^{-1}.\end{aligned} \tag{8.24}$$

This expression can be simplified by noting that

$$\begin{aligned}(s\mathbf{I} - \mathbf{A} + \mathbf{B}_u\mathbf{K})^{-1}\mathbf{B}_u &= \{(s\mathbf{I} - \mathbf{A})[\mathbf{I} + (s\mathbf{I} - \mathbf{A})^{-1}\mathbf{B}_u\mathbf{K}]\}^{-1}\mathbf{B}_u\\ &= [\mathbf{I} + (s\mathbf{I} - \mathbf{A})^{-1}\mathbf{B}_u\mathbf{K}]^{-1}(s\mathbf{I} - \mathbf{A})^{-1}\mathbf{B}_u\\ &= (s\mathbf{I} - \mathbf{A})^{-1}\mathbf{B}_u[\mathbf{I} + \mathbf{K}(s\mathbf{I} - \mathbf{A})^{-1}\mathbf{B}_u]^{-1},\end{aligned}$$

where the push through theorem is used in generating the final expression. Using this result in (8.24) yields

$$\begin{aligned}\mathbf{G}_c(s) &= -\mathbf{K}(s\mathbf{I} - \mathbf{A})^{-1}\mathbf{B}_u[\mathbf{I} + \mathbf{K}(s\mathbf{I} - \mathbf{A})^{-1}\mathbf{B}_u]^{-1}\\ &\quad\times \{\mathbf{C}_m(s\mathbf{I} - \mathbf{A})^{-1}\mathbf{B}_u[\mathbf{I} + \mathbf{K}(s\mathbf{I} - \mathbf{A})^{-1}\mathbf{B}_u]^{-1}\}^{-1};\end{aligned}$$

$$\mathbf{G}_c(s) = -\mathbf{K}(s\mathbf{I} - \mathbf{A})^{-1}\mathbf{B}_u\{\mathbf{C}_m(s\mathbf{I} - \mathbf{A})^{-1}\mathbf{B}_u\}^{-1}. \tag{8.25}$$

The loop transfer function (with the loop broken at the control input to the plant) is then[6]

$$\mathbf{G}_l(s) = \mathbf{K}(s\mathbf{I} - \mathbf{A})^{-1}\mathbf{B}_u,$$

since the term within the curly brackets in (8.25) is the plant transfer function. This limiting loop transfer function equals the loop transfer function of the linear quadratic regulator, which implies that the robustness of the LQR is recovered. From (8.25), the

[6] The minus sign in this loop transfer function is suppressed as a matter of convention.

limiting LTR controller is seen to include an inverse of the plant dynamics. Right half-plane plant zeros, therefore, yield unstable pole-zero cancellations in the loop. Note that this difficulty is not observed for unstable plant poles, since right half-plane poles are canceled within the controller equation and do not yield unstable pole-zero cancellations in the actual system.

In summary, the robustness properties of the LQR are obtained in the limit as the fictitious noise's spectral density approaches infinity provided that the plant is minimum phase and the number of measurements is greater than or equal to the number of controls. LTR may also work for nonminimal phase plants, especially when the nonminimal phase zeros are outside of the bandwidth of interest. But there is no guarantee of the effectiveness of LTR for nonminimal phase plants, and a try-it-and-see approach is required of the designer.

8.3.2 Robustness to Output-Multiplicative Perturbations

A dual of the loop transfer recovery procedure presented above can be used to recover the robustness of the Kalman filter to output-multiplicative perturbations. In this case, the state weighting matrix of the cost function is modified to increase robustness. This dual procedure can be applied when the number of measurements is less than or equal to the number of control inputs.

Loop transfer recovery at the output is obtained by adding an additional term, known as measurement weighting, to the cost function:

$$J = E[x^T(\infty)(\mathbf{Q} + \rho\mathbf{C}_m^T\mathbf{C}_m)x(\infty) + u^T(\infty)\mathbf{R}u(\infty)].$$

This additional term increases robustness, but results in a suboptimal controller relative to the original cost function. Again, a family of controllers is generated, and the selection of the desired controller represents a compromise between robustness and performance. Increasing ρ yields controllers with loop transfer functions (broken at the output) that are arbitrarily close to those obtained for the Kalman filter provided that

1. The number of measurements is less than or equal to the number of control inputs;
2. The plant transfer function,

   $$\mathbf{C}_m(s\mathbf{I} - \mathbf{A})^{-1}\mathbf{B}_u,$$

   is minimum phase.

The robustness properties of the Kalman filter with respect to output-multiplicative perturbations can, therefore, be recovered when these conditions are satisfied.

EXAMPLE 8.3 CONTINUED Loop transfer recovery at the output is applied to increase the stability margins of the LQG controller for this example. Note that the stability margins at the output and input are equal in this example, since the plant is SISO. Measurement weighting is added to the cost function with a weighting of $\rho = 10^4$, resulting in an LQG/LTR design. A second design is generated with a measurement weighting factor of 10^6. The Nyquist plots for these two designs are given in Figure 8.10e and 8.10f. Performance criteria and stability margins for these two LQG/LTR control systems (along with performance criteria and margins for the LQG and the margins for the Kalman filter) are given in Table 8.1. Note

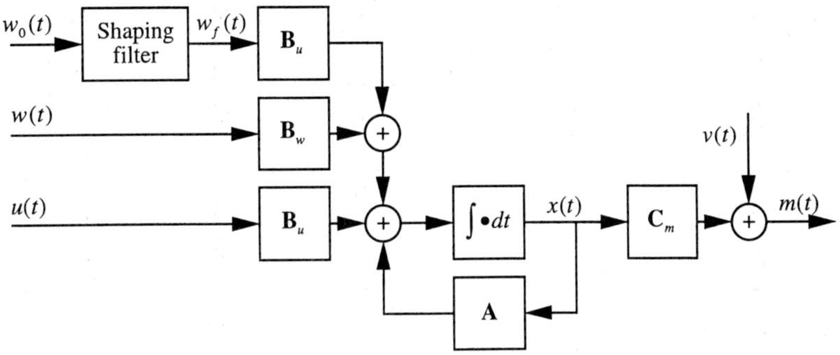

FIGURE 8.12 Frequency-weighted loop transfer recovery

that increasing the measurement weighting results in an increase in the stability margins of the system and also results in the Nyquist plot more closely approximating that of the Kalman filter (shown in Figure 8.10a). Further, increasing the measurement weighting results in a decrease in the standard deviation of the angle error. This improvement in regulation is obtained by increasing the control standard deviation and the cost. ◆

8.3.3 Frequency-Shaped LTR

The fictitious noise used in loop transfer recovery can be restricted to a particular frequency band, as shown in Figure 8.12. Frequency-shaped LTR at both the input and the output can be performed, but this section addresses only LTR at the input to simplify the discussion. In this case, the loop transfer function approaches that of the LQR in the frequency range selected, but remains largely unaffected at other frequencies. For example, lowpass fictitious noise can be used to increase the stability margins provided the cutoff frequency of the fictitious noise is greater than the phase and gain crossover frequencies. Since the fictitious noise is lowpass, the high-frequency roll-off of the LQG controller (40 dB/decade) is preserved. Note that the LQR high-frequency roll-off is 20 dB/decade. The more rapid high-frequency roll-off achieved using lowpass fictitious noise has the advantage of decreasing the response to high-frequency measurement noise, and decreasing the sensitivity of the system to high-frequency unmodeled dynamics. The resulting controller is higher-order than the original LQG controller due to the addition of the filter states. Low-order frequency-shaping filters are therefore desirable to maintain reasonable controller orders.

Bandpass fictitious noise can also be employed in loop transfer recovery. In particular, the gain margin can be increased by the addition of fictitious noise around the phase crossover frequency. The phase margin can be increased by the addition of fictitious noise around the gain crossover frequency. A more thorough discussion of the possible benefits of frequency-shaped LTR can be found in [3], page 241.

8.4 Tracking System Design

The LQG controller is a regulator; that is, it is designed to drive the states to zero. This is a very restrictive limitation, since most control systems are tracking systems that force the reference output to equal the reference input (not necessarily zero).

Fortunately, the LQG design methodology can be modified to generate tracking systems. Control systems for tracking constant reference inputs are initially considered, and these results are generalized to the tracking of more general inputs.

8.4.1 Tracking Constant Reference Inputs

Three approaches are presented for tracking constant reference inputs. Coordinate translations that result in the zero state yielding the desired reference output can be used when the system includes a pure integrator. A feedforward control input can be applied to force the system to exhibit the desired behavior. Integral control can also be applied to guarantee a zero steady-state tracking error. Applying the separation principle, these tracking systems are first developed assuming all of the states are measured exactly and then generalized to the case when only noisy partial state information is available.

Tracking via Coordinate Translation An intuitive approach to tracking system design is to replace state feedback with state error feedback:

$$u(t) = -\mathbf{K}\{x(t) - x_d\}, \tag{8.26}$$

where x_d is the desired state. The desired state is chosen so that the resulting reference output equals the desired reference input:

$$\mathbf{C}_y x_d = r. \tag{8.27}$$

Equation (8.27) is typically underdetermined; that is, there are multiple desired states that yield the same reference output. Desired state selection, in the underdetermined case, is discussed below after first developing an additional constraint on this state.

Using the control law (8.26) results in a steady-state error, in general, as can be seen by considering the coordinate translation:

$$\tilde{x}(t) = x(t) - x_d. \tag{8.28}$$

Applying this coordinate translation to the generic plant model with the control (8.26) gives

$$\dot{\tilde{x}}(t) = \dot{x}(t) = \mathbf{A}\{\tilde{x}(t) + x_d\} + \mathbf{B}_u\{-\mathbf{K}\tilde{x}(t)\} = (\mathbf{A} - \mathbf{B}_u\mathbf{K})\tilde{x}(t) + \mathbf{A}x_d.$$

This closed-loop state equation is driven by the constant input $\mathbf{A}x_d$, resulting in a constant steady-state value for $\tilde{x}(t)$ (i.e., $x(\infty) \neq x_d$) in general. Therefore, state error feedback is not desirable except in the following special case.

The closed-loop state equation for $\tilde{x}(t)$ becomes homogeneous, and zero steady-state error is obtained when

$$\mathbf{A}x_d = 0. \tag{8.29}$$

This condition can only be met if the state matrix is singular and x_d is in the null space of this matrix. The fact that the state matrix is singular indicates that this matrix has an eigenvalue at zero or, equivalently, that the plant contains a pure integrator. The desired state's being in the null space of the state matrix implies that x_d is an equilibrium state; that is, x_d can be maintained in steady state without the use of control.

The desired state must satisfy both (8.27) and (8.29). This is only possible for arbitrary reference inputs if the dimension of the null space of the state matrix is greater than or equal to the dimension of the reference input. The number of integrators in the plant must therefore be greater than or equal to the number of reference outputs to achieve arbitrary, nonzero set points. In addition, the row vectors associated with the reference outputs must not be orthogonal to the null space of the state matrix in order to ensure the existence of an appropriate desired state. When multiple desired states exist that satisfy both (8.27) and (8.29), the ambiguity is usually resolved by setting any unconstrained states to zero.

When (8.29) is satisfied, the open-loop state equation is invariant under the coordinate translation (8.28); that is, the state equation for $\tilde{x}(t)$ is identical to the state equation for $x(t)$:

$$\dot{\tilde{x}}(t) = \mathbf{A}\tilde{x}(t) + \mathbf{B}_u u(t).$$

LQR theory can then be applied to this new plant to yield an optimal control of the form in (8.26). This control is optimal for tracking constant inputs and is reasonable for tracking slowly changing or stepwise constant reference inputs.

EXAMPLE 8.4 The transfer function of an ac motor that contains an integrator is

$$\frac{\Theta(s)}{V(s)} = \frac{150}{s(s + 50)},$$

where $\theta(t)$ is the motor shaft angle, and $v(t)$ is the applied ac voltage. The state model of this system is

$$\begin{bmatrix} \dot{\theta}(t) \\ \ddot{\theta}(t) \end{bmatrix} = \begin{bmatrix} 0 & 1 \\ 0 & -50 \end{bmatrix} \begin{bmatrix} \theta(t) \\ \dot{\theta}(t) \end{bmatrix} + \begin{bmatrix} 0 \\ 150 \end{bmatrix} v(t).$$

A motor shaft angle of 0.5 radians is desired. The desired state is chosen to be the element of the null space of the state matrix that gives the desired motor shaft angle:

$$\begin{bmatrix} 0.5 \\ 0 \end{bmatrix}.$$

The coordinate translation is then

$$\tilde{x}(t) = \begin{bmatrix} \theta(t) \\ \dot{\theta}(t) \end{bmatrix} - \begin{bmatrix} 0.5 \\ 0 \end{bmatrix} = \begin{bmatrix} \theta(t) - 0.5 \\ \dot{\theta}(t) \end{bmatrix}.$$

The original state equation is recovered when this coordinate translation is performed, and the LQR can be applied to the transformed state model to yield an optimal control.

The control (8.26) is modified in the case of noisy partial state measurements by estimating $x(t)$ or, equivalently, $\tilde{x}(t)$, using the Kalman filter. The resulting control is

$$u(t) = -\mathbf{K}\{\hat{x}(t) - x_d\} = -\mathbf{K}\hat{\tilde{x}}(t).$$

The use of state error feedback or, equivalently, coordinate translation provides the simplest approach to optimal tracking system design when the plant contains an integrator and a desired state in the null space of the state matrix can be found.

Feedforward Control The control (8.26) is a special case of the control

$$u(t) = -\mathbf{K}x(t) + \mathbf{K}_r r, \tag{8.30}$$

where $\mathbf{K}_r r$ is termed a feedforward control. The feedforward control gain can be selected to drive the reference output to the desired value after first generating the state feedback gains for a regulator.

Applying the control (8.30), the closed-loop state model is

$$x(t) = (\mathbf{A} - \mathbf{B}_u\mathbf{K})x(t) + \mathbf{B}_u\mathbf{K}_r r;$$

$$y(t) = \mathbf{C}_y x(t).$$

Assuming the closed-loop system is stable, the feedforward control term generates the following steady-state reference output:

$$y(\infty) = -\mathbf{C}_y(\mathbf{A} - \mathbf{B}_u\mathbf{K})^{-1}\mathbf{B}_u\mathbf{K}_r r.$$

Note that this expression is simply the dc gain of the closed-loop system times the feedforward control. The inverse within this expression is guaranteed to exist, since all of the eigenvalues of the closed-loop state matrix have negative real parts, and therefore there is no eigenvalue at zero.

The feedforward gain is found by setting the steady-state reference output equal to the desired value:

$$r = y(\infty) = -\mathbf{C}_y(\mathbf{A} - \mathbf{B}_u\mathbf{K})^{-1}\mathbf{B}_u\mathbf{K}_r r. \tag{8.31a}$$

For arbitrary reference inputs, this expression can be solved for the feedforward gain provided that the dimension of the reference input (equal to the dimension of the reference output) is less than or equal to the dimension of the control input. This statement assumes that the matrix multiplying the feedforward control term in (8.31a) is full rank. When the dimension of the reference input is less than the dimension of the control input, there is typically an infinite number of solutions for the feedforward gain, any one yielding the desired reference output. A unique solution exists when these dimensions are equal:

$$\mathbf{K}_r = -\{\mathbf{C}_y(\mathbf{A} - \mathbf{B}_u\mathbf{K})^{-1}\mathbf{B}_u\}^{-1}. \tag{8.31b}$$

The use of feedback often greatly reduces the sensitivity of the output to plant errors (i.e., increases robustness). Feedforward control, on the other hand, does not share this desirable property, and the output may be quite sensitive to plant errors. Therefore, feedforward control should be used to achieve desired reference outputs only when the plant model is known to be accurate. In addition, a thorough analysis of the sensitivity of the output to plant errors should be conducted when using feedforward control.

EXAMPLE 8.5 A valve is constructed with a spring on the "flapper" so that if power is removed the valve closes. The control input is a torque applied to the flapper. The state equation for this valve is

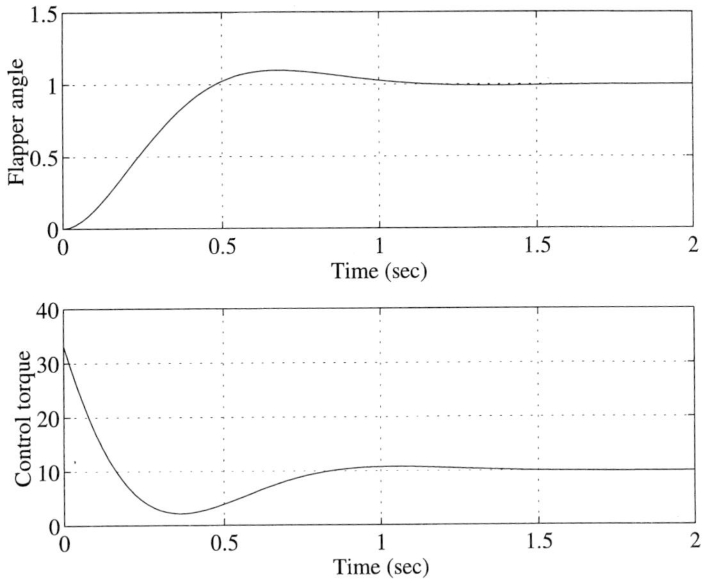

FIGURE 8.13 Tracking performance using feedforward control

$$\dot{x}(t) = \begin{bmatrix} 0 & 1 \\ -10 & -1 \end{bmatrix} x(t) + \begin{bmatrix} 0 \\ 1 \end{bmatrix} u(t),$$

and the output of interest is $y(t) = x_1(t)$. The LQR gain is found for the following cost function:

$$J = \int_0^\infty y^2(t) + 0.001u^2(t)dt.$$

The control required to drive the valve to an arbitrary position is then given by (8.30) and (8.31b):

$$u(t) = -[23 \quad 5.9]x(t) + 33r.$$

This tracking system is simulated and the results plotted in Figure 8.13. Zero steady-state error is achieved, since the plant model in the simulation is perfect. ◆

The state feedback plus feedforward controller is modified in the case of noisy, partial state measurements by utilizing the optimal state estimates:

$$\boxed{u(t) = -\mathbf{K}\hat{x}(t) + \mathbf{K}_r r.} \tag{8.32}$$

At first glance, the feedforward gain should be computed for the closed-loop system including the Kalman filter. But the same feedforward gain is computed with and without the Kalman filter in the feedback loop (see Exercise 8.6). Therefore, the feedforward gain should be computed for the simpler state feedback controller. The fact that

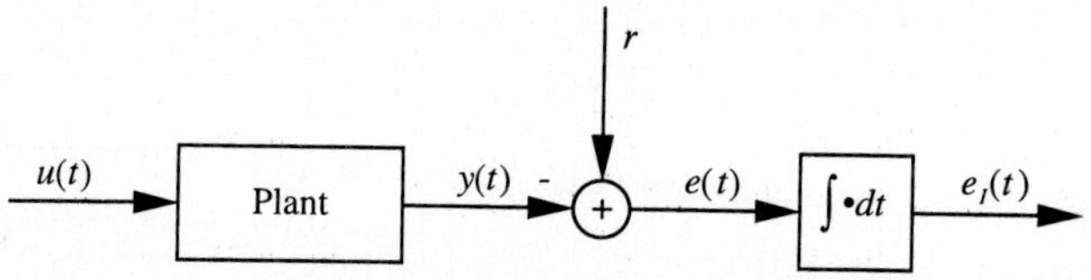

FIGURE 8.14 The augmented plant for integral control

this gain can be computed while ignoring the Kalman filter provides additional evidence for the generality of the separation principle.

Integral Control Integral feedback is used in classical control to zero out steady-state errors when tracking constant signals. Integral control can be generated in an LQR setting by appending an integrator to the plant (see Figure 8.14) before computation of the feedback gains. The integral of the error between the reference input and reference output is generated by the following differential equation:

$$\dot{e}_I(t) = r - y(t) = -\mathbf{C}_y x(t) + r. \tag{8.33}$$

Note that (8.33) requires the integration of each of the reference outputs, making the number of integrations equal to the number of reference outputs. The augmented state model is the combination of the plant state equation,

$$\dot{x}(t) = \mathbf{A}x(t) + \mathbf{B}_u u(t),$$

and the state equation for the integral of the error. The augmented state equation is then

$$\begin{bmatrix} \dot{x}(t) \\ \hline \dot{e}_I(t) \end{bmatrix} = \left[\begin{array}{c|c} \mathbf{A} & \mathbf{0} \\ \hline -\mathbf{C}_y & \mathbf{0} \end{array}\right] \begin{bmatrix} x(t) \\ \hline e_I(t) \end{bmatrix} + \begin{bmatrix} \mathbf{B}_u \\ \hline \mathbf{0} \end{bmatrix} u(t) + \begin{bmatrix} \mathbf{0} \\ \hline \mathbf{I} \end{bmatrix} r. \tag{8.34}$$

LQR theory can be used to generate a state feedback for the augmented plant. This controller is designed ignoring the constant reference input, and using a cost function that penalizes the integral of the error:

$$\begin{aligned} J &= \int_0^{t_f} e_I^T(t)e_I(t) + u^T(t)\mathbf{R}u(t)dt \\ &= \int_0^{t_f} [x^T(t) \mid e_I^T(t)] \left[\begin{array}{c|c} \mathbf{0} & \mathbf{0} \\ \hline \mathbf{0} & \mathbf{I} \end{array}\right] \begin{bmatrix} x(t) \\ \hline e_I(t) \end{bmatrix} + u^T(t)\mathbf{R}u(t)dt. \end{aligned} \tag{8.35}$$

The control weighting matrix is usually generated by trial and error, since the physical meaning of this cost function is often fairly nebulous. The optimal control is then

$$u(t) = -\mathbf{K}(t)\begin{bmatrix} x(t) \\ \hline e_I(t) \end{bmatrix} = -[\mathbf{K}_x(t) \mid \mathbf{K}_I(t)]\begin{bmatrix} x(t) \\ \hline e_I(t) \end{bmatrix}.$$

This control can be written explicitly in terms of the tracking error:

$$u(t) = -\mathbf{K}_x(t)x(t) - \mathbf{K}_I(t)\int_0^t \{r - y(t)\}dt, \tag{8.36}$$

which shows that the control includes integral feedback.

EXAMPLE 8.5 CONTINUED An integral control can be applied to track constant reference inputs for the valve in Example 8.5. In this case, the augmented system becomes

$$\begin{bmatrix} \dot{x}(t) \\ \hline \dot{e}_I(t) \end{bmatrix} = \left[\begin{array}{cc|c} 0 & 1 & 0 \\ -10 & -1 & 0 \\ \hline -1 & 0 & 0 \end{array}\right] \begin{bmatrix} x(t) \\ \hline e_I(t) \end{bmatrix} + \begin{bmatrix} 0 \\ 1 \\ \hline 0 \end{bmatrix} u(t) + \begin{bmatrix} 0 \\ 0 \\ \hline 1 \end{bmatrix} r.$$

Ignoring the reference input and applying LQR theory to minimize the cost function (8.35) yields the controller (8.36):

$$u(t) = -[31 \quad 7.0]x(t) + 100 \int_0^t \{r - y(t)\}dt.$$

The control weighting $R = 0.0001$ was chosen by trial and error to yield desirable pole locations. The integral control tracking system is simulated and the results plotted in Figure 8.15. Note that zero steady-state error is achieved. ◆

The separation principle can be used to generate a controller when only noisy, partial state measurements are available. In this case, minimum mean square estimates of $x(t)$ and $e_I(t)$ are required. The minimum mean square estimate of $x(t)$ is generated by the Kalman filter operating on the original plant equation. The minimum mean square estimate of $e_I(t)$ is then generated by integrating $\{r - \mathbf{C}_y\hat{x}(t)\}$. Note that this is the optimal estimate of $e_I(t)$, since minimum mean square estimation and linear operations commute. The optimal control is then

$$\boxed{u(t) = -\mathbf{K}_x(t)\hat{x}(t) - \mathbf{K}_I(t) \int_0^t \{r - \mathbf{C}_y\hat{x}(t)\}dt,} \qquad (8.37)$$

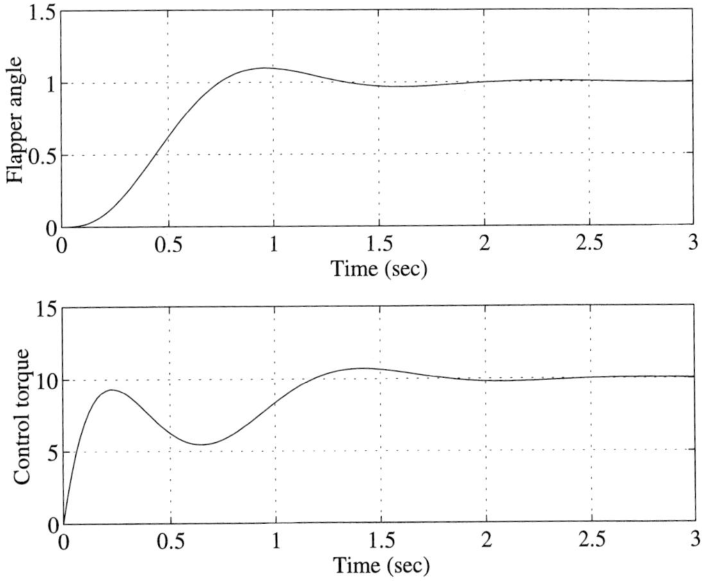

FIGURE 8.15 Tracking performance using integral control

where $\hat{x}(t)$ is the Kalman estimate of the original plant state. When using the steady-state gains, this controller also obeys the deterministic separation principle; that is, the poles of the closed-loop system are the poles of the state feedback plus the poles the Kalman filter.

8.4.2 Tracking Time-Varying Reference Inputs

Slowly varying and piecewise constant reference inputs can be tracked using the techniques presented in the previous subsection. Feedforward control can be used to track more general inputs by replacing the feedforward gain with a filter. This technique suffers from the lack of robustness typical of feedforward control and is not pursued further. The integral control technique can also be generalized to track reference inputs that are bandpass, sinusoidal, ramps, and so on.

Integral control works by increasing the loop gain in the frequency band of the reference input. This fact can be justified by considering the unity feedback system in Figure 8.16. For this system, the error is given in terms of the reference input:

$$E(s) = [\mathbf{I} - \mathbf{G}_l(s)]^{-1}R(s).$$

Clearly, a large loop gain results in a small error. A bandpass signal can be effectively tracked by appending a high-gain bandpass filter to the tracking error, as shown in Figure 8.17. The filter output is included within the cost function used in generating the LQR gain. The controller then implements this bandpass filter and uses the filter states for feedback along with the plant state feedback. The resulting loop gain is increased in the frequency range of the reference input, which yields better tracking. Note that the control weighting matrix may also require modification when adding the filter output to the cost.

Perfect tracking of sinusoidal reference inputs (with known frequency) can be obtained by appending an oscillator to the tracking error before generating the LQR gain. When the oscillator poles are at the known reference input frequency, an infinite loop gain is obtained, which yields zero steady-state error. Ramp inputs can be tracked perfectly by appending two integrators to the tracking error and then generating the LQR gain. These results for sinusoidal and ramp inputs are special cases of the internal model principle:

> Zero steady-state tracking error is obtained when the poles in the Laplace transform of the reference input are also poles of the loop transfer function.

This principle can be used in the design of tracking system for other simple reference inputs.

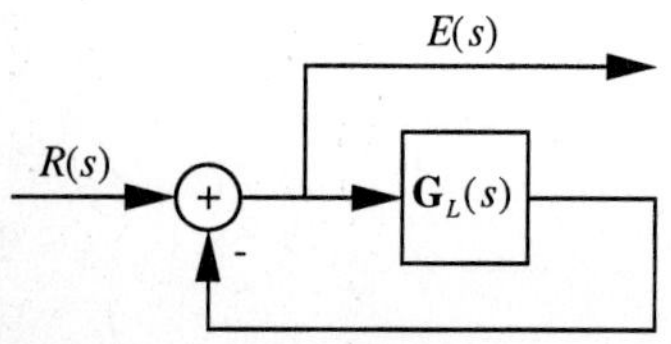

FIGURE 8.16 Unity feedback

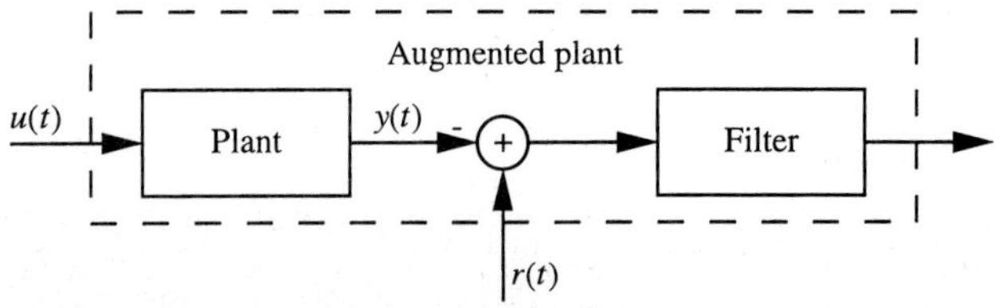

FIGURE 8.17 The augmented plant

8.5 Designing for Disturbance Rejection

The linear quadratic Gaussian control system is optimized to reject white noise disturbances, but may be suboptimal when applied to nonwhite disturbances. The LQG controller can be modified to improve rejection of nonwhite disturbances by using disturbance cancellation and shaping filters during controller design and implementation. These ideas are illustrated by application to the rejection of constant disturbances. The resulting design procedures can be extended to other disturbances in a straightforward manner.

8.5.1 Feedforward Disturbance Cancellation

The cancellation of disturbances is first addressed assuming that the state and the disturbance are both measured exactly. The plant is modeled as follows:

$$\dot{x}(t) = \mathbf{A}x(t) + \mathbf{B}_u u(t) + \mathbf{B}_{w_0} w_0;$$
$$y(t) = \mathbf{C}_y x(t),$$

where w_0 is a constant disturbance input. Ignoring the constant disturbance input, linear quadratic regulator theory can be used to generate state feedback gains for this system. Adding a feedforward control to the state feedback yields

$$u(t) = -\mathbf{K}x(t) + \mathbf{K}_w w_0. \tag{8.38}$$

Using this control, the closed-loop state equation is

$$\dot{x}(t) = (\mathbf{A} - \mathbf{B}_u\mathbf{K})x(t) + \mathbf{B}_u\mathbf{K}_w w_0 + \mathbf{B}_{w_0} w_0.$$

The feedforward gain is selected to cancel the effects of the disturbance input on the reference output. In steady state, the reference output due to the disturbance is

$$y_w(\infty) = -\mathbf{C}_y(\mathbf{A} - \mathbf{B}_u\mathbf{K})^{-1}\mathbf{B}_{w_0} w_0,$$

and the output due to the feedforward control is

$$y_u(\infty) = -\mathbf{C}_y(\mathbf{A} - \mathbf{B}_u\mathbf{K})^{-1}\mathbf{B}_u\mathbf{K}_w w_0.$$

The reference output is zero provided that these two terms cancel:

$$\begin{aligned} y_w(\infty) &= -\mathbf{C}_y(\mathbf{A} - \mathbf{B}_u\mathbf{K})^{-1}\mathbf{B}_{w_0} w_0 \\ &= \mathbf{C}_y(\mathbf{A} - \mathbf{B}_u\mathbf{K})^{-1}\mathbf{B}_u\mathbf{K}_w w_0 = -y_u(\infty). \end{aligned} \tag{8.39a}$$

A feedforward gain can be found that satisfies this condition provided that the dimension of the reference output is less than or equal to the dimension of the control input. It is assumed that the matrix multiplying the feedforward control term has full rank. When the reference output and control input have equal dimensions, a unique solution exists for the feedforward gain:

$$\mathbf{K}_w = -\{\mathbf{C}_y(\mathbf{A} - \mathbf{B}_u\mathbf{K})^{-1}\mathbf{B}_u\}^{-1}\mathbf{C}_y(\mathbf{A} - \mathbf{B}_u\mathbf{K})^{-1}\mathbf{B}_{w_0}. \tag{8.39b}$$

As in the case of reference input tracking, feedforward control is often not very robust. Therefore, feedforward control should be used only when the plant model is known to be accurate, and only after conducting a thorough analysis of the output sensitivity to plant errors.

EXAMPLE 8.6 The angular velocity of a field-controlled dc motor is required to equal a set point regardless of the load torque. The state equation of the motor is

$$\begin{bmatrix} \dot{\omega}(t) \\ \dot{\tau}_M(t) \end{bmatrix} = \begin{bmatrix} -1 & 1 \\ 0 & -0.1 \end{bmatrix} \begin{bmatrix} \omega(t) \\ \tau_M(t) \end{bmatrix} + \begin{bmatrix} 0 \\ 0.1 \end{bmatrix} u(t) + \begin{bmatrix} 1 \\ 0 \end{bmatrix} \tau_L(t)$$

where ω is the angular velocity of the motor, τ_M is the motor torque, u is the field voltage, and τ_L is the load torque. The states and load torque are assumed to be known. Steady-state LQR theory is used to generate state feedback gains. The cost function is

$$J = \int_0^\infty \omega^2(t) + 0.01u^2(t)dt,$$

which specifies the feedback gain. Using this gain in (8.39b) yields the following control:

$$u(t) = -[14.6 \quad 16.1] \begin{bmatrix} \omega(t) \\ \tau_M(t) \end{bmatrix} - 17.1\tau_L(t).$$

This system is simulated with a unit step load torque, and a set point of zero. The results are given in Figure 8.18. A maximum field voltage of 17 volts is used to maintain the angular velocity within 0.25 rad/sec of zero. In addition, the steady-state error is zero, since the model used in the simulation exactly matches the model used during design.

Disturbance cancellation can be applied when only noisy, partial state measurements are available. In this case, the plant state and the disturbance can be estimated using a Kalman filter. To estimate the disturbance, the plant model must be augmented with a filter that generates a nearly constant disturbance, as shown in Figure 8.19. Note

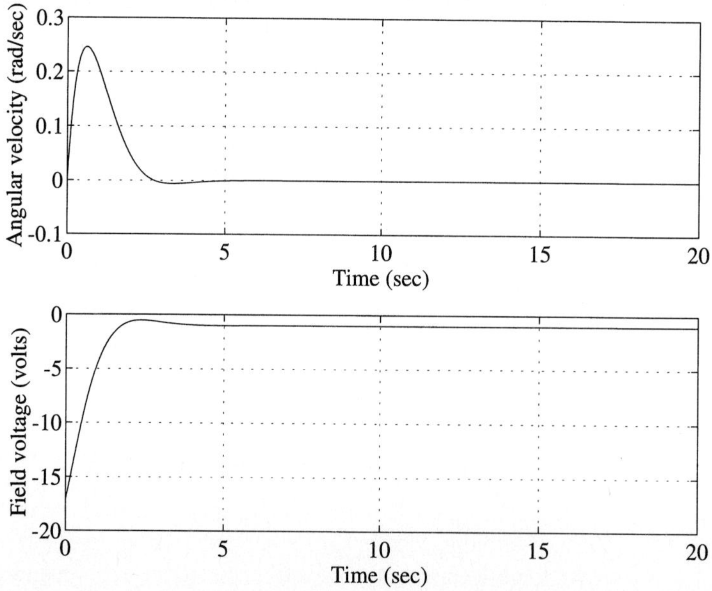

FIGURE 8.18 Feedforward disturbance rejection applied to Example 8.6

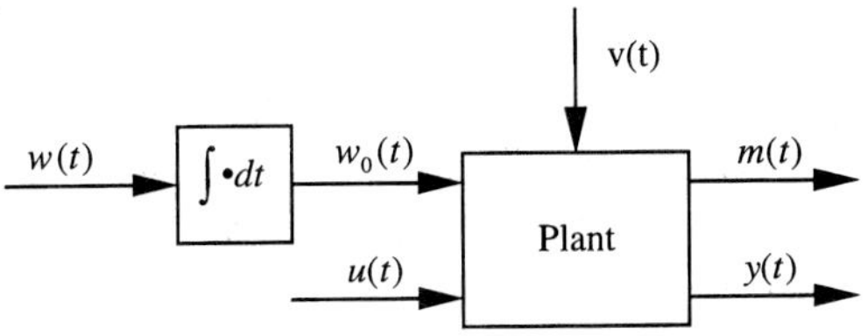

FIGURE 8.19 The augmented plant for estimating a constant disturbance

that the integral in this figure has a noise input that allows the disturbance input to change slowly. This input is included to keep the steady-state Kalman gain from being zero. Additional plant noise inputs can also be included in this model, but fictitious noise into the integrator is always required when using the steady-state Kalman gains. The plant state model used for Kalman filter design is then

$$\begin{bmatrix} \dot{x}(t) \\ \hline \dot{w}_0(t) \end{bmatrix} = \left[\begin{array}{c|c} \mathbf{A} & \mathbf{B}_{w0} \\ \hline \mathbf{0} & \mathbf{0} \end{array}\right] \begin{bmatrix} x(t) \\ \hline w_0(t) \end{bmatrix} + \begin{bmatrix} \mathbf{B}_u \\ \hline \mathbf{0} \end{bmatrix} u(t) + \begin{bmatrix} \mathbf{0} \\ \hline \mathbf{I} \end{bmatrix} w(t);$$

$$m(t) = [\mathbf{C}_m \mid 0] \begin{bmatrix} x(t) \\ \hline w_0(t) \end{bmatrix} + v(t).$$

The disturbance-canceling controller that utilizes noisy, partial state measurements is then

$$\boxed{u(t) = -\mathbf{K}\hat{x}(t) - \mathbf{K}_w \hat{w}(t),} \tag{8.40}$$

where the feedforward gain is given by (8.39). Note that the addition of the Kalman filter does not change this feedforward gain.

8.5.2 Integral Control

Integral control can be applied to mitigate the effects of constant disturbance inputs. A block diagram of an LQG integral control system is given in Figure 8.20. Integral control also allows tracking of constant reference inputs, so a nonzero set point is included in this block diagram.

The Kalman filter in the integral control system estimates both the plant state and the disturbance input. The estimated disturbance input is not used in generating the control, since the integral of the error is available to cancel the effects of the disturbance. The disturbance must still be estimated by the Kalman filter, since a bias in the estimation error results when the disturbance is not included. A steady-state tracking error is caused by bias in the reference output estimate, since the integral control drives the estimated reference output (not the actual reference output) to the reference input.

The LQG integral control yields a zero steady-state tracking error in the presence of a constant disturbance provided that the dimension of the control input is greater than or equal to both the dimension of the disturbance input and the dimension of the reference output. The reference output must also be physically consistent; that is, it can be achieved with a constant state.

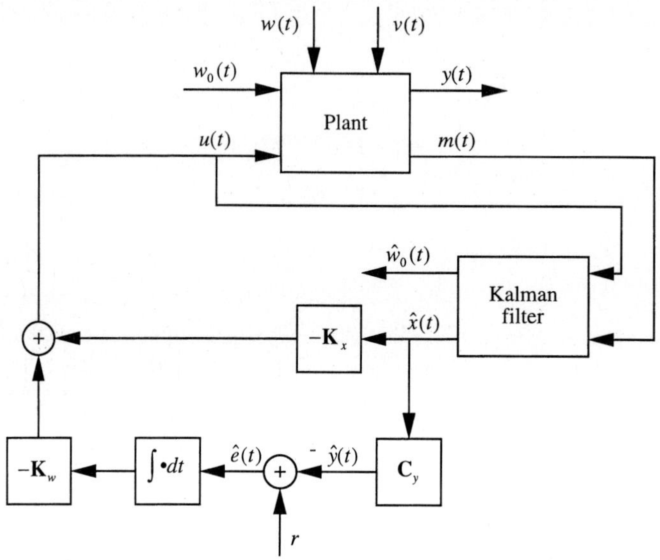

FIGURE 8.20 LQG integral control

EXAMPLE 8.6 CONTINUED An integral LQG controller is designed to maintain the motor velocity at the set point. The control system is implemented using a noisy tachometer measurement for input to the controller, and the load torque is assumed unknown. The state model of this plant is

$$\begin{bmatrix} \dot{\omega}(t) \\ \dot{\tau}_M(t) \end{bmatrix} = \begin{bmatrix} -1 & 1 \\ 0 & -0.1 \end{bmatrix} \begin{bmatrix} \omega(t) \\ \tau_M(t) \end{bmatrix} + \begin{bmatrix} 0 \\ 0.1 \end{bmatrix} u(t) + \begin{bmatrix} 0 \\ 1 \end{bmatrix} w(t) + \begin{bmatrix} 1 \\ 0 \end{bmatrix} \tau_L(t);$$

$$m(t) = [1 \quad 0] \begin{bmatrix} \omega(t) \\ \tau_M(t) \end{bmatrix} + v(t).$$

The output to be controlled is $\omega(t)$.

A steady-state Kalman filter is designed to estimate the augmented state. The augmented state model is

$$\begin{bmatrix} \dot{\omega}(t) \\ \dot{\tau}_M(t) \\ \dot{\tau}_L(t) \end{bmatrix} = \begin{bmatrix} -1 & 1 & 1 \\ 0 & -0.1 & 0 \\ 0 & 0 & 0 \end{bmatrix} \begin{bmatrix} \omega(t) \\ \tau_M(t) \\ \tau_L(t) \end{bmatrix} + \begin{bmatrix} 0 \\ 0.1 \\ 0 \end{bmatrix} u(t) + \begin{bmatrix} 0 & 0 \\ 1 & 0 \\ 0 & 1 \end{bmatrix} \begin{bmatrix} w(t) \\ w_T(t) \end{bmatrix};$$

$$m(t) = [1 \quad 0 \quad 0] \begin{bmatrix} \omega(t) \\ \tau_M(t) \\ \tau_L(t) \end{bmatrix} + v(t),$$

where w_T is the white noise that drives changes in the load torque. This noise must be nonzero for the steady-state load torque estimate to converge. The noise spectral densities are

$$\mathbf{S}_{\left[\begin{smallmatrix} w \\ w_T \end{smallmatrix}\right]} = \begin{bmatrix} 0.001 & 0 \\ 0 & 0.01 \end{bmatrix}; \ \mathbf{S}_v = 0.001,$$

where the spectral density of $w_T(t)$ was selected via trial and error.

A steady-state LQR integral state feedback is generated. The augmented state equation is

$$\begin{bmatrix} \dot{\omega}(t) \\ \dot{\tau}_M(t) \\ \dot{e}_I(t) \end{bmatrix} = \begin{bmatrix} -1 & 1 & 0 \\ 0 & -0.1 & 0 \\ -1 & 0 & 0 \end{bmatrix} \begin{bmatrix} \omega(t) \\ \tau_M(t) \\ e_I(t) \end{bmatrix} + \begin{bmatrix} 0 \\ 0.1 \\ 0 \end{bmatrix} u(t),$$

and the cost function is

$$J = \int_0^\infty e_I^2(t) + 0.0001u^2(t)dt.$$

The controller uses the original plant state estimates provided by the integral Kalman filter in the integral LQR. The resulting controller is fourth-order (an order is added for the integral Kalman and another for the integral LQR) with the following state model:

$$\begin{bmatrix} \dot{\hat{\omega}}(t) \\ \dot{\hat{\tau}}_M(t) \\ \dot{\hat{\tau}}_L(t) \\ \dot{\hat{e}}_I(t) \end{bmatrix} = \begin{bmatrix} -2.8 & 1 & 1 & 0 \\ -6.1 & -3.5 & 0 & 10 \\ -3.2 & 0 & 0 & 0 \\ -1 & 0 & 0 & 0 \end{bmatrix} \begin{bmatrix} \hat{\omega}(t) \\ \hat{\tau}_M(t) \\ \hat{\tau}_L(t) \\ \hat{e}_I(t) \end{bmatrix} + \begin{bmatrix} 1.8 \\ 0.1 \\ 3.2 \\ 0 \end{bmatrix} m(t);$$

$$u(t) = [-60 \quad -34 \quad 0 \quad 100] \begin{bmatrix} \hat{\omega}(t) \\ \hat{\tau}_M(t) \\ \hat{\tau}_L(t) \\ \hat{e}_I(t) \end{bmatrix}.$$

The closed-loop system is simulated with a unit step load torque, a set point of zero, measurement noise with the given spectral density, and plant noise with the spectral density $S_w = 0.001$. The additional plant noise term going into the load torque is not included, since this is not real and is only added during Kalman filter design to allow estimation of the disturbance torque. The simulation results are given in Figure 8.21.

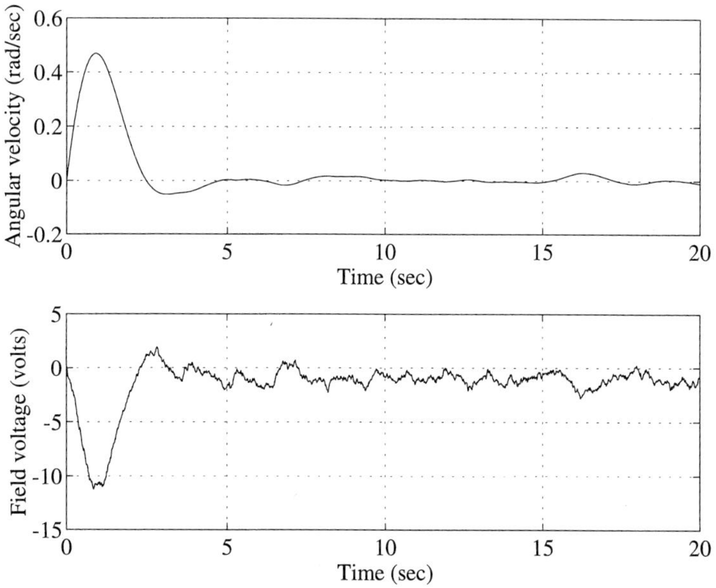

FIGURE 8.21 Integral LQG control applied to Example 8.6

The maximum field voltage is 12 volts, and the angular velocity is maintained within 0.5 rad/sec of zero. In addition, the average steady-state error is zero. The actual angular velocity varies around zero due to the noise inputs. ◆

8.6 Frequency-Shaped Control via LQG Methods

Trail-and-error cost function selection, loop transfer recovery, integral control for tracking and disturbance rejection, and the use of filters for tracking and disturbance rejection are all methods of shaping the frequency response of the controller and the loop transfer function. A much wider range of "tricks" can be applied to modify the frequency response of the LQG controller.

For example, consider a motor with a flexible shaft that has a resonant frequency within the controller bandwidth. In this case, it may be desirable to put a notch filter in the feedback loop so that the control does not excite motor shaft vibrations. This can be accomplished by appending a bandpass filter (with a passband around the resonant frequency) to the control, as shown in Figure 8.22. The cost function used to find the state feedback gains should include a large penalty on the output of this filter, that is, put a large penalty on the use of controls in the filter passband. When designing the Kalman filter for this controller, only the states of the original plant are estimated. The filter states x_f can be generated by inputting the known control input to the filter.

This example illustrates an additional use of shaping filters for modifying the frequency response of the basic LQG controller. The use of shaping filters in LQG design is fairly intuitive, but care must be taken to make sure that the LQR and Kalman filter problems are well posed. For example, the Kalman filter above was not required to estimate the states of the filter appended to the control, since the control is known. The development of other methods of modifying the controller frequency response are left to the imagination of the control system designer.

8.7 Equivalence of LQG and $\mathcal{H}_2$ Optimal Control

The steady-state, linear quadratic Gaussian optimal control problem is equivalent to an $\mathcal{H}_2$ optimization problem. This $\mathcal{H}_2$ optimization problem is formally stated as follows,

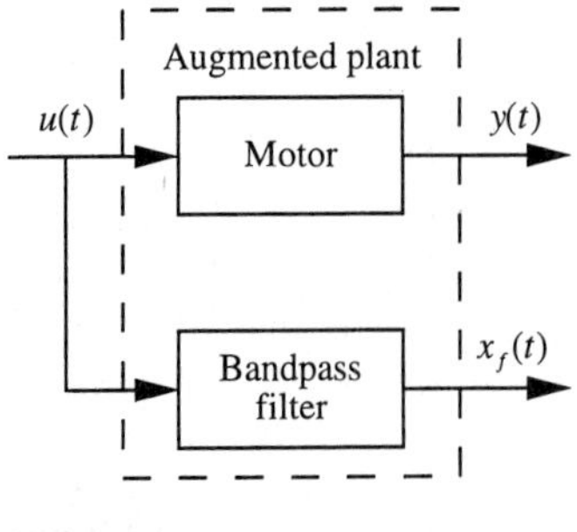

FIGURE 8.22 Adding a notch filter to the controller

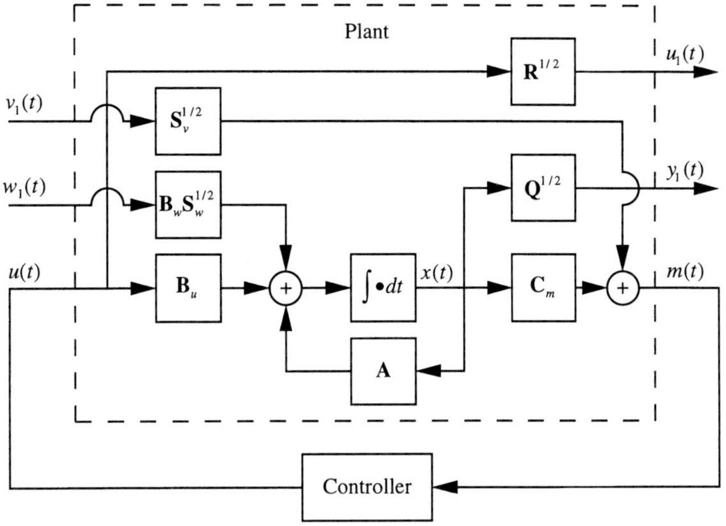

FIGURE 8.23 The LQG as an $\mathcal{H}_2$ optimization problem

given the plant with normalized disturbance inputs and reference outputs (see Figure 8.23):

$$\dot{x}(t) = \mathbf{A}x(t) + [\mathbf{B}_u \;\vdots\; \mathbf{B}_w\mathbf{S}_w^{1/2} \quad \mathbf{0}]\begin{bmatrix} u(t) \\ \hdashline w_1(t) \\ v_1(t) \end{bmatrix}; \tag{8.41a}$$

$$\begin{bmatrix} m(t) \\ \hdashline y_1(t) \\ u_1(t) \end{bmatrix} = \begin{bmatrix} \mathbf{C}_m \\ \hdashline \mathbf{Q}^{1/2} \\ \mathbf{0} \end{bmatrix} x(t) + \begin{bmatrix} \mathbf{0} & \mathbf{0} & \mathbf{S}_v^{1/2} \\ \hdashline \mathbf{0} & \mathbf{0} & \mathbf{0} \\ \mathbf{R}^{1/2} & \mathbf{0} & \mathbf{0} \end{bmatrix}\begin{bmatrix} u(t) \\ \hdashline w_1(t) \\ v_1(t) \end{bmatrix}, \tag{8.41b}$$

find a feedback controller that internally stabilizes the closed-loop system and minimizes the closed-loop system 2-norm:

$$J_2 = \|\mathbf{G}_{cl}\|_2. \tag{8.42}$$

To demonstrate this equivalence, the LQG cost function can be written in terms of the closed-loop system 2-norm:

$$\begin{aligned} J &= E[x^T(\infty)\mathbf{Q}x(\infty) + u^T(\infty)\mathbf{R}u(\infty)] \\ &= E\left\{[(\mathbf{Q}^{1/2}x(\infty))^T \;\vdots\; (\mathbf{R}^{1/2}u(\infty))^T]\begin{bmatrix} \mathbf{Q}^{1/2}x(\infty) \\ \hdashline \mathbf{R}^{1/2}u(\infty) \end{bmatrix}\right\} = \|\mathbf{G}_{cl}\|_2^2. \end{aligned}$$

Since the squaring operation is monotonic, minimizing the LQG cost function is equivalent to minimizing the closed-loop system 2-norm. The $\mathcal{H}_2$ optimization problem requirement that the closed-loop system be internally stable is also satisfied by the LQG controller provided the conditions for the existence of the associated steady-state LQR and Kalman filter solutions are satisfied.

The $\mathcal{H}_2$ formulation of the LQG problem indicates the possibility of generating alternative controller designs by using different system norms. This idea is pursued in the next chapter, where ∞-norm optimization is presented.

8.8 Summary

The linear quadratic Gaussian optimal control problem is defined and solved. The LQG problem is to find a feedback controller that utilizes noisy (white noise), partial state measurements to generate the control input. Further, the plant is subject to white noise disturbance inputs, the state initial condition is random, and a quadratic cost function must be minimized. The solution of this optimal control problem can be broken into two parts. The first part is to design a stochastic regulator (or LQR), where all the states are assumed to be measured exactly; The second part is to generate a Kalman filter for estimating the states from the noisy, partial state measurements. The use of the Kalman estimates in the stochastic regulator is then optimal in terms of the LQG cost function. This separation between controller design (assuming full state information) and estimator design is known as the stochastic separation principle.

A closed-loop state model is generated for use in evaluating LQG performance. The state in this model consists of the plant state and the estimation error. Note that this is not the only possible parameterization of the closed-loop system. Integral and differential equation formulas for the closed-loop state covariance matrix are given. This closed-loop state covariance matrix includes the covariance matrices of the plant state and the estimation error. In addition, formulas for the covariance matrices of the reference output and the control, in terms of the closed-loop state covariance matrix, are given. A formula for the cost in terms of the closed-loop state covariance matrix is also provided. All of these formulas are applicable to any state estimate feedback system, and can be used to compare the performance of the optimal controller with suboptimal controllers. In particular, comparing the optimal controller to the suboptimal steady-state controller is useful in determining when it is reasonable to use the steady-state controller.

The LQG optimal control is nearly time-invariant far from the initial and final times. This time-invariant controller is called the steady-state LQG controller and is optimal in terms of an appropriate cost function. Formulas are given for the cost and the steady-state covariance matrices of the state, estimation error, reference output, and control when using the steady-state LQG controller. In addition, the poles of the closed-loop system are shown to be the union of the poles of the LQR and the poles of the Kalman filter. This fact is often referred to as the deterministic separation principle.

The robustness of the LQR and the Kalman filter subject to input- and output-multiplicative perturbations, respectively, is very good. Unfortunately, it is shown that this robustness may be destroyed when combining the LQR and the Kalman filter. In general, there are no guarantees on the robustness of the LQG optimal control, and the robustness of each design should be carefully checked.

Loop transfer recovery is a procedure that generates a family of controllers with limiting controllers that recover the robustness of the LQR (or the Kalman filter). This family of controllers is generated by adding fictitious noise to the plant model used in Kalman filter design (or by modifying the LQR cost function). The resulting controllers are suboptimal in terms of the original cost function. The selection of which controller to use then represents a trade-off between robustness and performance. The fictitious noise used in LTR can also be frequency-shaped to allow loop transfer function recovery over a specified frequency range. Loop transfer recovery is guaranteed to recover the LQR robustness provided that the plant is minimal phase. While not guaranteed to work, LTR may also be applicable when the plant has right half-plane zeros (i.e., is not minimal phase).

A number of modifications are presented that allow the LQR and the LQG methodologies to be applied to tracking system design. Constant reference inputs can be tracked via coordinate translation, feedforward control, and integral control. Coordinate translation is accomplished by feeding back the error between the actual state and the desired state. The procedure works perfectly provided that the plant contains an integral and the desired state is selected to be in the null space of the plant state matrix. Feedforward control can also be applied to track constant reference inputs. Feedforward control allows the tracking of arbitrary reference inputs provided that the number of control inputs is greater than or equal to the number of reference inputs to be tracked. Unfortunately, feedforward control is often not very robust; it should be used with caution, and only after performing a thorough robustness analysis. Integral control can also be used to track constant inputs. In this case, the integral of the tracking error is appended to the plant before generating the state feedback gains. The resulting control uses integral feedback and can perfectly track arbitrary reference inputs. Similar modifications can be made to the LQG design to allow the tracking of time-varying reference inputs, and also for use with the time-varying LQG gains.

The LQG control system is optimized to reject white noise disturbances. The LQG can be modified to allow the rejection of constant disturbances via feedforward and integral control. The selection of a feedforward control gain is accomplished in a manner similar to that used for tracking system design. Again, this technique may suffer from poor robustness. Constant disturbance rejection via integral control requires the use of an integral Kalman filter and also integral state feedback. This integral LQG can be used simultaneously for both disturbance rejection and tracking. Similar modifications can be made to the LQG design to allow the tracking of time-varying reference inputs, and also for use with the time-varying LQG gains.

This chapter concludes by demonstrating that the LQG optimal control problem is equivalent to an $\mathcal{H}_2$ optimal control problem. The $\mathcal{H}_2$ optimal control problem is as follows: Find a feedback controller that internally stabilizes the closed-loop system and also minimizes a closed-loop system 2-norm. This formulation of the LQG problem suggests the possibility of using other system norms for optimal control.

The use of the stochastic separation principle in LQG design is applicable in both continuous time and discrete time. The formulas for the discrete-time LQR and Kalman filters are very similar to their continuous-time counterparts, and have already been given in previous chapters. The evaluation of performance in discrete time is also similar to that in continuous time. Further, the deterministic separation principle is directly applicable in discrete time. Loop transfer recovery can also be performed on the discrete-time LQG in a manner identical to that used for the continuous-time LQG.

Caution must be exercised in this application, since the discrete-time LQR and Kalman filter do not have the guaranteed robustness margins of their continuous-time analogs. Still, good results are usually obtained, especially when the sampling time is sufficiently short. The tricks presented for tracking system design and disturbance rejection are also applicable in discrete time provided that the shaping filters used are discretized. The fundamental equations presented in this chapter are summarized for both continuous-time and discrete-time systems in Table 8.2.

TABLE 8.2 Summary of Formulas for Both Continuous-Time and Discrete-Time Systems

<table>
<tr><th>Formula</th><th>Continuous-Time</th><th>Discrete-Time</th></tr>
<tr><td>The plant model</td><td>$\dot{x}(t) = \mathbf{A}x(t) + \mathbf{B}_u u(t) + \mathbf{B}_w w(t)$
$m(t) = \mathbf{C}_m x(t) + v(t)$</td><td>$x(k+1) = \mathbf{\Phi}x(k) + \mathbf{\Gamma}_u u(k) + \mathbf{\Gamma}_w w(k)$
$m(k) = \mathbf{C}_m x(k) + v(k)$</td></tr>
<tr><td>The noise model</td><td>$E[w(t)w^T(t+\tau)] = \mathbf{S}_w\delta(\tau)$
$E[v(t)v^T(t+\tau)] = \mathbf{S}_v\delta(\tau)$
$E[v(t)w^T(t+\tau)] = \mathbf{0}$</td><td>$E[w(k)w^T(k+p)] = \mathbf{\Sigma}_w\delta(p) = \frac{\mathbf{S}_w}{T}\delta(p)$
$E[v(k)v^T(k+p)] = \mathbf{\Sigma}_v\delta(p) = \frac{\mathbf{S}_v}{T}\delta(p)$
$E[v(k)w^T(k+p)] = \mathbf{0}$</td></tr>
<tr><td>The cost function</td><td>$J = E\left[\frac{1}{2}x^T(t_f)\mathbf{H}x(t_f) + \frac{1}{2}\int_0^{t_f} x^T(t)\mathbf{Q}x(t) + u^T(t)\mathbf{R}u(t)dt\right]$</td><td>$J = E\left[\frac{1}{2}x^T(k_f)\mathbf{H}_d x(k_f) + \frac{1}{2}\sum_{k=0}^{k_f-1} x^T(k)\mathbf{Q}_d x(k) + u^T(k)\mathbf{R}_d u(k)\right]$</td></tr>
<tr><td>The weighting matrices</td><td>$\mathbf{H}\,;\mathbf{Q}\,;\mathbf{R}$</td><td>$\mathbf{H}_d = \mathbf{H}\,;\mathbf{Q}_d = \mathbf{Q}T;\ \mathbf{R}_d = \mathbf{R}T$</td></tr>
<tr><td>The optimal control</td><td>$u(t) = -\mathbf{K}(t)\hat{x}(t)$
where $\hat{x}$ is the Kalman state estimate</td><td>$u(k) = -\mathbf{K}(k)\hat{x}(k)$
where $\hat{x}$ is the Kalman state estimate</td></tr>
<tr><td>The closed-loop state model</td><td>$\left[\begin{array}{c}\dot{x}(t)\\ \hline \dot{e}(t)\end{array}\right] = \mathbf{A}_{cl}(t)\left[\begin{array}{c}x(t)\\ \hline e(t)\end{array}\right] + \mathbf{B}_{cl}(t)\left[\begin{array}{c}w(t)\\ \hline v(t)\end{array}\right]$
$\mathbf{A}_{cl}(t) = \left[\begin{array}{c|c}\mathbf{A} - \mathbf{B}_u\mathbf{K}(t) & \mathbf{B}_u\mathbf{K}(t)\\ \hline \mathbf{0} & \mathbf{A} - \mathbf{G}(t)\mathbf{C}_m\end{array}\right]$
$\mathbf{B}_{cl}(t) = \left[\begin{array}{c|c}\mathbf{B}_w & \mathbf{0}\\ \hline \mathbf{B}_w & -\mathbf{G}(t)\end{array}\right]$</td><td>$\left[\begin{array}{c}x(k+1)\\ \hline e(k+1)\end{array}\right] = \mathbf{\Phi}_{cl}(k)\left[\begin{array}{c}x(k)\\ \hline e(k)\end{array}\right] + \mathbf{\Gamma}_{cl}(t)\left[\begin{array}{c}w(k)\\ \hline v(k)\end{array}\right]$
$\mathbf{\Phi}_{cl}(k) = \left[\begin{array}{c|c}\mathbf{\Phi} - \mathbf{\Gamma}_u\mathbf{K}(k) & \mathbf{\Gamma}_u\mathbf{K}(k)\\ \hline \mathbf{0} & \mathbf{\Phi} - \mathbf{G}_d(k)\mathbf{C}_m\end{array}\right]$
$\mathbf{\Gamma}_{cl}(k) = \left[\begin{array}{c|c}\mathbf{\Gamma}_w & \mathbf{0}\\ \hline \mathbf{\Gamma}_w & -\mathbf{G}_d(k)\end{array}\right]$</td></tr>
<tr><td>The closed-loop state covariance matrix</td><td>$\dot{\mathbf{\Sigma}}_{\left[\frac{x}{e}\right]}(t) = \mathbf{A}_{cl}(t)\mathbf{\Sigma}_{\left[\frac{x}{e}\right]}(t) + \mathbf{\Sigma}_{\left[\frac{x}{e}\right]}(t)\mathbf{A}_{cl}^T(t) + \mathbf{B}_{cl}(t)\left[\begin{array}{c|c}\mathbf{S}_w & \mathbf{0}\\ \hline \mathbf{0} & \mathbf{S}_v\end{array}\right]\mathbf{B}_{cl}^T(t);$
$\mathbf{\Sigma}_{\left[\frac{x}{e}\right]}(0) = E\left[\begin{array}{c|c}x(0)x^T(0) & x(0)e^T(0)\\ \hline e(0)x^T(0) & e(0)e^T(0)\end{array}\right]$</td><td>$\mathbf{\Sigma}_{\left[\frac{x}{e}\right]}(k+1) = \mathbf{\Phi}_{cl}(k)\mathbf{\Sigma}_{\left[\frac{x}{e}\right]}(k)\mathbf{\Phi}_{cl}^T(k) + \mathbf{\Gamma}_{cl}(k)\left[\begin{array}{c|c}\mathbf{\Sigma}_w & \mathbf{0}\\ \hline \mathbf{0} & \mathbf{\Sigma}_v\end{array}\right]\mathbf{\Gamma}_{cl}^T(k);$
$\mathbf{\Sigma}_{\left[\frac{x}{e}\right]}(0) = E\left[\begin{array}{c|c}x(0)x^T(0) & x(0)e^T(0)\\ \hline e(0)x^T(0) & e(0)e^T(0)\end{array}\right]$</td></tr>
<tr><td>Performance</td><td>$\mathbf{\Sigma}_y(t) = [\mathbf{C}_y \mid \mathbf{0}]\mathbf{\Sigma}_{\left[\frac{x}{e}\right]}(t)\left[\begin{array}{c}\mathbf{C}_y^T\\ \hline \mathbf{0}\end{array}\right] = \mathbf{C}_y\mathbf{\Sigma}_x(t)\mathbf{C}_y^T$
$\mathbf{\Sigma}_u(t) = [-\mathbf{K}(t) \mid \mathbf{K}(t)]\mathbf{\Sigma}_{\left[\frac{x}{e}\right]}(t)\left[\begin{array}{c}-\mathbf{K}^T(t)\\ \hline \mathbf{K}^T(t)\end{array}\right]$</td><td>$\mathbf{\Sigma}_y(k) = [\mathbf{C}_y \mid \mathbf{0}]\mathbf{\Sigma}_{\left[\frac{x}{e}\right]}(k)\left[\begin{array}{c}\mathbf{C}_y^T\\ \hline \mathbf{0}\end{array}\right] = \mathbf{C}_y\mathbf{\Sigma}_x(k)\mathbf{C}_y^T$
$\mathbf{\Sigma}_u(k) = [-\mathbf{K}(k) \mid \mathbf{K}(k)]\mathbf{\Sigma}_{\left[\frac{x}{e}\right]}(k)\left[\begin{array}{c}-\mathbf{K}^T(k)\\ \hline \mathbf{K}^T(k)\end{array}\right]$</td></tr>
<tr><td>Cost</td><td>$J = \frac{1}{2}trace\left\{\mathbf{\Sigma}_{\left[\frac{x}{e}\right]}(t_f)\left[\begin{array}{c|c}\mathbf{H} & \mathbf{0}\\ \hline \mathbf{0} & \mathbf{0}\end{array}\right] + \int_0^{t_f}\mathbf{\Sigma}_{\left[\frac{x}{e}\right]}(t) \times \left[\begin{array}{c|c}\mathbf{Q} + \mathbf{K}^T(t)\mathbf{R}\mathbf{K}(t) & -\mathbf{K}^T(t)\mathbf{R}\mathbf{K}(t)\\ \hline -\mathbf{K}^T(t)\mathbf{R}\mathbf{K}(t) & \mathbf{K}^T(t)\mathbf{R}\mathbf{K}(t)\end{array}\right]dt\right\}$</td><td>$J = \frac{1}{2}trace\left\{\mathbf{\Sigma}_{\left[\frac{x}{e}\right]}(k_f)\left[\begin{array}{c|c}\mathbf{H}_d & \mathbf{0}\\ \hline \mathbf{0} & \mathbf{0}\end{array}\right] + \sum_{k=0}^{k_f}\mathbf{\Sigma}_{\left[\frac{x}{e}\right]}(k) \times \left[\begin{array}{c|c}\mathbf{Q}_d + \mathbf{K}^T(k)\mathbf{R}_d\mathbf{K}(k) & -\mathbf{K}^T(k)\mathbf{R}_d\mathbf{K}(k)\\ \hline -\mathbf{K}^T(k)\mathbf{R}_d\mathbf{K}(k) & \mathbf{K}^T(k)\mathbf{R}_d\mathbf{K}(k)\end{array}\right]\right\}$</td></tr>
</table>

REFERENCES

[1] H. Kwakernaak and R. Sivan, *Linear Optimal Control Systems,* Wiley-Interscience, New York, 1972.

[2] H. Kwakernaak and R. Sivan, "The maximally achievable accuracy of linear optimal regulators and linear optimal filters," IEEE Transactions on Autom. Control, 17, no. 1 (1972): 79–86.

[3] B. D. O. Anderson and J. B. Moore, *Optimal Control: Linear Quadratic Methods,* Prentice-Hall, Englewood Cliffs, NJ, 1990.

[4] J. B. Moore, D. Gangsaas, and J. Blight, "Performance and Robustness Trades in LQG regulator designs," *Proceedings of the 20th IEEE Conference on Decision and Control,* San Diego, CA, Dec. 1981, pp. 1191–199.

Some additional references on linear quadratic Gaussian control follow:

[5] A. E. Bryson, Jr., and Y.-C. Ho, *Applied Optimal Control: Optimization, Estimation, and Control,* Hemisphere Publishing Co., Washington, DC, 1975.

[6] G. E. Franklin, J. D. Powell, and M. L. Workman, *Digital Control of Dynamic Systems,* 2d ed., Addison-Wesley, Reading, MA, 1990.

[7] B. Friedland, *Control System Design: An Introduction to State Space Methods,* McGraw-Hill, NY, 1986.

[8] D. E. Kirk, *Optimal Control Theory: An Introduction,* Prentice-Hall, Englewood Cliffs, NJ, 1970.

[9] F. L. Lewis, *Optimal Control,* John Wiley & Sons, New York, 1986.

[10] J. M. Maciejowski, *Multivariable Feedback Design,* Addison-Wesley, Reading, MA, 1989.

[11] A. P. Sage and C. C. White, III, *Optimum Systems Control,* 2d ed., Prentice-Hall, Englewood Cliffs, NJ, 1977.

EXERCISES

8.1 For a time-varying LQG control system, do the following.

a. Generate a state model for this system using the following state:

$$\begin{bmatrix} x(t) \\ \hline \hat{x}(t) \end{bmatrix}.$$

b. Generate an expression for the cost function of the LQG control system in terms of the covariance matrix of this state.

8.2 The plant

$$\dot{x}(t) = \begin{bmatrix} 0 & 1 \\ 1 & -1 \end{bmatrix} x(t) + \begin{bmatrix} 0 \\ 1 \end{bmatrix} u(t) + \begin{bmatrix} 0 \\ 1 \end{bmatrix} w(t) \; ; \; m(t) = [1 \quad 0]x(t) + v(t)$$

is controlled with the following observer feedback:

$$\dot{\hat{x}}(t) = \begin{bmatrix} 0 & 1 \\ 1 & -1 \end{bmatrix} \hat{x}(t) + \begin{bmatrix} 0 \\ 1 \end{bmatrix} u(t) + \begin{bmatrix} 15 \\ 114 \end{bmatrix} \{m(t) - [1 \quad 0]\hat{x}(t)\} \; ; \; u(t) = -[9 \quad 3]\hat{x}(t).$$

a. Compute the steady-state covariance matrix of the closed-loop state, given $S_w = 10$, and $S_v = 0.01$. What is the steady-state standard deviation of $x_1(t)$?

b. For this system, compute the cost:

$$J = E[x_1^2(\infty) + 0.001u^2(\infty)].$$

c. Generate an LQG controller for this system, and compute the cost. How does this cost compare to the cost associated with the observer feedback system?

8.3 Generate an expression for the LQR gain of a SISO plant, with the following cost function:

$$J = E[x^T(\infty)(\mathbf{Q} + \rho \mathbf{C}_m^T \mathbf{C}_m)x(\infty) + u^T(\infty)Ru(\infty)],$$

in the limit as ρ approaches infinity. The limiting gain should be a function of the ratio (R/ρ). Using this gain in an LQG controller, verify that the loop transfer function (broken at the output, and with the sign of the negative feedback removed) reduces to

$$G_L(s) = \mathbf{C}_m(sI - A)^{-1}\mathbf{G},$$

which is the loop transfer function of a Kalman filter.

8.4 Generate a state model for the controller in Figure 8.20. Note that the input to the controller is $m(t)$ and the output of the controller is $u(t)$.

8.5 Given the plant

$$\dot{x}(t) = \begin{bmatrix} 0 & 1 \\ 1 & -1 \end{bmatrix} x(t) + \begin{bmatrix} 0 \\ 1 \end{bmatrix} u(t),$$

do the following.

a. Generate the optimal control for this plant with the following cost function:

$$J = \int_0^{\infty} x_1^2(t) + 0.1u^2(t)dt,$$

assuming all of the states are measured exactly.

b. Using the optimal feedback gains just found, compute the feedforward control gain required to drive $x_1(t)$ to 4 in the steady state.

c. Generate an integral state feedback to drive $x_1(t)$ to 4 in the steady state.

8.6 Show that the feedforward gain required to drive the reference output to a desired value when using state feedback equals the feedforward gain required when using the estimated states in place of the actual states. *Hint:* Generate expressions for the steady-state gains of the closed-loop LQR and the closed-loop LQG, and compare the results.

8.7 Generate a steady-state LQG integral control system for the plant in Exercise 8.5. Assume that the plant noise is input with the control input, and the measured output is $m(t) = x_1(t) + v(t)$. The spectral densities of the plant and measurement noises are $S_w = 1$ and $S_v = 10$, respectively. Use fictitious noise with the spectral density $S_{w_0} = 0.001$ as an input to the integrator that generates the nearly constant disturbance. Compute the poles of the closed-loop system, and show that these poles equal the poles of the Kalman filter plus the poles of the state feedback.

8.8 Given the system

$$\dot{x}(t) = \begin{bmatrix} 1 & 2 \\ 1 & 1 \end{bmatrix} x(t) + \begin{bmatrix} 0 \\ 1 \end{bmatrix} u(t) + \begin{bmatrix} 1 \\ 1 \end{bmatrix} w(t) + \begin{bmatrix} 1 \\ 1 \end{bmatrix} w_0(t);$$

$$m(t) = [1 \quad 1]x(t) + v(t),$$

where $S_w = 1$, $S_v = 2$, and the cost function is

$$J = E[x_1^2(t) + 0.1u^2(t)],$$

do the following.

a. Generate a state model for the optimal controller (controller only, not the closed-loop system) assuming that $w_0(t)$ is zero.

b. Generate a feedforward gain to remove the effects of $w_0(t) = 3$.

8.9 For the system in Exercise 8.8, with $y(t) = x_1(t)$, do the following.

a. Generate a controller using integral state feedback and the standard Kalman filter. Generate the loop transfer function for this control system (break the loop at the input to the plant). Does this transfer function contain an integral?

b. Generate a controller using ordinary state feedback and an integral Kalman filter. Generate the loop transfer function for this control system (break the loop at the input to the plant). Does this transfer function contain an integral?

c. Generate a controller using integral state feedback and an integral Kalman filter. Generate the loop transfer function for this control system (break the loop at the input to the plant). Does this transfer function contain an integral?

8.10 Show that the integral Kalman filter for estimating the plant state, generated by appending an integral disturbance model to the plant, can be written as the ordinary Kalman filter with an integral gain.

COMPUTER EXERCISES

8.1 Optimal Control of the Altitude of an Unmanned Surveillance Vehicle

Design a system for controlling the altitude of an unmanned surveillance vehicle. The available information consists of noisy radar altimeter measurements, which can be filtered (using the Kalman filter) to yield estimates of the vehicle's altitude, vertical velocity, and the vertical acceleration generated by the engine. These estimates will be fed back to the throttle to yield a desired altitude near 1000 feet.

A block diagram for the vertical motion of the vehicle is given in Figure P8.1, where $h(t)$ is the height of the vehicle, $a(t)$ is the vehicle's acceleration due to the engine, $u(t)$ is the throttle input, $\tau = 2$ is a time constant indicating that acceleration does not change instantaneously, $w_1(t)$ is a random acceleration (disturbance input), and $w_2(t)$ is a random throttle component (disturbance input). The disturbance $[w_1(t) + 10]$ is a white, Gaussian random process with spectral density $S_{w_1} = S_w$. Note that the disturbance $w_1(t)$ has a mean value of -10. The disturbance $w_2(t)$ is a white, Gaussian random process with spectral density $S_{w_2} = 3S_w$. The random processes $w_1(t)$ and $w_2(t)$ are uncorrelated. The radar altimeter generates measurements as follows:

$$y(t) = h(t) + v(t),$$

where $v(t)$ is a Gaussian random signal with a mean of zero, a variance σ_v^2, and a correlation time of 0.1 seconds. This random signal can be approximated by white noise.

a. Design a control system to minimize the following cost function:

$$J = E\left[[h(t) - 1000 \quad \dot{h}(t) \quad a(t)]\mathbf{Q}\begin{bmatrix} h(t) - 1000 \\ \dot{h}(t) \\ a(t) \end{bmatrix} + Ru^2(t) \right],$$

with the baseline parameters

$$\mathbf{Q} = \begin{bmatrix} 1 & 0 & 0 \\ 0 & 0 & 0 \\ 0 & 0 & 0 \end{bmatrix} ; R = 200 \; ; S_w = 200 \; ; \sigma_v^2 = 400.$$

Add a reference input to cancel the effects of the disturbance on the vehicle height. Simulate the closed-loop system and plot the plant state, the Kalman filter estimates of the

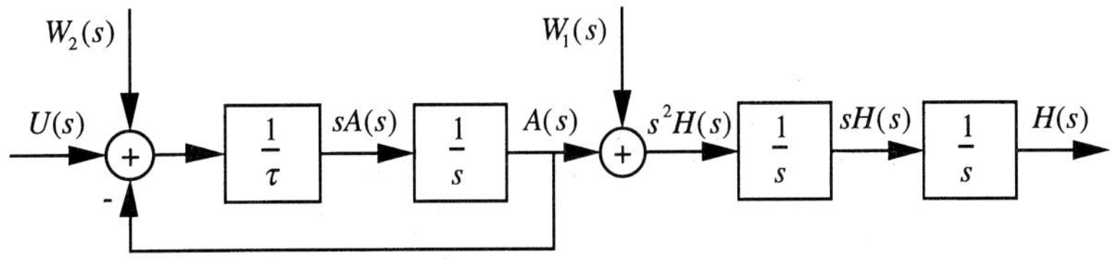

FIGURE P8.1 Block diagram for Computer Exercise 8.1

plant state, and the control input. Design additional controllers with the baseline parameters modified as follows:

i. $$R = 2000.$$

ii. $$\mathbf{Q} = \begin{bmatrix} 10 & 0 & 0 \\ 0 & 0 & 0 \\ 0 & 0 & 0 \end{bmatrix}.$$

iii. $$\mathbf{Q} = \begin{bmatrix} 1 & 0 & 0 \\ 0 & 100 & 0 \\ 0 & 0 & 0 \end{bmatrix}.$$

iv. $$S_w = 2000.$$

v. $$\sigma_v^2 = 4000.$$

Simulate the closed-loop system for each of these controllers and plot $h(t)$ and the estimate of $h(t)$. Comment on the effect of changing these parameters on the plots.

b. Design a control system to minimize the following finite-time cost function:

$$J = E\left[\int_0^{20}\left\{[h(t) - 1000 \quad \dot{h}(t) \quad a(t)]\mathbf{Q}\begin{bmatrix} h(t) - 1000 \\ \dot{h}(t) \\ a(t) \end{bmatrix} + Ru^2(t)\right\}dt\right],$$

where $\mathbf{Q}, R, S_w$, and σ_v^2 are as given in the baseline parameter set. Again, add a reference input to cancel the effects of the disturbance on the vehicle height. Simulate the closed-loop system and plot the plant state, the Kalman filter estimates of the plant state, the control input, the Kalman gains, and the control gains for this design. Generate the cost, the standard deviation of $h(t)$, the standard deviation of $u(t)$, and the standard deviation of $[h(t) - \hat{h}(t)]$ for this design. Use the steady-state baseline controller and reevaluate these quantities (for finite-time control). Compare the results obtained with the steady-state design to those obtained with the time-varying design.

8.2 Linear Quadratic Gaussian/Loop Transfer Recovery Control System Design
The plant consists of a simple servo in parallel with a vibrational mode.[7] This plant is subject to a colored noise disturbance, which is modeled as the output of a filter subjected to white noise. The block diagram of the augmented plant is shown in Figure P8.2. This augmented plant can also be described by the following state model:

$$\dot{x}(t) = \begin{bmatrix} -1 & 0 & 0 & 1 \\ 0 & 0 & 1 & 0 \\ 0 & -100 & -0.2 & 0 \\ 0 & 0 & 0 & -1 \end{bmatrix} x(t) + \begin{bmatrix} 1 \\ 0 \\ 100 \\ 0 \end{bmatrix} u(t) + \begin{bmatrix} 0 \\ 0 \\ 0 \\ .45 \end{bmatrix} w(t);$$

$$m(t) = [1 \quad 10 \quad 0 \quad 1]x(t) + v(t),$$

where

$$E[w(t)] = E[v(t)] = 0 \; ; \; E[w(t)v(\tau)] = 0;$$

$$E[w(t)w(\tau)] = \delta(t - \tau) \; ; \; E[v(t)v(\tau)] = 0.01\delta(t - \tau).$$

a. Design a control system for the above plant that minimizes the following performance function:

$$J = E[4x_1^2(\infty) + u^2(\infty)].$$

Generate a Nyquist plot for the resulting system. What are the phase and gain margins of this system. Find the cost and the steady-state standard deviations of $x_1(t)$ and the control.

[7] This plant is drawn from [4].

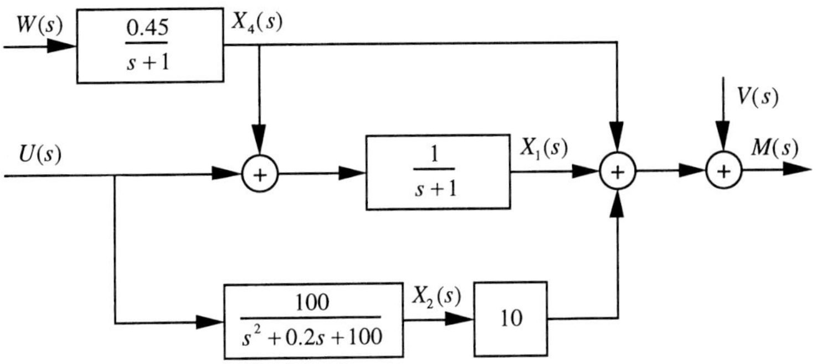

FIGURE P8.2 Block diagram for Computer Exercise 8.2

b. Apply loop transfer recovery to improve the phase and gain margins of the baseline design (use three different levels of fictitious noise). For each level of fictitious noise, generate a Nyquist plot for the resulting system; compute the phase and gain margins of this system; and evaluate the cost and the steady-state standard deviations of $x_1(t)$ and the control. Comment on the effects of the fictitious noise on the control system robustness, the cost, and the steady-state standard deviations of $x_1(t)$ and the control. Specify which design you consider to be best.

8.3 Integral Control of Blood Gases During Extracorporeal Support

Design a feedback system for controlling the partial pressure of oxygen and the partial pressure of carbon dioxide during extracorporeal support. This system should be capable of setting these partial pressures at any desired value.

The state model of the oxygenator is

$$\begin{bmatrix}\dot{x}_1(t)\\ \dot{x}_2(t)\\ \dot{x}_3(t)\\ \dot{x}_4(t)\end{bmatrix} = \begin{bmatrix}-10 & 0 & 0 & 0\\ 0 & -10 & 0 & 0\\ 6 & -3 & -5 & 0\\ 0 & 0.5 & 0 & -5\end{bmatrix}\begin{bmatrix}x_1(t)\\ x_2(t)\\ x_3(t)\\ x_4(t)\end{bmatrix} + \begin{bmatrix}10 & 0\\ 0 & 10\\ 0 & 0\\ 0 & 0\end{bmatrix}\begin{bmatrix}u_1(t)\\ u_2(t)\end{bmatrix} + \begin{bmatrix}10 & 0\\ 0 & 10\\ 0 & 0\\ 0 & 0\end{bmatrix}\begin{bmatrix}w_1(t)\\ w_2(t)\end{bmatrix};$$

$$\begin{bmatrix}m_1(t)\\ m_2(t)\end{bmatrix} = \begin{bmatrix}0 & 0 & 1 & 0\\ 0 & 0 & 0 & 1\end{bmatrix}\begin{bmatrix}x_1(t)\\ x_2(t)\\ x_3(t)\\ x_4(t)\end{bmatrix} + \begin{bmatrix}v_1(t)\\ v_2(t)\end{bmatrix},$$

where $x_1(t)$ is the flow rate of oxygen, $x_2(t)$ is the flow rate of carbon dioxide, $x_3(t)$ is the arterial partial pressure of oxygen, $x_4(t)$ is the arterial partial pressure of carbon dioxide, $u_1(t)$ is the commanded oxygen flow rate, $u_2(t)$ is the commanded carbon dioxide flow rate, $w_1(t)$ is an error in the oxygen valve position, and $w_2(t)$ is an error in the carbon dioxide valve position. The outputs of interest are $x_3(t)$ and $x_4(t)$:

$$\begin{bmatrix}y_1(t)\\ y_2(t)\end{bmatrix} = \begin{bmatrix}0 & 0 & 1 & 0\\ 0 & 0 & 0 & 1\end{bmatrix}\begin{bmatrix}x_1(t)\\ x_2(t)\\ x_3(t)\\ x_4(t)\end{bmatrix}.$$

The spectral densities of the plant noise and measurement noise are

$$\mathbf{S}_w = \begin{bmatrix}0.02 & 0\\ 0 & 0.01\end{bmatrix}; \ \mathbf{S}_v = \begin{bmatrix}10^{-4} & 0\\ 0 & 10^{-5}\end{bmatrix},$$

respectively.

a. Design an LQG control system to minimize the following cost function:

$$J = E\left[y^T(\infty)\begin{bmatrix} 1 & 0 \\ 0 & 10 \end{bmatrix} y(\infty) + \frac{1}{10} u^T(\infty)u(\infty) \right].$$

Use the resulting gains and estimates to form the control system:

$$\begin{bmatrix} u_1(t) \\ u_2(t) \end{bmatrix} = -\mathbf{K}\begin{bmatrix} \hat{x}_1(t) \\ \hat{x}_2(t) \\ \hat{x}_3(t) - r_1 \\ \hat{x}_4(t) - r_2 \end{bmatrix}.$$

Simulate the closed-loop system with an initial state $x(0) = 0$ and the following reference input:

$$\begin{bmatrix} r_1 \\ r_2 \end{bmatrix} = \begin{bmatrix} 50 \\ 5 \end{bmatrix}.$$

Plot the outputs and the control inputs, and comment on the performance of this system.

b. Design a second controller using an integral LQR operating on the Kalman estimates. The integral Kalman filter is not required, since there are no significant constant disturbances in this problem. Simulate the system with the given test conditions, and plot the outputs and the control inputs. Note that the weighting on the integral of the errors should be adjusted to yield control inputs comparable in size to those obtained in part *a*. Comment on the performance of this system.

PART 3
$\mathcal{H}_\infty$ Control

9 Full Information Control and Estimation

The linear quadratic regulator, Kalman filter, and linear quadratic Gaussian problems can all be posed as 2-norm optimization problems. These optimization problems can also be posed using the system ∞-norm as a cost function. The ∞-norm is the worst-case gain of the system and therefore provides a good match to engineering specifications, which are typically given in terms of bounds on errors and controls.

The small-gain theorem states that, for unstructured perturbations, robust stability depends on the ∞-norm of the closed-loop system from the perturbation input to the perturbation output. The minimization of the closed-loop ∞-norm can therefore also be used as a means of maximizing robustness.

Robust stability in the presence of structured perturbations, and robust performance both depend on the supremum (over frequency) of the structured singular value from the perturbation input to the perturbation output. Note that a performance block is added to the perturbation (forming a structured perturbation) for robust performance analysis. The supremum of the SSV is bounded by the ∞-norm of the diagonal-scaled, closed-loop system. In fact, this bound is typically used in place of the SSV in robustness analysis. Therefore, it is reasonable to assume that minimization of the system ∞-norm plays a role in maximizing both robust stability (for structured perturbations) and robust performance (in general).

In summary, the optimization of the ∞-norm has applications in both maximizing performance and robustness. Control and estimation problems, with the goal of minimizing of the system ∞-norm, are termed $\mathcal{H}_\infty$ optimizations problems.[1] This chapter develops some basic results on the design of systems that minimize the system ∞-norm, in particular, full information control and output estimation. Full information control and output estimation are combined in the next chapter to form $\mathcal{H}_\infty$ output feedback. The results in this chapter are derived solely within a performance optimization framework. The maximization of performance robustness is discussed in the next chapter.

9.1 Differential Games

An $\mathcal{H}_\infty$ optimal controller minimizes the worst-case gain of the system. This problem can be thought of as a game with two participants: the designer, who is seeking a control that minimizes the gain; and nature, which is seeking a disturbance that maximizes the

[1] The $\mathcal{H}$ in $\mathcal{H}_\infty$ refers the Hardy space of all stable systems, and the ∞ refers to the system norm. While stability is not an issue in finite-time problems, these problems are also referred to as $\mathcal{H}_\infty$ optimization problems in keeping with standard terminology in the literature.

gain (the worst-case input). Games of this type are termed differential games when the dynamics of the game are described by differential equations. Some fundamental results from the field of differential game theory are presented in this section for application to $\mathcal{H}_\infty$ optimization problems.

A differential game is described by the game dynamics and the objective function.[2] The game dynamics are modeled by a generic state equation:

$$\dot{x}(t) = \mathbf{A}x(t) + \mathbf{B}_u u(t) + \mathbf{B}_w w(t), \tag{9.1}$$

where $u(t)$ is the control input, which is selected by the designer, and $w(t)$ is the disturbance input, which is selected by nature. The objective function is a real function (not necessarily positive definite) of the state, the control, and the disturbance:

$$J\{x(t),\ u(t),\ w(t)\}.$$

The solution of the differential game consists of the optimal control trajectory $u^*(t)$ and the worst-case disturbance input $w^*(t)$. This solution is necessarily a saddle point of the objective function, which is defined by the following inequalities:

$$J\{x(u^*,\ w),\ u^*(t),\ w(t)\} \le J\{x(u^*,\ w^*),\ u^*(t),\ w^*(t)\} \le J\{x(u,\ w^*),\ u(t),\ w^*(t)\}.$$

The state is shown as a function of the inputs in this expression to emphasize that the state is completely specified by the inputs (provided the state initial condition is zero). A saddle point can also be defined as the argument of the mini-max problem:

$$\min_u (\max_w J\{x(u,\ v),\ u(t),\ w(t)\}).$$

Lagrange multipliers can be used to convert a constrained mini-max problem into an unconstrained mini-max problem of higher order. Appending the constraint equation (9.1) to the objective function yields the augmented objective function:

$$J_a(x,\ u,\ w,\ p) = J(x,\ u,\ w) + \int_0^{t_f} p^T(t)\{\mathbf{A}x(t) + \mathbf{B}_u u(t) + \mathbf{B}_w w(t) - \dot{x}(t)\}dt.$$

A necessary condition for a saddle point is that the variation of this augmented objective function (with respect to perturbations in the state, the control, the disturbance, and the Lagrange multiplier) must be zero. This result can be intuitively justified by noting that the mini-max problem consists of two optimization problems, or by following the procedure given in Subsection 6.1.2.

9.2 Full Information Control

An optimal full information controller minimizes the infinity norm of the closed-loop system from the disturbance input to the reference output, assuming that all of the plant states and all of the disturbance inputs are available for feedback. In the stochastic regulator and in LQG control, the disturbance input is white noise. This disturbance input has an infinite variance, is totally unpredictable, and is not measurable. Using this in-

[2] Constraints on the controls and disturbances are typically included in a differential game description. These constraints are not included in this presentation, since they are not used in the following treatment of $\mathcal{H}_\infty$ optimization.

put for feedback is then not practical. In $\mathcal{H}_\infty$ optimization problems, the disturbance input may be measured or predicted and is included as a possible feedback input. The term *full information* is used to reflect this change with respect to the state feedback controller, which is the solution of the linear quadratic regulator problem.

The $\mathcal{H}_\infty$ full information control problem is formally stated below. Let the plant be described by the following state model:

$$\dot{x}(t) = \mathbf{A}x(t) + \mathbf{B}_u u(t) + \mathbf{B}_w w(t); \tag{9.2a}$$

$$y(t) = \mathbf{C}_y x(t) + \mathbf{D}_{yu} u(t), \tag{9.2b}$$

where

$$\mathbf{D}_{yu}^T \mathbf{C}_y = \mathbf{0}; \tag{9.2c}$$

$$\mathbf{D}_{yu}^T \mathbf{D}_{yu} = \mathbf{I}. \tag{9.2d}$$

The plant is assumed to be observable from the reference output $y(t)$ and controllable from the control input $u(t)$.[3] The zero in condition (9.2c) specifies that the output consists of two distinct components: linear combinations of the state and linear combinations of the control. This separation between the state and control is equivalent to the absence of cross terms between the state and control in the LQR cost function. The identity matrix in (9.2d) indicates that the coupling matrix between the control and the output is normalized; that is, the control is normalized so that all controls are equally weighted in the cost. This normalization is performed to simplify subsequent derivations, and so that the final results match those in the current literature. Normalization of this coupling matrix can always be accomplished provided $\mathbf{D}_{yu}$ has full rank.

The $\mathcal{H}_\infty$ full information control problem is to find a feedback controller, utilizing the state and the disturbance, that minimizes the closed-loop system ∞-norm:

$$J = \|\mathbf{G}_{yw}\|_{\infty,[0,t_f]} = \sup_{\|w(t)\|_{2,[0,t_f]} \neq 0} \frac{\|y(t)\|_{2,[0,t_f]}}{\|w(t)\|_{2,[0,t_f]}}. \tag{9.3}$$

A block diagram of the full information control system is given in Figure 9.1 where the controller is a linear system (not necessarily time-invariant) denoted $\mathcal{K}(\bullet)$. Note that defining the cost in terms of the ∞-norm implies that the initial state is zero:[4]

$$x(0) = 0.$$

[3] These assumptions can be relaxed. In general, the plant need only be detectable from the reference output and stabilizable from the control input.

[4] In applications, the initial state may not be zero. The $\mathcal{H}_\infty$ full information controller typically yields a reasonable transient response, but this is not guaranteed and must be verified. Applying weights to the various terms in the reference output can be used to modify the transient response when necessary.

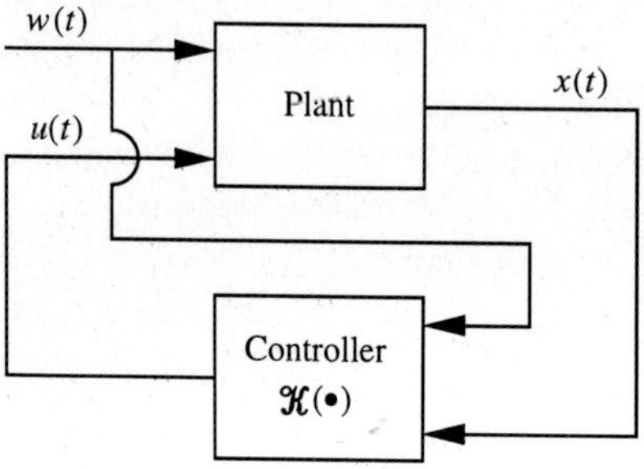

FIGURE 9.1 Full information control

The ∞-norm can be defined on either a finite or infinite ($t_f = \infty$) time interval. When the time interval is infinite, the controller must also internally stabilize the closed-loop system.

A suitable objective function is required in order to apply differential game theory to the solution of the full information control problem. The ∞-norm cost function (9.3) is not acceptable as an objective function, since this cost only depends on the controller; that is, the supremum makes this function independent of a particular disturbance input. Dropping the supremum from the cost function (9.3) yields a suitable objective function. While theoretically acceptable, the solution of the differential game with this objective function is intractable, in general.

A quadratic objective function, which yields tractable solutions of the differential game, can be obtained by considering the bound on the closed-loop ∞-norm:

$$\|\mathbf{G}_{yw}\|_{\infty,[0,t_f]} = \sup_{\|w(t)\|_{2,[0,t_f]}\neq 0} \frac{\|y(t)\|_{2,[0,t_f]}}{\|w(t)\|_{2,[0,t_f]}} < \gamma, \tag{9.4}$$

where γ is called the *performance bound.* A controller that satisfies this bound is called a suboptimal solution to the $\mathcal{H}_\infty$ full information control problem, or simply a suboptimal controller. The suboptimal controller must also satisfy the bound obtained by squaring (9.4):

$$\|\mathbf{G}_{yw}\|^2_{\infty,[0,t_f]} = \sup_{\|w(t)\|_{2,[0,t_f]}\neq 0} \left\{ \frac{\|y(t)\|^2_{2,[0,t_f]}}{\|w(t)\|^2_{2,[0,t_f]}} \right\} < \gamma^2.$$

For the supremum to satisfy this strict inequality, the term within the curly brackets must be bounded away from γ^2; that is, for some ε,

$$\frac{\|y(t)\|^2_{2,[0,t_f]}}{\|w(t)\|^2_{2,[0,t_f]}} \leq \gamma^2 - \varepsilon^2.$$

Multiplying both sides by the denominator, and grouping the resulting terms yields

$$\|y(t)\|^2_{2,[0,t_f]} - \gamma^2\|w(t)\|^2_{2,[0,t_f]} \leq -\varepsilon^2\|w(t)\|^2_{2,[0,t_f]}. \tag{9.5}$$

Note that satisfaction of this inequality for all disturbance inputs and some ε is equivalent to the bound on the closed-loop ∞-norm (9.4). The expression on the left of (9.5) can be used as an objective function:

$$J_\gamma(x,\, u,\, w) \;=\; \|y(t)\|_2^2 \;-\; \gamma^2\|w(t)\|_2^2. \tag{9.6}$$

Differential game theory can then be applied to generate a control that minimizes this objective function in the presence of the worst-case disturbance. If the minimum of the objective function satisfies (9.5), then the bound on the closed-loop ∞-norm is achieved.

This differential game is posed as an open-loop mini-max problem; that is, the control input is not required to be generated by a feedback controller. Therefore, this problem allows more general control inputs than those allowed in the $\mathcal{H}_\infty$ suboptimal control problem. As will be seen, the solution of the differential game does yield a full information feedback control, making it a valid candidate solution of the $\mathcal{H}_\infty$ suboptimal control problem. In fact, the saddle point of the differential game is an $\mathcal{H}_\infty$ suboptimal full information controller. This fact is demonstrated after first deriving the solution of the differential game, which is referred to as an $\mathcal{H}_\infty$ suboptimal full information controller in anticipation of demonstrating this equivalence.

General conditions can be given for the existence of $\mathcal{H}_\infty$ suboptimal controllers. The $\mathcal{H}_\infty$ optimal controller can then be approximated by decreasing the performance bound until a suboptimal controller no longer exists. As a matter of practice, a binary search with a Cauchy convergence criteria can be used when decreasing this bound, allowing arbitrarily close approximation of the $\mathcal{H}_\infty$ optimal solution.

9.2.1 The Hamiltonian Equations

The differential game specified by the objective function (9.6) and the game dynamics (9.2) yields a constrained mini-max problem. An unconstrained mini-max problem of higher dimension can be obtained by appending the constraint to the objective function:

$$\begin{aligned} J_{a,\gamma}(u,\, w,\, p) \;=\; \int_0^{t_f} & y^T(t)y(t) \;-\; \gamma^2 w^T(t)w(t) \\ & +\; 2p^T(t)\{\mathbf{A}x(t) \;+\; \mathbf{B}_u u(t) \;+\; \mathbf{B}_w w(t) \;-\; \dot{x}(t)\}dt, \end{aligned}$$

where the Lagrange multiplier term $p(t)$ is referred to as the *costate* (following the terminology presented for the LQR). The factor of 2 multiplying the constraint equation is used to simplify the final results. Note that 2 times a constraint equation is also a constraint equation.

A necessary condition for a saddle point is that the variation of the augmented objective function equal zero. The increment of the augmented cost function is

$$\begin{aligned} &\Delta J_{a,\gamma}(u,\, w,\, p,\, \delta u,\, \delta w,\, \delta p) \\ &= J_{a,\gamma}(u + \delta u,\, w + \delta w,\, p + \delta p) - J_{a,\gamma}(u,\, w,\, p) \\ &= \int_0^{t_f} (x + \delta x)^T\mathbf{C}_y^T\mathbf{C}_y(x + \delta x) + (u + \delta u)^T(u + \delta u) - \gamma^2(w + \delta w)^T(w + \delta w) \\ &\quad + 2(p + \delta p)^T\{\mathbf{A}(x + \delta x) + \mathbf{B}_u(u + \delta u) + \mathbf{B}_w(w + \delta w) - (\dot{x} + \delta\dot{x})\}dt \\ &\quad - \int_0^{t_f} x^T\mathbf{C}_y^T\mathbf{C}_y x + u^T u - \gamma^2 w^T w + 2p^T\{\mathbf{A}x + \mathbf{B}_u u + \mathbf{B}_w w - \dot{x}\}dt, \end{aligned}$$

where (9.2b) has been used to substitute for $y(t)$, the fact that $\mathbf{D}_{yu}^T\mathbf{D}_{yu} = \mathbf{I}$ has been used, and the time indexes have been omitted to simplify the notation. Expanding this expression and grouping terms, we have

$$\begin{aligned}\Delta J_{a,\gamma} = \int_0^{t_f} & \delta x^T \mathbf{C}_y^T \mathbf{C}_y \delta x + \delta u^T \delta u - \gamma^2 \delta w^T \delta w \\ & + 2\delta p^T\{\mathbf{A}\delta x + \mathbf{B}_u \delta u + \mathbf{B}_w \delta w - \delta\dot{x}\} + 2u^T\delta u \\ & + 2x^T\mathbf{C}_y^T\mathbf{C}_y\delta x - 2\gamma^2 w^T\delta w + 2\delta p^T\{\mathbf{A}x + \mathbf{B}_u u + \mathbf{B}_w w - \dot{x}\} \\ & + 2p^T\{\mathbf{A}\delta x + \mathbf{B}_u\delta u + \mathbf{B}_w \delta w - \delta\dot{x}\}dt.\end{aligned}$$

A necessary condition for the trajectory $x(t), p(t), u(t)$, and $w(t)$ to be a saddle point is that the variation of the objective function equal zero:

$$\begin{aligned}\delta J_{a,\gamma}(u, w, p, \delta u, \delta w, \delta p) = \int_0^{t_f} & 2x^T\mathbf{C}_y^T\mathbf{C}_y\delta x + 2u^T\delta u - 2\gamma^2 w^T \delta w \\ & + 2\delta p^T\{\mathbf{A}x + \mathbf{B}_u u + \mathbf{B}_w w - \dot{x}\} \\ & + 2p^T\{\mathbf{A}\delta x + \mathbf{B}_u \delta u + \mathbf{B}_w\delta w - \delta\dot{x}\}dt = 0.\end{aligned} \tag{9.7}$$

Integration by parts yields

$$\int_0^{t_f} p^T(t)\delta\dot{x}(t)dt = p^T(t_f)\delta x(t_f) - p^T(0)\delta x(0) - \int_0^{t_f} \dot{p}^T(t)\delta x(t)dt, \tag{9.8}$$

which can be used to eliminate the variation of the state derivative in (9.7). Further, the state initial condition is fixed, so $\delta x(0) = 0$. Substituting this result and (9.8) into (9.7), and regrouping terms yields the necessary condition for a saddle point:

$$\begin{aligned}\delta J_{a,\gamma}(u, w, p, \delta u, \delta w, \delta p) = -2p^T(t_f)\delta x(t_f) + \int_0^{t_f} & (2\dot{p}^T + 2x^T\mathbf{C}_y^T\mathbf{C}_y + 2p^T\mathbf{A})\delta x \\ & + (2p^T\mathbf{B}_w - 2\gamma^2 w^T)\delta w + (2u^T + 2p^T\mathbf{B}_u)\delta u \\ & + 2\delta p^T(\mathbf{A}x + \mathbf{B}_u u + \mathbf{B}_w w - \dot{x})dt = 0.\end{aligned}$$

Since the variations $\delta x(t_f)$, δx, δu, and δp are all arbitrary, this expression is only equal to zero if

$$p(t_f) = 0; \tag{9.9a}$$

$$\dot{p}(t) = -\mathbf{C}_y^T\mathbf{C}_y x(t) - \mathbf{A}^T p(t); \tag{9.9b}$$

$$u(t) = -\mathbf{B}_u^T p(t); \tag{9.9c}$$

$$w(t) = \gamma^{-2}\mathbf{B}_w^T p(t); \tag{9.9d}$$

$$\dot{x}(t) = \mathbf{A}x(t) + \mathbf{B}_u u(t) + \mathbf{B}_w w(t). \tag{9.9e}$$

Eliminating $u(t)$ and $w(t)$ from (9.9b) and (9.9e), and combining the resulting equations into a single state equation yields

$$\begin{bmatrix} \dot{x}(t) \\ \dot{p}(t) \end{bmatrix} = \begin{bmatrix} \mathbf{A} & -\mathbf{B}_u\mathbf{B}_u^T + \gamma^{-2}\mathbf{B}_w\mathbf{B}_w^T \\ -\mathbf{C}_y^T\mathbf{C}_y & -\mathbf{A}^T \end{bmatrix} \begin{bmatrix} x(t) \\ p(t) \end{bmatrix} = \mathfrak{X}_\infty \begin{bmatrix} x(t) \\ p(t) \end{bmatrix}. \tag{9.10}$$

This is the Hamiltonian system for the full information $\mathcal{H}_\infty$ control problem, and the matrix $\mathfrak{X}_\infty$ is called the Hamiltonian.

The $\mathcal{H}_\infty$ suboptimal control can be found by solving the Hamiltonian system subject to the final condition

$$p(t_f) = 0.$$

The solution of the Hamiltonian system (9.10), at the final time, given an initial condition at time t, is

$$\begin{bmatrix} x(t_f) \\ \hline p(t_f) \end{bmatrix} = e^{\mathfrak{X}_\infty(t_f - t)} \begin{bmatrix} x(t) \\ \hline p(t) \end{bmatrix} = \left[\begin{array}{c:c} \mathbf{\Phi}_{11}(t_f - t) & \mathbf{\Phi}_{12}(t_f - t) \\ \hline \mathbf{\Phi}_{21}(t_f - t) & \mathbf{\Phi}_{22}(t_f - t) \end{array}\right] \begin{bmatrix} x(t) \\ \hline p(t) \end{bmatrix}.$$

Substituting the final condition (9.9a) into this equation yields

$$\begin{bmatrix} x(t_f) \\ \hline 0 \end{bmatrix} = \left[\begin{array}{c:c} \mathbf{\Phi}_{11}(t_f - t) & \mathbf{\Phi}_{12}(t_f - t) \\ \hline \mathbf{\Phi}_{21}(t_f - t) & \mathbf{\Phi}_{22}(t_f - t) \end{array}\right] \begin{bmatrix} x(t) \\ \hline p(t) \end{bmatrix}.$$

The lower block of this matrix equation can be solved for the costate in terms of the following state:

$$p(t) = -\{\mathbf{\Phi}_{22}(t_f - t)\}^{-1}\mathbf{\Phi}_{21}(t_f - t)x(t) = \mathbf{P}(t)x(t), \tag{9.11}$$

where $\mathbf{P}(t)$ is the matrix of proportionality between the costate and the state. This matrix of proportionality is fully specified by the state-transition matrix of the Hamiltonian system, provided the inverse in (9.11) exists at all times between the initial time and the final time.

The $\mathcal{H}_\infty$ suboptimal control is then found from (9.9c):

$$\boxed{u(t) = -\mathbf{B}_u^T\mathbf{P}(t)x(t) = \mathbf{B}_u^T\{\mathbf{\Phi}_{22}(t_f - t)\}^{-1}\mathbf{\Phi}_{21}(t_f - t)x(t) = -\mathbf{K}(t)x(t),} \tag{9.12}$$

where $\mathbf{K}(t)$ is termed the $\mathcal{H}_\infty$ suboptimal feedback gain matrix. While this control is generated from the necessary conditions for the solution of the differential game, it in fact specifies a solution of the differential game (provided the inverse exists at all times). Further, this control is given by a linear, time-varying, state feedback law, making it a candidate solution of the $\mathcal{H}_\infty$ full information suboptimal control problem. Indeed, this feedback law is a solution of this suboptimal control problem, which will be demonstrated after first presenting an example, and showing that the matrix $\mathbf{P}(t)$ can be found as the solution of a Riccati equation.

EXAMPLE 9.1 Let the plant be modeled as follows:

$$\dot{x}(t) = x(t) + u(t) + 2w(t);$$

$$y(t) = \begin{bmatrix} 10 \\ 0 \end{bmatrix} x(t) + \begin{bmatrix} 0 \\ 1 \end{bmatrix} u(t),$$

and the objective function be

$$J_\gamma\{x(t), u(t), w(t)\} = \|y(t)\|_2^2 - \gamma^2\|w(t)\|_2^2.$$

Note that state and control weightings are incorporated into the output equation, which is scaled to yield a unity weight on the control. The Hamiltonian system is

$$\begin{bmatrix} \dot{x}(t) \\ \dot{p}(t) \end{bmatrix} = \begin{bmatrix} 1 & -1 + 4\gamma^{-2} \\ -100 & -1 \end{bmatrix} \begin{bmatrix} x(t) \\ p(t) \end{bmatrix},$$

which has the state-transition matrix

$$\begin{aligned} \mathbf{\Phi}(t) &= e^{\mathfrak{X}_\infty t} = \mathcal{L}^{-1}\{(s\mathbf{I} - \mathfrak{X}_\infty)^{-1}\} \\ &= \mathcal{L}^{-1}\left\{ \frac{1}{s^2 - 101 + 400\gamma^{-2}} \begin{bmatrix} s + 1 & -1 + 4\gamma^{-2} \\ -100 & s - 1 \end{bmatrix} \right\}. \end{aligned}$$

A saddle point of this differential game exists (and an $\mathcal{H}_\infty$ full information suboptimal controller exists) for a given performance bound provided $\mathbf{\Phi}_{22}(t_f - t)$ has an inverse throughout the time interval; that is, the scalar $\mathbf{\Phi}_{22}(t_f - t) \neq 0$. This element of the state transition matrix can be computed:

$$\mathbf{\Phi}_{22}(t) = \mathcal{L}^{-1}\left\{ \frac{s - 1}{s^2 - 101 + 400\gamma^{-2}} \right\} = \begin{cases} \dfrac{a - 1}{2a} e^{at} + \dfrac{a + 1}{2a} e^{-at} & \gamma > \sqrt{\dfrac{400}{101}} \\ 1 - t & \gamma = \sqrt{\dfrac{400}{101}} \\ \sqrt{\dfrac{\omega^2 + 1}{\omega}} \sin(\omega t + \theta) & \gamma < \sqrt{\dfrac{400}{101}} \end{cases}$$

where $a = \sqrt{101 - 400\gamma^{-2}}$, $\omega = \sqrt{-101 + 400\gamma^{-2}}$, and $\theta = -\tan^{-1}(\omega)$. Note that as γ gets smaller, a gets smaller, ω gets bigger, and θ gets more negative. When $\gamma < \sqrt{400/101}$, the eigenvalues of the Hamiltonian matrix are imaginary, and the gain does not exist for long time intervals, since $\mathbf{\Phi}_{22}(t)$ periodically crosses zero. Further, the fact that the feedback gain does not exist implies that there is no solution of the differential game (and no solution of the $\mathcal{H}_\infty$ suboptimal control problem) for the given bound. Less obvious from these expressions is that the time interval over which a feedback gain exists becomes monotonically larger as γ is increased, reaching infinity for $\gamma > 2$.

For the bound $\gamma = 2.03$ ($a = 2$), and the final time $t_f = 3$, the $\mathcal{H}_\infty$ suboptimal feedback gain is

$$K(t) = \frac{100e^{2(3-t)} - 100e^{-2(3-t)}}{e^{2(3-t)} + 3e^{-2(3-t)}},$$

which is plotted in Figure 9.2. Note that the gain exhibits a transient and then approaches a steady-state value far from the final time. In situations where the time interval is long compared to the settling time of this transient, it may be reasonable to use only the steady-state gain. ◆

An $\mathcal{H}_\infty$ suboptimal controller fails to exist over long time intervals, in the above example, when the Hamiltonian has purely imaginary eigenvalues. This result is a general property of $\mathcal{H}_\infty$ suboptimal control:

> The feedback gain exists for arbitrary time intervals, and there is a solution to the $\mathcal{H}_\infty$ suboptimal control problem, only if the Hamiltonian matrix has no purely imaginary eigenvalues.

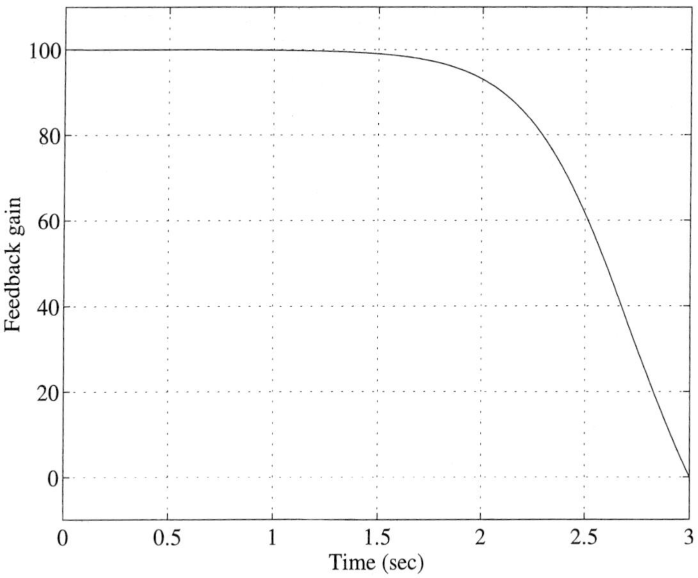

FIGURE 9.2 The feedback gain for Example 9.1

Note that this is an "only if" statement. The fact that the Hamiltonian matrix has no purely imaginary eigenvalues is not sufficient to guarantee existence of a solution to the steady-state $\mathcal{H}_\infty$ suboptimal control problem. Additional conditions for the existence of this steady-state control are presented in subsection 9.2.4.

9.2.2 The Riccati Equation

The solution of the Hamiltonian system for the linear quadratic regulator was shown in Chapter 6 to be related to a nonlinear matrix differential equation, known as a Riccati equation. The Hamiltonian system for the $\mathcal{H}_\infty$ suboptimal control problem can also be related to a Riccati equation. This Riccati equation has only final conditions and can be solved backward in time using any numerical integration package.

The solution of the $\mathcal{H}_\infty$ suboptimal control problem can be reduced to finding the matrix $\mathbf{P}(t)$, since the feedback gain (9.12) only depends on this matrix and $\mathbf{B}_u$. A differential equation for $\mathbf{P}(t)$ can be generated by taking the derivative of (9.11):

$$\dot{p}(t) = \dot{\mathbf{P}}(t)x(t) + \mathbf{P}(t)\dot{x}(t).$$

Substituting for $\dot{p}(t)$ and $\dot{x}(t)$, using the Hamiltonian system (9.10) yields

$$-\mathbf{C}_y^T\mathbf{C}_y x(t) - \mathbf{A}^T p(t) = \dot{\mathbf{P}}(t)x(t) + \mathbf{P}(t)\{\mathbf{A}x(t) - (\mathbf{B}_u\mathbf{B}_u^T - \gamma^{-2}\mathbf{B}_w\mathbf{B}_w^T)p(t)\}.$$

Then, substituting for $p(t)$ using (9.11) and rearranging yields

$$\{\dot{\mathbf{P}}(t) + \mathbf{P}(t)\mathbf{A} + \mathbf{A}^T\mathbf{P}(t) + \mathbf{C}_y^T\mathbf{C}_y - \mathbf{P}(t)(\mathbf{B}_u\mathbf{B}_u^T - \gamma^{-2}\mathbf{B}_w\mathbf{B}_w^T)\mathbf{P}(t)\}x(t) = 0.$$

This equation is valid for any state $x(t)$; therefore, the matrix in curly brackets must equal zero:

$$\dot{\mathbf{P}}(t) = -\mathbf{P}(t)\mathbf{A} - \mathbf{A}^T\mathbf{P}(t) - \mathbf{C}_y^T\mathbf{C}_y + \mathbf{P}(t)(\mathbf{B}_u\mathbf{B}_u^T - \gamma^{-2}\mathbf{B}_w\mathbf{B}_w^T)\mathbf{P}(t). \tag{9.13}$$

This equation is the Riccati differential equation for the $\mathcal{H}_\infty$ suboptimal control problem.

The Riccati solution $\mathbf{P}(t)$ is a symmetric matrix (if it exists), which can be found by solving (9.13) backward in time from the final condition:

$$\mathbf{P}(t_f) = \mathbf{0}. \tag{9.14}$$

This final condition is obtained by letting $p(t_f) = 0$ in (9.11) and recognizing that the result is valid for any final state. The fact that the Riccati solution is symmetric is shown by noting that the derivative is symmetric whenever the Riccati solution is symmetric:

$$\begin{aligned}\dot{\mathbf{P}}^T(t) &= \{-\mathbf{P}(t)\mathbf{A} - \mathbf{A}^T\mathbf{P}(t) - \mathbf{C}_y^T\mathbf{C}_y + \mathbf{P}(t)(\mathbf{B}_u\mathbf{B}_u^T - \gamma^{-2}\mathbf{B}_w\mathbf{B}_w^T)\mathbf{P}(t)\}^T \\ &= -\mathbf{A}^T\mathbf{P}(t) - \mathbf{P}(t)\mathbf{A} - \mathbf{C}_y^T\mathbf{C}_y + \mathbf{P}(t)(\mathbf{B}_u\mathbf{B}_u^T - \gamma^{-2}\mathbf{B}_w\mathbf{B}_w^T)\mathbf{P}(t) = \dot{\mathbf{P}}(t).\end{aligned}$$

Therefore, integrating this derivative backward in time from the final condition yields a symmetric Riccati solution.

EXAMPLE 9.1 CONTINUED The Riccati equation associated with the differential game (and $\mathcal{H}_\infty$ suboptimal control problem) in Example 9.1 is:

$$\dot{P}(t) = -2P(t) + (1 - 4\gamma^{-2})P^2(t) - 100.$$

The final condition for this differential equation is $P(3) = 0$. The solution to this final value problem is

$$P(t) = K(t) = \frac{100e^{2(3-t)} - 100e^{-2(3-t)}}{e^{2(3-t)} + 3e^{-(3-t)}}.$$

This solution can be verified by substitution into the Riccati equation and by evaluating the solution at the final time. Note that the feedback gain equals the Riccati solution, in this example, since $\mathbf{B}_u = 1$. ◆

9.2.3 The Value of the Objective Function

A feedback controller and a worst-case disturbance result when applying Lagrange multipliers and variation theory to the differential game with dynamics (9.2) and objective function (9.6). The feedback gain is found from the solution of either a Hamiltonian or a Riccati equation. All of these results are based on the variation of the objective function, which only specifies necessary conditions for the existence of a saddle point. In fact, the given control and worst-case disturbance form a saddle point of the differential game, and this control is an $\mathcal{H}_\infty$ suboptimal controller. These two results can be verified by cleverly rewriting the objective function.

To begin, note that

$$\int_0^{t_f} \frac{dx^T(t)\mathbf{P}(t)x(t)}{dt}dt = x^T(t_f)\mathbf{P}(t_f)x(t_f) - x^T(0)\mathbf{P}(0)x(0) = 0,$$

since $\mathbf{P}(t_f) = 0$ and $x(0) = 0$. This integral can be added to the objective function (to complete the square) without changing the value of this function:

$$\begin{aligned} J_\gamma &= \int_0^{t_f} x^T(t)\mathbf{C}_y^T\mathbf{C}_y x(t) + u^T(t)u(t) - \gamma^2 w^T(t)w(t) + \frac{dx^T(t)\mathbf{P}(t)x(t)}{dt}dt \\ &= \int_0^{t_f} x^T(t)\mathbf{C}_y^T\mathbf{C}_y x(t) + u^T(t)u(t) - \gamma^2 w^T(t)w(t) \\ &\quad + \dot{x}^T(t)\mathbf{P}x(t) + x^T(t)\dot{\mathbf{P}}(t)x(t) + x^T(t)\mathbf{P}(t)\dot{x}(t)dt. \end{aligned}$$

Substituting for the derivative of the state using (9.2a) and regrouping terms yields

$$\begin{aligned} J_\gamma &= \int_0^{t_f} x^T(\dot{\mathbf{P}} + \mathbf{C}_y^T\mathbf{C}_y + \mathbf{A}^T\mathbf{P} + \mathbf{P}\mathbf{A})x + u^Tu - \gamma^2 w^Tw + (\mathbf{B}_u u + \mathbf{B}_w w)^T\mathbf{P}x \\ &\quad + x^T\mathbf{P}(\mathbf{B}_u u + \mathbf{B}_w w)dt, \end{aligned}$$

where the time argument has been dropped to simplify the notation.

Substituting for $(\dot{\mathbf{P}} + \mathbf{C}_y^T\mathbf{C}_y + \mathbf{A}^T\mathbf{P} + \mathbf{P}\mathbf{A})$ using the Riccati equation (9.13) gives

$$\begin{aligned} J_\gamma &= \int_0^{t_f} x^T\mathbf{P}(\mathbf{B}_u\mathbf{B}_u^T - \gamma^{-2}\mathbf{B}_w\mathbf{B}_w^T)\mathbf{P}x + u^Tu - \gamma^2 w^Tw \\ &\quad + (\mathbf{B}_u u + \mathbf{B}_w w)^T\mathbf{P}x + x^T\mathbf{P}(\mathbf{B}_u u + \mathbf{B}_w w)dt. \end{aligned}$$

Regrouping again yields

$$\begin{aligned} J_\gamma &= \int_0^{t_f} (u + \mathbf{B}_u^T\mathbf{P}x)^T(u + \mathbf{B}_u^T\mathbf{P}x) - \gamma^2(w - \gamma^{-2}\mathbf{B}_w^T\mathbf{P}x)^T(w - \gamma^{-2}\mathbf{B}_w^T\mathbf{P}x)dt \\ &= \int_0^{t_f} \|u + \mathbf{B}_u^T\mathbf{P}x\|_E^2 - \gamma^2\|w - \gamma^{-2}\mathbf{B}_w^T\mathbf{P}x\|_E^2 dt \\ &= \|u + \mathbf{B}_u^T\mathbf{P}x\|_{2,[0,t_f]}^2 - \gamma^2\|w - \gamma^{-2}\mathbf{B}_w^T\mathbf{P}x\|_{2,[0,t_f]}^2, \end{aligned} \tag{9.15}$$

where the norms in the final expression are signal norms. The objective function has a value of zero when the control and disturbance inputs are given by the necessary conditions:

$$u(t) = -\mathbf{B}_u^T\mathbf{P}(t)x(t); \tag{9.16}$$

$$w(t) = \gamma^{-2}\mathbf{B}_w^T\mathbf{P}(t)x(t). \tag{9.17}$$

This point is a saddle point since any other disturbance input, with control (9.16), yields a decrease in the objective value, and any other control input, with disturbance (9.17) yields an increase in the objective value.

To verify that the controller (9.16) is an $\mathcal{H}_\infty$ suboptimal controller, it must be shown (9.5) that

$$\begin{aligned} J_\gamma &= \|u(t) + \mathbf{B}_u^T\mathbf{P}x(t)\|_{2,[0,t_f]}^2 - \gamma^2\|w(t) - \gamma^{-2}\mathbf{B}_w^T\mathbf{P}x(t)\|_{2,[0,t_f]}^2 \\ &\le -\varepsilon^2\|w(t)\|_{2,[0,t_f]}^2 \end{aligned} \tag{9.18}$$

for some positive ε. When using the controller (9.16), (9.18) reduces to

$$J_\gamma = -\gamma^2\|w(t) - \gamma^{-2}\mathbf{B}_w^T\mathbf{P}x(t)\|^2_{2,[0,t_f]} \leq -\varepsilon^2\|w(t)\|^2_{2,[0,t_f]},$$

or, equivalently,

$$\|w(t)\|^2_{2,[0,t_f]} \leq \frac{\gamma^2}{\varepsilon^2}\|w(t) - \gamma^{-2}\mathbf{B}_w^T\mathbf{P}x(t)\|^2_{2,[0,t_f]}. \tag{9.19}$$

The fact that a positive ε exists that satisfies this equation can be developed by noting that the disturbance input can be generated by a time-varying state model from the input $[w(t) - \gamma^{-2}\mathbf{B}_w^T\mathbf{P}(t)x(t)]$:[5]

$$\dot{x}(t) = [\mathbf{A} - \mathbf{B}_u\mathbf{B}_u^T\mathbf{P}(t) + \gamma^{-2}\mathbf{B}_w\mathbf{B}_w^T\mathbf{P}(t)]x(t) + \mathbf{B}_w[w(t) - \gamma^{-2}\mathbf{B}_w^T\mathbf{P}(t)x(t)]; \tag{9.20a}$$

$$w(t) = -\gamma^{-2}\mathbf{B}_w^T\mathbf{P}(t)x(t) + [w(t) - \gamma^{-2}\mathbf{B}_w^T\mathbf{P}(t)x(t)]. \tag{9.20b}$$

Since all matrices in this state model are finite, the output is bounded; see Appendix equation (A9.6):

$$\|w(t)\|^2_{2,[0,t_f]} \leq \|\mathbf{G}_{w\Delta w}\|^2_{\infty,[0,t_f]}\|w(t) - \gamma^{-2}\mathbf{B}_w^T\mathbf{P}(t)x(t)\|^2_{2,[0,t_f]},$$

where $\mathbf{G}_{w\Delta w}$ denotes the state model (9.20). Comparing this result with (9.19), a positive ε that satisfies (9.19) is $\varepsilon = \gamma/\|\mathbf{G}_{w\Delta w}\|_{\infty,[0,t_f]}$ and the control law (9.16) is therefore a solution of the suboptimal $\mathcal{H}_\infty$ control problem.

The suboptimal control exists provided the Riccati equation (9.13) has a solution over the entire time interval from 0 to t_f. This control depends on the performance bound selected. Note that the Riccati equation (9.13) reduces to the LQR Riccati equation (6.20) when the performance bound approaches infinity. The LQR Riccati equation is guaranteed to have a solution, indicating that the $\mathcal{H}_\infty$ Riccati equation has a solution for sufficiently large performance bounds.

The control law (9.16) has been shown to satisfy the infinity norm bound (9.6), and to exist whenever the Riccati equation (9.13) has a solution. The following question immediately comes to mind: If this Riccati equation has no solution, is it possible to find a full information controller that satisfies the bound (9.6)? The answer to this question is no! The existence of a solution to the Riccati equation (9.13) is both necessary and sufficient for the existence of a solution to the $\mathcal{H}_\infty$ suboptimal control problem. Therefore, the $\mathcal{H}_\infty$ optimal solution can be approximated to an arbitrary degree of closeness by decreasing the bound until a Riccati solution no longer exists. A formal derivation of the fact that the Riccati solution is required for the existence of a suboptimal full information control can be found in [1], page 224.

The $\mathcal{H}_\infty$ suboptimal controller presented is not unique. A family of controllers can be generated that all satisfy the bound (9.6), whenever a solution exists. The controller

[5] This model can be obtained by inverting the state model for the system that generates $(w - \gamma^{-2}\mathbf{B}_w^T\mathbf{P}x)$ from w (see Exercise 9.7).

presented above is often referred to as the central, or minimum entropy, controller. The generation of this family of controllers is discussed in [1], page 232, and [2], page 431.

9.2.4 Steady-State Full Information Control

The feedback gain in Example 9.1 approaches a steady-state value far from the final time. When operating over infinite time intervals, all finite times are infinitely far from the final time. The time-invariant controller that utilizes the steady-state gain is therefore a solution to the $\mathcal{H}_\infty$ suboptimal control problem when the time interval is infinite (provided this gain internally stabilizes the system). This time-invariant, full information control can also be utilized to simplify controller implementation when operating over finite time intervals. In this case, the steady-state controller may not satisfy the ∞-norm bound, but is probably reasonably close provided the time interval is long compared to the gain settling time. These observations, made concerning Example 9.1, can be applied to general $\mathcal{H}_\infty$ suboptimal control problems.

The existence of a steady-state solution to the $\mathcal{H}_\infty$ suboptimal control problem is not guaranteed for all values of the ∞-norm bound (as seen in Example 9.1). A solution does exists for ∞-norm bounds sufficiently large, given that the plant is controllable from the control input and observable from the reference output (conditions included in the $\mathcal{H}_\infty$ suboptimal control problem specification). This fact can be deduced by noting that the $\mathcal{H}_\infty$ Riccati equation approaches the LQR Riccati equation as the performance bound approaches infinity. The above observability and controllability conditions are sufficient to guarantee the existence of a steady-state solution to the LQR and therefore also to guarantee the existence of a steady-state $\mathcal{H}_\infty$ suboptimal control for sufficiently large performance bounds.

For smaller bounds, the question immediately arises: Under what conditions does a steady-state solution to the $\mathcal{H}_\infty$ suboptimal control problem exist? The answer follows:

> A suboptimal solution that internally stabilizes the closed-loop system and bounds the closed-loop ∞-norm (9.4) exists if and only if there exists a positive semidefinite solution of the algebraic Riccati equation,
>
> $$\boxed{\mathbf{PA} + \mathbf{A}^T\mathbf{P} - \mathbf{P}(\mathbf{B}_u\mathbf{B}_u^T - \gamma^{-2}\mathbf{B}_w\mathbf{B}_w^T)\mathbf{P} + \mathbf{C}_y^T\mathbf{C}_y = \mathbf{0},} \tag{9.21}$$
>
> such that
>
> $$\boxed{\mathbf{A} - (\mathbf{B}_u\mathbf{B}_u^T - \gamma^{-2}\mathbf{B}_w\mathbf{B}_w^T)\mathbf{P}} \tag{9.22}$$
>
> is stable, that is, all of the eigenvalues of this matrix have negative real parts. The suboptimal controller is then given as
>
> $$\boxed{u(t) = -\mathbf{B}_u^T\mathbf{P}x(t) = -\mathbf{K}x(t).} \tag{9.23}$$

Note that condition (9.22) is included to guarantee the internal stability of the feedback system.

Equation (9.15) was used to show that the infinity norm is bounded as in (9.4), in the finite time case. The derivation of (9.15) requires the existence of the Riccati solution at all times during the interval of operation. While the above conditions are sufficient to guarantee the existence of the Riccati solution at all times, this fact is certainly not obvious. Simply taking the limit of (9.15) as the final time approaches infinity is therefore not sufficient to verify that the performance bound is achieved. An alternative method of demonstrating that (9.23) represents a solution of the $\mathcal{H}_\infty$ suboptimal control problem is presented below.

The fact that (9.23) is a solution of the $\mathcal{H}_\infty$ suboptimal control problem can be verified by forming the closed-loop state model, verifying that the closed-loop system is stable, and applying the bounded real lemma to verify that the gain satisfies the performance bound. The closed-loop state model is

$$\dot{x}(t) = (\mathbf{A} - \mathbf{B}_u\mathbf{B}_u^T\mathbf{P})x(t) + \mathbf{B}_w w(t) = \mathbf{A}_{cl}x(t) + \mathbf{B}_{cl}w(t);$$

$$y(t) = \begin{bmatrix} \mathbf{C}_y \\ \hline -\mathbf{D}\mathbf{B}_u^T\mathbf{P} \end{bmatrix} x(t) + \mathbf{0}w(t) = \mathbf{C}_{cl}x(t) + \mathbf{D}_{cl}w(t).$$

To show that this system is internally stable, add and subtract $(\mathbf{P}\mathbf{B}_u\mathbf{B}_u^T\mathbf{P})$ to the algebraic Riccati equation (9.21) and rearrange to give

$$\begin{aligned} &\mathbf{P}(\mathbf{A} - \mathbf{B}_u\mathbf{B}_u^T\mathbf{P}) + (\mathbf{A} - \mathbf{B}_u\mathbf{B}_u^T\mathbf{P})^T\mathbf{P} \\ &+ \gamma^{-2}\mathbf{P}\mathbf{B}_w\mathbf{B}_w^T\mathbf{P} + \mathbf{P}\mathbf{B}_u\mathbf{B}_u^T\mathbf{P} + \mathbf{C}_y^T\mathbf{C}_y = \mathbf{0}. \end{aligned} \tag{9.24}$$

Note that this is a Lyapunov equation:

$$\mathbf{P}\mathbf{A}_{cl} + \mathbf{A}_{cl}^T\mathbf{P} + [\gamma^{-1}\mathbf{P}\mathbf{B}_w \mid \mathbf{P}\mathbf{B}_u \mid \mathbf{C}_y^T] \begin{bmatrix} \gamma^{-1}\mathbf{B}_w^T\mathbf{P} \\ \hline \mathbf{B}_u^T\mathbf{P} \\ \hline \mathbf{C}_y \end{bmatrix} = \mathbf{0}.$$

The existence of a positive semidefinite solution $\mathbf{P}$ for this equation implies that

$$\lim_{t\to\infty} \begin{bmatrix} \gamma^{-1}\mathbf{B}_w^T\mathbf{P} \\ \hline \mathbf{B}_u^T\mathbf{P} \\ \hline \mathbf{C}_y \end{bmatrix} e^{\mathbf{A}_{cl}t} = \mathbf{0}, \tag{9.25}$$

by Appendix equation (A7.7).

Since the plant is observable from $y(t)$ (the output associated with $\mathbf{C}_y$), and the controller contains no states, the closed-loop system is observable from $y(t)$. Therefore, the closed-loop system is internally stable, since any unstable modes, modes that do not decay to zero, will appear in the output. A formal proof that (9.25) and observability imply stability is given in the Appendix; see Appendix equation (A7.8).

The bounded real lemma is used to demonstrate that the closed-loop system ∞-norm satisfies the bound (9.4). This lemma states,

> The infinity norm of the generic stable system (2.4) is bounded,
>
> $$\|\mathbf{G}(s)\|_\infty < \gamma, \tag{9.26}$$
>
> if and only if
>
> $$\bar{\sigma}(\mathbf{D}) < \gamma, \tag{9.27a}$$

and there exists a symmetric matrix $\mathbf{P}$ that satisfies the algebraic Riccati equation

$$\mathbf{P}(\mathbf{A} + \mathbf{B}\mathbf{R}^{-1}\mathbf{D}^T\mathbf{C}) + (\mathbf{A} + \mathbf{B}\mathbf{R}^{-1}\mathbf{D}^T\mathbf{C})^T\mathbf{P} \tag{9.27b}$$
$$+ \mathbf{P}\mathbf{B}\mathbf{R}^{-1}\mathbf{B}^T\mathbf{P} + \mathbf{C}^T(\mathbf{I} + \mathbf{D}\mathbf{R}^{-1}\mathbf{D}^T)\mathbf{C} = \mathbf{0},$$

such that

$$\mathbf{A} + \mathbf{B}\mathbf{R}^{-1}\mathbf{D}^T\mathbf{C} + \mathbf{B}\mathbf{R}^{-1}\mathbf{B}^T\mathbf{P} \tag{9.27c}$$

is stable, that is, has all eigenvalues with negative real parts, where $\mathbf{R} = \gamma^2\mathbf{I} + \mathbf{D}^T\mathbf{D}$.

A proof of this lemma can be found in Section A10 of the Appendix. To apply this result, it must be shown that the closed-loop state model satisfies the conditions given in (9.27).

Condition (9.27a) requires that the input-to-output coupling matrix be bounded:

$$\bar{\sigma}(\mathbf{D}_{cl}) = \bar{\sigma}(\mathbf{0}) = 0 < \gamma.$$

Further, conditions (9.21) and (9.22) are equivalent to conditions (9.27b) and (9.27c) when applied to the closed-loop system. To see this, substitute the matrices from the closed-loop state model into (9.27b):

$$\mathbf{P}(\mathbf{A} - \mathbf{B}_u\mathbf{B}_u^T\mathbf{P}) + (\mathbf{A} - \mathbf{B}_u\mathbf{B}_u^T\mathbf{P})^T\mathbf{P}$$
$$+ \gamma^{-2}\mathbf{P}\mathbf{B}_w\mathbf{B}_w^T\mathbf{P} + [\mathbf{C}^T \;\vdots\; -\mathbf{P}\mathbf{B}_u\mathbf{D}^T]\begin{bmatrix}\mathbf{C} \\ \hdashline -\mathbf{D}\mathbf{B}_u^T\mathbf{P}\end{bmatrix} = \mathbf{0}.$$

Expanding the last term in this equation and remembering that $\mathbf{D}^T\mathbf{D} = \mathbf{I}$ yields (9.24), which is equivalent to (9.21), as demonstrated above. For the closed-loop system, the matrix in (9.27c) reduces to (9.22):

$$\mathbf{A}_{cl} + \mathbf{B}_{cl}\mathbf{R}^{-1}\mathbf{D}_{cl}^T\mathbf{C}_{cl} + \mathbf{B}_{cl}\mathbf{R}^{-1}\mathbf{B}_{cl}^T\mathbf{P} = \mathbf{A} - \mathbf{B}_u\mathbf{B}_u^T\mathbf{P} + \gamma^{-2}\mathbf{B}_w\mathbf{B}_w^T\mathbf{P},$$

which is required to be stable. Summarizing, for the closed-loop system, (9.27a) is satisfied, and there exists a solution to (9.27b) such that (9.27c) is stable. Together these results imply that the closed-loop system ∞-norm is bounded as given in (9.26) and equivalently in (9.4). Since the closed-loop system is stable, this implies that the control (9.23) is an $\mathcal{H}_\infty$ suboptimal controller.

The above presentation does not demonstrate that the conditions relating to (9.21) and (9.22) are necessary for the existence of an $\mathcal{H}_\infty$ suboptimal controller.[6] But in fact, these conditions are both necessary and sufficient for the existence of such a controller. A formal proof of the necessity of these conditions can be found in [1], page 242.

EXAMPLE 9.2 An antenna is required to remain pointed at a satellite in the presence of disturbance torques caused by gravity, wind, squirrels climbing on the dish, and so on. A state model for this antenna is

$$\dot{x}(t) = \begin{bmatrix} 0 & 1 \\ 0 & -1 \end{bmatrix} x(t) + \begin{bmatrix} 0 \\ 1 \end{bmatrix} u(t) + \begin{bmatrix} 0 \\ 1 \end{bmatrix} w(t);$$
$$e(t) = [1 \quad 0]x(t),$$

[6] Actually, given the result in Section A10 of the Appendix, necessity is demonstrated for the state feedback control, but not general full information control.

where $u(t)$ is the control torque, $w(t)$ is the disturbance torque, and $e(t)$ is the tracking error of the antenna. The tracking error of less than 0.1 rad is desired, even in the presence of disturbance torques of up to 2 N-m. The control magnitude is also required to be less than 10 N-m. These specifications can be appended to the plant as weighting functions to yield the model in standard form:

$$\dot{x}(t) = \begin{bmatrix} 0 & 1 \\ 0 & -1 \end{bmatrix} x(t) + \begin{bmatrix} 0 \\ 10 \end{bmatrix} u_1(t) + \begin{bmatrix} 0 \\ 2 \end{bmatrix} w_1(t);$$

$$y(t) = \begin{bmatrix} 10 & 0 \\ 0 & 0 \end{bmatrix} x(t) + \begin{bmatrix} 0 \\ 1 \end{bmatrix} u_1(t),$$

where $w_1(t)$ and $u_1(t)$ are the normalized disturbance and control inputs, respectively. In this form, $w_1(t)$ is less than 1 and both elements of $y(t)$ are required to be less than 1. The original control input can be recovered from the normalized control:

$$u(t) = 10u_1(t).$$

A full information controller is generated for $\gamma = 1$:

$$u_1(t) = -[10.2 \quad 1.36]x(t) \Rightarrow u(t) = -[102 \quad 13.6]x(t).$$

The magnitude of the closed-loop system frequency response with this control is given in Figure 9.3. This frequency response is for the unnormalized system. Note that the maximum gain from the disturbance input to the error is 0.01, which is less than the gain of 0.05 required to meet the tracking error specification (provided the input is sinusoidal). In addition, the maximum gain from the disturbance input to the control is 1.26, which is less than the gain of 5 required to maintain a control less than 10 (again, for sinusoidal disturbance inputs). Based on these results, it is expected that the tracking error and control magnitude will meet the specified bounds even in the presence of reasonable nonsinusoidal disturbance inputs, but this is not guaranteed.

The given controller is designed solely to achieve the specified bounds on the closed-loop system gains. No specifications were included on the transient response of the control system.[7] Therefore, the transient response should be checked to determine if it is acceptable. The closed-loop poles are $-7.3 \pm 7.0j$, which should yield a reasonably good transient response.

Smaller values of the performance bound (with $\gamma > 0.2$) can be used to generate additional controllers that meet the specifications. This limit on the performance bound can be obtained by decreasing the bound, solving (9.21), and checking to see if this solution is positive definite and (9.22) is stable. Alternatively, this bound can be generated by noting that the control and the disturbance enter the system at the same point. A full information controller,

$$u(t) = -w(t),$$

would then yield perfect tracking and satisfy the bound on the control input only for values of $\gamma > 0.2$.

[7] Transient response can be mandated, to a degree, in an $\mathcal{H}_\infty$ setting by specifying the desired closed-loop frequency response.

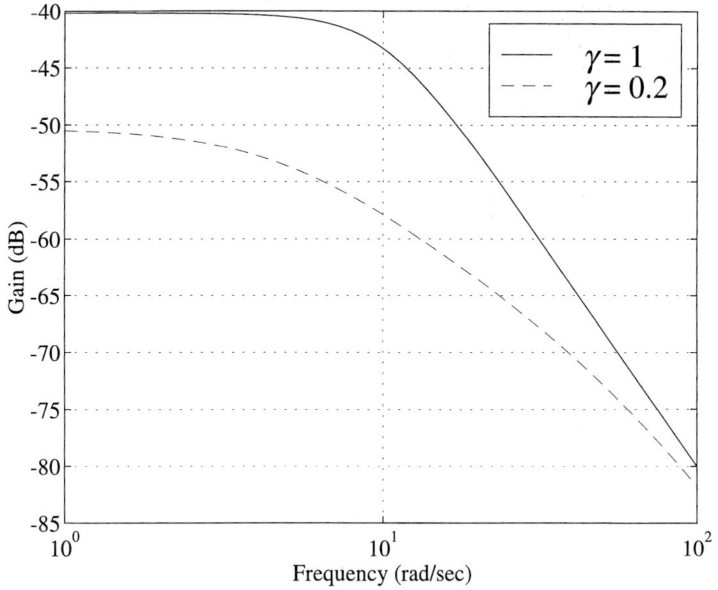

(a)

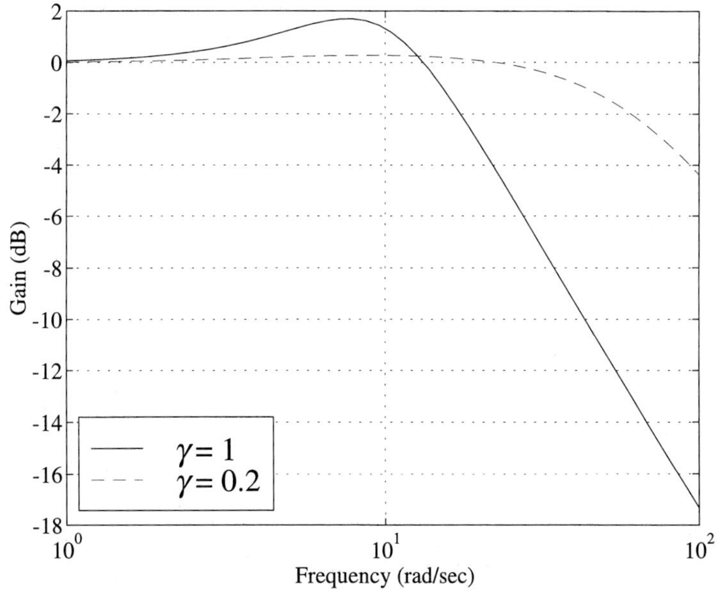

(b)

FIGURE 9.3 Frequency responses for Example 9.2: (a) tracking error; (b) control

Another $\mathcal{H}_\infty$ suboptimal controller is generated for $\gamma = 0.21$:

$$u_1(t) = -[32.8 \quad 7.39]x(t) \Rightarrow u(t) = -[328 \quad 73.9]x(t).$$

The closed-loop frequency response when using this controller is also shown in Figure 9.3. Comparing the two controllers, the use of the smaller performance bound increases the feedback gains, significantly decreases the maximum tracking error, and increases the bandwidth of the transfer function from the disturbance input to the control. This increase in bandwidth is also reflected in the closed-loop poles $\{-4.7, -70\}$. These poles are indicative of an overdamped, but probably acceptable, transient response.

The weights on the control and tracking errors can be adjusted individually to trade off tracking accuracy and control magnitude. The results of these changes are analogous to the results obtained when changing the weighting matrices in the LQR. ◆

The bandwidth of the gain between the disturbance input and the control input was observed to increase when approaching the $\mathcal{H}_\infty$ optimal solution in this example. This tends to be a general property of $\mathcal{H}_\infty$ control. This increase in bandwidth is often detrimental to the overall control system, since many actuators have finite bandwidths, and controller dynamics (which may have been ignored) become more significant at higher frequencies. Therefore, it is often desirable to use an $\mathcal{H}_\infty$ suboptimal controller that just meets (or exceeds by a safe margin of error) the specifications rather than attempt to optimize the infinity norm.

Computation of the Steady-State $\mathcal{H}_\infty$ Full Information Control The steady-state solution of the algebraic Riccati equation can be found from the eigensystem decomposition of the Hamiltonian matrix:

$$\mathbf{P} = \mathbf{\Psi}_{21}(\mathbf{\Psi}_{11})^{-1},$$

where

$$\begin{bmatrix} \mathbf{\Psi}_{11} \\ \hline \mathbf{\Psi}_{21} \end{bmatrix}$$

is a matrix whose columns are the eigenvectors of the Hamiltonian associated with the stable eigenvalues. The derivation of this result is analogous to that presented for the LQR and is therefore omitted. The steady-state Riccati solution is then used to generate the steady-state full information controller as given in (9.23).

Existence Results in Terms of the Hamiltonian The existence of an $\mathcal{H}_\infty$ suboptimal control can also be specified in terms of the Hamiltonian matrix:

> A suboptimal solution that stabilizes the system and bounds the ∞-norm of the closed-loop transfer function (9.4) exists if and only if the Hamiltonian matrix,

$$\mathfrak{X}_\infty = \left[\begin{array}{c|c} \mathbf{A} & -\mathbf{B}_u\mathbf{B}_u^T + \gamma^{-2}\mathbf{B}_w\mathbf{B}_w^T \\ \hline -\mathbf{C}_y^T\mathbf{C}_y & -\mathbf{A}^T \end{array}\right], \tag{9.28}$$

has no eigenvalues on the imaginary axis, $\mathbf{\Psi}_{11}$ is invertible, and

$$\boxed{\mathbf{P} = \mathbf{\Psi}_{21}(\mathbf{\Psi}_{11})^{-1} \geq \mathbf{0}.} \tag{9.29}$$

The suboptimal controller is then given as

$$\boxed{u(t) = -\mathbf{B}_u^T\mathbf{\Psi}_{21}(\mathbf{\Psi}_{11})^{-1}x(t) = -\mathbf{K}x(t).} \tag{9.30}$$

The fact that no $\mathcal{H}_\infty$ suboptimal control exists when the Hamiltonian has eigenvalues on the imaginary axis can be intuitively understood by considering (9.12). Purely imaginary eigenvalues give rise to undamped oscillations in the state-transition matrix of the Hamiltonian. These oscillations, in turn, yield oscillations in the time-varying Riccati solution, thus guaranteeing that a limit does not exist.

The above result is simply a restatement of the previous existence result for the $\mathcal{H}_\infty$ suboptimal control problem based on the relationship between the Hamiltonian and the steady-state Riccati solution. This formulation of the existence result is presented for completeness and to aid in understanding related results in the literature.

9.2.5 Generalizations

Several assumptions have been made in the development of the full information control to simplify the mathematics. Specifically, the control was assumed to be normalized so that the input-to-output coupling matrix between the control and the reference output satisfied

$$\mathbf{D}_{yu}^T\mathbf{D}_{yu} = \mathbf{I}.$$

This condition can be achieved by proper definition (normalization) of the control input. Alternatively, the $\mathcal{H}_\infty$ suboptimal controller formulas could be modified to incorporate the use of a non-normalized control input, as given in [2], page 424. Most computer-aided design packages that allow the generation of $\mathcal{H}_\infty$ suboptimal controllers do not require normalization of the control. In particular, normalization is not required when using the μ-Analysis and Synthesis Toolbox of MATLAB.

The assumption that

$$\mathbf{D}_{yu}^T\mathbf{C}_y = \mathbf{0}$$

is also made in the derivation of the $\mathcal{H}_\infty$ suboptimal controller formulas. This assumption specifies that there are no cross terms between the control and the state in the cost function. These cross terms can be incorporated into the problem statement, yielding an increase in the complexity of the $\mathcal{H}_\infty$ suboptimal controller formulas. These more general formulas can be found in [2], page 449, and software for finding the associated controllers can be found in the μ-Analysis and Synthesis Toolbox of MATLAB.

Terminal state weighting can be employed in the differential game used to generate the $\mathcal{H}_\infty$ full information control:

$$J_\gamma(x, u, w) = \|y\|_2^2 - \gamma^2\|w\|_2^2 + x^T(t_f)\mathbf{H}x(t_f).$$

The resulting controller (if it exists) is still an $\mathcal{H}_\infty$ full information suboptimal controller. This controller can be found by solving the Riccati equation with the final condition

$$\mathbf{P}(t_f) = \mathbf{H},$$

and using this Riccati solution to find the feedback gain. Note that the final condition can influence the existence of a Riccati equation solution, in general. The use of final state weighting has a similar effect on the $\mathcal{H}_\infty$ full information controller as on the LQR; that is, it forces the final state closer to zero.

The $\mathcal{H}_\infty$ suboptimal full information controller is not unique. The set of all controllers that satisfy the bound can be constructed by adding terms to the given state feedback controller. A second optimization may then be performed, if desired, over this set to yield controllers that have additional properties, for example, robustness. Parameterizations for the set of all $\mathcal{H}_\infty$ suboptimal full information controllers that satisfy a given bound can be found in [1], page 250, and [2], page 425.

9.3 $\mathcal{H}_\infty$ Estimation

The linear quadratic Gaussian control is generated by an optimal state feedback control law operating on estimates of the state. The states, in this case, are estimated using the Kalman filter. A similar structure exists for the $\mathcal{H}_\infty$ output feedback controller, as presented in the next chapter. Before discussing this structure in detail, the $\mathcal{H}_\infty$ estimation problem is posed and solved.

Two fundamental differences exist between the Kalman filter and the $\mathcal{H}_\infty$ optimal estimator (or filter). First, the $\mathcal{H}_\infty$ filter is optimal in terms of minimizing the ∞-norm of the gain between a set of disturbance inputs, and the estimation error. This performance criteria specifies that the worst-case gain be minimized. In contrast, the Kalman filter minimizes the mean square estimation error, or equivalently, minimizes the mean square gain between the disturbances and the estimation error.

The second difference stems from the fact that minimum mean square estimation commutes with linear operations; that is, the minimum mean square estimate of any linear combination of the state is simply the same linear combination of the optimal state estimate. The Kalman filter, which provides optimal estimates of the state, can therefore also be used to estimate any linear combination of the state. Minimal ∞-norm estimation does not possess this property, and the optimal $\mathcal{H}_\infty$ estimator depends on the plant output being estimated.

The specification of the $\mathcal{H}_\infty$ estimation problem requires a model of the plant and a cost function. The plant is modeled as follows:

$$\dot{x}(t) = \mathbf{A}x(t) + \mathbf{B}_u u(t) + \mathbf{B}_w w(t); \tag{9.31a}$$

$$m(t) = \mathbf{C}_m x(t) + \mathbf{D}_{mw} w(t), \tag{9.31b}$$

where $u(t)$ is a known input, and $w(t)$ is an unknown (but not necessarily random) disturbance input. The plant is assumed to be observable from the measurement and controllable from the disturbance input.[8] The matrices $\mathbf{B}_w$ and $\mathbf{D}_{mw}$ are assumed to satisfy the following conditions:

$$\mathbf{D}_{mw}\mathbf{B}_w^T = \mathbf{0}; \tag{9.31c}$$

$$\mathbf{D}_{mw}\mathbf{D}_{mw}^T = \mathbf{I}. \tag{9.31d}$$

Condition (9.31c) specifies that the disturbances entering the plant via the state equation (similar to plant noise) and the disturbances entering the plant via the measurement equation (similar to measurement noise) must be distinct. This condition is akin to requiring the measurement and plant noises be independent in the Kalman filter setting. Condition (9.31d) specifies that the output equation must be scaled to normalize the input-to-output coupling matrix between the disturbance and the measurement. This normalization can always be accomplished provided $\mathbf{D}_{mw}$ has full rank. Assuming that the output equation is normalized is not necessary to the theory, but simplifies the subsequent derivations.

The $\mathcal{H}_\infty$ filter estimates linear combinations of the state

$$y(t) = \mathbf{C}_y x(t), \tag{9.32}$$

given the measured output of the plant. An optimal $\mathcal{H}_\infty$ estimator generates estimates that minimize the worst-case gain between the disturbance input and the estimation error $e(t) = y(t) - \hat{y}(t)$:

$$J = \|\mathbf{G}_{ew}\|_{\infty[0,t_f]} = \sup_{w \neq 0} \frac{\|y - \hat{y}\|_{2,[0,t_f]}}{\|w\|_{2,[0,t_f]}}. \tag{9.33}$$

This infinity norm can be defined over either a finite or an infinite time interval. The estimator is required to be stable when operating over an infinite time interval. The plant and estimator are shown in Figure 9.4, where the estimator is a linear system (not necessarily time-invariant) denoted $\mathcal{G}(\bullet)$.

The $\mathcal{H}_\infty$ estimation problem is solved by utilizing the duality between estimation and control. Duality was first presented to explain the similarity between the equations for the linear quadratic regulator and the Kalman filter. The following section provides a more detailed treatment of duality based on the adjoint system.

9.3.1 The Adjoint System

The duality between the control and estimation problems can be explained by noting that these problems are related via the adjoint operation. The adjoint system is a

[8] In general, the plant need only be detectable from the measurement and stabilizable from the disturbance input.

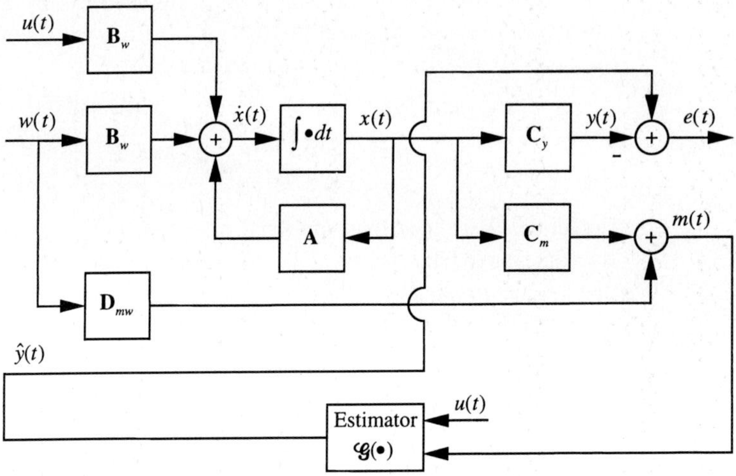

FIGURE 9.4 The estimator problem in standard form

modification of another system that has the same ∞-norm and 2-norm as the original system.[9]

A time-varying linear system (referred to below as the original system) is given:

$$\dot{x}(t) = \mathbf{A}(t)x(t) + \mathbf{B}(t)u(t);$$

$$y(t) = \mathbf{C}(t)x(t) + \mathbf{D}(t)u(t).$$

The adjoint system (associated with the original system) is defined as

$$\dot{\tilde{x}}(\tau) = \mathbf{A}^T(t_f - \tau)\tilde{x}(\tau)\mathbf{C}^T(t_f - \tau)\tilde{y}(\tau);$$

$$\tilde{u}(\tau) = \mathbf{B}^T(t_f - \tau)\tilde{x}(\tau) + \mathbf{D}^T(t_f - \tau)\tilde{y}(\tau).$$

Note that the input, output, and state of the adjoint system are distinct from the input, output, and state of the original system. Block diagrams of the original system and the adjoint system are shown in Figure 9.5. Comparing these block diagrams, note that the adjoint system is the "reverse" of the original system.[10] In general, the adjoint of a collection of subsystems is the adjoint of each subsystem connected in reverse order.

Two important properties of the adjoint system are that it has the same 2-norm and the same ∞-norm as the original system. These facts are rigorously demonstrated for time-varying systems in Section A11 of the Appendix. A simpler demonstration of the equivalence between the ∞-norms of the original system and the adjoint system is given

---

[9] Adjoints are typically defined by the relationship where $< \mathcal{G}x, y > = < x, \tilde{\mathcal{G}}y >$, where $\mathcal{G}$ is a linear operator, and $< \bullet, \bullet >$ is an inner product. The adjoint system defined in this section is not a true adjoint, but is closely related to the true adjoint. Some authors refer to the system defined in this section as the modified adjoint system.

[10] The time variable in the adjoint system is typically defined as $\tau = t_f - t$, which reinforces the idea of the adjoint system being the reverse of the original system.

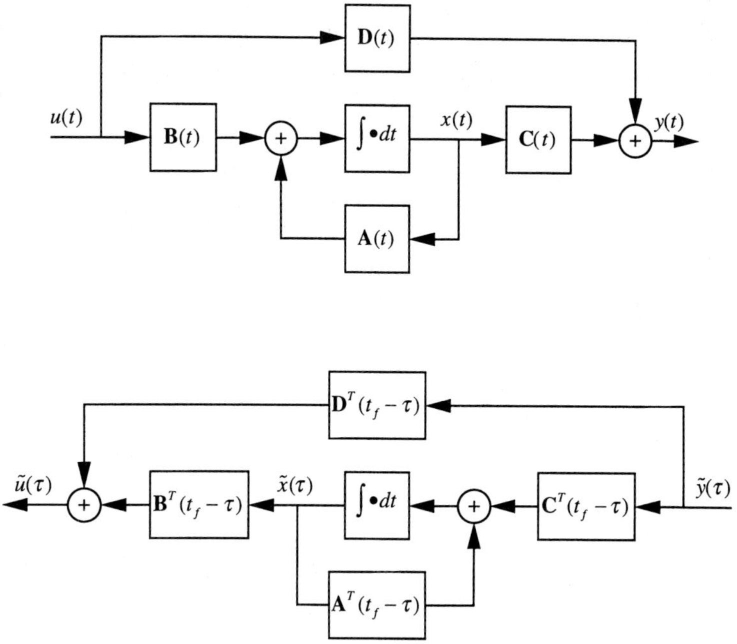

FIGURE 9.5 Relationship between the adjoint system and the original system: (a) the original system; (b) the adjoint system

below. For this demonstration, the systems are assumed to be time-invariant, and the ∞-norm is defined over an infinite time interval.

The transfer function of a time-invariant system can be related to the transfer function of the adjoint system. The transfer function of the original system is

$$\mathbf{G}(s) = \mathbf{C}(s\mathbf{I} - \mathbf{A})^{-1} + \mathbf{D}.$$

Taking the transpose of this transfer function yields the transfer function of the adjoint system:

$$\mathbf{G}^T(s) = \mathbf{B}^T(s\mathbf{I} - \mathbf{A}^T)^{-1}\mathbf{C}^T + \mathbf{D}^T = \tilde{\mathbf{G}}(s).$$

The infinity norm of the adjoint system is then given:

$$\|\tilde{\mathbf{G}}\|_\infty = \sup_\omega\{\bar{\sigma}[\tilde{\mathbf{G}}(j\omega)]\} = \sup_\omega\{\bar{\sigma}[\mathbf{G}^T(j\omega)]\} = \sup_\omega\{\bar{\sigma}[\mathbf{G}(j\omega)]\} = \|\mathbf{G}\|_\infty.$$

The fact that the maximum singular value of a matrix equals the maximum singular value of the matrix transposed can be easily understood by noting that the nonzero singular values of a matrix $\mathbf{M}$ are the nonzero eigenvalues of $\mathbf{M}\mathbf{M}^T$, which equal the nonzero eigenvalues of $\mathbf{M}^T\mathbf{M}$.

9.3.2 Finite-Time Estimation

The $\mathcal{H}_\infty$ estimation problem is solved by using the adjoint to convert the estimation problem into an equivalent control problem. The full information results generated in the previous section can be used to solve this equivalent control problem. The $\mathcal{H}_\infty$ estimator is then obtained by taking the adjoint of the resulting controller.

Paralleling the full information results, suboptimal estimators are generated that yield a given bound on the infinity norm from the disturbance input to the estimation error. General conditions can be given for the existence of these suboptimal estimators. The optimal estimator can then be approximated to an arbitrary degree of closeness by decreasing the bound until a solution no longer exists.

The equations for the plant and estimator (see Figure 9.4) are summarized as follows:

$$\dot{x}(t) = \mathbf{A}x(t) + \mathbf{B}_w w(t); \tag{9.34a}$$

$$m(t) = \mathbf{C}_m x(t) + \mathbf{D}_{mw} w(t); \tag{9.34b}$$

$$e(t) = \mathbf{C}_y x(t) - \hat{y}(t); \tag{9.34c}$$

$$\hat{y}(t) = \mathcal{G}\{m(t)\}, \tag{9.34d}$$

where $\mathcal{G}(\bullet)$ denotes a linear (not necessarily time-invariant) estimator. The input to the model (9.34) is $w(t)$, and the output is $e(t)$. The known input $u(t)$ has been removed from this formulation of the estimation problem to simplify the subsequent derivations. The known input can be ignored without loss of generality, since the effects of this input on the output are easily computed. Superposition can then be invoked to add the effects of the known input into the estimator.

The adjoint of this system is obtained by taking the adjoint of the plant (9.34a) through (9.34c), taking the adjoint of the estimator (9.34d), and combining the results:

$$\dot{\tilde{x}}(\tau) = \mathbf{A}^T\tilde{x}(\tau) + \mathbf{C}_m^T\tilde{m}(\tau) + \mathbf{C}_y^T\tilde{e}(\tau); \tag{9.35a}$$

$$\tilde{w}(\tau) = \mathbf{B}_w^T\tilde{x}(\tau) + \mathbf{D}_{mw}^T\tilde{m}(\tau); \tag{9.35b}$$

$$\tilde{y}(\tau) = -\tilde{e}(\tau); \tag{9.35c}$$

$$\tilde{m}(\tau) = \tilde{\mathcal{G}}\{\tilde{y}(\tau)\}. \tag{9.35d}$$

A block diagram of the adjoint system is shown in Figure 9.6.

The system in Figure 9.6 is a control system. For this system, the controller $\tilde{\mathcal{G}}\{\bullet\}$ can be selected to bound the cost:

$$J = \|\mathbf{G}_{\tilde{w}\tilde{e}}\|_{\infty,[0,t_f]} = \sup_{\|\tilde{e}\|_{2,[0,t_f]}\neq 0} \frac{\|\tilde{w}\|_{2,[0,t_f]}}{\|\tilde{e}\|_{2,[0,t_f]}} < \gamma. \tag{9.36}$$

This suboptimal control problem is very similar to the full information control problem. Note that the plant model (9.35a) and (9.35b) has exactly the form of (9.2a) and (9.2b) when $\tilde{m}(\tau)$ is the "control input." Further, the conditions (9.30c) and (9.30d) are equivalent to (9.2c) and (9.2d) for the adjoint plant:

$$(\mathbf{D}_{mw}^T)^T\mathbf{B}_w^T = \mathbf{0};$$

$$(\mathbf{D}_{mw}^T)^T\mathbf{D}_{mw}^T = \mathbf{I}.$$

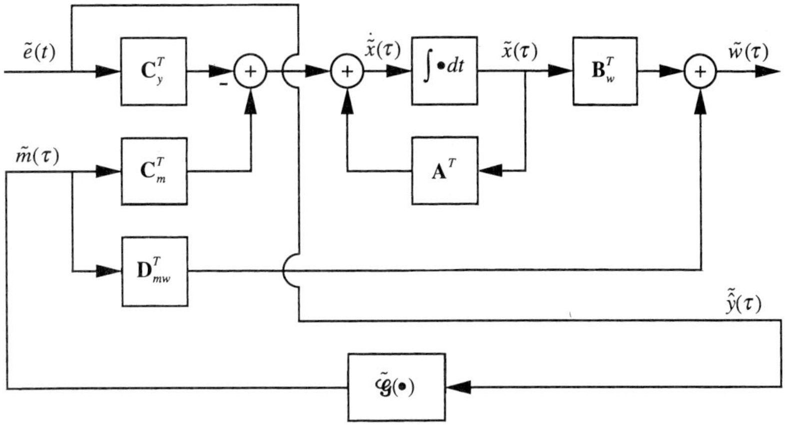

FIGURE 9.6 The adjoint of the estimator problem in standard form

Lastly, the initial condition for the adjoint system can be assumed to equal zero, since the ∞-norm only depends on the forced response.

The control problem (9.35) and (9.36) differs from the full information control problem only in the fact that the measurement available for feedback (9.35c) is the disturbance input. In contrast, both the disturbance input and the state are assumed to be available in the full information control problem. A controller that utilizes only the disturbance input is referred to as a disturbance feedforward controller.

The existence of an $\mathcal{H}_\infty$ full information suboptimal control for the plant (9.35) is necessary for the existence of a solution to the disturbance feedforward control problem. This observation is based on the fact that a disturbance feedforward controller is a special case of a full information controller. Less obvious is the fact that the full information control problem is mathematically equivalent to the disturbance feedforward control problem.

The equivalence of the full information and disturbance feedforward control problems is demonstrated by noting that the state of a plant can be perfectly reconstructed from the initial condition and the inputs. Since the initial condition for the plant (9.35) is zero, the state can be perfectly reconstructed from the disturbance input and the control input. Since the controller always has knowledge of the control input, knowledge of the disturbance input is equivalent to full information.

The full information control results can be used to generate a suboptimal solution to the control problem in Figure 9.6. The estimator $\mathcal{G}(\cdot)$ can then be recovered from the resulting controller $\tilde{\mathcal{G}}(\cdot)$ by taking the adjoint. The resulting estimator is a suboptimal solution of the $\mathcal{H}_\infty$ estimation problem, since the ∞-norm of the closed-loop adjoint system equals the ∞-norm of the original system; see Appendix equation (A10.6).

The Riccati Equation The disturbance feedforward controller is found by utilizing the full information results to yield a state feedback controller. The state used in this controller is reconstructed by applying the control and disturbance inputs to the adjoint model. The resulting controller is given:

$$\dot{\tilde{x}}(\tau) = \mathbf{A}^T\tilde{x}(\tau) + \mathbf{C}_m^T\tilde{m}(\tau) + \mathbf{C}_y^T\tilde{e}(\tau);$$
$$\tilde{m}(\tau) = -\tilde{\mathbf{G}}(\tau)\tilde{x}(\tau),$$

or, equivalently,

$$\dot{\tilde{x}}(\tau) = [\mathbf{A}^T - \mathbf{C}_m^T\tilde{\mathbf{G}}(\tau)]\tilde{x}(\tau) + \mathbf{C}_y^T\tilde{e}(\tau); \tag{9.37a}$$
$$\tilde{m}(\tau) = \tilde{\mathbf{G}}(\tau)\tilde{x}(\tau). \tag{9.37b}$$

The state feedback gain in (9.37b) is equal to the full information control:

$$\tilde{\mathbf{G}}(\tau) = -\mathbf{C}_m\tilde{\mathbf{Q}}(\tau). \tag{9.38}$$

The matrix $\tilde{\mathbf{Q}}(\tau)$ is found by solving the following Riccati differential equation:

$$\dot{\tilde{\mathbf{Q}}}(\tau) = -\tilde{\mathbf{Q}}(\tau)\mathbf{A}^T - \mathbf{A}\tilde{\mathbf{Q}}(\tau) - \mathbf{B}_w\mathbf{B}_w^T + \tilde{\mathbf{Q}}(\tau)(\mathbf{C}_m^T\mathbf{C}_m - \gamma^{-2}\mathbf{C}_y^T\mathbf{C}_y)\tilde{\mathbf{Q}}(\tau) \tag{9.39a}$$

backward in time from the final condition

$$\tilde{\mathbf{Q}}(t_f) = \mathbf{0}. \tag{9.39b}$$

The $\mathcal{H}_\infty$ suboptimal estimator is the adjoint of the controller (9.37):

$$\dot{\hat{x}}(\tau) = [\mathbf{A} - \tilde{\mathbf{G}}^T(t_f - t)\mathbf{C}_m]\hat{x}(t) + \tilde{\mathbf{G}}^T(t_f - t)m(t);$$
$$e(t) = \mathbf{C}_y\tilde{x}(t).$$

Defining,

$$\mathbf{G}(t) = \tilde{\mathbf{G}}^T(t_f - t), \tag{9.40}$$

adding the known input into the estimator equation, and rearranging yields the $\mathcal{H}_\infty$ suboptimal estimator:

$$\dot{\hat{x}}(t) = \mathbf{A}\hat{x}(t) + \mathbf{B}_u u(t) + \mathbf{G}(t)\{m(t) - \mathbf{C}_y\hat{x}(t)\}; \tag{9.41a}$$
$$\hat{y}(t) = \mathbf{C}_y\hat{x}(t). \tag{9.41b}$$

A block diagram of this estimator is shown in Figure 9.7. Note that this estimator has the structure of the Kalman filter, but the estimator gain is selected to minimize the ∞-norm criterion.

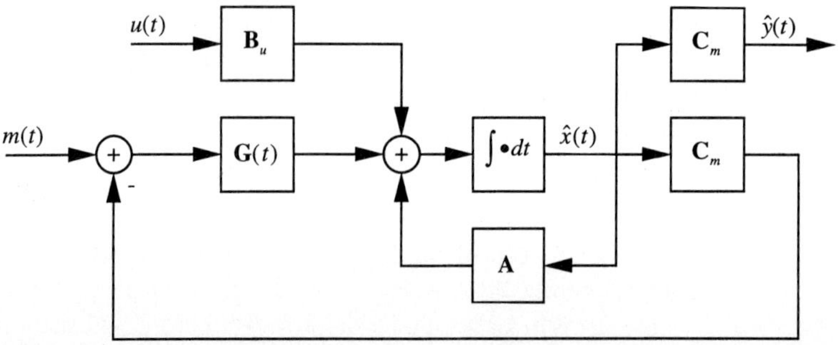

FIGURE 9.7 The $\mathcal{H}_\infty$ optimal estimator

The Riccati equation (9.39) can be placed in a more convenient form by performing the change of variables $\tau = t_f - t$, and defining

$$\mathbf{Q}(t) = \tilde{\mathbf{Q}}(t_f - \tau).$$

A Riccati equation for this new matrix is then

$$\boxed{\dot{\mathbf{Q}}(t) = \mathbf{Q}(t)\mathbf{A}^T + \mathbf{A}\mathbf{Q}(t) + \mathbf{B}_w\mathbf{B}_w^T - \mathbf{Q}(t)(\mathbf{C}_m^T\mathbf{C}_m - \gamma^{-2}\mathbf{C}_y^T\mathbf{C}_y)\mathbf{Q}(t).} \tag{9.42a}$$

This Riccati equation is solved forward in time from the initial condition

$$\boxed{\mathbf{Q}(0) = \mathbf{0}.} \tag{9.42b}$$

The estimator gain can be written in terms of this new Riccati solution:

$$\boxed{\mathbf{G}(t) = \mathbf{Q}(t)\mathbf{C}_m^T.} \tag{9.43}$$

Equations (9.41) through (9.43) completely specify the suboptimal $\mathcal{H}_\infty$ estimator. Note that the Riccati equation and therefore the estimator gain depend on the output being estimated, since this equation contains $\mathbf{C}_y$.

EXAMPLE 9.3 An $\mathcal{H}_\infty$ estimator can be applied to the radar range tracking of an aircraft (first presented in Example 7.2). A state equation for the range $r(t)$ and radial velocity of an aircraft is

$$\begin{bmatrix} \dot{r}(t) \\ \ddot{r}(t) \end{bmatrix} = \begin{bmatrix} 0 & 1 \\ 0 & 0 \end{bmatrix}\begin{bmatrix} r(t) \\ \dot{r}(t) \end{bmatrix} + \begin{bmatrix} 0 \\ 1 \end{bmatrix} w(t).$$

The range measurements are given,

$$m(t) = [1 \quad 0]\begin{bmatrix} r(t) \\ \dot{r}(t) \end{bmatrix} + v(t),$$

and the output to be estimated is the entire state:

$$y(t) = \begin{bmatrix} 1 & 0 \\ 0 & 1 \end{bmatrix}\begin{bmatrix} r(t) \\ \dot{r}(t) \end{bmatrix}.$$

The plant input and measurement error are assumed to be bounded:

$$\begin{bmatrix} |w(t)| \\ |v(t)| \end{bmatrix} \le \begin{bmatrix} 4 \ m/\text{sec}^2 \\ 20 \ m \end{bmatrix}.$$

A state equation of the form in (9.31a) can be generated by including the measurement error as part of the disturbance input:

$$\begin{bmatrix} \dot{r}(t) \\ \ddot{r}(t) \end{bmatrix} = \begin{bmatrix} 0 & 1 \\ 0 & 0 \end{bmatrix}\begin{bmatrix} r(t) \\ \dot{r}(t) \end{bmatrix} + \begin{bmatrix} 0 & 0 \\ 4 & 0 \end{bmatrix}\begin{bmatrix} w_1(t) \\ v_1(t) \end{bmatrix}.$$

The inputs are normalized, and the plant input bound is included as a weight in this state equation. Appending the measurement error bound to the measurement equation yields

$$m(t) = [1 \quad 0]\begin{bmatrix} r(t) \\ \dot{r}(t) \end{bmatrix} + [0 \quad 20]\begin{bmatrix} w_1(t) \\ v_1(t) \end{bmatrix}.$$

Normalizing this equation so that the input-to-output coupling matrix satisfies the constraint (9.31d) gives

$$m_1(t) = [\tfrac{1}{20} \quad 0]\begin{bmatrix} r(t) \\ \dot{r}(t) \end{bmatrix} + [0 \quad 1]\begin{bmatrix} w_1(t) \\ v_1(t) \end{bmatrix},$$

where $m(t) = 20m_1(t)$. This measurement equation and the state equation (including the input bound) are in a form appropriate for applying (9.41) through (9.43) for $\mathcal{H}_\infty$ estimator synthesis.

The resulting $\mathcal{H}_\infty$ estimator is given as

$$\dot{\hat{x}}(t) = \mathbf{A}\hat{x}(t) + \mathbf{G}_1(t)\{m_1(t) - [\tfrac{1}{20} \quad 0]\hat{x}(t)\};$$

$$\hat{y}(t) = \mathbf{C}_y\hat{x}(t).$$

The state equation for this estimator can be written in terms of the original measurement:

$$\begin{aligned}\dot{\hat{x}}(t) &= \mathbf{A}\hat{x}(t) + \tfrac{1}{20}\mathbf{G}_1(t)\{20m_1(t) - 20[\tfrac{1}{20} \quad 0]\hat{x}(t)\} \\ &= \mathbf{A}\hat{x}(t) + \mathbf{G}(t)\{m(t) - [1 \quad 0]\hat{x}(t)\},\end{aligned}$$

where $\mathbf{G}(t) = \mathbf{G}_1(t)/20$.[11]

The plant and the $\mathcal{H}_\infty$ estimator are simulated. The disturbance input and the measurement error are both assumed to be discrete-time white noise, uniformly distributed within the given bounds. An initial estimation error is included in the simulation to display the transient performance of the estimator, even though this initial error is assumed to be zero during estimator development.

The estimator gains and the estimated outputs (the states in this case) are shown in Figure 9.8, where the estimator performance bound is: $\gamma = 22.5$. This particular performance bound is selected because it is roughly 10% larger than the optimal bound.[12] The gains converge to steady-state values. Also, the estimates converge in slightly over 10 seconds for both position and velocity, and the estimator tracks the actual state. Figure 9.9 shows the estimator gains and the estimated output when the bound on the measurement error is decreased to 2. In this case, an estimator performance bound of $\gamma = 3.5$ (~10% over optimal) is used. The estimates are seen to converge in less than

[11] The normalization of the measurement equation and conversion of the estimator state equation to utilize the original measurements can typically be handled by CAD software packages. For example, normalization of the measurement is not required when using the μ-Synthesis and Analysis Toolbox of MATLAB.

[12] Numerical difficulties are frequently encountered when doing discrete-time implementations of near-optimal $\mathcal{H}_\infty$ estimators. These difficulties are mostly avoided by making the bound slightly larger than optimal.

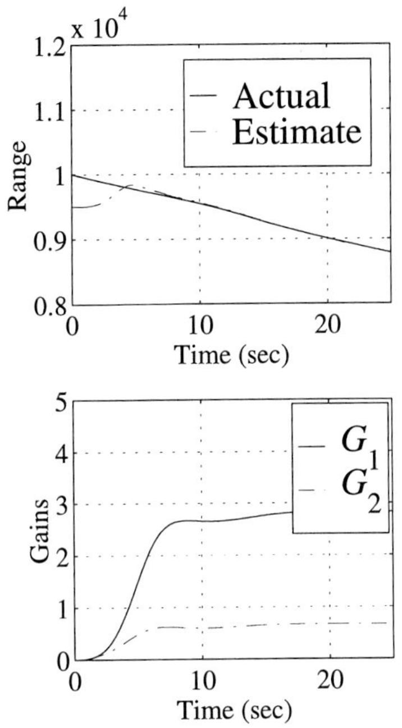

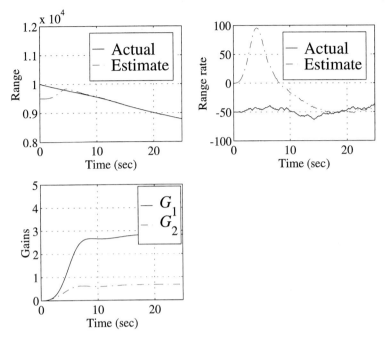

FIGURE 9.8 Baseline results for Example 9.3

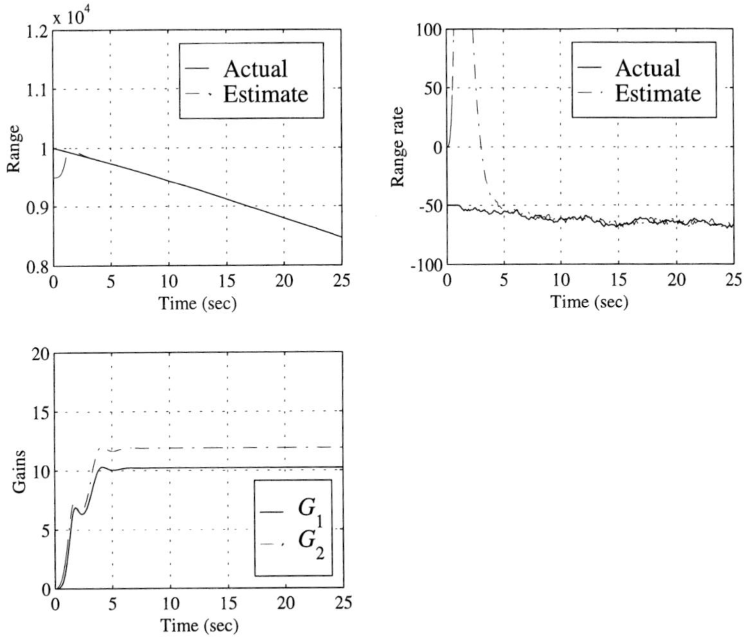

FIGURE 9.9 Results for Example 9.3 with decreased measurement error

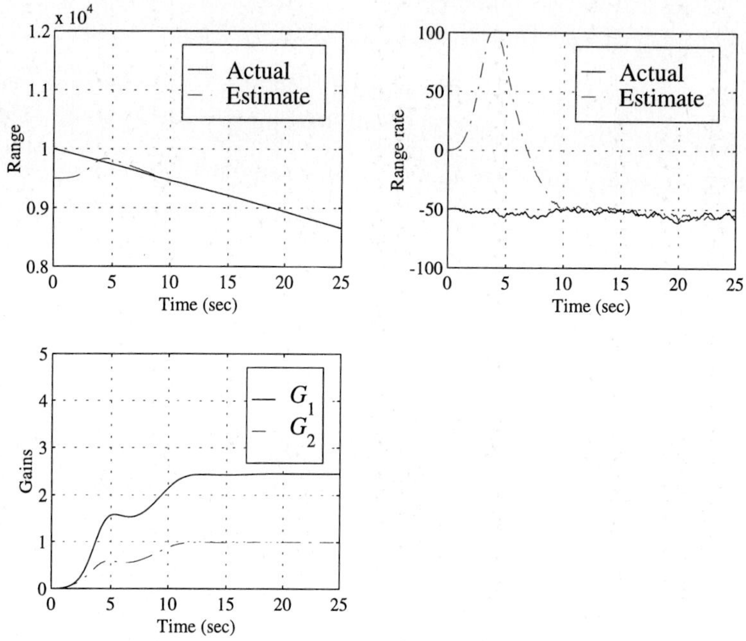

FIGURE 9.10 Results for Example 9.3 with only range rate estimation

5 seconds, about twice as fast as in the baseline case. This faster rate of convergence results from the larger estimator gains, and also increases the overshoot in the velocity estimate. This is expected from the Kalman filtering results, where faster convergence occurs when the spectral density of the measurement noise is decreased.

Figure 9.10 shows the estimator gains and the estimated states when the reference output is only the range rate:

$$y(t) = [0 \quad 1]\begin{bmatrix} r(t) \\ \dot{r}(t) \end{bmatrix}.$$

Note both range and range rate are still available as the estimator states and are therefore included in the figure. In this case, an estimator performance bound of $\gamma = 10$ ($\sim$10% over optimal) was used. When exclusively estimating range rate, the gains change slightly. Again, it is worth noting that the $\mathcal{H}_\infty$ estimator depends on the output of interest. ◆

The Riccati equation (9.42) approaches the Kalman Riccati equation as the performance bound γ approaches infinity.[13] A solution is therefore guaranteed to exist for

[13] The covariance matrices of the plant driving noise and the measurement noise are assumed to both equal identity matrices to yield the associated Kalman filtering problem. This can be accomplished, in general, by absorbing these covariance matrices into the input matrix for the plant noise, and also by appropriately scaling the measurement equation (normalizing the measurement noise).

sufficiently large performance bounds. For finite performance bounds, this equation may or may not have a solution. In general, solutions exist for performance bounds above a limiting value equal to the optimal value, where optimal is defined in terms of minimizing the norm (9.33).

The Hamiltonian Equations The suboptimal full information controller, and therefore the suboptimal estimator, can be generated by solving a Riccati equation or by solving the Hamiltonian equations. The Hamiltonian equations for the estimator can be combined to form the Hamiltonian system:

$$\begin{bmatrix} \dot{\tilde{x}}(t) \\ \dot{\tilde{p}}(t) \end{bmatrix} = \begin{bmatrix} \mathbf{A}^T & -\mathbf{C}_m^T\mathbf{C}_m + \gamma^{-2}\mathbf{C}_y^T\mathbf{C}_y \\ -\mathbf{B}_w\mathbf{B}_w^T & -\mathbf{A} \end{bmatrix} \begin{bmatrix} \tilde{x}(t) \\ \tilde{p}(t) \end{bmatrix} = \mathcal{Y}_\infty \begin{bmatrix} \tilde{x}(t) \\ \tilde{p}(t) \end{bmatrix}, \tag{9.44}$$

where the matrix $\mathcal{Y}_\infty$ is called the Hamiltonian.

The $\mathcal{H}_\infty$ suboptimal estimator gain can be found by solving the Hamiltonian system subject to the final condition

$$\tilde{p}(t_f) = 0.$$

This gain is then given as

$$\mathbf{G}(t) = -\mathbf{Q}(t)\mathbf{C}_m^T = -\mathbf{\Phi}_{21}^T(t)\{\mathbf{\Phi}_{22}(t)\}^{-T}\mathbf{C}_m^T = -\{\mathbf{\Phi}_{22}(t)\}^{-1}\mathbf{\Phi}_{21}(t)\mathbf{C}_m^T, \tag{9.45}$$

where the terms $\mathbf{\Phi}_{ij}$ are blocks in the state-transition matrix of the Hamiltonian system:

$$e^{\mathcal{Y}_\infty t} = \begin{bmatrix} \mathbf{\Phi}_{11}(t) & \mathbf{\Phi}_{12}(t) \\ \mathbf{\Phi}_{21}(t) & \mathbf{\Phi}_{22}(t) \end{bmatrix}.$$

9.3.3 Steady-State Estimation

The $\mathcal{H}_\infty$ suboptimal estimator becomes time-invariant far from the initial time provided the bound is chosen sufficiently large that a steady-state solution to the Riccati equation exists. When operating over long time intervals, it is often desirable to use the steady-state estimator to simplify implementation. Mathematically, the steady-state estimator must be stable for the ∞-norm (9.33) to be defined over infinite time intervals. In practice, stable estimators are also desirable for operation over moderate to long time intervals, especially when initial errors and unmodeled dynamics are present. Thus the steady-state $\mathcal{H}_\infty$ suboptimal estimator problem is to find a stable, time-invariant estimator that satisfies an ∞-norm (defined over an infinite time interval) bound on the transfer function from the disturbance input to the estimation error.

Conditions for the existence of an $\mathcal{H}_\infty$ suboptimal estimator are summarized below. When these conditions are satisfied, the steady-state estimator is given by (9.41), with the gain found by using the steady-state Riccati solution in (9.42). These existence results and estimator equations are given without proof, since they can be related to the full information results via duality.

An estimator that is stable and bounds the ∞-norm of the closed-loop transfer function (9.33) exists if and only if there exists a positive semidefinite solution of the algebraic Riccati equation

$$\boxed{\mathbf{AQ} + \mathbf{QA}^T - \mathbf{Q}(\mathbf{C}_m^T\mathbf{C}_m - \gamma^{-2}\mathbf{C}_y^T\mathbf{C}_y)\mathbf{Q} + \mathbf{B}_w\mathbf{B}_w^T = \mathbf{0},} \tag{9.46}$$

such that the matrix

$$\boxed{\mathbf{A} - \mathbf{Q}(\mathbf{C}_m^T\mathbf{C}_m - \gamma^{-2}\mathbf{C}_y^T\mathbf{C}_y)} \tag{9.47}$$

is stable.

The steady-state Riccati solution can be found from the eigensolution of the Hamiltonian:

$$\mathbf{Q} = \mathbf{\Psi}_{21}(\mathbf{\Psi}_{21})^{-1},$$

where

$$\begin{bmatrix} \mathbf{\Psi}_{11} \\ \hdashline \mathbf{\Psi}_{21} \end{bmatrix}$$

is a matrix whose columns are the eigenvectors of the Hamiltonian associated with the stable eigenvalues. As observed previously, the $\mathcal{H}_\infty$ optimal estimator can be approximated to an arbitrary degree of closeness by iteration of the bound and testing for the existence of suboptimal estimators.

EXAMPLE 9.4 An ac motor is described by the following state equation:

$$\dot{x}(t) = -x(t) + 100u(t) + 100w(t),$$

where $x(t)$ is the rotational velocity of the shaft, $u(t)$ is the nominal input of 10 volts rms, and $w(t)$ is the error between the actual applied voltage and the nominal value. This error is assumed to be caused by modulation of the ac envelope due to harmonics on the power line. This modulation is assumed to be less than 1%, implying the disturbance input can be bounded:

$$|w(t)| \leq 0.1 \text{ volts rms.}$$

An $\mathcal{H}_\infty$ estimator is used to estimate the true rotational velocity of the shaft from the noisy tachometer measurements:

$$m(t) = x(t) + v(t),$$

where $v(t)$ is the measurement error. This error consists of a possible dc bias and sinusoidal interference, and is bounded:

$$|v(t)| \leq 1 \text{ volts rms.}$$

The output of interest is simply the state.

The Hamiltonian for this estimation problem, with the bound on the disturbance input appended to the plant as a weight, is

$$\mathcal{Y}_\infty = \begin{bmatrix} -1 & -1 + \gamma^{-2} \\ -100 & 1 \end{bmatrix}.$$

Letting $\gamma = 1.1$, the eigenvalues and eigenvectors of the Hamiltonian are

$$\lambda_1 = -4.28 \ ; \ \lambda_2 = 4.28 \ ; \ \phi_1 = \begin{bmatrix} -0.0528 \\ -0.9986 \end{bmatrix} \ ; \ \phi_2 = \begin{bmatrix} 0.0328 \\ -0.9995 \end{bmatrix}.$$

The resulting normalized estimator gain is

$$G_1 = 18.92,$$

yielding the estimator

$$\begin{aligned} \dot{\hat{x}}(t) &= -\hat{x}(t) + 100u(t) + 1.892\{m(t) - \hat{x}(t)\} \\ &= -2.892\hat{x}(t) + 100u(t) + 1.892m(t). \end{aligned}$$

The output of interest is the state in this example, so no output equation is required.

The plant and the $\mathcal{H}_\infty$ estimator are simulated. The disturbance input is a 2 Hz sinusoid of amplitude 0.1. The measurement error is a dc bias with an amplitude of 1. The initial state and initial estimate are both set equal to 1000 rad/sec, which is the nominal motor shaft velocity, since only steady-state performance is of interest. The plant state and the estimated state are shown in Figure 9.11. Note that the maximum error exceeds the performance bound. This is reasonable since the performance bound

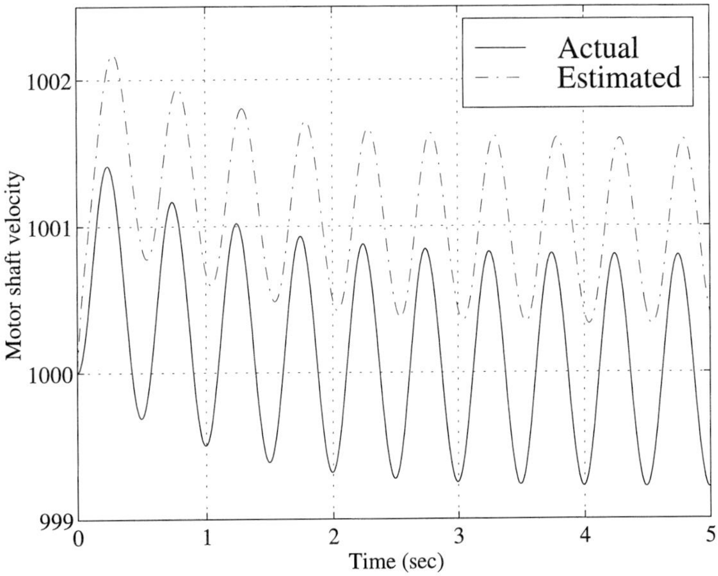

FIGURE 9.11 $\mathcal{H}_\infty$ estimation of motor shaft velocity

assumes that both inputs have the same frequency and the sum of the squares of their amplitudes is bounded by 1 (for the normalized state equation). In reality, the inputs have distinct frequencies and each achieves the maximum amplitude. Still, the performance bound provides a good indication of the observed estimation error. ◆

9.3.4 Generalizations

Normalizing the measurement error simplifies the formulas for the $\mathcal{H}_\infty$ suboptimal estimator. More general formulas can be derived that do not require this normalization. Generalizations also exist to the estimator formulas, which allow a particular disturbance input to enter both the state equation and the measurement equation. This generalization is equivalent to allowing correlation between the plant and measurement noises in the Kalman filter problem. These more general $\mathcal{H}_\infty$ suboptimal estimator formulas are derived in [2], page 449, and software for finding the associated estimators can be found in the μ-Analysis and Synthesis Toolbox of MATLAB.

The initial condition on the Riccati equation can be modified to reflect uncertainty in the initial state, with a large initial Riccati solution being used when the state is very uncertain. The initial condition on the Riccati equation has an effect on the $\mathcal{H}_\infty$ estimator that is similar to the effect of the initial condition covariance matrix on the Kalman filter. In fact, it is reasonable to use the initial condition covariance matrix when initializing the $\mathcal{H}_\infty$ Riccati equation. The resulting estimator (if it exists) is still an $\mathcal{H}_\infty$ suboptimal estimator, but the initial condition can influence the existence of a Riccati equation solution.

A parameterization exists for the set of all possible $\mathcal{H}_\infty$ suboptimal estimators that satisfy a given performance bound; see [1], page 266, and [2], page 433. This parameterization allows the designer to perform a second optimization to maximize or minimize an additional performance criteria. Note that the resulting estimator still satisfies the original estimation error bound.

Smoothing involves estimating the output based on both past and future measurements. While smoothing cannot be applied in real time, it is useful for off-line data analysis. $\mathcal{H}_\infty$ smoothing algorithms can be developed on the basis of the $\mathcal{H}_\infty$ output estimation results given in this chapter. $\mathcal{H}_\infty$ smoothing is presented in [3].

9.4 Summary

The $\mathcal{H}_\infty$ optimal full information control problem is to find a feedback controller that utilizes the plant state and the disturbance input, and minimizes the closed-loop system ∞-norm. This feedback controller is also required to internally stabilize the system in the infinite-time case. Differential game theory is used to generate a solution to the full information problem. Unfortunately, differential game theory does not directly yield a solution to the optimal full information problem. Instead, a suboptimal controller is generated that satisfies a bound on the closed-loop system ∞-norm. This suboptimal controller is state feedback with a gain that is given in terms of the solution of a Hamiltonian equation or, equivalently, in terms of the solution of a Riccati equation. General conditions are given for the existence of a suboptimal controller that satisfies a given ∞-norm bound. Therefore, the optimal solution can be approximated arbitrarily closely by decreasing the bound until a suboptimal controller no longer exists.

A suboptimal, steady-state full information controller is generated as a limit of the finite-time full information controller. The bounded real lemma is used to demonstrate that this controller satisfies the desired ∞-norm bound. The existence results for the steady-state full information controller are given both in terms of a Riccati equation and in terms of a Hamiltonian equation. In both cases, an additional condition is required to guarantee internal stability of the feedback system.

The $\mathcal{H}_\infty$ estimation problem is to estimate a linear combination of the plant state, given measurements and known plant inputs. The resulting estimator should minimize the ∞-norm of the transfer function between the disturbance input (including measurement error) and the estimation error. This estimator is also required to be stable in the infinite-time case. The $\mathcal{H}_\infty$ estimator, while technically not the dual of the full information controller, is still derived using the full information results and duality arguments. The resulting suboptimal estimators have the observer structure of the Kalman filter. The observer gain is given in terms of the solution of either a Riccati equation or a Hamiltonian equation. This gain is dependent on the output being estimated, which is a fundamental distinction between $\mathcal{H}_\infty$ estimation and Kalman filtering. General conditions that guarantee the existence of an $\mathcal{H}_\infty$ suboptimal estimator are also given. The optimal estimator can be approximated by decreasing the bound on the ∞-norm until a solution no longer exists.

Game theory can be applied directly to the discrete-time plant model to yield a suboptimal, $\mathcal{H}_\infty$ full information controller [4]. An alternative approach to discrete-time $\mathcal{H}_\infty$ controller synthesis is presented in [5]. The fundamental equations for the steady-state $\mathcal{H}_\infty$ full information controller and the steady-state $\mathcal{H}_\infty$ estimator are summarized for both continuous-time and discrete-time systems in Table 9.1.

TABLE 9.1 Summary of Formulas for Both Continuous-Time and Discrete-Time Systems

Formula	Continuous-Time	Discrete-Time
The $\mathcal{H}_\infty$ full information control problem	Given the plant $\dot{x}(t) = \mathbf{A}x(t) + \mathbf{B}_u u(t) + \mathbf{B}_w w(t)$; $y(t) = \mathbf{C}_y x(t) + \mathbf{D}_{yu} u(t)$, where $\mathbf{D}_{yu}^T\mathbf{C}_y = \mathbf{0}$; $\mathbf{D}_{yu}^T\mathbf{D}_{yu} = \mathbf{I}$, generate a feedback controller such that $\|\mathbf{G}_{yw}\|_\infty < \gamma$.	Given the plant $\dot{x}(t) = \mathbf{\Phi}x(t) + \mathbf{\Gamma}_u u(t) + \mathbf{\Gamma}_w w(t)$; $y(t) = \mathbf{C}_y x(t) + \mathbf{D}_{yu} u(t)$, where $\mathbf{D}_{yu}^T\mathbf{C}_y = \mathbf{0}$; $\mathbf{D}_{yu}^T\mathbf{D}_{yu} = \mathbf{I}$, generate a feedback controller such that[1] $\|\mathbf{G}_{yw}\|_\infty < \gamma$.
Existence conditions for a suboptimal control	There exists a solution $\mathbf{P} \geq \mathbf{0}$ of the algebraic Riccati equation $\mathbf{PA} + \mathbf{A}^T\mathbf{P} - \mathbf{P}(\mathbf{B}_u\mathbf{B}_u^T - \gamma^{-2}\mathbf{B}_w\mathbf{B}_w^T)\mathbf{P} + \mathbf{C}_y^T\mathbf{C}_y = \mathbf{0}$, where $\mathbf{A} - (\mathbf{B}_u\mathbf{B}_u^T - \gamma^{-2}\mathbf{B}_w\mathbf{B}_w^T)\mathbf{P}$ is stable.	There exists a solution $\mathbf{P} \geq \mathbf{0}$ of the algebraic Riccati equation $\mathbf{P} = \mathbf{A}^T\mathbf{PA} - \mathbf{A}^T\mathbf{P}[\mathbf{B}_w \mid \mathbf{B}_u]\mathbf{R}^{-1}\begin{bmatrix}\mathbf{B}_w^T \\ \mathbf{B}_u^T\end{bmatrix}\mathbf{PA} + \mathbf{C}_y^T\mathbf{C}_y$, where $\mathbf{A} - [\mathbf{B}_w \mid \mathbf{B}_u]\mathbf{R}^{-1}\begin{bmatrix}\mathbf{B}_w^T \\ \mathbf{B}_u^T\end{bmatrix}\mathbf{PA}$ is stable, and $\mathbf{R} = \begin{bmatrix}-\gamma^2\mathbf{I} + \mathbf{B}_w^T\mathbf{PB}_w & \mathbf{B}_w^T\mathbf{PB}_u \\ \mathbf{B}_u^T\mathbf{PB}_w & \mathbf{I} + \mathbf{B}_u^T\mathbf{PB}_u\end{bmatrix}$.
The suboptimal control	$u(t) = -\mathbf{B}_u^T\mathbf{P}x(t) = -\mathbf{K}x(t)$	$u(k) = -(\mathbf{I} + \mathbf{B}_u^T\mathbf{PB}_u)^{-1}\mathbf{B}_u^T\mathbf{PA}x(k) - (\mathbf{I} + \mathbf{B}_u^T\mathbf{PB}_u)^{-1}\mathbf{B}_u^T\mathbf{PB}_w w(k)$

continued

TABLE 9.1 *Continued*

Formula	Continuous-Time	Discrete-Time
The $\mathcal{H}_\infty$ estimation problem	Given the plant $\dot{x}(t) = \mathbf{A}x(t) + \mathbf{B}_u u(t) + \mathbf{B}_w w(t)$; $m(t) = \mathbf{C}_m x(t) + \mathbf{D}_{mw} w(t)$, $Y(t) = \mathbf{C}_y x(t)$ where $\mathbf{D}_{mw}\mathbf{B}_w^T = \mathbf{0}$; $\mathbf{D}_{mw}\mathbf{D}_{mw}^T = \mathbf{I}$, generate an estimator such that $\Vert\mathbf{G}_{ew}\Vert_\infty < \gamma$.	Given the plant $\dot{x}(t) = \mathbf{\Phi} x(t) + \mathbf{\Gamma}_u u(t) + \mathbf{\Gamma}_w w(t)$; $m(t) = \mathbf{C}_m x(t) + \mathbf{D}_{mw} w(t)$, $Y(t) = \mathbf{C}_y x(t)$ where $\mathbf{D}_{mw}\mathbf{B}_w^T = \mathbf{0}$; $\mathbf{D}_{mw}\mathbf{D}_{mw}^T = \mathbf{I}$, generate an estimator such that $\Vert\mathbf{G}_{ew}\Vert_\infty < \gamma$.
Existence conditions for a suboptimal estimator	There exists a solution $\mathbf{Q} \geq \mathbf{0}$ of the algebraic Riccati equation $\mathbf{AQ} + \mathbf{QA}^T - \mathbf{Q}(\mathbf{C}_m^T\mathbf{C}_m - \gamma^{-2}\mathbf{C}_y^T\mathbf{C}_y)\mathbf{Q} + \mathbf{B}_w\mathbf{B}_w^T = \mathbf{0}$, where $\mathbf{A} - \mathbf{Q}(\mathbf{C}_m^T\mathbf{C}_m - \gamma^{-2}\mathbf{C}_y^T\mathbf{C}_y)$ is stable.	There exists a solution $\mathbf{Q} \geq \mathbf{0}$ of the algebraic Riccati equation $\mathbf{Q} = \mathbf{AQA}^T - \mathbf{AQ}[\mathbf{C}_y^T \mid \mathbf{C}_m^T]\mathbf{S}^{-1}\begin{bmatrix}\mathbf{C}_y \\ \mathbf{C}_m\end{bmatrix}\mathbf{QA}^T + \mathbf{B}_w\mathbf{B}_w^T$, where $\mathbf{A} - \mathbf{AQ}[\mathbf{C}_y^T \mid \mathbf{C}_m^T]\mathbf{S}^{-1}\begin{bmatrix}\mathbf{C}_y \\ \mathbf{C}_m\end{bmatrix}$ is stable, and $\mathbf{S} = \begin{bmatrix}-\gamma^2\mathbf{I} + \mathbf{C}_y\mathbf{QC}_y^T & \mathbf{C}_y\mathbf{QC}_m^T \\ \mathbf{C}_m\mathbf{QC}_y^T & \mathbf{I} + \mathbf{C}_m\mathbf{QC}_m^T\end{bmatrix}$.
The suboptimal estimator	$\dot{\hat{x}}(t) = \mathbf{A}\hat{x}(t) + \mathbf{B}_u u(t) + \mathbf{G}\{m(t) - \mathbf{C}_m\hat{x}(t)\}$ $\hat{y}(t) = \mathbf{C}_y\hat{x}(t)$ $\mathbf{G} = \mathbf{QC}_m^T$	$\hat{x}(k+1) = \mathbf{\Phi}\hat{x}(k) + \mathbf{\Gamma}_u u(k) + \mathbf{G}\{m(k) - \mathbf{C}_m\hat{x}(k)\}$ $\hat{y}(t) = \mathbf{C}_y\hat{x}(t) + \mathbf{G}_r\{m(t) - \mathbf{C}_m\hat{x}(t)\}$ $\mathbf{G} = \mathbf{AQC}_m^T(\mathbf{I} + \mathbf{C}_m\mathbf{QC}_m^T)^{-1}$ $\mathbf{G}_r = \mathbf{C}_y\mathbf{QC}_m^T(\mathbf{I} + \mathbf{C}_m\mathbf{QC}_m^T)^{-1}$

[1] For discrete-time systems, the signal 2-norm and the system ∞-norm are defined as

$$\Vert w(k)\Vert_2 = \left\{\sum_{k=-\infty}^{\infty} w^T(k)w(k)\right\}; \ \Vert\mathbf{G}\Vert_\infty = \sup_{w(k)\neq 0} \frac{\Vert\mathbf{g}(k) \otimes w(k)\Vert_2}{\Vert w(k)\Vert_2}.$$

REFERENCES

[1] M. Greene and D. J. N. Limebeer, *Linear Robust Control,* Prentice-Hall, Englewood Cliffs, NJ, 1995.

[2] K. Zhou, with J. C. Doyle and K. Glover, *Robust and Optimal Control,* Prentice-Hall, Englewood Cliffs, NJ, 1996.

[3] K. M. Nagpal and P. P. Khargonekar, "Filtering and smoothing in an $\mathcal{H}_\infty$ setting," *IEEE Transactions on Automatic Control,* 36, no. 2 (1991): 152–66.

[4] T. Basar, "A dynamic games approach to controller design: Disturbance rejection in discrete-time," *Proceedings of the IEEE Conference on Decision and Control,* pp. 407–14, 1989.

[5] D. J. N. Limebeer, M. Greene, and D. Walker, "Discrete-time $\mathcal{H}_\infty$ control," *Proceedings of the IEEE Conference on Decision and Control,* pp. 392–96, 1989.

Some additional references on $\mathcal{H}_\infty$ control follow:

[6] T. Basar and P. Bernhard, *$\mathcal{H}_\infty$-Optimal Control and Related Minimax Design Problems: A Dynamic Game Approach,* Systems and Control: Foundations and Applications, Boston: Birkhauser, 1991.

[7] K. Zhou, with J. C. Doyle, *Essentials of Robust Control,* Prentice-Hall, Englewood Cliffs, NJ, 1998.

[8] J. C. Doyle, K. Glover, P. P. Khargonekar, and B. A. Francis, "State-space solution to standard $\mathcal{H}_2$ and $\mathcal{H}_\infty$ control problems," *IEEE Transactions on Automatic Control,* 34, no. 8 (1989): 831–47.

[9] A. Elsayed and M. J. Grimble, "A new approach to $\mathcal{H}_\infty$ design of optimal digital filters," *IMA Journal of Mathematical Control and Information,* 6, no. 8 (1989): 233–51.

[10] B. A. Francis, *A Course in $\mathcal{H}_\infty$ Control Theory,* Lecture Notes in Control and Information Sciences, vol. 88, Berlin: Springer-Verlag, 1987.

[11] D. J. N. Limebeer, B. D. O. Anderson, P. P. Khargonekar, and M. Green, "A game theoretic approach to $\mathcal{H}_\infty$ control for time-varying systems," *SIAM J. Control and Optimization,* 30, no. 2 (1992): 262–83.

[12] J. M. Maciejowski, *Multivariable Feedback Design,* Addison-Wesley, Reading, MA, 1989.

EXERCISES

9.1 Using the Lagrange multiplier method, generate the necessary conditions for a saddle point of

$$f(x, y, z) = 2x^2 + 42x + 4xy - 5y^2 + z^2,$$

given the following constraint:

$$x + y + z = 0.$$

Show that this point is a saddle point for (x, y) by eliminating z from f using the constraint, and alternatively adding perturbations to x and y.

9.2 Generate a suboptimal full information controller for the following plant:

$$\dot{x}(t) = -x(t) + 5u(t) + w(t) \; ; \; y(t) = \begin{bmatrix} 1 \\ 0 \end{bmatrix} x(t) + \begin{bmatrix} 0 \\ 1 \end{bmatrix} u(t),$$

where the performance bound γ is a variable. This suboptimal control gain should be an analytic function of the performance bound. For what values of the performance bound does a suboptimal controller exist for an infinite time interval?

9.3 The discrete-time plant is

$$x(k+1) = \mathbf{\Phi} x(k) + \mathbf{\Gamma}_u u(k) + \mathbf{\Gamma}_w w(k) \; ; \; y(k) = \mathbf{C}_y x(k) + \mathbf{D}_{yu} u(k),$$

where

$$\mathbf{D}_{yu}^T \left[\begin{array}{c} \mathbf{C}_y \\ \hline \mathbf{D}_{yu} \end{array} \right] = \left[\begin{array}{c} \mathbf{0} \\ \hline \mathbf{I} \end{array} \right].$$

Generate necessary conditions for a saddle point of the objective function

$$J(x(k), u(k), w(k)) = \sum_{k=0}^{k_f - 1} y^T(k) y(k) - \gamma^2 w^T(k) w(k).$$

9.4 Show that the poles of the Hamiltonian system

$$\left[\begin{array}{c} \dot{x}(t) \\ \hline \dot{p}(t) \end{array} \right] = \left[\begin{array}{c|c} \mathbf{A} & -\mathbf{B}_u \mathbf{B}_u^T + \gamma^{-2} \mathbf{B}_w \mathbf{B}_w^T \\ \hline -\mathbf{C}_y^T \mathbf{C}_y & -\mathbf{A}^T \end{array} \right] \left[\begin{array}{c} x(t) \\ \hline p(t) \end{array} \right]$$

are symmetric around the imaginary axis.

9.5 Show that

$$\lim_{t \to \infty} \mathbf{K}(t) = -\lim_{t \to \infty} \mathbf{B}_u^T \{ \mathbf{\Phi}_{22}(t_f - t) \}^{-1} \mathbf{\Phi}_{21}(t_f - t)$$

does not exist when the Hamiltonian system (9.10) has a pole on the imaginary axis. Assume all the poles of the Hamiltonian are distinct. Note that $\mathbf{\Phi}(t)$ is the state-transition matrix of the Hamiltonian system.

9.6 Generate a suboptimal $\mathcal{H}_\infty$ estimator of the plant state for

$$\dot{x}(t) = -x(t) + 5u(t) + w(t) \; ; \; m(t) = x(t) + v(t),$$

where the performance bound γ is a variable. The estimator gain should be an analytic function of the performance bound. For what values of the performance bound does a suboptimal estimator exist for an infinite time interval?

9.7 Show that the transfer function of

$$\dot{\overline{x}}(t) = (\mathbf{A} - \mathbf{B}_u\mathbf{K} + \gamma^{-2}\mathbf{B}_w\mathbf{B}_w^T\mathbf{P})\overline{x}(t) + \mathbf{B}_w\overline{w}(t);$$
$$\overline{y}(t) = \gamma^{-2}\mathbf{B}_w^T\mathbf{P}\overline{x}(t) + \overline{w}(t)$$

is the inverse of the transfer function of

$$\dot{x}(t) = (\mathbf{A} - \mathbf{B}_u\mathbf{K})x(t) + \mathbf{B}_w w(t);$$
$$y(t) = -\gamma^{-2}\mathbf{B}_w^T\mathbf{P}x(t) + w(t).$$

Hint: Use the matrix inversion lemma of Appendix equation (A2.8) to show that

$$[\mathbf{I} + \mathbf{C}(s\mathbf{I} - \mathbf{A})^{-1}\mathbf{B}]^{-1} = \mathbf{I} - \mathbf{C}(s\mathbf{I} - \mathbf{A} + \mathbf{BC})^{-1}\mathbf{B},$$

and then use this result to derive the relationship between the two systems.

9.8 You are given the following plant:

$$\dot{x}(t) = \mathbf{A}x(t) + [\mathbf{B}_w \mid \mathbf{0}]\begin{bmatrix} w(t) \\ \hdashline v(t) \end{bmatrix};\ m(t) = \mathbf{C}_m x(t) + [\mathbf{0} \mid \mathbf{I}]\begin{bmatrix} w(t) \\ \hdashline v(t) \end{bmatrix};\ y(t) = \mathbf{C}_y x(t).$$

The estimator of $y(t)$ is

$$\dot{\hat{x}}(t) = \mathbf{A}\hat{x}(t) + \mathbf{G}[m(t) - \mathbf{C}_m\hat{x}(t)]\ ;\ \hat{y}(t) = \mathbf{C}_y\hat{x}(t),$$

and the estimator gain is given as

$$\mathbf{G} = \mathbf{PC}_m^T,$$

where $\mathbf{P} \geq \mathbf{0}$ is the solution of the algebraic Riccati equation

$$\mathbf{AP} + \mathbf{PA}^T - \mathbf{P}[\mathbf{C}_m^T\mathbf{C}_m - \gamma^{-2}\mathbf{C}_y^T\mathbf{C}_y]\mathbf{P} + \mathbf{B}_w\mathbf{B}_w^T = \mathbf{0},$$

and $(\mathbf{A} - \mathbf{PC}_m^T\mathbf{C}_m + \gamma^{-2}\mathbf{PC}_y^T\mathbf{C}_y)$ is assumed to be stable. Show that this estimator is stable and that the transfer function $\mathbf{T}_{ew}(s)$,

$$e(s) = y(s) - \hat{y}(s) = \mathbf{T}_{ew}(s)\begin{bmatrix} w(s) \\ \hdashline v(s) \end{bmatrix}$$

is bounded:

$$\|\mathbf{T}_{ew}(s)\|_\infty < \gamma.$$

Hint: Take the adjoint of the error equation, and use the bounded real lemma.

9.9 Generate an example system, with input $w(t)$ and output $y(t)$, to show that

$$\|y(t)\|_2^2 - \gamma^2\|w(t)\|_2^2 < 0$$

is not sufficient to guarantee that

$$\sup_{w(t)\neq 0}\frac{\|y(t)\|_2^2}{\|w(t)\|_2^2} < \gamma.$$

Note that all norms are over infinite time intervals.

COMPUTER EXERCISES

9.1 Full Information Control of Blood Gases During Extracorporeal Support
Design a system for controlling the partial pressure of oxygen and the partial pressure of carbon dioxide during extracorporeal support. The entire state is measured and available for feedback to control the oxygenator.

The state model of the oxygenator is

$$\begin{bmatrix} \dot{x}_1(t) \\ \dot{x}_2(t) \\ \dot{x}_3(t) \\ \dot{x}_4(t) \end{bmatrix} = \begin{bmatrix} -10 & 0 & 0 & 0 \\ 0 & -10 & 0 & 0 \\ 6 & -3 & -5 & 0 \\ 0 & 0.5 & 0 & -5 \end{bmatrix}\begin{bmatrix} x_1(t) \\ x_2(t) \\ x_3(t) \\ x_4(t) \end{bmatrix} + \begin{bmatrix} 10 & 0 \\ 0 & 10 \\ 0 & 0 \\ 0 & 0 \end{bmatrix}\begin{bmatrix} u_1(t) \\ u_2(t) \end{bmatrix} + \begin{bmatrix} 0.1 & 0 \\ 0 & 0.1 \\ 0 & 0 \\ 0 & 0 \end{bmatrix}\begin{bmatrix} w_1(t) \\ w_2(t) \end{bmatrix}$$

where x_1 is the flow rate of oxygen, x_2 is the flow rate of carbon dioxide, x_3 is the arterial partial pressure of oxygen, x_4 is the arterial partial pressure of carbon dioxide, u_1 is the commanded oxygen flow rate, u_2 is the commanded carbon dioxide flow rate, w_1 is a disturbance that generates errors in the oxygen flow rate, and w_2 is a disturbance that generates errors in the carbon dioxide flow rate. The outputs of interest are x_3 and x_4:

$$\begin{bmatrix} y_1(t) \\ y_2(t) \end{bmatrix} = \begin{bmatrix} 0 & 0 & 1 & 0 \\ 0 & 0 & 0 & 1 \end{bmatrix} \begin{bmatrix} x_1(t) \\ x_2(t) \\ x_3(t) \\ x_4(t) \end{bmatrix}.$$

a. Design a control system such that the gains from the disturbance inputs to the outputs of interest are bounded by 0.001. Further, the gain from the disturbance inputs to the control inputs should be bounded by 1. Compute the frequency responses of the closed-loop system from the disturbance inputs to both the outputs of interest and the controls. Simulate the closed-loop system with the disturbance inputs both equal to unit step functions and the initial state equal to zero. Plot the reference outputs and the control inputs. Simulate the closed-loop system with zero disturbance inputs and the following initial state:

$$x(0) = [0 \quad 0 \quad 50 \quad 20]^T.$$

Plot the reference outputs and the control inputs. Compute the norm of the smallest destabilizing, input-multiplicative perturbation.

b. Design a control system such that the gains from the disturbance inputs to the outputs of interest are bounded by 0.1. Further, the gains from the disturbance inputs to the control inputs should be bounded by 1. Compute the frequency responses, simulate the closed-loop system, and plot as described in part *a*. Also compute the norm of the smallest destabilizing, input-multiplicative perturbation. Comment on the differences between the results obtained in part *b* and those obtained in part *a*.

9.2 Attitude Estimation for a Flexible Spacecraft

The pitch, yaw, and roll angles can be used to describe a spacecraft's attitude. A linear dynamic model can be used for the spacecraft, provided these angles (and the associated angle rates) are small. The linear model has no coupling between the axes. The pitch-plane dynamics are

$$\begin{bmatrix} \dot{\theta}_p(t) \\ \ddot{\theta}_p(t) \end{bmatrix} = \begin{bmatrix} 0 & 1 \\ 0 & 0 \end{bmatrix} \begin{bmatrix} \theta_p(t) \\ \dot{\theta}_p(t) \end{bmatrix} + \begin{bmatrix} 0 \\ 0.005 \end{bmatrix} u(t) + \begin{bmatrix} 0 \\ 0.005 \end{bmatrix} w(t),$$

where $\theta_p(t)$ is the pitch angle in degrees and $u(t)$ is an applied force. The disturbance input is mainly caused by vibration of a flexible sensor boom. This boom generates forces that are bounded by 100 N. This error could be removed if the vibration were known or estimated. Unfortunately, the modal frequencies for this vibration are not well known, since the position of the sensor boom is variable; that is, the natural frequencies of the boom are only known to lie between 0.5 and 5 Hz. Note that a reasonable disturbance input can be generated from a normalized disturbance input w_1 by the following filter:

$$\frac{w(s)}{w_1(s)} = \frac{s^2 + 20.05s + 1}{s^2 + 0.2s + 1.01}.$$

The pitch angle and pitch rate can be estimated from pitch and pitch rate measurements:

$$m(t) = \begin{bmatrix} 1 & 0 \\ 0 & 1 \end{bmatrix} \begin{bmatrix} \theta_p(t) \\ \dot{\theta}_p(t) \end{bmatrix} + v(t),$$

where $v(t)$ is the measurement error. The measurement errors are bounded by $\pm 0.1°$ and $\pm 3°$/sec, respectively. The correlation time for these errors is 0.001.

a. Generate an $\mathcal{H}_\infty$ filter to estimate the pitch angle and pitch rate of this spacecraft. This filter should minimize the gains from the normalized disturbance input $w_1(t)$ and the measurement error $v(t)$ to the estimation error for the pitch and the pitch rate. Simulate the plant and $\mathcal{H}_\infty$ filter with a disturbance input,

$$w(t) = 100\cos(6t),$$

and a discrete-time white noise measurement error. Use a sampling time of 0.001, and make the measurement errors uniformly distributed between the given bounds. Assume the initial conditions on the plant and the filter are zero. Plot the plant states and the estimated plant states, and use the simulation results to generate a sample standard deviation for the estimation errors.

b. Generate a Kalman filter for estimating the spacecraft states. Let the spectral densities for the disturbance input and the plant noise equal the squares of the respective bounds on these variables. This is reasonable, since the Kalman gains only depend on the relative amplitudes of the plant and measurement noises. Simulate the plant and Kalman filter with the input given in part *a*. Plot the plant states and the estimated plant states, and use the simulation results to generate a sample standard deviation for the estimation errors. Contrast the performance of the Kalman filter and the $\mathcal{H}_\infty$ estimator.

c. Generate an $\mathcal{H}_\infty$ filter to estimate the pitch rate only of this spacecraft. Simulate the plant and filter with the input given in part *a*. Plot the plant states and the estimated plant states, and use the simulation results to generate a sample standard deviation for the estimation error. Contrast the performance of this filter and the $\mathcal{H}_\infty$ estimator developed in part *a*.

10 $\mathcal{H}_\infty$ Output Feedback

The $\mathcal{H}_\infty$ output feedback controller (or simply the $\mathcal{H}_\infty$ controller) utilizes partial state measurements, corrupted by disturbances, to generate the control. This controller can be synthesized by combining an $\mathcal{H}_\infty$ full information controller with an $\mathcal{H}_\infty$ estimator. At first glance it appears appropriate to estimate the state from the measurements and apply the full information feedback gain, as was done in the case of the LQG optimal controller. This turns out to be not exactly correct, since unlike the Kalman filter, the $\mathcal{H}_\infty$ estimator gain depends on what linear combination of the states is being estimated. Instead of estimating the state, it is more appropriate to estimate the desired control input. In addition, the worst-case disturbance input that appears in the full information mini-max problem must be included in the $\mathcal{H}_\infty$ estimator equations. The $\mathcal{H}_\infty$ output feedback controller therefore has a structure like that of the LQG controller, which consists of an estimator and a full information controller. But this structure technically violates the separation principle as presented for the LQG controller, since the estimator now depends on the full information controller design.

The suboptimal $\mathcal{H}_\infty$ control problem is defined by the plant and the cost function. The plant is given by the following state model:

$$\dot{x}(t) = \mathbf{A}x(t) + [\mathbf{B}_u \;\vdots\; \mathbf{B}_w]\begin{bmatrix} u(t) \\ \hdashline w(t) \end{bmatrix}; \tag{10.1a}$$

$$\begin{bmatrix} m(t) \\ \hdashline y(t) \end{bmatrix} = \begin{bmatrix} \mathbf{C}_m \\ \hdashline \mathbf{C}_y \end{bmatrix} x(t) + \begin{bmatrix} \mathbf{0} & \mathbf{D}_{mw} \\ \hdashline \mathbf{D}_{yu} & \mathbf{0} \end{bmatrix}\begin{bmatrix} u(t) \\ \hdashline w(t) \end{bmatrix}. \tag{10.1b}$$

The matrices $\mathbf{B}_w$ and $\mathbf{D}_{mw}$ are assumed to satisfy the following conditions:

$$\mathbf{D}_{mw}\mathbf{B}_w^T = \mathbf{0}; \tag{10.2a}$$

$$\mathbf{D}_{mw}\mathbf{D}_{mw}^T = \mathbf{I}. \tag{10.2b}$$

These conditions require that the disturbances entering the plant and the measurement be distinct and that the output equation of the plant be scaled to normalize the measurement noise. The matrices $\mathbf{C}_y$ and $\mathbf{D}_{yu}$ are assumed to satisfy the following conditions:

$$\mathbf{D}_{yu}^T \mathbf{C}_y = \mathbf{0}; \qquad (10.2c)$$

$$\mathbf{D}_{yu}^T \mathbf{D}_{yu} = \mathbf{I}. \qquad (10.2d)$$

These conditions require that the reference output consist of an output dependent only on the state and a distinct output dependent only on the control input. Further, the portion of the output that depends on the control is simply equal to the control input or an orthogonally transformed version of this input. The plant is assumed to be controllable from the control input and observable from the measured output.[1] These conditions guarantee that the plant can be stabilized using output feedback, a necessity when operating over infinite time intervals and always desirable. Further, the plant is assumed to be controllable from the disturbance input and observable from the reference output. These conditions guarantee the existence of a steady-state $\mathcal{H}_\infty$ suboptimal output feedback for sufficiently large performance bounds.[2]

The suboptimal $\mathcal{H}_\infty$ control problem is to find a feedback controller for the above plant such that the ∞-norm of the closed-loop system is bounded:

$$\|\mathbf{G}_{yw}\|_{\infty,[0,t_f]} = \sup_{\|w(t)\|_{2,[0,t_f]} \neq 0} \frac{\|y(t)\|_{2,[0,t_f]}}{\|w(t)\|_{2,[0,t_f]}} < \gamma. \qquad (10.3)$$

The closed-loop system is also required to be internally stable when the final time is infinite. The solution of the optimal $\mathcal{H}_\infty$ control problem (minimizing the closed-loop ∞-norm) is discussed after first presenting a suboptimal solution.

10.1 Controller Structure

The suboptimal $\mathcal{H}_\infty$ controller can be synthesized by combining a full information controller and an output estimator. This structure for the suboptimal $\mathcal{H}_\infty$ controller is derived in this subsection, assuming that the suboptimal full information controller exists.

The ∞-norm bound (10.3), that defines the suboptimal $\mathcal{H}_\infty$ controller, is equivalent to requiring that the inequality

$$J_\gamma = \|y(t)\|_{2,[0,t_f]}^2 - \gamma^2\|w(t)\|_{2,[0,t_f]}^2 \leq -\varepsilon^2\|w(t)\|_{2,[0,t_f]}^2 \qquad (10.4)$$

be satisfied for some positive ε and for all disturbance inputs. The expression on the left in this inequality is the objective function for the differential game used when solving the full information control problem. A general expression for this objective function with arbitrary inputs was derived in Subsection 9.2.3. This expression is derived assuming that a solution of the Riccati equation (9.13) exists (i.e., that a suboptimal $\mathcal{H}_\infty$ full information controller exists).

A suboptimal full information controller exists whenever a suboptimal output feedback controller exists, since all possible outputs can be generated from the state and

[1] In general, these conditions can be relaxed. The plant must be stabilizable from the control input and detectable from the measured output.

[2] These conditions can be relaxed significantly (see, for example, [1], page 449).

disturbance input. Therefore, the inequality (10.4) can be written in terms of the full information Riccati solution by substituting (9.15) into (10.4):

$$\begin{aligned} J_\gamma &= \|u(t) + \mathbf{B}_u^T\mathbf{P}(t)x(t)\|^2_{2,[0,t_f]} - \gamma^2\|w(t) - \gamma^{-2}\mathbf{B}_w^T\mathbf{P}(t)x(t)\|^2_{2,[0,t_f]} \\ &\leq -\varepsilon^2\|w(t)\|^2_{2,[0,t_f]}. \end{aligned} \tag{10.5}$$

This inequality is equivalent to the bound on the ∞-norm:

$$\|\mathbf{G}_\Delta\|_\infty = \sup_{w(t)-\gamma^{-2}\mathbf{B}_w^T\mathbf{P}(t)x(t)\neq 0} \frac{\|u(t) + \mathbf{B}_u^T\mathbf{P}(t)x(t)\|_{2,[0,t_f]}}{\|w(t) - \gamma^{-2}\mathbf{B}_w^T\mathbf{P}(t)x(t)\|_{2,[0,t_f]}} < \gamma. \tag{10.6}$$

This bound can be used in formulating a suboptimal $\mathcal{H}_\infty$ output estimation problem. The solution of this problem leads directly to a suboptimal $\mathcal{H}_\infty$ controller.

This output estimation problem is stated as follows: Estimate the full information control input,

$$u(t) = -\mathbf{B}_u^T\mathbf{P}(t)x(t), \tag{10.7}$$

given the measurement $m(t)$, such that the ∞-norm of the transfer function between the disturbance input $\Delta w(t) = w(t) - \gamma^{-2}\mathbf{B}_w^T\mathbf{P}(t)x(t)$, and the estimation error is bounded:

$$\begin{aligned} \|\mathbf{G}_\Delta\|_\infty &= \sup_{\Delta w(t)\neq 0} \frac{\|y(t) - \hat{y}(t)\|_{2,[0,t_f]}}{\|w(t) - \gamma^{-2}\mathbf{B}_w^T\mathbf{P}(t)x(t)\|_{2,[0,t_f]}} \\ &= \sup_{\Delta w(t)\neq 0} \frac{\|-\mathbf{B}_u^T\mathbf{P}(t)x(t) - u(t)\|_{2,[0,t_f]}}{\|w(t) - \gamma^{-2}\mathbf{B}_w^T\mathbf{P}(t)x(t)\|_{2,[0,t_f]}} < \gamma. \end{aligned}$$

Note that this equation is equivalent to (10.6), since $\|-x\| = \|x\|$.

The state model of the "plant" in this estimation problem has $\Delta w(t)$ as the disturbance input and $m(t)$ as the measured output. The state equation for this model can be generated by adding plus and minus $\gamma^{-2}\mathbf{B}_w\mathbf{B}_w^T\mathbf{P}(t)x(t)$ to the original state equation (10.1a):

$$\begin{aligned} \dot{x}(t) &= \mathbf{A}x(t) + \mathbf{B}_u u(t) + \mathbf{B}_w w(t) + \gamma^{-2}\mathbf{B}_w\mathbf{B}_w^T\mathbf{P}(t)x(t) - \gamma^{-2}\mathbf{B}_w\mathbf{B}_w^T\mathbf{P}(t)x(t) \\ &= [\mathbf{A} + \gamma^{-2}\mathbf{B}_w\mathbf{B}_w^T\mathbf{P}(t)]x(t) + \mathbf{B}_u u(t) + \mathbf{B}_w\Delta w(t). \end{aligned} \tag{10.8a}$$

The measurement equation for this model can be generated by subtracting $\gamma^{-2}\mathbf{D}_{mw}\mathbf{B}_w^T\mathbf{P}x$, which equals zero because $\mathbf{D}_{mw}\mathbf{B}_w^T = \mathbf{0}$, from the original measurement equation (10.1b):

$$\begin{aligned} m(t) &= \mathbf{C}_m x(t) + \mathbf{D}_{mw}w(t) - \gamma^{-2}\mathbf{D}_{mw}\mathbf{B}_w^T\mathbf{P}(t)x(t) \\ &= \mathbf{C}_m x(t) + \mathbf{D}_{mw}\Delta w(t). \end{aligned} \tag{10.8b}$$

The plant model (10.8) has the form of the plant model (9.31), which appears in the $\mathcal{H}_\infty$ estimation problem statement.

The suboptimal $\mathcal{H}_\infty$ estimator for the problem specified by (10.6) through (10.8) generates estimates of the full information control. The inequality (10.6), which is equivalent to (10.3), is satisfied when applying these estimates as control inputs to the original plant. In this case, the estimator becomes an output feedback controller, since it generates control inputs from the measurements. Therefore, the estimator is a suboptimal $\mathcal{H}_\infty$ controller.

In summary, the suboptimal (or optimal) $\mathcal{H}_\infty$ output feedback control law is the $\mathcal{H}_\infty$ suboptimal (or optimal) estimate of the full information control.† The controller is therefore designed in two stages: a full information controller is synthesized, and an output estimator is synthesized. The final output feedback controller is generated by combining these two components. This controller has a structure similar to the LQG controller, but technically violates the separation principle, since the estimator design depends on the full information controller design.

10.2 Finite-Time Control

The suboptimal $\mathcal{H}_\infty$ output estimation problem specified by (10.6) through (10.8) has a solution if the Riccati equation

$$\begin{aligned}\dot{\mathbf{Q}}_m(t) = &-\mathbf{Q}_m(t)[\mathbf{A} + \gamma^{-2}\mathbf{B}_w\mathbf{B}_w^T\mathbf{P}(t)]^T - [\mathbf{A} + \gamma^{-2}\mathbf{B}_w\mathbf{B}_w^T\mathbf{P}(t)]\mathbf{Q}_m(t) \\ &- \mathbf{B}_w\mathbf{B}_w^T + \mathbf{Q}_m(t)[\mathbf{C}_m^T\mathbf{C}_m - \gamma^{-2}\mathbf{P}(t)\mathbf{B}_u\mathbf{B}_u^T\mathbf{P}(t)]\mathbf{Q}_m(t)\end{aligned} \tag{10.9}$$

has a solution, given the initial condition $\mathbf{Q}_m(0) = \mathbf{0}$. The suboptimal estimator is

$$\begin{aligned}\dot{\hat{x}}(t) &= [\mathbf{A} + \gamma^{-2}\mathbf{B}_w\mathbf{B}_w^T\mathbf{P}(t)]\hat{x}(t) + \mathbf{B}_u u(t) + \mathbf{G}(t)[m(t) - \mathbf{C}_m\hat{x}(t)]; \\ \hat{u}(t) &= -\mathbf{B}_u^T\mathbf{P}(t)\hat{x}(t),\end{aligned}$$

and the gain is

$$\mathbf{G}(t) = \mathbf{Q}_m(t)\mathbf{C}_m^T. \tag{10.10}$$

This estimator contains two known inputs: $u(t)$ which enters through $\mathbf{B}_u$; and $\gamma^{-2}\mathbf{B}_w^T\mathbf{P}(t)\hat{x}(t)$ which enters through $\mathbf{B}_w$. The second input is the worst-case disturbance encountered during full information controller optimization. This estimator can then be described as estimating the full information control in the presence of the worst case disturbance.

This estimator can be used as a feedback controller. Substituting for the gain (10.10) in the estimator and setting $u(t) = \hat{u}(t)$ yields the controller:

$$\begin{aligned}\dot{\hat{x}}(t) = &[\mathbf{A} + \gamma^{-2}\mathbf{B}_w\mathbf{B}_w^T\mathbf{P}(t) - \mathbf{B}_u\mathbf{B}_u^T\mathbf{P}(t) - \mathbf{Q}_m(t)\mathbf{C}_m^T\mathbf{C}_m]\hat{x}(t) \\ &+ \mathbf{Q}_m(t)\mathbf{C}_m^T m(t);\end{aligned} \tag{10.11a}$$

$$u(t) = -\mathbf{B}_u^T\mathbf{P}(t)\hat{x}(t). \tag{10.11b}$$

Solutions to the Riccati equation (10.9) and the full information Riccati equation (9.13) are sufficient to guarantee the existence of this suboptimal $\mathcal{H}_\infty$ controller.

10.2.1 An Alternative Estimator Riccati Equation

The generation of the $\mathcal{H}_\infty$ output feedback controller (as given above) proceeds by synthesizing the full information control and then estimating this control. This process results in solving the full information Riccati equation for the plant. In addition, an estimator Riccati equation for a modified plant is solved. This Riccati solution can be related to the Riccati solution associated with reference output estimation for the original plant. Developing this correspondence produces a symmetric (the symmetry is between the control and estimation Riccati equations) pair of Riccati equations that can be solved to

†The full information control estimator includes the worst case disturbance as a known input. This fact will be discussed in Section 10.2.

generate the $\mathcal{H}_\infty$ output feedback controller. The resulting equations are in the form most frequently encountered in the research literature.

The solution of the estimator Riccati equation (10.9) for the modified plant (if it exists) can be related to the solution of the output estimation Riccati equation (9.42) with $\mathbf{Q}(0) = \mathbf{0}$:

$$\mathbf{Q}_m(t) = \mathbf{Q}(t)[\mathbf{I} - \gamma^{-2}\mathbf{P}(t)\mathbf{Q}(t)]^{-1}. \tag{10.12}$$

This fact is demonstrated after first presenting an additional condition that is necessary to guarantee the existence of $\mathbf{Q}_m(t)$.

The existence of $\mathbf{Q}_m(t)$ in (10.12) hinges on the existence of the matrix inverse in this equation on the interval from 0 to t_f. This matrix is invertible if it is positive definite, that is, all the eigenvalues of $[\mathbf{I} - \gamma^{-2}\mathbf{P}(t)\mathbf{Q}(t)]$ are positive. The eigenvalues of this matrix can be related to the eigenvalues of $\mathbf{P}(t)\mathbf{Q}(t)$. If λ is an eigenvalue of $\mathbf{P}(t)\mathbf{Q}(t)$, then

$$1 - \gamma^{-2}\lambda(t)$$

is an eigenvalue of $(\mathbf{I} - \gamma^{-2}\mathbf{P}(t)\mathbf{Q}(t))$; see Appendix equation (A4.1). Therefore, the inverse in (10.12) exists, and $\mathbf{Q}_m(t)$ exists, provided $\mathbf{P}(t)$ exists, $\mathbf{Q}(t)$ exists, and all of the eigenvalues of $\mathbf{P}(t)\mathbf{Q}(t)$ are less than γ^2; that is,

$$\rho[\mathbf{P}(t)\mathbf{Q}(t)] < \gamma^2, \tag{10.13}$$

where $\rho(\bullet)$ is the spectral radius.

It remains to show that $\mathbf{Q}_m(t)$ is a solution of the Riccati equation (10.9), with the initial condition $\mathbf{Q}_m(0) = \mathbf{0}$. The initial condition on $\mathbf{Q}_m(t)$ is easily verified:

$$\mathbf{Q}_m(0) = \mathbf{Q}(0)[\mathbf{I} - \gamma^{-2}\mathbf{P}(0)\mathbf{Q}(0)]^{-1} = \mathbf{0},$$

since $\mathbf{Q}(0) = \mathbf{0}$. The fact that $\mathbf{Q}_m(t)$ is a solution of the Riccati equation (10.9) can be verified by considering the Hamiltonian system for the estimator (9.44):

$$\dot{x}(t) = \left[\begin{array}{c:c} \mathbf{A}^T & -\mathbf{C}_m^T\mathbf{C}_m + \gamma^{-2}\mathbf{C}_y^T\mathbf{C}_y \\ \hdashline -\mathbf{B}_w\mathbf{B}_w^T & -\mathbf{A} \end{array}\right] x(t).$$

The generic state $x(t)$ is used in this equation to simplify the notation. Using the transformation matrix

$$\mathbf{T}(t) = \left[\begin{array}{c:c} \mathbf{I} & -\gamma^{-2}\mathbf{P}(t) \\ \hdashline \mathbf{0} & \mathbf{I} \end{array}\right],$$

a time-varying similarity transformation, as in Appendix equation (A6.2), can be performed on the Hamiltonian system:

$$\begin{aligned}
\dot{\tilde{x}} &= \left(\left[\begin{array}{c:c} \mathbf{I} & \gamma^{-2}\mathbf{P} \\ \hdashline \mathbf{0} & \mathbf{I} \end{array}\right] \left[\begin{array}{c:c} \mathbf{A}^T & -\mathbf{C}_m^T\mathbf{C}_m + \gamma^{-2}\mathbf{C}_y^T\mathbf{C}_y \\ \hdashline -\mathbf{B}_w\mathbf{B}_w^T & -\mathbf{A} \end{array}\right] \left[\begin{array}{c:c} \mathbf{I} & -\gamma^{-2}\mathbf{P} \\ \hdashline \mathbf{0} & \mathbf{I} \end{array}\right] \right. \\
&\quad \left. - \left[\begin{array}{c:c} \mathbf{I} & \gamma^{-2}\mathbf{P} \\ \hdashline \mathbf{0} & \mathbf{I} \end{array}\right] \left[\begin{array}{c:c} \mathbf{0} & -\gamma^{-2}\dot{\mathbf{P}} \\ \hdashline \mathbf{0} & \mathbf{0} \end{array}\right] \right) \tilde{x} \\
&= \left[\begin{array}{c:c} (\mathbf{A} + \gamma^{-2}\mathbf{B}_w\mathbf{B}_w^T\mathbf{P})^T & \gamma^{-2}(\dot{\mathbf{P}} + \mathbf{PA} + \mathbf{A}^T\mathbf{P} + \gamma^{-2}\mathbf{PB}_w\mathbf{B}_w^T\mathbf{P} + \mathbf{C}_y^T\mathbf{C}_y) - \mathbf{C}_m^T\mathbf{C}_m \\ \hdashline -\mathbf{B}_w\mathbf{B}_w^T & -(\mathbf{A} + \gamma^{-2}\mathbf{B}_w\mathbf{B}_w^T\mathbf{P}) \end{array}\right] \tilde{x} \\
&= \tilde{\mathfrak{Y}}_m \tilde{x}
\end{aligned}$$

where time index has been deleted to simplify the notation. Adding and subtracting $\gamma^{-2}\mathbf{PB}_u\mathbf{B}_u^T\mathbf{P}$ in the (1, 2) block of the transformed Hamiltonian matrix yields

$$\begin{aligned}\tilde{\mathcal{Y}}_{12} &= \gamma^{-2}\{\dot{\mathbf{P}} + \mathbf{PA} + \mathbf{A}^T\mathbf{P} - \mathbf{P}(\mathbf{B}_u\mathbf{B}_u^T - \gamma^{-2}\mathbf{B}_w\mathbf{B}_w^T)\mathbf{P} + \mathbf{C}_y^T\mathbf{C}_y\}\\ &\quad + \gamma^{-2}\mathbf{PB}_u\mathbf{B}_u^T\mathbf{P} - \mathbf{C}_m^T\mathbf{C}_m\\ &= \gamma^{-2}\mathbf{PB}_u\mathbf{B}_u^T\mathbf{P} - \mathbf{C}_m^T\mathbf{C}_m.\end{aligned}$$

This expression has been simplified by noting that the term in curly brackets is the difference between the two sides of the full information Riccati equation (9.13) and is therefore equal to zero. The transformed Hamiltonian system is then

$$\dot{\tilde{x}}(t) = \left[\begin{array}{c|c} (\mathbf{A} + \gamma^{-2}\mathbf{B}_w\mathbf{B}_w^T\mathbf{P})^T & \gamma^{-2}\mathbf{PB}_u\mathbf{B}_u^T\mathbf{P} - \mathbf{C}_m^T\mathbf{C}_m \\ \hline -\mathbf{B}_w\mathbf{B}_w^T & -(\mathbf{A} + \gamma^{-2}\mathbf{B}_w\mathbf{B}_w^T\mathbf{P}) \end{array}\right]\tilde{x}(t) \qquad (10.14)$$

This equation is the Hamiltonian system associated with the Riccati equation (10.9). The solution of this Riccati equation can be given in terms of the state-transition matrix $\mathbf{\Phi}_m(t)$ of the Hamiltonian system (10.14):

$$\mathbf{Q}_m(t) = [\mathbf{\Phi}_{m22}(t)]^{-1}\mathbf{\Phi}_{m21}(t),$$

where

$$\mathbf{\Phi}_m(t) = \left[\begin{array}{c|c} \mathbf{\Phi}_{m11}(t) & \mathbf{\Phi}_{m12}(t) \\ \hline \mathbf{\Phi}_{m21}(t) & \mathbf{\Phi}_{m22}(t) \end{array}\right].$$

Further, this state-transition matrix of (10.14) can be related to the state-transition matrix of (10.12):

$$\begin{aligned}\left[\begin{array}{c|c} \mathbf{\Phi}_{m11} & \mathbf{\Phi}_{m12} \\ \hline \mathbf{\Phi}_{m21} & \mathbf{\Phi}_{m22} \end{array}\right] &= \left[\begin{array}{c|c} \mathbf{I} & -\gamma^{-2}\mathbf{P} \\ \hline \mathbf{0} & \mathbf{I} \end{array}\right]\left[\begin{array}{c|c} \mathbf{\Phi}_{11} & \mathbf{\Phi}_{12} \\ \hline \mathbf{\Phi}_{21} & \mathbf{\Phi}_{22} \end{array}\right]\\ &= \left[\begin{array}{c|c} \mathbf{\Phi}_{11} - \gamma^{-2}\mathbf{P}\mathbf{\Phi}_{21} & \mathbf{\Phi}_{12} - \gamma^{-2}\mathbf{P}\mathbf{\Phi}_{22} \\ \hline \mathbf{\Phi}_{21} & \mathbf{\Phi}_{22} \end{array}\right],\end{aligned}$$

since these systems are related by a time-varying similarity transform; see Appendix equation (A11.3). Note that the time indexes have again been dropped to simplify the notation. The solution of the Riccati equation (10.9) is then

$$\begin{aligned}\mathbf{Q}_m &= \mathbf{\Phi}_{21}(\mathbf{\Phi}_{22} - \gamma^{-2}\mathbf{P}\mathbf{\Phi}_{21})^{-1}\\ &= \mathbf{\Phi}_{21}(\mathbf{\Phi}_{22})^{-1}[(\mathbf{I} - \gamma^{-2}\mathbf{P}\mathbf{\Phi}_{21}(\mathbf{\Phi}_{22})^{-1}]^{-1} = \mathbf{Q}(\mathbf{I} - \gamma^{-2}\mathbf{PQ})^{-1},\end{aligned}$$

as given in (10.12).

10.2.2 Summary

The solution of the finite-time, suboptimal $\mathcal{H}_\infty$ output feedback control problem can be given in terms of the Riccati solutions $\mathbf{P}$ and $\mathbf{Q}$: A suboptimal $\mathcal{H}_\infty$ controller exists if and only if the following conditions are satisfied:

1. There is a solution of the Riccati equation

$$\boxed{\dot{\mathbf{P}}(t) = \mathbf{P}(t)\mathbf{A} + \mathbf{A}^T\mathbf{P}(t) - \mathbf{P}(t)(\mathbf{B}_u\mathbf{B}_u^T - \gamma^{-2}\mathbf{B}_w\mathbf{B}_w^T)\mathbf{P}(t) + \mathbf{C}_y^T\mathbf{C}_y} \qquad (10.15)$$

on the interval from 0 to t_f, given $\mathbf{P}(t_f) = \mathbf{0}$.

2. There is a solution of the Riccati equation

$$\dot{\mathbf{Q}}(t) = \mathbf{A}\mathbf{Q}(t) + \mathbf{Q}(t)\mathbf{A}^T - \mathbf{Q}(t)[\mathbf{C}_m^T\mathbf{C}_m - \gamma^{-2}\mathbf{C}_y^T\mathbf{C}_y]\mathbf{Q}(t) + \mathbf{B}_w\mathbf{B}_w^T \tag{10.16}$$

on the interval from 0 to t_f, given $\mathbf{Q}(0) = \mathbf{0}$.

3. On the interval from 0 to t_f,

$$\rho[\mathbf{P}(t)\mathbf{Q}(t)] < \gamma^2. \tag{10.17}$$

Note that the Hamiltonian formulations of the first two conditions are given in the sections on $\mathcal{H}_\infty$ full information control and $\mathcal{H}_\infty$ estimation, respectively.

A suboptimal controller that satisfies this bound is given:

$$\dot{x}_c(t) = \mathbf{A}_c(t)x_c(t) + \mathbf{B}_c(t)m(t); \tag{10.18a}$$

$$u(t) = \mathbf{C}_c(t)x_c(t), \tag{10.18b}$$

where the matrices $\mathbf{A}_c(t)$, $\mathbf{B}_c(t)$, and $\mathbf{C}_c(t)$ are found using the solutions of the above algebraic Riccati equations:

$$\begin{aligned}\mathbf{A}_c(t) = \mathbf{A} + \gamma^{-2}\mathbf{B}_w\mathbf{B}_w^T\mathbf{P}(t) - \mathbf{B}_u\mathbf{B}_u^T\mathbf{P}(t)\\ - [\mathbf{I} - \gamma^{-2}\mathbf{Q}(t)\mathbf{P}(t)]^{-1}\mathbf{Q}(t)\mathbf{C}_m^T\mathbf{C}_m;\end{aligned} \tag{10.18c}$$

$$\mathbf{B}_c(t) = [\mathbf{I} - \gamma^{-2}\mathbf{Q}(t)\mathbf{P}(t)]^{-1}\mathbf{Q}(t)\mathbf{C}_m^T; \tag{10.18d}$$

$$\mathbf{C}_c(t) = -\mathbf{B}_u^T\mathbf{P}(t). \tag{10.18e}$$

The existence conditions for the suboptimal $\mathcal{H}_\infty$ controller imply the existence of both a suboptimal $\mathcal{H}_\infty$ full information controller and a suboptimal $\mathcal{H}_\infty$ filter for estimating the reference output. The implied existence of the full information controller can be understood by noting that full information includes information on all possible outputs. Therefore, output feedback is a special case of full information control. The existence of the $\mathcal{H}_\infty$ suboptimal filter for estimating the reference output can be understood by noting that the reference output cannot be controlled to a greater degree of accuracy than that at which the reference output can be estimated.

The optimal controller can be approximated to an arbitrary degree of accuracy by decreasing the performance bound until the conditions (10.15) through (10.17) are no longer satisfied. Note that a solution always exists for a sufficiently large performance bound, since the above Riccati equations reduce to those for the LQR, and the Kalman filter in the limit as this bound approaches infinity.

10.3 Steady-State Control

The estimator gains and the state feedback gains of the $\mathcal{H}_\infty$ controller typically approach steady-state values far from the initial time and the final time, respectively. In applications where the control system is designed to operate for time periods that are long compared to the transient times of these gains, it is reasonable to ignore the transients and use the steady-state gains, exclusively. The use of the steady-state gains simplifies controller implementation and results in a time-invariant, closed-loop system. Time-invariance of the closed-loop system allows the use of many robustness and performance analysis techniques that are not applicable to time-varying systems.

The steady-state $\mathcal{H}_\infty$ controller is the solution of the following suboptimal control problem: Find a linear, time-invariant controller system, described in the Laplace domain as follows:

$$u(s) = \mathbf{K}(s)m(s),$$

that internally stabilizes the closed-loop system and bounds the ∞-norm of the closed-loop system:

$$\|\mathbf{G}_{yw}\|_\infty < \gamma.$$

The steady-state $\mathcal{H}_\infty$ suboptimal control can be obtained by combining the steady-state $\mathcal{H}_\infty$ full information controller and the steady-state $\mathcal{H}_\infty$ estimator of this control. The existence of this steady-state controller is predicated on the existence of both the full information controller and the estimator. As in the finite-time case, the existence of an estimator of the full information control can be related to the existence of an estimator of the reference output. Combining the existence results for the full information controller and the output estimator, we find as follows: A solution exists for the suboptimal $\mathcal{H}_\infty$ control problem if and only if the following conditions are satisfied:

1. There is a positive semidefinite solution of the algebraic Riccati equation

$$\boxed{\mathbf{PA} + \mathbf{A}^T\mathbf{P} - \mathbf{P}(\mathbf{B}_u\mathbf{B}_u^T - \gamma^{-2}\mathbf{B}_w\mathbf{B}_w^T)\mathbf{P} + \mathbf{C}_y^T\mathbf{C}_y = \mathbf{0},}$$

   such that $\{\mathbf{A} - (\mathbf{B}_u\mathbf{B}_u^T - \gamma^{-2}\mathbf{B}_w\mathbf{B}_w^T)\mathbf{P}\}$ is stable (i.e., has only eigenvalues with negative real parts).
2. There is a positive semidefinite solution of the algebraic Riccati equation

$$\boxed{\mathbf{AQ} + \mathbf{QA}^T - \mathbf{Q}(\mathbf{C}_m^T\mathbf{C}_m - \gamma^{-2}\mathbf{C}_y^T\mathbf{C}_y)\mathbf{Q} + \mathbf{B}_w\mathbf{B}_w^T = \mathbf{0}}$$

   such that $\{\mathbf{A} - \mathbf{Q}(\mathbf{C}_m^T\mathbf{C}_m - \gamma^{-2}\mathbf{C}_y^T\mathbf{C}_y)\}$ is stable.
3. The spectral radius of the product of these Riccati solutions is bounded:

$$\boxed{\rho(\mathbf{PQ}) < \gamma^2.}$$

Note that Hamiltonian formulations for the first two conditions are given in Chapter 9 for $\mathcal{H}_\infty$ full information control and $\mathcal{H}_\infty$ estimation. A state model for the steady-state $\mathcal{H}_\infty$ suboptimal controller is obtained by using the algebraic Riccati solutions in (10.18).

EXAMPLE 10.1 A satellite tracking antenna, containing measurement errors and subject to wind torques, (as in Examples 6.11 and 8. 1), can be modeled as follows:

$$\begin{bmatrix} \dot{\theta}(t) \\ \ddot{\theta}(t) \end{bmatrix} = \begin{bmatrix} 0 & 1 \\ 0 & -0.1 \end{bmatrix} \begin{bmatrix} \theta(t) \\ \dot{\theta}(t) \end{bmatrix} + \begin{bmatrix} 0 \\ 0.001 \end{bmatrix} u(t) + \begin{bmatrix} 0 \\ 0.001W_b \end{bmatrix} w_1(t),$$

where $\theta(t)$ is the pointing error of the antenna in degrees, $u(t)$ is the control torque, $w(t)$ is the normalized wind torque (disturbance input), and W_b is the bound on the wind torque. The normalized wind torque is assumed to be sinusoidal with a bounded amplitude $|w_1(t)| \le 1$, implying that the true wind torque is bounded: $|w(t)| = |W_b w_1(t)| \le W_b$. The pointing error is measured as

$$m(t) = [1 \quad 0]\begin{bmatrix}\theta(t)\\ \dot{\theta}(t)\end{bmatrix} + v(t),$$

where the measurement error is assumed to be sinusoidal with a bounded amplitude: $|v(t)| \le 1$ degree. The outputs of interest are the pointing error and the control input

$$y(t) = \begin{bmatrix}W_\theta & 0\\ 0 & 0\end{bmatrix}\begin{bmatrix}\theta(t)\\ \dot{\theta}(t)\end{bmatrix} + \begin{bmatrix}0\\ 1\end{bmatrix}u(t).$$

The weight on the control input is selected to be 1 in order to satisfy condition (10.2d). A variable weight W_θ is added to the pointing error to allow tuning of the controller. Note that the bound on the disturbance torque can also be used as a tuning parameter for this controller.

The steady-state controller is synthesized for $W_b = 70$ and $W_\theta = 13$ (these numbers roughly match those selected for the LQG controller in Example 8.1). A performance bound $\gamma = 90$ (approximately 10% over the optimal) is used in this design. The system is simulated with a true disturbance input $w(t) = 70$. The measurement error is sampled white noise (the sampling time is 0.1), uniformly distributed between -1 to 1. The results are shown in Figure 10.1. The system is also simulated for controllers synthesized with $W_b = 700$ and $W_\theta = 13$ (Figure 10.2), and $W_b = 70$ and $W_\theta = 130$ (Figure 10.3). Note that the actual disturbance torque has an amplitude of 70 in all

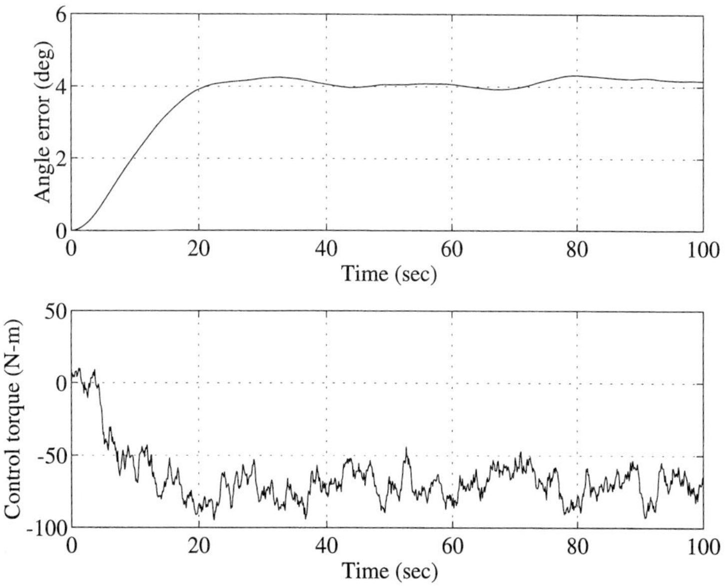

FIGURE 10.1 Baseline results for Example 10.1

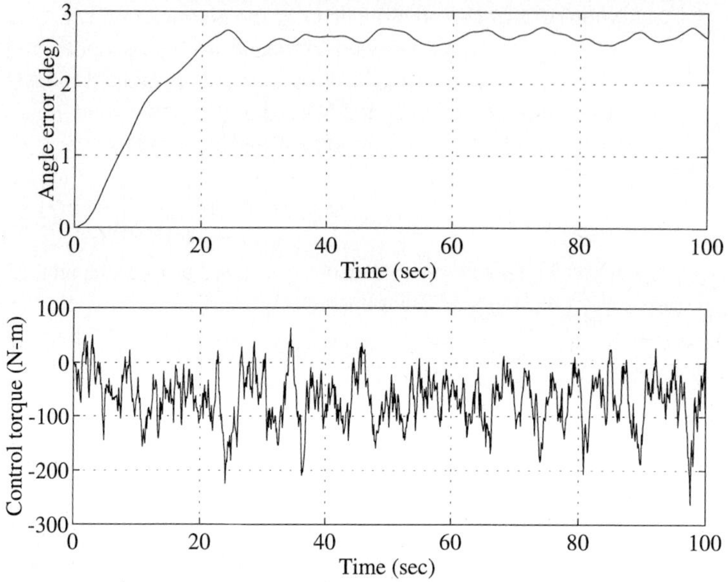

FIGURE 10.2 Results for Example 10.1 with increased disturbance input bound

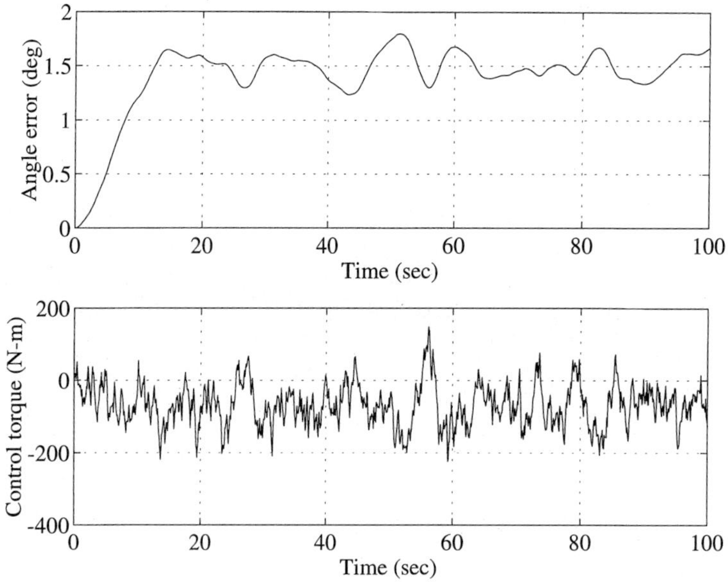

FIGURE 10.3 Results for Example 10.1 with increased angle error weighting

cases, and all controllers are generated with performance bounds approximately 10% over optimal.

The steady-state angle error is composed of a dc component due to the disturbance torque and a random component due to the measurement error. Increasing the bound on the disturbance torque in the design process (as in Figure 10.2) results in a closed-loop response that uses more control, decreases the constant angle error due to the disturbance torque, and increases the random component of the angle error due to the measurement noise. Increasing the angle error weighting (as in Figure 10.3) also yields a closed-loop response that uses more control, decreases the constant angle error due to the disturbance torque, and increases the random component of the angle error due to the measurement noise. ◆

10.4 Application of $\mathcal{H}_\infty$ Control

$\mathcal{H}_\infty$ control can be used as an alternative to LQG optimal control. Both cost functions are reasonable for a wide range of problems, and in many applications the choice of a quadratic versus an ∞-norm cost function is arbitrary. In these applications, the LQG controller is typically selected, since controller optimization is simpler and yields a unique solution. But the LQG control system may have undesirable properties; that is, it may not be robust, it may have an undesirable frequency response, and so on. In these cases, it is reasonable to try an $\mathcal{H}_\infty$ optimal (or suboptimal) controller. The results obtained with the $\mathcal{H}_\infty$ controller may be better or worse than those obtained with the LQG controller, since the $\mathcal{H}_\infty$ controller has no magic robustness or frequency-domain properties. Still, it is worth trying when the LQG controller is not performing adequately. An example of reformulating an LQG problem in an $\mathcal{H}_\infty$ control setting is provided by Example 10.1, which is a reformulation of the LQG problem in Example 8.1.

$\mathcal{H}_\infty$ control is a natural for applications where the specifications are given in terms of bounds on the outputs (both output errors and controls). Requiring the outputs to remain below prescribed levels is typical of engineering design specifications. Output bounds are given by the closed-loop system ∞-norm provided the disturbance inputs are sinusoidal (or constant), all inputs are at the same frequency, the inputs are normalized, and the outputs are normalized.

The inputs should be normalized so that

$$|w_i(t)| \leq \frac{1}{\sqrt{\text{\# of inputs}}}, \tag{10.19}$$

when designing to achieve a given output bound. This normalization assures that the sum of the squares of the contributions from all inputs remains below the specification. The outputs should also be normalized so that the specifications require a bound of 1 for each output. The desired output bounds are then achieved provided the closed-loop ∞-norm is less than 1. Note that these bounds are only guaranteed for sinusoidal disturbance inputs.[3]

$\mathcal{H}_\infty$ control can be used to generate systems that meet output bound specifications when the inputs are sinusoidal, but with different frequencies. In this case, the ∞-norm provides a worst-case gain from each input to each output. The output is then bounded by the sum, over all inputs, of the output amplitude bounds. In this case, the inputs should be normalized as in (10.19), except the number of inputs is replaced by the number of inputs at a given frequency. The resulting bound is typically conservative

[3] The system 1-norm provides an absolute bound on the size of the output of a SISO system for arbitrary, bounded inputs. This result can be generalized to MIMO systems. Unfortunately, optimization of the system 1-norm is not trivial.

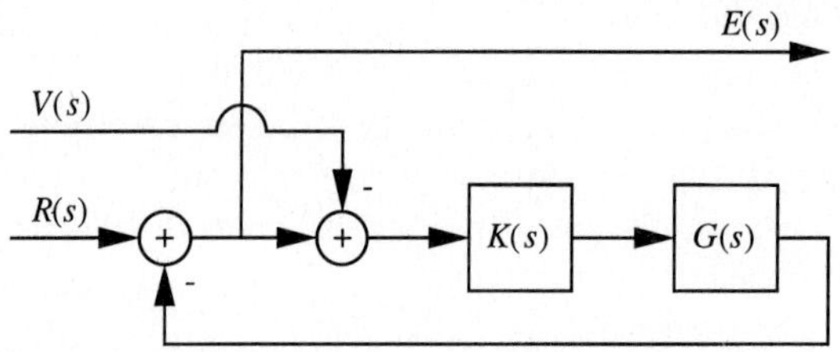

FIGURE 10.4 Unity feedback system with measurement noise

since, at each frequency, the input "direction" is constrained, whereas the ∞-norm gives the bound on the gain for unconstrained input directions. $\mathcal{H}_\infty$ control can also be used when the inputs are not sinusoidal. For nonsinusoidal inputs, the closed-loop system ∞-norm provides an indication of the maximum size of the outputs, but does not provide a formal bound on the outputs.

The existence conditions for the suboptimal $\mathcal{H}_\infty$ controller are very useful when performing trade-offs between competing control objectives. These existence conditions can be used to determine when a given set of specifications are consistent with a reasonable design and when the system is overspecified.

10.4.1 Performance Limitations

The specification of performance using the ∞-norm can easily result in unobtainable requirements. As an example, consider the generic SISO unity feedback system shown in Figure 10.4. A controller is desired for this system so that

$$\|\mathbf{G}_{cl}\|_\infty = \max_\omega\{\bar{\sigma}[\mathbf{G}_{cl}(j\omega)]\} < 0.1,$$

where[4]

$$\mathbf{G}_{cl}(s) = [G_{er}(s) \quad G_{ev}(s)] = \begin{bmatrix} \dfrac{1}{1 + G(s)K(s)} & \dfrac{G(s)K(s)}{1 + G(s)K(s)} \end{bmatrix}.$$

While this specification seems quite reasonable, it is not achievable with any plant and controller unless the inputs are separated in frequency.

The minimum closed-loop system ∞-norm that can be achieved for this system is 0.707. This bound is a consequence of the relationship between the elements in the closed-loop transfer function:

$$G_{er}(s) + G_{ev}(s) = \frac{1}{1 + G(s)K(s)} + \frac{G(s)K(s)}{1 + G(s)K(s)} = 1. \tag{10.20}$$

To derive the ∞-norm bound, note that the singular value (there is only one) of the closed-loop transfer function is

$$\bar{\sigma}[G_{cl}(j\omega)] = \sqrt{G_{cl}(j\omega)G_{cl}^T(j\omega)} = \sqrt{|G_{er}(j\omega)|^2 + |G_{ev}(j\omega)|^2}.$$

[4] These two transfer functions are often called (1) the sensitivity and (2) the closed-loop transfer function (or the complimentary sensitivity) in classical control.

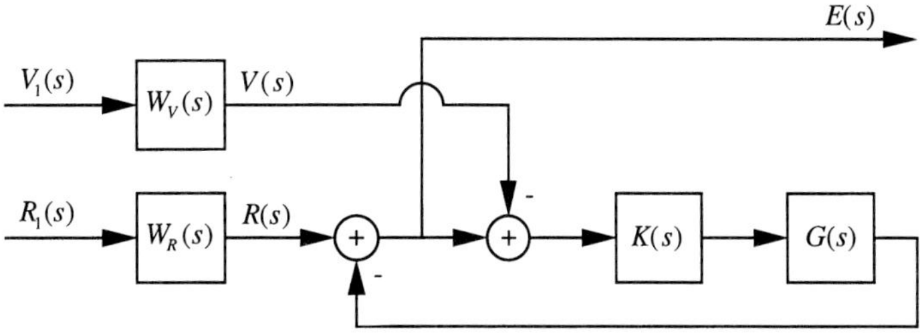

FIGURE 10.5 Weighting functions for the unity feedback system with measurement noise

Minimizing this singular value with respect to the constraint (10.20) yields[5]

$$\min_{G_{er}+G_{ev}=1} \bar{\sigma}[G_{cl}(j\omega)] = \min_{G_{er}+G_{ev}=1} \sqrt{|G_{er}(j\omega)|^2 + |G_{ev}(j\omega)|^2} = \frac{1}{\sqrt{2}}.$$

Since this is the minimum possible singular value, the closed-loop system ∞-norm is bounded:

$$\|G_{cl}\|_\infty = \max_{\omega}\{\bar{\sigma}[G_{cl}(j\omega)]\} \geq \frac{1}{\sqrt{2}}.$$

The constraint (10.20) imposes a fundamental limitation on tracking performance in the presence of measurement error. This is also true for MIMO control systems, where (10.20) becomes

$$\mathbf{G}_{er}(s) + \mathbf{G}_{ev}(s) = [\mathbf{I} + \mathbf{G}(s)\mathbf{K}(s)]^{-1} + [\mathbf{I} + \mathbf{G}(s)\mathbf{K}(s)]^{-1}\mathbf{G}(s)\mathbf{K}(s) = \mathbf{I}.$$

This performance limitation can be intuitively understood by noting that large loop gain is required for good tracking performance. But large loop gain results in the plant tracking the measurement noise. Therefore, the controller cannot simultaneously reduce tracking errors due to both reference inputs and measurement errors unless these signals are separated in frequency.

The specifications can be made more reasonable by separating the reference inputs and the measurement noise in frequency. For example, the reference input may be assumed to be slowly varying or, equivalently, to be a lowpass signal. The measurement error may be assumed to be rapidly varying (typical of noise signals) or, equivalently, to be a highpass signal. The transfer function from the reference input to the tracking error can then be made small over the frequency range allotted to the reference input. In addition, the transfer function from the measurement noise to the tracking error can be made small over the separate frequency band allotted to the measurement noise.

The ∞-norm can be used to specify performance over frequency bands by appending weighting functions to the plant inputs, as shown in Figure 10.5. A controller that yields an ∞-norm less than 0.1 is then possible, at least in theory, provided the weighting functions $W_R(j\omega)$ and $W_V(j\omega)$ are appropriately selected; that is, these weighting functions cannot contain significant overlap.

[5] The minimum occurs when $G_{er}(j\omega) = G_{ev}(j\omega) = 1/2$. This result is easily obtained using the Lagrange multiplier method for constrained optimization.

EXAMPLE 10.2 An ac motor is described by the following state equation:

$$\dot{x}(t) = -x(t) + u(t),$$

where $x(t)$ is the rotational velocity of the shaft, and $u(t)$ is the control input. Noisy measurements of the rotational velocity error are available:

$$m(t) = r(t) - x(t) - v(t),$$

where $r(t)$ is the reference input.

The measurement noise and reference input are both assumed to be bounded by 1. The normalized reference output is then defined as[6]

$$y(t) = \begin{bmatrix} -1 \\ 0 \end{bmatrix} x(t) + \begin{bmatrix} 1 \\ 0 \end{bmatrix} r(t) + \begin{bmatrix} 0 \\ 0.001 \end{bmatrix} u(t).$$

The control is included as a reference output to keep the control finite. But the control weight is small, which means that this term should not significantly affect the performance.

Performing $\mathcal{H}_\infty$ optimization, we find that the minimum achievable ∞-norm for this system is 1. This bound is achievable in this example by using the controller $u(t) = 0$. Obviously, this bound on the error and this controller are not acceptable.

A more appropriate controller can be generated by $\mathcal{H}_\infty$ optimization, after adding frequency weighting to the reference input and measurement error, as shown in Figure 10.5. Specifying that the reference input has a bandwidth of 0.1 rad/sec yields the following weighting function:

$$W_R(s) = \frac{0.1}{s + 0.1}.$$

Specifying that the measurement error is highpass with a stop band below 10 rad/sec, yields the following weighting function:

$$W_V(s) = \frac{s}{s + 10}.$$

Note that both of these weighting functions have a nearly unity gain in the band of interest. As required, the gain at the crossover frequency of the weighting functions is small. This crossover gain limits the achievable closed-loop ∞-norm due to the constraint (10.20). Appending the weighting functions to the plant yields

$$\begin{bmatrix} \dot{x}(t) \\ \dot{x}_r(t) \\ \dot{x}_v(t) \end{bmatrix} = \begin{bmatrix} -1 & 0 & 0 \\ 0 & -0.1 & 0 \\ 0 & 0 & -10 \end{bmatrix} \begin{bmatrix} x(t) \\ x_r(t) \\ x_v(t) \end{bmatrix} + \begin{bmatrix} 1 \\ 0 \\ 0 \end{bmatrix} u(t) + \begin{bmatrix} 0 & 0 \\ 0.1 & 0 \\ 0 & 10 \end{bmatrix} \begin{bmatrix} w_1(t) \\ w_2(t) \end{bmatrix};$$

$$\begin{bmatrix} m(t) \\ \hline y(t) \end{bmatrix} = \begin{bmatrix} -1 & 1 & -1 \\ \hline -1 & 1 & 0 \\ 0 & 0 & 0 \end{bmatrix} \begin{bmatrix} x(t) \\ x_r(t) \\ x_v(t) \end{bmatrix} + \left[\begin{array}{c|cc} 0 & 0 & 1 \\ \hline 0 & 0 & 0 \\ 0.001 & 0 & 0 \end{array}\right] \begin{bmatrix} u(t) \\ \hline w_1(t) \\ w_2(t) \end{bmatrix}.$$

[6] This reference output is not consistent with the $\mathcal{H}_\infty$ optimization problem statement (10.1b) due to the inclusion of the disturbance input in the reference output. More general $\mathcal{H}_\infty$ suboptimal control formulas that allow the inclusion of disturbance inputs in the reference output can be found in [1], page 449.

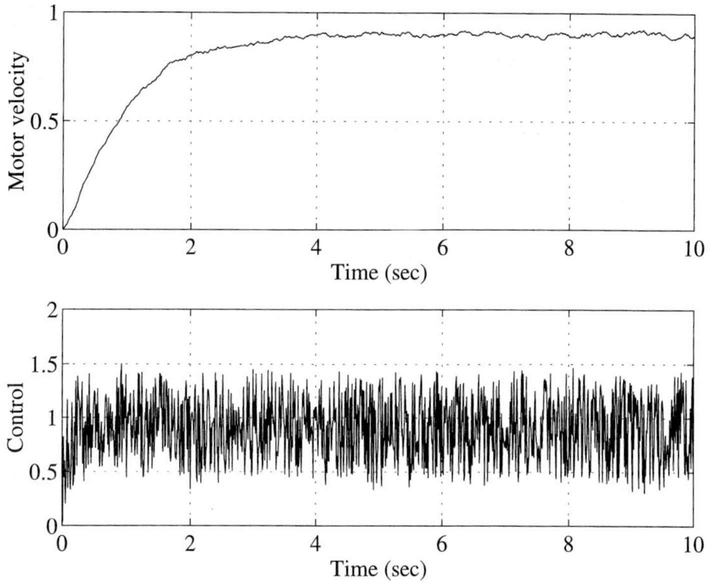

FIGURE 10.6 Results for Example 10.2

This model can be placed in the form [see (10.1) and (10.2)] appropriate for $\mathcal{H}_\infty$ suboptimal controller synthesis by normalizing the control input.

An $\mathcal{H}_\infty$ suboptimal controller is generated for this plant using the performance bound $\gamma = 0.1$ (approximately 10% over the optimal). The closed-loop system is then simulated with the reference input equal to 1. The measurement error is formed by passing discrete-time white noise (uniformly distributed from -1 to 1 and with a sampling time of 0.01) through the highpass filter $W_V(s)$. The results are plotted in Figure 10.6. Observe that the steady-state tracking error due to the reference input is less than 0.1 as guaranteed by the performance bound. Also, the random tracking error due to measurement error is less than 0.1. This error is not guaranteed by the theory to be less than 0.1, since the measurement error is not sinusoidal. Still, the closed-loop ∞-norm provides a reasonable gauge of the output size.

The control input is considerably less than the guaranteed bound of 100, because the control input is not limited by the specified bound on the control. Instead, the control input is limited by the bound on the gain from the measurement error to the tracking error. This bound limits the control input, since large feedback gains result in a large coupling between the measurement noise and the tracking error. ◆

An achievable set of specifications was obtained in the previous example by using frequency weighting. In this example, we found that good tracking and good measurement noise rejection place contradictory requirements on the controller. A trade-off between these two factors is therefore required during controller design. Additional trade-offs may also be encountered during controller design.

A number of control system trade-offs can be described by considering the generic feedback system shown in Figure 10.7. The inputs to this system are the reference input

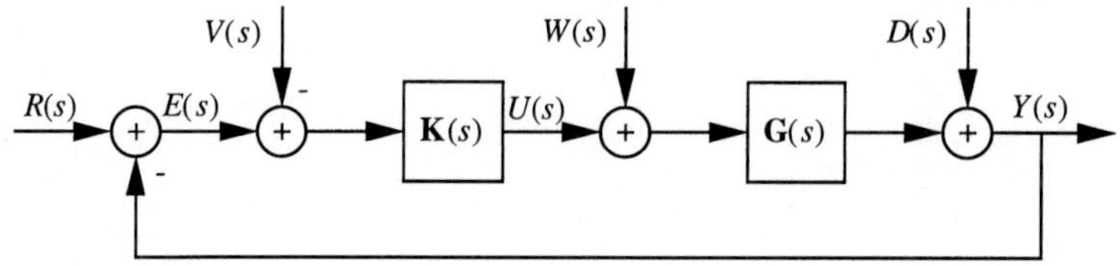

FIGURE 10.7 A generic unity feedback control system

$r(t)$, measurement error $v(t)$, actuator error $w(t)$, and output disturbance $d(t)$. The output disturbance is typically used to describe plant modeling errors. The various transfer functions in this block diagram are given:

$$E(s) = [\mathbf{I} + \mathbf{GK}]^{-1}[R(s) - D(s)] + [\mathbf{I} + \mathbf{GK}]^{-1}\mathbf{GK}V(s) - [\mathbf{I} + \mathbf{GK}]^{-1}\mathbf{G}W(s);$$

$$Y(s) = [\mathbf{I} + \mathbf{GK}]^{-1}\mathbf{GK}[R(s) - V(s)] + [\mathbf{I} + \mathbf{GK}]^{-1}\mathbf{G}W(s) + [\mathbf{I} + \mathbf{GK}]^{-1}D(s);$$

$$U(s) = [\mathbf{I} + \mathbf{KG}]^{-1}\mathbf{K}[R(s) - D(s) - V(s)] - [\mathbf{I} + \mathbf{KG}]^{-1}\mathbf{KG}W(s).$$

Note, for example, that

$$\mathbf{G}_{ed}(s) - \mathbf{G}_{ev}(s) = \mathbf{I}.$$

Therefore, there is an inherent trade-off between making the system tolerant of output disturbances (i.e., providing good robustness to plant modeling errors), and making the system tolerant of measurement noise. Other trade-offs can be obtained by looking for additional constraints between the various transfer functions and by looking at the effects of large and/or small loop gains, controller gains, and plant gains on the various transfer functions.

Careful consideration of the trade-offs inherent in a control system design is also beneficial when iteration of the design is required. For example, changing the control bound in Example 10.2 had little effect of the controller because the control gains were being constrained by the measurement error gain. Therefore, using this parameter for design iteration is not very useful.

In summary, $\mathcal{H}_\infty$ specifications should be selected after careful consideration of the desired control objectives. Constraints among the various closed-loop transfer functions should be identified before defining the specifications. In addition, trade-offs between the various control objectives should be defined, and redundant control objectives should be identified. Using this information, the specifications should be frequency-weighted to avoid conflicts caused by competing control objectives. It is typically a good idea to avoid allpass specifications when multiple disturbance inputs are included in the model. In addition, it is often a good idea to avoid redundant specifications whenever possible. Note that these redundant specifications can be added back into the model when evaluating performance, if desired. A good deal of art is involved in setting specifications for $\mathcal{H}_\infty$ controller design. This art is facilitated by a thorough understanding of the plant and by experience. The setting of $\mathcal{H}_\infty$ specifications is discussed further in subsequent examples.

10.4.2 Integral Control

Integral control is used to remove steady-state errors due to constant reference inputs and disturbances. Unfortunately, controllers with integral terms cannot be generated directly using the $\mathcal{H}_\infty$ synthesis theory presented. This limitation is typically not a problem since "nearly" integral controllers can be generated. In practice, these controllers can usually be replaced by controllers with true integral terms, if desired.

A reasonable approach to designing integral controllers is to append an integral to the plant before controller synthesis. This approach fails (as it also did in the LQG case) because the integral state is uncontrollable if the integral is on the plant input, and the integral state is unobservable if the integral is on the plant output. Since the integral is unstable, the closed-loop system is not stabilizable, implying that no suboptimal $\mathcal{H}_\infty$ controller exists. This limitation was circumvented in LQG design by using a separate model for synthesizing the state feedback controller and for synthesizing the Kalman filter. This approach is not valid in this case, since the general separation principle is technically not valid in the $\mathcal{H}_\infty$ setting. Instead, integral controllers are synthesized in an *ad hoc* manner. This *ad hoc* approach to integral controller design can also be applied in LQG design.

A nearly integral controller can be generated by appending filters of the form

$$W(s) = \frac{1}{s + \varepsilon} \tag{10.21}$$

to the plant inputs (either reference or disturbance) where dc signals are to be rejected.[7] For small ε, this filter has a large dc gain and approximates an integrator. The loop gain at dc for the feedback system must therefore also be large in order to avoid a large dc closed-loop gain. This large dc loop gain is obtained by placing a pole of the controller near the origin in the complex plane.[8] When true integral action is desired, this controller pole can be replaced by a pole at the origin. This replacement typically has very little impact on the closed-loop poles and stability of the feedback system. But caution must be used when making this substitution, since there is no mathematical guarantee of performance or even stability. This design procedure is illustrated in the following example.

EXAMPLE 10.3 A valve is constructed with a spring on the "flapper" so that if power is removed the valve closes (see Example 8.5). The control input is a torque applied to the flapper. The state equation for this valve is

$$\dot{x}(t) = \begin{bmatrix} 0 & 1 \\ -10 & -1 \end{bmatrix} x(t) + \begin{bmatrix} 0 \\ 1 \end{bmatrix} u(t) + \begin{bmatrix} 0 & 0 \\ 0 & 0 \end{bmatrix} \begin{bmatrix} r(t) \\ v(t) \end{bmatrix},$$

where $r(t)$ is the reference input and $v(t)$ is measurement error. The reference input is assumed to be generated by application of a normalized input to the filter (10.21). The

[7] Integral controllers can also be obtained by appending these filters to the outputs, but this is typically less desirable, since the filters then operate on the measurement error gain. This makes specification of the measurement error more difficult.

[8] A pole near the origin in the plant also yields good dc tracking. Such a pole is assumed not to exist.

measurement consists of the difference between the position of the valve and the desired position, plus a measurement error. The measurement noise is assumed to be the output of a highpass filter:

$$W_v(s) = \frac{s}{s + 10}.$$

Appending this filter and (10.21) to the plant yields the following state model:

$$\dot{x}(t) = \begin{bmatrix} 0 & 1 & 0 & 0 \\ -10 & -1 & 0 & 0 \\ 0 & 0 & -\varepsilon & 0 \\ 0 & 0 & 0 & -10 \end{bmatrix} \begin{bmatrix} \dot{x}_1(t) \\ \dot{x}_2(t) \\ \dot{r}(t) \\ \dot{v}(t) \end{bmatrix} + \left[\begin{array}{c|cc} 0 & 0 & 0 \\ 1 & 0 & 0 \\ 0 & 1 & 0 \\ 0 & 0 & 10 \end{array}\right] \begin{bmatrix} u(t) \\ \hline r_0(t) \\ v_0(t) \end{bmatrix};$$

$$\begin{bmatrix} m(t) \\ \hline y(t) \end{bmatrix} = \left[\begin{array}{cccc} -1 & 0 & 1 & -1 \\ \hline -1 & 0 & 1 & 0 \\ 0 & 0 & 0 & 0 \end{array}\right] \begin{bmatrix} x_1(t) \\ x_2(t) \\ r(t) \\ v(t) \end{bmatrix} + \left[\begin{array}{c|cc} 0 & 0 & 1 \\ \hline 0 & 0 & 0 \\ 0.001 & 0 & 0 \end{array}\right] \begin{bmatrix} u(t) \\ \hline r_0(t) \\ v_0(t) \end{bmatrix}.$$

A suboptimal $\mathcal{H}_\infty$ controller is generated with $\varepsilon = 0.01$ and $\gamma = 1100$ (approximately 10% over optimal). This controller is described by the following transfer function:

$$G_c(s) = \frac{99(s + 10)(s^2 + s + 10)}{(s + 0.03)(s + 4.4)(s^2 + 13.6s + 112)}.$$

Integral action is then added by replacing the pole at 0.03 with a pole at the origin:

$$G_c(s) = \frac{99(s + 10)(s^2 + s + 10)}{s(s + 4.4)(s^2 + 13.6s + 112)}.$$

The feedback system formed using this controller and the original plant is stable. This system is simulated with a unit step reference input. The measurement error is generated by putting discrete-time white noise into the highpass filter given above. This measurement noise has a maximum amplitude approximately equal to 1. The results of this simulation are plotted in Figure 10.8 and show that zero steady-state error is achieved. Note that this controller inverts the stable plant dynamics, which is typical when the measurement errors are small. ◆

This example illustrates the *ad hoc* integral control design procedure. The approximate integrator could also have been appended at the plant output. In a MIMO setting, it may be necessary to append approximate integrators to multiple inputs.

10.4.3 Designing for Robustness

Stability robustness to an unstructured perturbation is guaranteed when the ∞-norm of the transfer function from the perturbation input to the perturbation output is bounded by 1 (the perturbation bound is assumed to be normalized to 1). This transfer function can be included within the closed-loop transfer function by appending the perturbation input and output to the disturbance input and reference output, respectively (see Figure 10.9). Adding weights at the input and output of the perturbation (as in Figure 10.9) allows the designer to generate a family of controllers that trade off performance with robustness. Note that the original plant results when both of these weights equal zero.

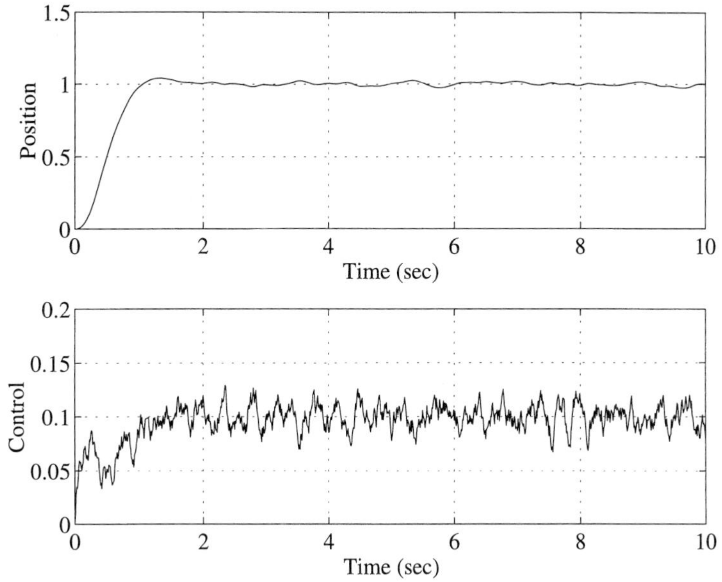

FIGURE 10.8 Integral control for Example 10.3

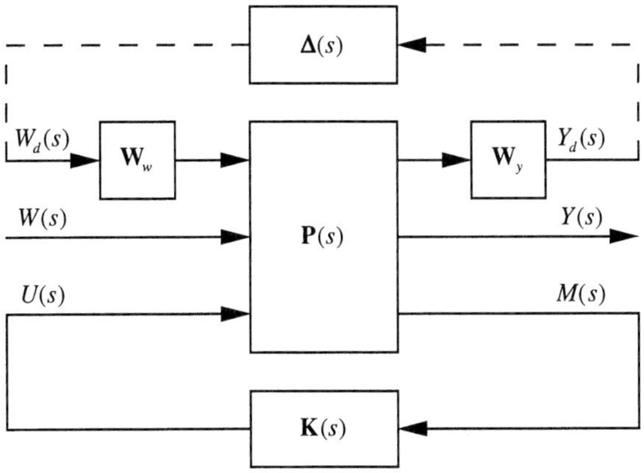

FIGURE 10.9 Block diagram for robust performance design

EXAMPLE 10.4 A single link robot arm is being held vertically to perform a task located overhead, as first presented in Example 8.3 (see Figure 8.9). A state model for this system is

$$\begin{bmatrix} \dot{\theta}(t) \\ \ddot{\theta}(t) \end{bmatrix} = \begin{bmatrix} 0 & 1 \\ 0.5 & 0 \end{bmatrix} \begin{bmatrix} \theta(t) \\ \dot{\theta}(t) \end{bmatrix} + \begin{bmatrix} 0 \\ 0.1 \end{bmatrix} u(t) + \begin{bmatrix} 0 & 0 \\ 0.1 & 0 \end{bmatrix} \begin{bmatrix} w(t) \\ v(t) \end{bmatrix}.$$

The disturbance torque $w(t)$ is caused by interference in the torquer circuit and is bounded by 1. The angular position of the arm is measured. The reference output consists of both the angular position of the arm and the weighted control input:

$$\begin{bmatrix} m(t) \\ \hdashline y(t) \end{bmatrix} = \begin{bmatrix} 1 & 0 \\ \hdashline 1 & 0 \\ 0 & 0 \end{bmatrix} \begin{bmatrix} \theta(t) \\ \dot{\theta}(t) \end{bmatrix} + \begin{bmatrix} 0 \\ \hdashline 0 \\ 4 \end{bmatrix} u(t) + \begin{bmatrix} 0 & \sqrt{10} \\ \hdashline 0 & 0 \\ 0 & 0 \end{bmatrix} \begin{bmatrix} w(t) \\ v(t) \end{bmatrix}.$$

The parameters in this equation are chosen to match the LQG control problem posed in Example 8.3. A steady-state, suboptimal $\mathcal{H}_\infty$ controller is synthesized for this plant model with a performance bound $\gamma = 275$ (approximately 10% over optimal). The Nyquist plot of this system is shown in Figure 10.10a. The stability margins of this system, along with the closed-loop system norm, are given in Table 10.1 (the row labeled original plant).

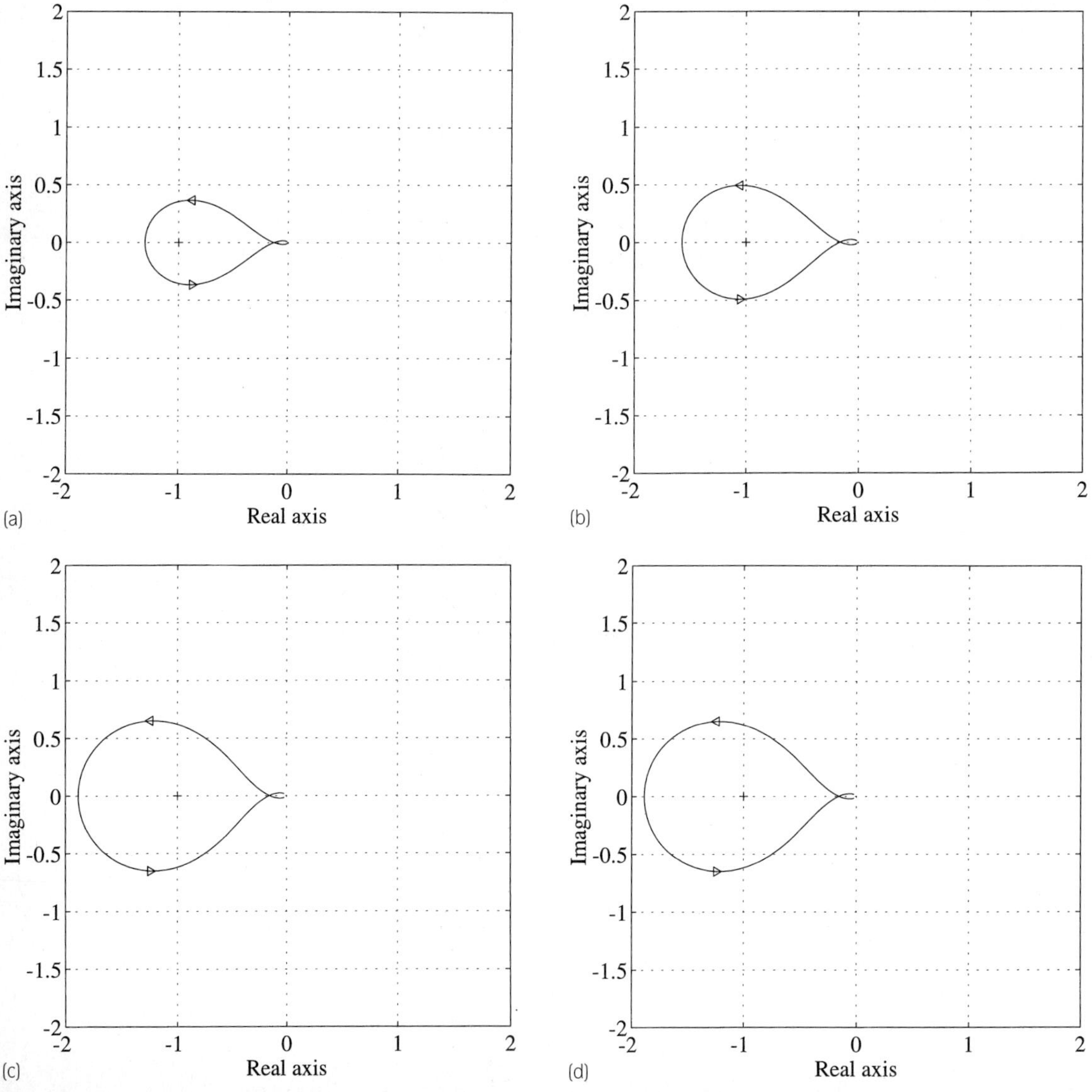

FIGURE 10.10 Nyquist plots for Example 10.4

TABLE 10.1 Performance and Robustness of the Closed-Loop System

	GM^+	GM^-	PM	$\|\mathbf{G}_{cl}\|_\infty$
Original plant	7.6	0.77	21°	275
$W_w = \frac{275}{8}$; $W_y = \frac{4}{275}$	6.2	0.63	28°	390
$W_w = \frac{275}{4}$; $W_y = \frac{4}{275}$	6.5	0.53	34°	552
$W_w = \frac{275}{4}$; $W_y = \frac{2}{275}$	6.5	0.53	34°	552

Note that the downside gain margin and phase margin are unacceptably small. Three additional designs are then generated after appending an input-multiplicative perturbation to the plant:

$$\begin{bmatrix} \dot{\theta}(t) \\ \ddot{\theta}(t) \end{bmatrix} = \begin{bmatrix} 0 & 1 \\ 0.5 & 0 \end{bmatrix} \begin{bmatrix} \theta(t) \\ \dot{\theta}(t) \end{bmatrix} + \begin{bmatrix} 0 & 0 & 0 & 0 \\ 0.1 & 0.1 & 0 & W_w 0.1 \end{bmatrix} \begin{bmatrix} u(t) \\ w(t) \\ v(t) \\ w_d(t) \end{bmatrix};$$

$$\begin{bmatrix} m(t) \\ y(t) \end{bmatrix} = \begin{bmatrix} 1 & 0 \\ \frac{1}{275} & 0 \\ 0 & 0 \\ 0 & 0 \end{bmatrix} \begin{bmatrix} \theta(t) \\ \dot{\theta}(t) \end{bmatrix} + \begin{bmatrix} 0 & 0 & \sqrt{10} & 0 \\ 0 & 0 & 0 & 0 \\ \frac{4}{275} & 0 & 0 & 0 \\ W_y & 0 & 0 & 0 \end{bmatrix} \begin{bmatrix} u(t) \\ w(t) \\ v(t) \\ w_d(t) \end{bmatrix}.$$

The closed-loop infinity norm is normalized to 1 before adding the perturbation. The weights appearing in this plant are positioned as shown in Figure 10.9, and are given in Table 10.1. The weight W_y was initially selected to equal 4/275, since the output y_d is proportional to the control portion of the original reference output. Therefore, the contribution of y_d to the system norm is comparable to the contribution due to the control. The weight W_w was then selected to yield a product $W_y W_w = 0.5$. This value was selected because a downside gain margin of 0.67 and a phase margin of 30° are guaranteed when the system possesses stability robustness to input-multiplicative perturbations bounded by 0.5. The stability margins and the closed-loop ∞-norm of the resulting system are given in Table 10.1. The Nyquist plot of this system is shown in Figure 10.10b. A performance bound of approximately 10% over the optimal was used in this design. The downside gain margin and the phase margin are improved over the original design. The closed-loop ∞-norm is increased, demonstrating that there is a trade-off between performance and robustness. The gain margin is decreased but still very acceptable. Note that robustness to the perturbation guarantees a gain margin of at least 2. It should also be noted that the final system is not guaranteed to be robust to the given input-multiplicative uncertainty unless the ∞-norm of the augmented system is less than 1 (not the case in this example).

A third design is generated by doubling the weight W_w. The Nyquist plot for the resulting system is shown in Figure 10.10c. The stability margins and the closed-loop ∞-norm for this design are given in Table 10.1. Increasing this weight increases the downside gain margin, phase margin, and the closed-loop ∞-norm when compared to the previous design. The gain margin is slightly decreased, but still very large.

A fourth design is generated by returning W_w to the original value and doubling the weight W_y. The Nyquist plot for the resulting system is shown in Figure 10.10d. The stability margins and the closed-loop ∞-norm for this design are given in Table 10.1. Doubling this weight yields a controller with performance very similar to the third design. The fact that this control system has better stability margins than the second design underscores the fact that the location of the weight is important (i.e., not simply the product $W_y W_w$). This fact is due to the coupling between the other inputs and the perturbation output, and the coupling between the perturbation input and other outputs. This coupling does not exist when using the maximum structured singular value as a performance measure. The optimization of the maximum structured singular value is discussed in detail in the next section.

Increasing the weight on the perturbation input W_w is roughly analogous to increasing fictitious noise in LQG loop transfer recovery. Using this analogy, the perturbation input corresponds to the fictitious noise input used in LTR, and increasing the weight is equivalent to increasing the fictitious noise spectral density. The perturbation output scales the control weighting in the cost, resulting in a minor divergence of the two procedures. ◆

The final comments in the example state that adding an input-multiplicative perturbation to the cost function used in $\mathcal{H}_\infty$ controller design is roughly analogous to LQG loop transfer recovery. Note that other types of perturbations can be included within the $\mathcal{H}_\infty$ controller design framework. This provides a very general method of increasing robustness to the given perturbations; that is, it results in a control system that is tailored to tolerate specific perturbations. This generality is useful in many applications, but caution is necessary when using a very restricted set of perturbations. This caution results from the fact that the use of general perturbations in $\mathcal{H}_\infty$ controller design does not yield the guaranteed asymptotic robustness properties of LQG/LTR. The resulting system may still be sensitive to unmodeled perturbations.

The minimization of the maximum structured singular value (not the ∞-norm, which is the maximum singular value) is the mathematically precise method of generating controllers that meet robust performance specifications. Adding perturbation inputs and outputs to the plant as shown in Figure 10.9 yields the transfer function required for structured singular value analysis. A procedure to minimize the SSV of this closed-loop transfer function is called μ-synthesis. $\mathcal{H}_\infty$ controller design forms the first step in the μ-synthesis design procedure.

10.5 μ-Synthesis

Robust performance can be analyzed using the structured singular value for systems containing both structured and unstructured perturbations. A system in standard form, with normalized performance criteria and perturbations (see Figure 10.11a), performs robustly if and only if

$$\sup_{\omega}\{\mu_{\bar{\Delta}}[\mathbf{N}(j\omega)]\} < 1, \tag{10.22}$$

where $\mathbf{N}(s)$ is the nominal closed-loop system formed by combining $\mathbf{P}(s)$ and $\mathbf{K}(s)$ and the block structure is implied by Figure 10.11b. Minimization of the cost function

$$J = \sup_{\omega}\{\mu_{\bar{\Delta}}[\mathbf{N}(j\omega)]\} \tag{10.23}$$

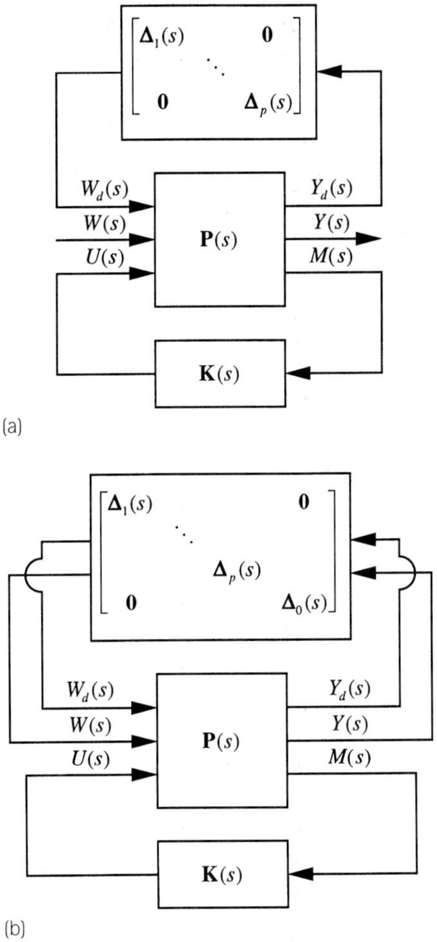

FIGURE 10.11 Robust performance analysis: (a) standard form for robust performance analysis; (b) robust performance as a robust stability problem

provides a means of obtaining robust performance, or demonstrating that robust performance is not possible, for the given specifications. The μ-synthesis controller design procedure addresses the minimization of the cost function (10.23).

10.5.1 $\mathfrak{D}$-Scaling and the Structured Singular Value

The direct computation of the structured singular value is intractable in all but the simplest cases. Therefore, bounds on the SSV are typically used in place of the actual SSV during robust performance analysis. In particular,

$$\mu_{\bar{\Delta}}(\mathbf{N}) \leq \min_{\substack{\{d_1,d_2,\cdots,d_n\} \\ d_i \in (0,\infty)}} \bar{\sigma}(\mathfrak{D}_R \mathbf{N} \mathfrak{D}_L^{-1}) \tag{10.24a}$$

provides a tight upper bound on the SSV and can be reliably computed.[9] The matrices

$$\mathfrak{D}_L = \begin{bmatrix} d_1\mathbf{I}_{l_1} & \mathbf{0} & \cdots & \mathbf{0} \\ \mathbf{0} & d_2\mathbf{I}_{l_2} & & \mathbf{0} \\ \vdots & & \ddots & \vdots \\ \mathbf{0} & \mathbf{0} & \cdots & d_n\mathbf{I}_{l_n} \end{bmatrix}; \mathfrak{D}_R = \begin{bmatrix} d_1\mathbf{I}_{n_1} & \mathbf{0} & \cdots & \mathbf{0} \\ \mathbf{0} & d_2\mathbf{I}_{n_2} & & \mathbf{0} \\ \vdots & & \ddots & \vdots \\ \mathbf{0} & \mathbf{0} & \cdots & d_n\mathbf{I}_{n_n} \end{bmatrix} \tag{10.24b}$$

are referred to as $\mathfrak{D}$-scaling matrices. Note that the identity matrices in (10.24b) are appropriately dimensioned to match the block structure of the perturbations. The scalar terms $\{d_i\}$ are referred to as $\mathfrak{D}$-scales.

Robust performance is evaluated by calculating the SSV bound for the nominal closed-loop transfer function $\mathbf{N}(j\omega)$. Frequency sampling is necessitated when generating this bound because the minimization of (10.24) is performed numerically. The resulting data are samples of the frequency-dependent SSV and the frequency-dependent $\mathfrak{D}$-scales.

The $\mathfrak{D}$-$\mathcal{K}$ iteration algorithm, discussed in detail in the next subsection, attempts to minimize the upper bound (10.24). This algorithm utilizes $\mathcal{H}_\infty$ optimization operating on the augmented plant $\mathfrak{D}_R(s)\mathbf{N}(s)\mathfrak{D}_L^{-1}(s)$. A state model for the augmented plant is required to apply the $\mathcal{H}_\infty$ optimization results given earlier in this chapter. Such a model can be obtained by appending $\mathfrak{D}_R(s)$ and $\mathfrak{D}_L^{-1}(s)$ to the nominal closed-loop transfer function $\mathbf{N}(s)$. This approach requires that the numerical $\mathfrak{D}$-scales, obtained when computing the SSV bound, be approximated by finite-order Laplace transfer functions.

Fitting Transfer Functions to the $\mathfrak{D}$-Scales Weighted least squares can be used to generate stable, minimum-phase transfer functions that approximate sampled, complex, frequency response data. Frequency response data can be approximated by either stable or unstable transfer functions. Stable transfer functions are preferred because instability of the $\mathfrak{D}$-scales needlessly complicates the controller design. In addition, the frequency response of an unstable system should always be viewed with caution because persistent transients typically overwhelm the sinusoidal response.

The $\mathfrak{D}$-scales and the $\mathfrak{D}$-scale inverses are appended to the nominal closed-loop system. The $\mathfrak{D}$-scale inverses should also be stable for the given reasons. Stable $\mathfrak{D}$-scale inverses are obtained provided the transfer functions are minimum-phase and the numerator and denominator orders are equal. Note that the zeros of a minimum-phase transfer function are all in the left half-plane.

The numerical $\mathfrak{D}$-scale data contains only magnitude information, but the bound (10.24) is unaffected by the inclusion of phase shifts into the $\mathfrak{D}$-scales. Therefore, phase information can be added to the magnitude $\mathfrak{D}$-scale data to yield the complex $\mathfrak{D}$-scale data used for least squares parameter identification.

The phase of a stable, minimum-phase transfer function (normalized such that the dc gain is positive) is uniquely determined by the magnitude of the transfer function. This fact can be used to generate phase information from the magnitude-only $\mathfrak{D}$-scale

[9] The computation of the $\mathfrak{D}$-scales, that is, performing the optimization of (10.22a), is discussed in Section 5.5.3.

data. The phase of the complex $\mathfrak{D}$-scale $\tilde{d}_k(j\omega)$, at a sample frequency ω_l, is (see Section A13 of the Appendix)

$$\angle\tilde{d}_k(j\omega_l) = \frac{2\omega_l}{\pi}\int_0^\infty \frac{\ln\{d_k(\omega)\} - \ln\{d_k(\omega_l)\}}{\omega^2 - \omega_l^2}d\omega,$$

where $d_k(\omega)$ is the magnitude of the $\mathfrak{D}$-scale. This integral can be evaluated numerically using the samples $d_k(\omega_l)$. The complex $\mathfrak{D}$-scale is then given as

$$\tilde{d}_k(j\omega) = d_k(\omega)e^{j\angle\tilde{d}_k(j\omega)}.$$

Least squares can now be used to generate a transfer function model of order n_o,

$$G(s) = \frac{b_{n_o}s^{n_o} + \cdots + b_1 s + b_0}{s^{n_o} + a_{(n_o-1)}s^{(n_o-1)} + \cdots + a_1 s + a_0},$$

that approximates each complex $\mathfrak{D}$-scale. The order is assumed to be fixed during the subsequent least squares development. A discussion of order selection is provided at the end of this subsection.

The frequency response of the transfer function model should approximate the complex $\mathfrak{D}$-scale. At each sample frequency, this implies that

$$\tilde{d}_k(\omega_l) \approx \frac{b_{n_o}\{j\omega_l\}^{n_o} + \cdots + b_1\{j\omega_l\} + b_0}{\{j\omega_l\}^{n_o} + a_{(n_o-1)}\{j\omega_l\}^{(n_o-1)} + \cdots + a_1\{j\omega_l\} + a_0}.$$

The error is then

$$\begin{aligned} e_k(\omega_l) = {} & \tilde{d}_k(\omega_l)\{j\omega_l\}^{n_0} + \tilde{d}_k(\omega_l)a_{(n_0-1)}\{j\omega_l\}^{(n_o-1)} \\ & + \cdots + \tilde{d}_k(\omega_l)a_1\{j\omega_l\} + \tilde{d}_k(\omega_l)a_0 \\ & - b_{n_0}\{j\omega_l\}^{n_0} - \cdots - b_1\{j\omega_l\} - b_0. \end{aligned}$$

The errors at all sample frequencies can be combined into a vector:

$$E_k = [e_k(\omega_1) \quad e_k(\omega_2) \quad \cdots \quad e_k(\omega_{n_\omega})]^T,$$

where n_ω is the number of frequencies. This error vector can be computed for a given transfer function model,

$$E_k = Y_k - \mathbf{M}_k\theta_k,$$

where

$$Y_k = [\tilde{d}_k(\omega_1)\{j\omega_1\}^{n_o} \quad \tilde{d}_k(\omega_2)\{j\omega_2\}^{n_o} \cdots \tilde{d}_k(\omega_{n_\omega})\{j\omega_{n_\omega}\}^{n_o}]^T;$$

$$\mathbf{M}_k = \begin{bmatrix} \tilde{d}_k(\omega_1)\{j\omega_1\}^{(n_o-1)} & \cdots & \tilde{d}_k(\omega_1) & \{j\omega_1)\}^{n_o} & \cdots & 1 \\ \vdots & & \vdots & \vdots & & \vdots \\ \tilde{d}_k(\omega_{n_\omega})\{j\omega_{n_\omega}\}^{(n_o-1)} & \cdots & \tilde{d}_k(\omega_{n_\omega}) & \{j\omega_{n_\omega}\}^{n_o} & \cdots & 1 \end{bmatrix};$$

$$\theta_k = [a_{(n_o-1)} \cdots a_0 \quad b_{n_o} \cdots b_0]^T.$$

The transfer function parameters, the elements of θ_k, are selected to minimize the weighted square error:

$$J_k = \sum_{l=1}^{n_\omega} W_l|e_k(\omega_l)|^2 = E_k^\dagger \mathbf{W} E_k = (Y_k - \mathbf{M}_k\theta_k)^\dagger \mathbf{W}(Y_k - \mathbf{M}_k\theta_k),$$

where $\mathbf{W}$ is a diagonal matrix with the elements $W_l \in [0, \infty)$. The weights W_l can be used to emphasize or deemphasize the fit at a given frequency or frequencies.

The transfer function parameters that minimize the weighted square error can be found by expanding the cost function:

$$\begin{aligned} J_k &= (Y_k - \mathbf{M}_k\theta_k)^\dagger \mathbf{W}(Y_k - \mathbf{M}_k\theta_k) \\ &= Y_k^\dagger \mathbf{W} Y_k - \theta_k^\dagger \mathbf{M}_k^\dagger \mathbf{W} Y_k - Y_k^\dagger \mathbf{W}\mathbf{M}_k\theta_k + \theta_k^\dagger \mathbf{M}_k^\dagger \mathbf{W}\mathbf{M}_k\theta_k. \end{aligned}$$

Taking the derivative of this scalar with respect to the parameter vector θ_k [see Appendix equations (A1.8) and (A1.10)] yields

$$\frac{\partial J_k(\theta_k)}{\partial \theta_k} = -2Y_k^\dagger \mathbf{W}\mathbf{M}_k + 2\theta_k^\dagger \mathbf{M}_k^\dagger \mathbf{W}\mathbf{M}_k.$$

Setting this derivative equal to zero and solving for θ_k yields the parameter vector that minimizes the weighted square error:

$$\hat{\theta} = (\mathbf{M}_k^\dagger \mathbf{W}\mathbf{M}_k)^{-1}\mathbf{M}_k^\dagger \mathbf{W} Y_k,$$

provided the inverse exists. This inverse typically exists when the number of frequencies is large compared to the number of parameters that are being identified (i.e., the model order is small compared to the number of sample frequencies used in computing the SSV).

Frequency weighting is included in the squared error criterion. The frequency weights can be selected to provide an accurate match at frequencies where the bound (10.24) is particularly sensitive to the $\mathfrak{D}$-scale. The sensitivity of the bound (10.24) to the $\mathfrak{D}$-scale can be computed by perturbing the $\mathfrak{D}$-scale and finding the change in the bound. The frequency weighting should then be set equal to this change. Applying this frequency weighting results in an approximation that is most accurate at frequencies where the sensitivity of the bound to the $\mathfrak{D}$-scale is high.

The order of the transfer function approximation is selected so that the weighted frequency error is small. This can be accomplished by plotting the magnitude frequency response of the approximations for several orders along with the numerical $\mathfrak{D}$-scale data. The user is then free to make a decision. Alternatively, an automated algorithm can be used that increases the order of the transfer function until the weighted frequency error is less than a given bound at all frequencies. Regardless of the method used to select order, the order should be maintained as small as is reasonable because high-order $\mathfrak{D}$-scales increase computation time and result in higher-order controllers when applying the $\mathfrak{D}$-$\mathcal{K}$ iteration algorithm.

10.5.2 $\mathfrak{D}$-$\mathcal{K}$ Iteration

The μ-synthesis design methodology attempts to minimize the supremum of the closed-loop system's structured singular value as given by the cost function (10.23). Direct minimization of this cost function is typically not tractable. As an alternative, it is reasonable to minimize the upper bound on the SSV:

$$J = \sup_{\omega} \min_{\substack{\{d_1, d_2, \cdots, d_p\} \\ d_i \in (0,\infty)}} \bar{\sigma}[\mathfrak{D}_R(j\omega)\mathbf{N}(j\omega)\mathfrak{D}_L^{-1}(j\omega)], \tag{10.25}$$

where the $\mathfrak{D}$-scales are frequency-dependent. Note that robust performance is achieved when the supremum of the SSV is appropriately bounded. Therefore, achieving this bound guarantees robust performance, but the design may be conservative if the $\mathfrak{D}$-scaled bound (10.26) is not tight.

The optimization of the cost function (10.25) is still intractable in most cases, but an *ad hoc* algorithm known as $\mathfrak{D}$-$\mathcal{K}$ iteration has been found to work well in many applications. This algorithm is based on the observation that for a given set of $\mathfrak{D}$-scales, the cost function (10.25) is simply an ∞-norm optimization problem:

$$J_{\mathfrak{D}} = \sup_{\omega} \bar{\sigma}[\mathfrak{D}_R(j\omega)\mathbf{N}(j\omega)\mathfrak{D}_L^{-1}(j\omega)] = \|\mathfrak{D}_R\mathbf{N}\mathfrak{D}_L^{-1}\|_\infty. \tag{10.26}$$

The solution to this problem given earlier in this chapter is valid provided that $\mathfrak{D}_R\mathbf{N}\mathfrak{D}_L^{-1}$ is described by a state model that satisfies the conditions (10.2). Unfortunately, the $\mathfrak{D}$-scales in (10.25) are not fixed, since they depend on the closed-loop model which, in turn, depends on the controller. $\mathfrak{D}$-$\mathcal{K}$ iteration seeks to overcome this problem by alternatively performing ∞-norm optimization and $\mathfrak{D}$-scale optimization.

The $\mathfrak{D}$-$\mathcal{K}$ iteration algorithm is summarized as follows.

1. Model the plant. The plant model should include disturbance inputs, control inputs, reference outputs, measured outputs, and perturbations. Append the performance block to the uncertainty matrix.
2. Generate a control system to minimize the ∞-norm of the transfer function from the augmented perturbation input to the augmented perturbation output.
3. Compute the structured singular values for the closed-loop system (with both uncertainty and performance blocks). Save the $\mathfrak{D}$-scales used in computing the structured singular value.
4. Fit a low-order transfer function to each frequency-dependent $\mathfrak{D}$-scale, as discussed in the previous subsection.
5. Append these transfer functions to the plant. The rational transfer function approximations for the $\mathfrak{D}$-scales and the inverse $\mathfrak{D}$-scales are appended to the nominal closed-loop system. This is typically accomplished by generating state models for the $\mathfrak{D}$-scales and the inverse $\mathfrak{D}$-scales, and appending these state models to the nominal closed-loop system.
6. For this augmented plant, generate a controller to minimize the ∞-norm of the transfer function from the augmented perturbation input to the augmented perturbation output.
7. Return to step 4, until the structured singular value of the closed-loop system fails to improve.

This algorithm has typically been found to converge to a minimum cost in a few iterations. Caution must be exercised in interpreting the meaning of this minimum, since the $\mathfrak{D}$-$\mathcal{K}$ iteration algorithm is not guaranteed to converge to the global minimum of the cost function (10.25). Further, this global minimum is not guaranteed to equal the global minimum of the cost function (10.23), except when the number of performance and perturbation blocks is less than or equal to 3.

EXAMPLE 10.5 For the single-link robot arm described in Example 8.3, the linearized plant model is

$$\begin{bmatrix} \dot{\theta}(t) \\ \ddot{\theta}(t) \end{bmatrix} = \begin{bmatrix} 0 & 1 \\ 0.5 & 0 \end{bmatrix} \begin{bmatrix} \theta(t) \\ \dot{\theta}(t) \end{bmatrix} + \begin{bmatrix} 0 \\ 0.1 \end{bmatrix} u(t) + \begin{bmatrix} 0 & 0 \\ 0.1 & 0 \end{bmatrix} \begin{bmatrix} w(t) \\ v(t) \end{bmatrix}.$$

The disturbance input consists of a torque caused by noise and biases in the torquer circuit. This torque is bounded: $|w(t)| \leq 0.5$. The angular position of the arm is the output of interest, and the measured output is a corrupted version of this angular position:

$$m(t) = [1 \quad 0] \begin{bmatrix} \theta(t) \\ \dot{\theta}(t) \end{bmatrix} + [0 \quad 1] \begin{bmatrix} w(t) \\ v(t) \end{bmatrix}.$$

The error in this measurement is bounded: $|v(t)| \leq 0.05$. The desired bounds on the output and the control input are 0.2 and 100, respectively. In order to obtain good stability margins for this system, an input-multiplicative perturbation is added to the plant. This perturbation is bounded: $\|\mathbf{\Delta}_i\|_\infty \leq 0.5$. A block diagram of this plant in standard form is shown in Figure 10.12, where $w_1(t)$ and $y_1(t)$ are the normalized disturbance input and reference output, respectively. The $\mathfrak{D}$-$\mathcal{K}$ iteration algorithm is used to obtain a near optimum controller for this system.

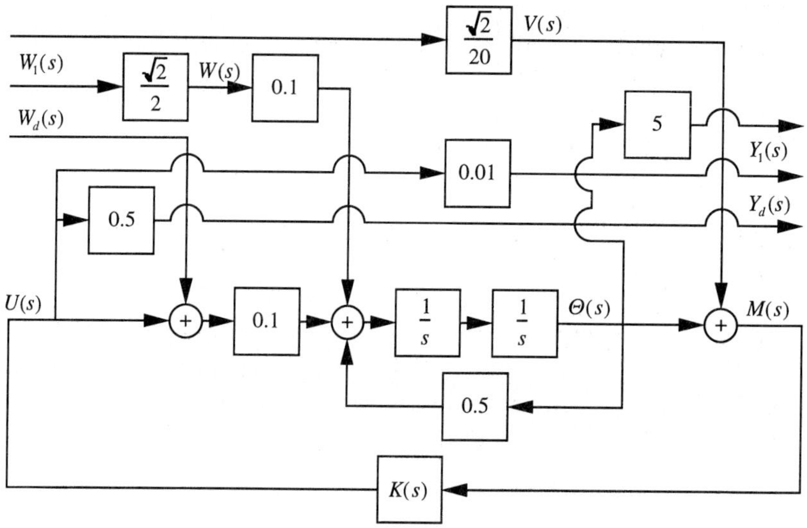

FIGURE 10.12 The plant in standard form for Example 10.5

Table 10.2 summarizes the relevant information on the $\mathfrak{D}$-$\mathcal{K}$ iteration process. From this table, the algorithm is seen to nearly converge in four iterations, and to yield a cost less than half of that obtained using the ∞-norm optimal controller (the results of iteration 1). The disadvantage of this algorithm is that the resulting controller has order 10 compared to the second-order controller resulting from ∞-norm optimization.

TABLE 10.2 Summary of $\mathfrak{D}$-$\mathcal{K}$ Iteration Results from Example 10.5

Iteration #	1	2	3	4
Controller order	2	10	10	10
Maximum μ	1.66	0.88	0.79	0.76

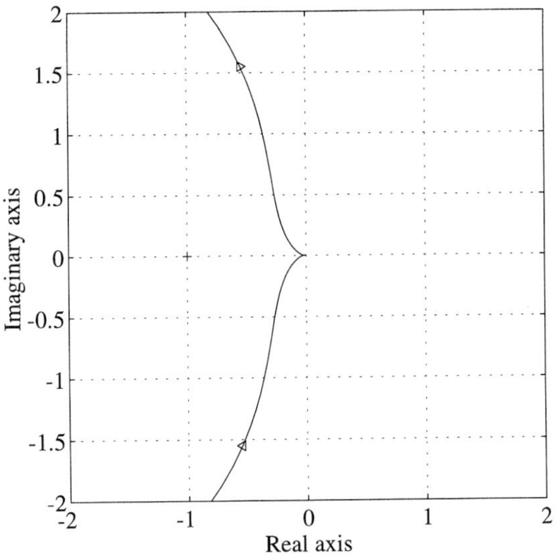

FIGURE 10.13 Nyquist plot for the μ-synthesis controller in Example 10.5

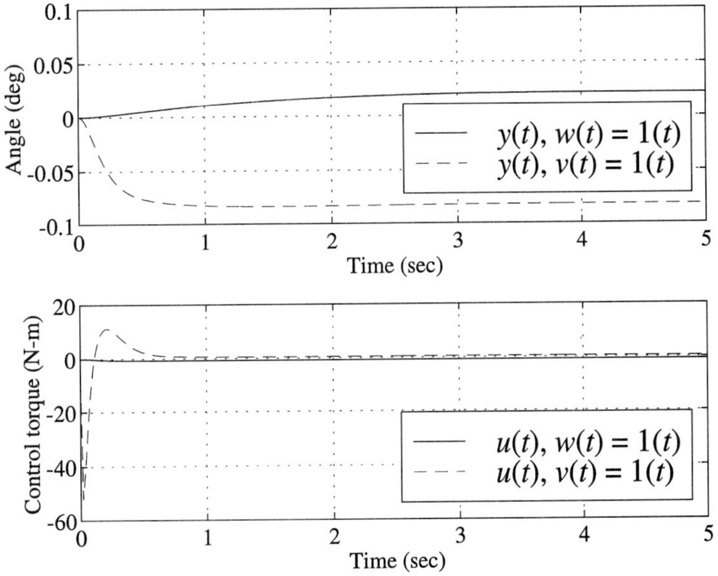

FIGURE 10.14 Step response for Example 10.5 with the μ-synthesis controller

The controller achieves the desired bound of one for the maximum SSV. The resulting controller yields excellent stability margins, as shown by the Nyquist plot in Figure 10.13. Figure 10.14 shows the step response of the nominal closed-loop system. Note that the nominal system achieves the given performance bounds for the unit step input. ◆

10.6 Comparison of Design Methodologies

Linear quadratic Gaussian control and μ-synthesis have both been presented as means of designing output feedback controllers. Both methodologies have been successfully applied in many control applications, even though these approaches represent very distinct design philosophies. The following example is presented to illustrate the similarities and differences of these two approaches, and to help illustrate the strengths and weaknesses of the various methods. In this example, both LQG control and μ-synthesis are applied to the design of a missile autopilot.

EXAMPLE 10.6 A tail-fin-controlled missile is shown in Figure 10.15. The control system for this missile is divided into two sections: the guidance law and the autopilot. The guidance law utilizes external measurements to generate desired missile accelerations. The autopilot produces these desired accelerations. This example is directed at the design of an autopilot for controlling the vertical acceleration of this missile.

The missile's pitch-plane dynamics (pertaining to the vertical motion) are nonlinear, in general. A block diagram of a linearized version of these dynamics is shown in Figure 10.16.[10] In this diagram, $\delta_c(t)$ is the fin-deflection command, $\delta(t)$ is the actual fin

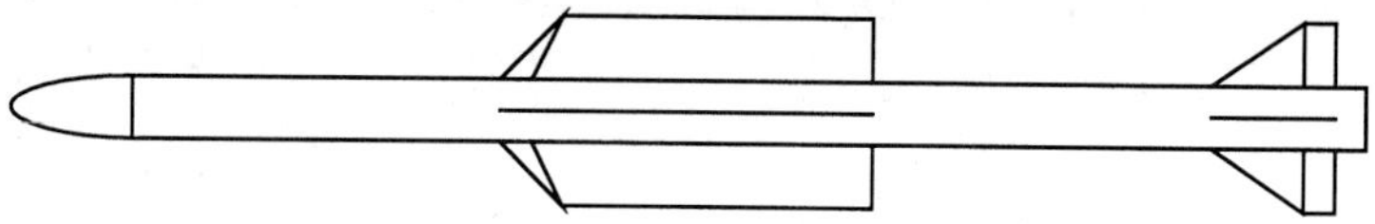

FIGURE 10.15 A generic tail-fin-controlled missile

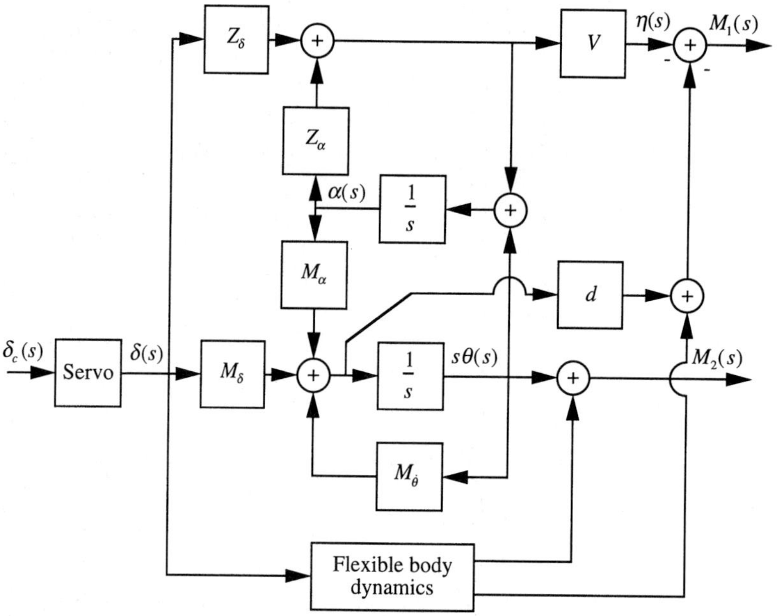

FIGURE 10.16 Block diagram of a tail-fin-controlled missile

[10] Autopilot design for this missile has previously been addressed in [2].

TABLE 10.3 Missile Parameters and Uncertainties

Parameter	Nominal	Uncertainty
Z_α	−5.24	±5.00
M_α	−47.0	±7.05
$M_{\dot\theta}$	−4.69	±1.41
Z_δ	−0.73	±0.20
M_δ	−11.3	±1.70
V	1.82	±0
d	8.55×10^{-4}	±0

deflection, $\theta(t)$ is the angular position, $\alpha(t)$ is the angle of attack, $\gamma(t)$ is the flight-path angle, and $\eta(t)$ is the missile's vertical acceleration. Numerical values for the parameters in the missile model are given in Table 10.3.

The aerodynamic parameters M_α, $M_{\dot\theta}$, and M_δ give the angular acceleration of the missile as a linear function of the angle of attack, the body rotation rate, and the fin deflection, respectively. These coefficients are generated by linearizing the nonlinear functions that relate the angular acceleration to the angle of attack and so on. The uncertainties, given in Table 10.3, provide upper and lower bounds for the nonlinear gain over the operating region. The generation of these bounds is illustrated in Figure 10.17. Similarly, the aerodynamic parameters Z_α and Z_δ are generated by linearizing the functions relating the flight-path angle rate to the angle of attack and the fin deflection, respectively. Uncertainty bounds for these parameters are also included in Table 10.3. The parameter V is the missile's speed, which is assumed to be constant and accurately known (i.e., the uncertainty is zero). The parameter d is the separation of the accelerometer and the missile's center of gravity, which is accurately known. A servo is used to

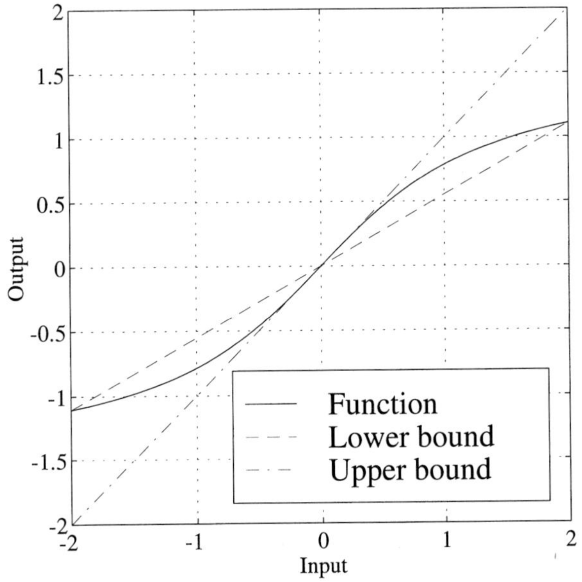

FIGURE 10.17 Gain bounds for a nonlinear function

position the tail fins at the commanded position. This servo is modeled as a second-order transfer function with a damping ratio $\zeta = 0.7$ and a natural frequency of 189:

$$G_{Servo}(s) = \frac{35{,}721}{s^2 + 265s + 35{,}721}.$$

The control input to this plant is the commanded tail-fin deflection. The measurements from the plant consist of the missile's vertical acceleration and pitch rate. The missile body is slightly flexible and vibrates when excited. Bending of the flexible body generates spurious accelerations and body rates at the sensors. These spurious accelerations and body rates appear in the block diagram (Figure 10.16) as additive measurement errors.

The autopilot is tasked with maintaining the desired vertical acceleration up to the maximum acceleration of 30 g. The response to a step vertical acceleration command should be fast, accurate, and use a reasonable amount of control. The steady-state error (due to a step in commanded acceleration) should be less than 2%, and the settling time should be as small as possible. The maximum tail-fin deflection is 40° (0.7 rad), and the rate of change of the tail-fin deflection is limited to 600°/sec (10.5 rad/sec).

The disturbance inputs consist of the commanded acceleration and an error (bias and/or noise) on the pitch rate measurement. This error is assumed to be less than 0.001 rad/sec. Note that the accelerometer is so accurate that the measurement noise on this sensor is ignored.

Robust performance of the autopilot can be analyzed using the structured singular value. The uncertainty and performance weights used to normalize the plant are shown in Figure 10.18. This block diagram has not been placed in standard form to improve clarity.

Five perturbations appear in this block diagram. The maximum perturbations on the three parameters Z_α, M_α, and $M_{\dot\theta}$ are given in Table 10.3. These maximum perturbation values are used as the weights on the appropriate perturbation feedback loop. An input-multiplicative perturbation $\mathbf{\Delta}_i$ is included to guarantee good stability margins for the system. The weight on this perturbation is

$$W_i = 0.5.$$

Note that the uncertainty on the parameters Z_δ and M_δ is ignored because the input-multiplicative perturbation allows for larger variations in the loop gain than is generated by the uncertainties in these parameters. The fifth perturbation $\mathbf{\Delta}_f$ is at the input to the flexible body dynamics. These dynamics are not well known, and are therefore modeled as a high-frequency additive uncertainty. The weight on the flexible body dynamics is

$$\mathbf{W}_{fb}(s) = \frac{1.75 \times 10^{-4}(s/0.0025 + 1)(s/25 + 1)}{(s/250 + 1)^2}\begin{bmatrix} 1 \\ 1 \end{bmatrix}.$$

The performance is specified by bounds on the disturbance inputs and the reference outputs. The commanded acceleration is limited to 30, which generates the weight

$$W_\eta = 30.$$

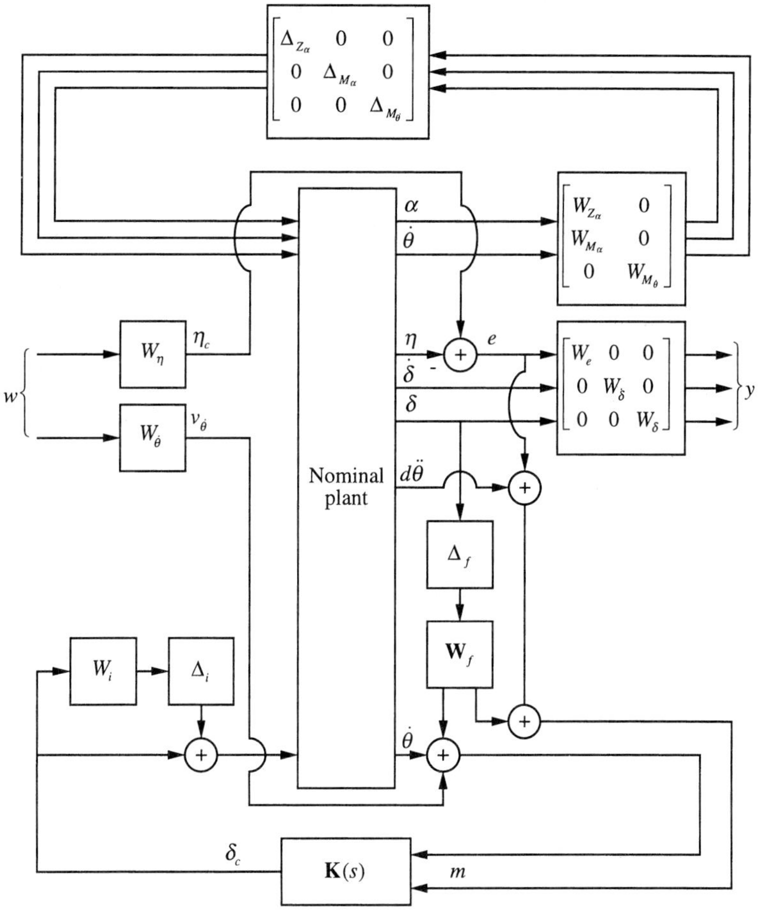

FIGURE 10.18 Performance and uncertainty weights for missile autopilot design

The error in the pitch rate measurement $v_{\dot\theta}$ is bounded by 0.001, yielding the weight

$$W_{\dot\theta} = 0.001.$$

Note that zero error on the acceleration measurement is modeled. This is typically not a good idea, but is acceptable in this case since the commanded acceleration is mathematically equivalent to a measurement error (see Figure 10.18). The tracking error is required to be less than 2% at low frequencies. This specification is incorporated into the error weight:

$$W_e(s) = \frac{50(0.025s + 1)}{30(10s + 1)}.$$

The additional weighting functions,

$$W_\delta = \frac{1}{0.7} \quad \text{and} \quad W_{\dot\delta} = \frac{1}{10.5},$$

are the inverses of the bounds on the tail-fin deflection and the tail-fin deflection rate, respectively.

The robust performance specifications fully demarcate a μ-synthesis design problem that can be solved using $\mathfrak{D}$-$\mathfrak{K}$ iteration. The results of applying $\mathfrak{D}$-$\mathfrak{K}$ iteration to this problem are summarized in Table 10.4.

TABLE 10.4 Summary of $\mathfrak{D}$-$\mathfrak{K}$ Iteration Results from Example 10.6

Iteration #	1	2	3	4
Controller order	7	15	11	17
Maximum μ	14.96	2.18	1.33	1.26

A controller that meets the robust performance specifications is not obtained for this example. Therefore, the specifications must be relaxed in order to obtain a controller that performs robustly.[11] For purposes of this example, the controller from the third iteration is used even though it does not perform robustly. This controller is selected because it nearly meets the robust performance specification, and the six additional controller states in iteration 4 do not significantly reduce the maximum structured singular value.

The response of the resulting system to a 30 g acceleration step command is shown in Figure 10.19. The acceleration settling time is approximately 0.4 seconds. Both the fin-deflection angle and the fin rate remain within the given limits. This system has good stability margins, as shown in the Nyquist plot in Figure 10.20. The structured singular values of $\mathbf{N}(j\omega)$ and $\mathbf{N}_{y_d w_d}(j\omega)$ are given in Figure 10.21, along with the maximum singular value of $\mathbf{N}_{yw}(j\omega)$. These plots indicate that the nominal system meets the performance objectives, and the system is robustly stable, but (as mentioned previously) the system does not robustly meet the performance objectives.

A second controller is designed using linear quadratic Gaussian techniques. The specifications require tracking a slowly varying reference input. Therefore, a controller that utilizes integral control is desirable. Integral action is incorporated in the LQR design, but not in the Kalman filter design, since there are no expected constant disturbance inputs. Note that a bias term appears in the actual controller to offset the effects of gravity. This term is easily computed and added after controller design and is therefore ignored in this example.

The feedback gains are found using LQR theory. The augmented state equation used in the LQR design is

$$\begin{bmatrix} \dot{\alpha}(t) \\ \ddot{\theta}(t) \\ \dot{\delta}(t) \\ \ddot{\delta}(t) \\ \dot{\eta}_I(t) \end{bmatrix} = \begin{bmatrix} Z_\alpha & 1 & Z_\delta & 0 & 0 \\ M_\alpha & M_{\dot{\theta}} & M_\delta & 0 & 0 \\ 0 & 0 & 0 & 1 & 0 \\ 0 & 0 & -\omega_n^2 & -2\zeta\omega_n & 0 \\ VZ_\alpha & 0 & VZ_\delta & 0 & 0 \end{bmatrix} \begin{bmatrix} \alpha(t) \\ \dot{\theta}(t) \\ \delta(t) \\ \dot{\delta}(t) \\ \eta_I(t) \end{bmatrix} + \begin{bmatrix} 0 \\ 0 \\ 0 \\ \omega_n^2 \\ 0 \end{bmatrix} \delta_c(t).$$

[11] Robust performance may be obtainable using a little trial-and-error modification of the controller, since $\mathfrak{D}$-$\mathfrak{K}$ iteration is not guaranteed to be optimal. Alternatively, uncertainty reduction before controller design [3] often yields a 15%–50% decrease in the SSV, and may yield a controller that performs robustly.

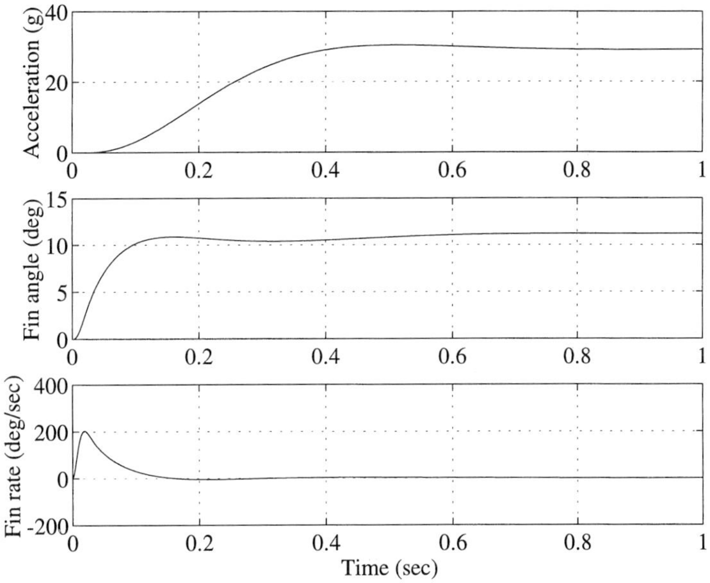

FIGURE 10.19 Maximum acceleration step response using the μ-synthesis controller

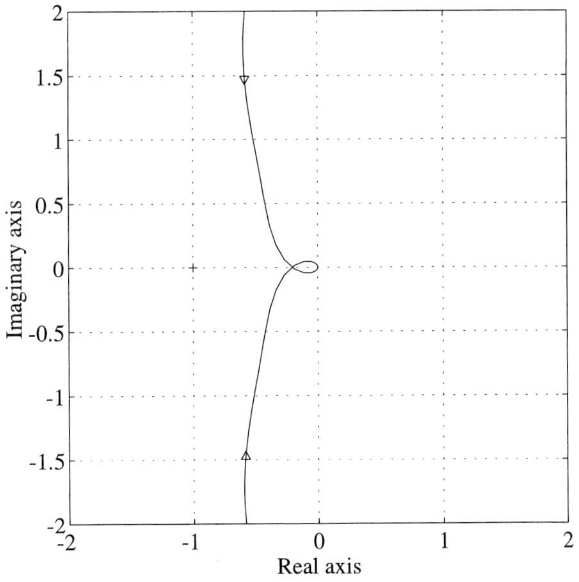

FIGURE 10.20 Nyquist plot with the μ-synthesis controller

where $\eta_I(t)$ is the integral of the missile's vertical acceleration. The cost function used in the LQR design is

$$J = \int_0^\infty 50\eta_I^2(t) + 1.4\delta^2(t) + 0.1\dot{\delta}^2(t) + \delta_c^2(t)dt.$$

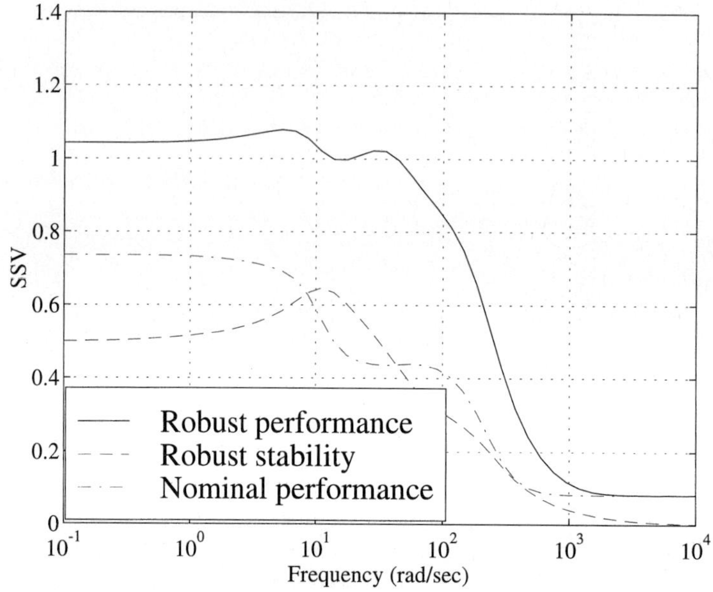

FIGURE 10.21 Structured singular value plot with the μ-synthesis controller

The weightings on the fin deflection and the fin rate are set to the inverse of the bounds on these parameters. The weights on the control input and the integral of the acceleration are then generated by trial and error, with the goal of achieving an appropriate closed-loop step response.

A Kalman filter is used to estimate the states. The problem statement given above does not include any specifications on the plant noise. Therefore, the form of the plant noise can be selected at the convenience of the designer. The state model used in the Kalman filter design is given:

$$\begin{bmatrix} \dot{\alpha}(t) \\ \ddot{\theta}(t) \\ \dot{\delta}(t) \\ \ddot{\delta}(t) \end{bmatrix} = \begin{bmatrix} Z_\alpha & 1 & Z_\delta & 0 \\ M_\alpha & M_{\dot{\theta}} & M_\delta & 0 \\ 0 & 0 & 0 & 1 \\ 0 & 0 & -\omega_n^2 & -2\zeta\omega_n \end{bmatrix} \begin{bmatrix} \alpha(t) \\ \dot{\theta}(t) \\ \delta(t) \\ \dot{\delta}(t) \end{bmatrix} + \begin{bmatrix} 0 \\ 0 \\ 0 \\ \omega_n^2 \end{bmatrix} \delta_c(t) + \begin{bmatrix} 1000 & 0 & 0 \\ 0 & 1000 & 0 \\ 0 & 0 & 0 \\ 0 & 0 & \omega_n^2 \end{bmatrix} \begin{bmatrix} w_1(t) \\ w_2(t) \\ w_3(t) \end{bmatrix};$$

$$\begin{bmatrix} m_1(t) \\ m_2(t) \end{bmatrix} = \begin{bmatrix} VZ_\alpha + dM_\alpha & dM_{\dot{\theta}} & VZ_\delta + dM_\delta & 0 \\ 0 & 1 & 0 & 0 \end{bmatrix} \begin{bmatrix} \alpha(t) \\ \dot{\theta}(t) \\ \delta(t) \\ \dot{\delta}(t) \end{bmatrix} + \begin{bmatrix} v_\eta(t) \\ v_{\dot{\theta}}(t) \end{bmatrix}.$$

The three plant noises in this model are selected to improve the system robustness. These noises are injected where the uncertainties introduce error in the state model.

Including these noises during controller design forces the control system to be insensitive to perturbation inputs caused by parameter uncertainties. This is a generalization of loop transfer recovery. The noise spectral densities are chosen as follows:

$$\mathbf{S}_w = \mathbf{I};\ \mathbf{S}_v = \begin{bmatrix} 10^{-6} & 0 \\ 0 & 10^{-3} \end{bmatrix}.$$

The gains in the input matrix for the plant noise are then generated by trial and error, with the goal of achieving a desirable closed-loop step response. Adjusting these gains is equivalent to varying the spectral densities of the various plant noises.

The response of the resulting system to a 30 g acceleration step command is shown in Figure 10.22. The acceleration settling time is approximately 0.35 seconds, slightly faster than the μ-synthesis design. Both the fin-deflection angle and the fin rate remain within the given limits. This system has good stability margins, as shown in the Nyquist plot in Figure 10.23, but the stability margins are not as good as those obtained with the μ-synthesis controller. The structured singular values of $\mathbf{N}(j\omega)$ and $\mathbf{N}_{y_d w_d}(j\omega)$ are given in Figure 10.24, along with the maximum singular value of $\mathbf{N}_{yw}(j\omega)$. These plots indicate that the nominal system does not meet the performance objectives, the system is not robustly stable, and it does not robustly meet the performance objectives. Additional robustness, at the cost of decreased performance, can be obtained by increasing the plant noise. Even though the LQG design does not meet the robust performance specifications, this design yields a nice step response and is also fairly robust, as indicated by the stability margins. ◆

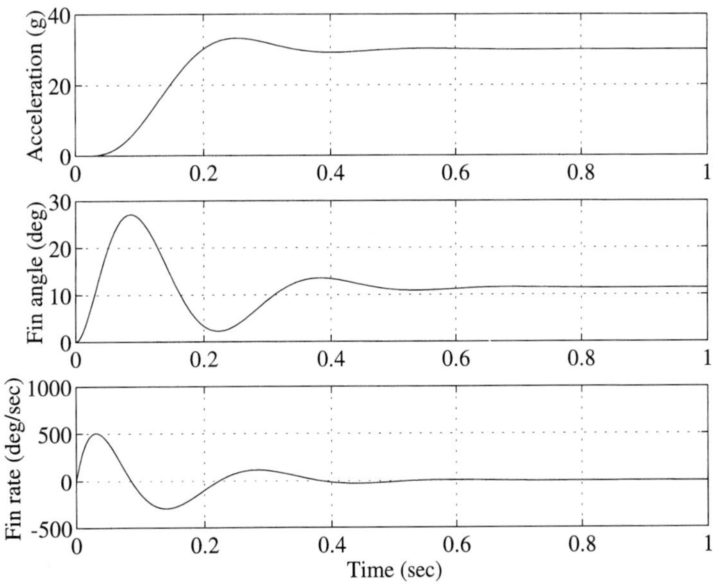

FIGURE 10.22 Maximum acceleration step response using the LQG controller

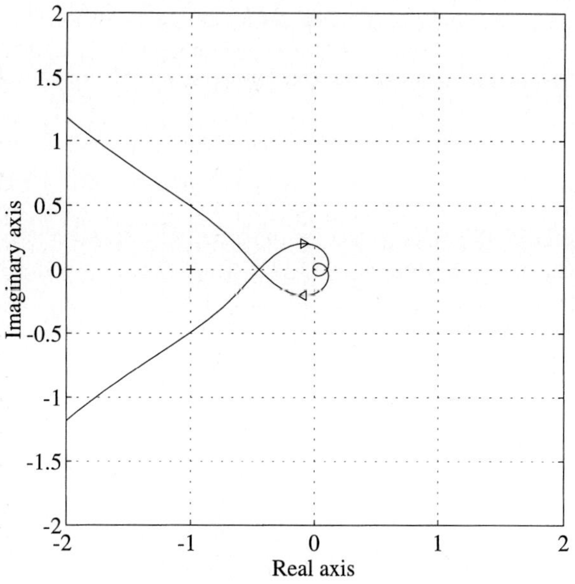

FIGURE 10.23 Nyquist plot with the LQG controller

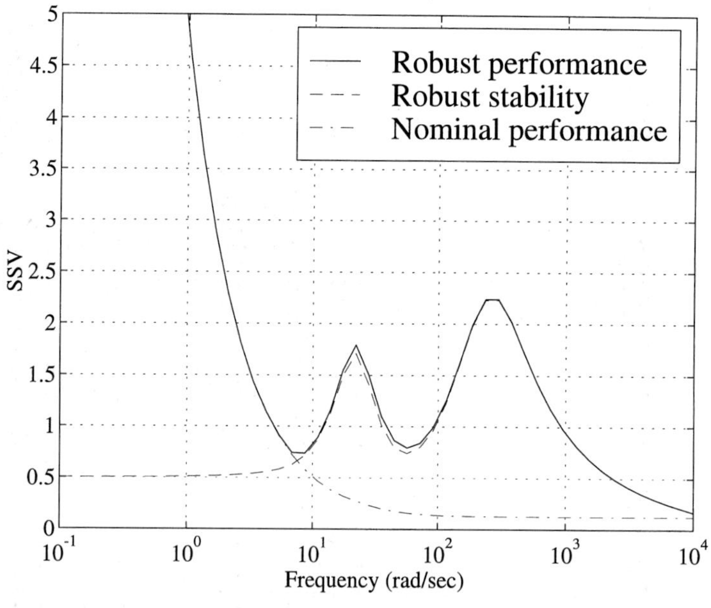

FIGURE 10.24 Structured singular value plot with the LQG controller

The subjectivity of the cost criteria is the largest impediment to the direct comparison of the two design methods. In this example, controllers designed with both methods performed very well, but neither controller was perfect. The LQG design requires iteration of some parameters in the cost function and the spectral density matrices in order to achieve the desired performance and robustness. The μ-synthesis design procedure can avoid this iteration by precisely specifying the desired level of robustness. But it is very easy to overspecify the μ-synthesis problem (as was done in this example) such that no controller achieves the robust performance objectives. Then, iteration of the specifications is again required. The μ-synthesis controller is of higher order, quite possibly requiring controller order reduction. In addition, the computer time required for the μ-synthesis design is considerably higher, making iteration more tedious. The bottom line is that both control design methods may yield high-performance, robust controls when used by skilled engineers with a solid understanding of the plant, the relevant performance limitations, and the performance objectives.

10.7 Summary

The $\mathcal{H}_\infty$ optimal control problem is to find a feedback controller that minimizes the closed-loop system ∞-norm. This feedback controller is also required to internally stabilize the system in the infinite-time case. The solution of this control problem can be divided into the synthesis of full information controller and the synthesis of an output estimator. This structure for the $\mathcal{H}_\infty$ controller design is similar to that for the LQG controller, but technically violates the separation principle. The estimator in this design depends on the full information control and also incorporates the worst-case disturbance (defined during full information controller design). Still, $\mathcal{H}_\infty$ controller design is simplified by dividing the overall design into full information controller design and estimator design.

$\mathcal{H}_\infty$ optimal and suboptimal control can be used as an alternative to LQG optimal control. Both cost functions are reasonable in a wide variety of applications. Quadratic cost functions are more often used in applications due to the ease of generating optimal solutions. But an $\mathcal{H}_\infty$ control can be used as an alternative when LQG controllers have undesirable properties (i.e., poor robustness, tracking, disturbance rejection, etc.). The use of an $\mathcal{H}_\infty$ controller is not guaranteed to improve the situation when the LQG controller fails, but it can be tried as an alternative.

$\mathcal{H}_\infty$ control is often desirable when the specifications are given in terms of bounds on the outputs. The ∞-norm can only be used to generate absolute output bounds when the inputs are sinusoidal, but it can be used as an approximate guide for a wider range of inputs. The designer must be aware of the performance limitations and trade-offs inherent in the control system design to effectively use $\mathcal{H}_\infty$ control theory to generate practical designs. As a general rule, the various inputs should be band-limited via shaping filters to avoid overspecification of the performance.

Tracking systems can be designed in a straightforward manner using $\mathcal{H}_\infty$ control. The reference input is included as a disturbance input, and the tracking error is included as a reference output. The gain from the reference input to the tracking error is then minimized or bounded (subject to the constraints and trade-offs imposed by other disturbance inputs and reference outputs) during $\mathcal{H}_\infty$ controller synthesis. Tracking systems using integral control are generated by using shaping filters on the reference inputs that have poles near the origin. The resulting controller then typically has a pole near the origin. Approximating this pole near the origin by a pole at the origin yields an integral controller. This is an *ad hoc* procedure for generating integral control but yields good results in many applications. Still, the designer is cautioned to evaluate the resulting control system thoroughly since this approximation invalidates all guarantees of stability and performance.

The structured singular value of the closed-loop plant can be used to specify robust performance. The maximum of this SSV can also be used as a cost function for an optimal control problem. Minimizing this cost function (a process known as μ-synthesis) then maximizes the robust performance of the system. The direct computation of the structured singular value, let alone minimization of this quantity, is typically intractable. An *ad hoc* approach to μ-synthesis known as $\mathfrak{D}$-$\mathcal{K}$ iteration is therefore presented.

The $\mathfrak{D}$-$\mathcal{K}$ iteration method is based on the facts that (1) the SSV of a matrix can be approximated by the maximum singular value of the $\mathfrak{D}$-scaled matrix; and (2) the supremum over frequency of the maximum singular value of a transfer function equals the system ∞-norm. Therefore, minimizing the system ∞-norm of the $\mathfrak{D}$-scaled plant should also minimize the supremum over frequency of the SSV. This is a reasonable approach to μ-synthesis because ∞-norm optimization has previously been developed. Unfortunately, the $\mathfrak{D}$-scales change when modifying the controller during $\mathcal{H}_\infty$ optimization. This necessitates an iterative approach to μ-synthesis, where $\mathfrak{D}$-scale computation and $\mathcal{H}_\infty$ optimization are alternately performed.

The $\mathfrak{D}$-$\mathcal{K}$ iteration method proceeds by iteratively performing $\mathcal{H}_\infty$ optimization, generating $\mathfrak{D}$-scales while computing the SSV, and appending these $\mathfrak{D}$-scales to the plant. There are no guarantees that $\mathfrak{D}$-$\mathcal{K}$ iteration will converge, and if it converges, there are no guarantees concerning the true optimality of the solution. Still, this procedure typically converges in a few iterations to a reasonable approximation of the optimal solution of the μ-synthesis problem.

The discrete-time $\mathcal{H}_\infty$ controller has the same structure as the continuous-time $\mathcal{H}_\infty$ controller. Therefore, discrete-time $\mathcal{H}_\infty$ output feedback controllers can be synthesized using the discrete-time full information and estimation results presented in Table 9.1. The resulting formulas for the discrete-time, time-varying output feedback controller are quite complex and are therefore not included in this summary. These formulas can be found in [4], page 475, [5], and [6].

A discrete-time controller can also be obtained by discretization of the continuous-time $\mathcal{H}_\infty$ suboptimal controller. $\mathcal{H}_\infty$ suboptimal controllers often incorporate fast subsystems, which makes discretization particularly problematic. Therefore, it is usually desirable to remove these fast subsystems from the $\mathcal{H}_\infty$ suboptimal controller prior to discretization. The removal of fast subsystems is discussed in the next chapter on controller order reduction.

REFERENCES

[1] K. Zhou, with J. C. Doyle, and K. Glover, *Robust and Optimal Control,* Prentice-Hall, Englewood Cliffs, NJ, 1996.

[2] J. Bible and H. Stalford, "Mu-synthesis autopilot design for a flexible missile," *Twenty-Ninth Aerospace Sciences Meeting,* Reno, NV, 1991.

[3] J. B. Burl and D. L. Krueger, "Parametric uncertainty reduction in robust missile autopilot design," *AIAA Journal of Guidance, Control, and Dynamics,* 19, no. 3 (1996): pp. 733–36.

[4] M. Greene and D. J. N. Limebeer, *Linear Robust Control*, Prentice-Hall, Englewood Cliffs, NJ, 1995.

[5] T. Basar, "A dynamic games approach to controller design: Disturbance rejection in discrete-time," *Proceedings of the IEEE Conference on Decision and Control,* pp. 407–14, 1989.

[6] D. J. N. Limebeer, M. Greene, and D. Walker, "Discrete-time $\mathcal{H}_\infty$ control," *Proceedings of the IEEE Conference on Decision and Control,* pp. 392–96, 1989.

[7] J. B. Moore, D. Gangsaas, and J. Blight, "Performance and robustness trades in LQG regulator designs," *Proceedings of the 20th IEEE Conference on Decision and Control,* San Diego, CA, Dec. 1981, pp. 1191–99.

Some additional references on $\mathcal{H}_\infty$ control follow:

[8] T. Basar and P. Bernhard, *$\mathcal{H}_\infty$-Optimal Control and Related Minimax Design Problems: A Dynamic Game Approach,* Systems and Control: Foundations and Applications, Birkhauser, Boston, 1991.

[9] J. C. Doyle, K. Glover, P. P. Khargonekar, and B. A. Francis, "State-space solution to standard $\mathcal{H}_2$ and $\mathcal{H}_\infty$ control problems," *IEEE Transactions on Automatic Control,* 34, no. 8 (1989): 831–47.

[10] B. A. Francis, *A Course in $\mathcal{H}_\infty$ Control Theory,* Lecture Notes in Control and Information Sciences, vol. 88, Springer-Verlag, Berlin, 1987.

[11] D. J. N. Limebeer, B. D. O. Anderson, P. P. Khargonekar, and M. Green, "A game theoretic approach to $\mathcal{H}_\infty$ control for time-varying systems," *SIAM J. Control and Optimization,* 30, no. 2 (1992): 262–83.

[12] J. M. Maciejowski, *Multivariable Feedback Design,* Addison-Wesley, Reading, MA, 1989.

An additional reference on μ-synthesis follows:

[13] G. J. Balas, J. C. Doyle, K. Glover, A. Packard, and R. Smith, *μ-Analysis and Synthesis Toolbox for use with Matlab,* The Math Works, Natick, MA, 1995.

EXERCISES

10.1 You are given the following plant:

$$\dot{x}(t) = -x(t) + u(t) + w(t) \; ; \; m(t) = x(t) + v(t) \; ; \; y(t) = \begin{bmatrix} x(t) \\ \hdashline 0.1u(t) \end{bmatrix}.$$

Generate a family of $\mathcal{H}_\infty$ suboptimal controllers for a variable performance bound γ. What is the limiting cost? Find the optimal $\mathcal{H}_\infty$ controller for this system by taking the limit of the suboptimal controllers as the performance bound approaches the optimal value.

10.2 A dc motor is modeled as follows:

$$\dot{x}(t) = \begin{bmatrix} 0 & 1 \\ 0 & -1 \end{bmatrix} x(t) + \begin{bmatrix} 0 \\ 1 \end{bmatrix} u(t) \; ; \; m(t) = [1 \quad 0]x(t) + v(t) \; ; \; y(t) = [1 \quad 0]x(t),$$

where $m(t)$ is the measured output, $y(t)$ is the reference output, and $v(t)$ is the measurement noise. The states are estimated using the following observer:

$$\dot{\hat{x}}(t) = (\mathbf{A} - \mathbf{G}\mathbf{C}_m)\hat{x}(t) + \mathbf{B}_u u(t) + \mathbf{G}m(t),$$

where $\mathbf{G} = [15 \quad 113]$ is the observer gain.

a. Compute the ∞-norm of the transfer function from the input $[r(t) \,\vdots\, v(t)]$ to the tracking error $e(t) = r(t) - y(t)$ for each of the following controllers:

$$u(t) = -[8 \quad 3]\hat{x}(t) + 8r(t);$$
$$u(t) = -[32 \quad 7]\hat{x}(t) + 32r(t).$$

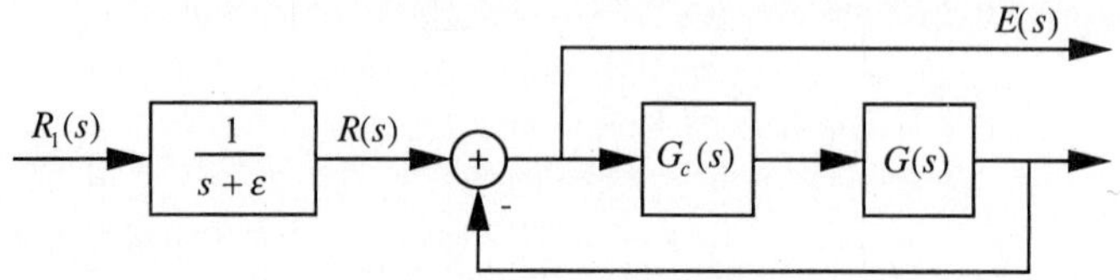

FIGURE P10.1 Frequency-weighted unity feedback system

b. Append first-order frequency weighting functions to the closed-loop plant. Assume that the reference input is lowpass with a cutoff frequency of 1 rad/sec, and the measurement noise is highpass with a cutoff frequency of 100 rad/sec. Compute the ∞-norms of the frequency weighted transfer functions obtained with each of the controllers.

10.3 A scalar tracking system with near integral weighting on the reference input is shown in Figure P10.1. Assume that a suboptimal controller exists such that

$$\left|\frac{E(j\omega)}{R_1(j\omega)}\right| \le 1.$$

a. Show that the steady-state error due to a unit step input $r(t)$ [not $r_1(t)$] approaches zero as ε goes to zero.

b. If the plant has finite dc gain, show that a pole of the controller must converge to ε as ε goes to zero.

10.4 Given the function

$$f(x, y) = ax^2 + by^2 + cxy + dx + ey + f,$$

and the initial guess $x = y = 0$, alternately minimize over x and y under the following conditions.

a. $a = b = c = d = e = f = 1$. Does the result converge to the values of x and y that minimize $f(x, y)$?

b. $a = b = c = e = f = 1$ and $d = -10$. Does the result converge to the values of x and y that minimize $f(x, y)$? In general, under what circumstances do you expect to get convergence of this algorithm?

COMPUTER EXERCISES

10.1 Servo System Design Using $\mathcal{H}_\infty$ Control

The plant consists of a simple servo in parallel with a vibrational mode (see Computer Exercise 8.2).[12] This plant is subject to a low-frequency disturbance that is modeled as the output of a filter. The block diagram of the augmented plant is shown in Figure P10.2. This augmented plant can also be described by the following state model:

$$\dot{x}(t) = \begin{bmatrix} -1 & 0 & 0 & 1 \\ 0 & 0 & 1 & 0 \\ 0 & -100 & -0.2 & 0 \\ 0 & 0 & 0 & -1 \end{bmatrix} x(t) + \begin{bmatrix} 1 \\ 0 \\ 100 \\ 0 \end{bmatrix} u(t) + \begin{bmatrix} 0 \\ 0 \\ 0 \\ .45 \end{bmatrix} w(t);$$

$$m(t) = [1 \quad 10 \quad 0 \quad 1]x(t) + 0.1v(t).$$

a. Design a control system for the above plant that minimizes the ∞-norm of the transfer function from

$$\begin{bmatrix} w(t) \\ v(t) \end{bmatrix} \text{ to } \begin{bmatrix} 4x_1(t) \\ 2u(t) \end{bmatrix}.$$

[12] This plant is drawn from [7].

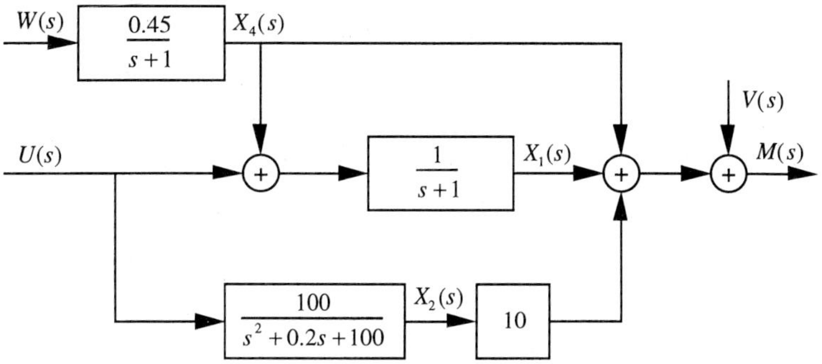

FIGURE P10.2 Plant model for Computer Exercise 10.1

Generate a Nyquist plot for the resulting system. What are the phase and gain margins of this system? Compute the frequency response and ∞-norm of the closed-loop transfer function.

b. Add an input-multiplicative perturbation to the plant model, where

$$\|\Delta_i\|_\infty \leq \Delta_{\max} = 0.0001.$$

Design a control system for the above plant that minimizes the ∞-norm of the transfer function from

$$\begin{bmatrix} w(t) \\ v(t) \\ w_d(t) \end{bmatrix} \quad \text{to} \quad \begin{bmatrix} 4x_1(t) \\ 2u(t) \\ y_d(t) \end{bmatrix},$$

where $w_d(t)$ and $y_d(t)$ are the input and output associated with the normalized perturbation, as shown in Figure P10.3. Note that this very small perturbation is sufficient to alter robustness due to the interactions of the perturbation inputs and outputs with the other plant inputs and outputs. Generate a Nyquist plot for the resulting system. What are the phase and gain margins? Using this controller, compute the frequency response and ∞-norm of the transfer function defined in part *a*. Compare these results with the previous design.

c. Let the bound on the input-multiplicative perturbation be increased to

$$\|\Delta_i\|_\infty \leq \Delta_{\max} = 0.01.$$

Design a control system for the above plant that minimizes the ∞-norm of the transfer function given in part *b*. Generate a Nyquist plot for the resulting system. What are the phase and gain margins? Using this controller, compute the frequency response and ∞-norm of the transfer function defined in part *a*. Compare these results with the previous designs.

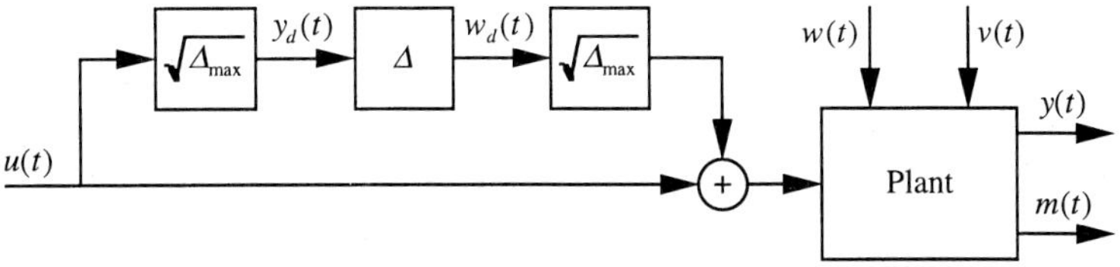

FIGURE P10.3 Input-multiplicative uncertainty for Computer Exercise 10.1

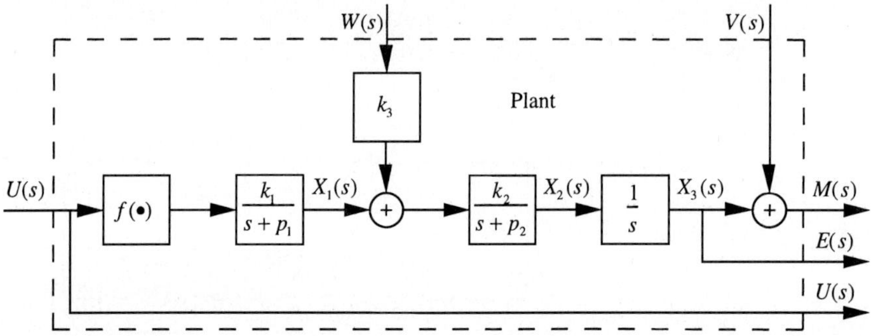

FIGURE P10.4 Block diagram of a dc motor

d. Let the bound on the input-multiplicative perturbation be increased to

$$\|\Delta_i\|_\infty \le \Delta_{\max} = 1.$$

Design a control system for the above plant that minimizes the ∞-norm of the transfer function given in part *b*. Generate a Nyquist plot for the resulting system. What are the phase and gain margins? Using this controller, compute the frequency response and ∞-norm of the transfer function defined in part *a*. Compare these results with the previous designs.

10.2 Control System Design for a DC Motor Using μ-Synthesis

The block diagram of a dc motor is shown in Figure P10.4, where $w(t)$ is a disturbance torque, $v(t)$ is measurement error, and $e(t)$ is the motor shaft angle error. Note that the coordinate system is chosen such that the desired motor shaft angle is zero (i.e., a regulator is desired). The gains are $k_1 = 3500, k_2 = 1, k_3 = 50$, and the poles have uncertain locations:

$$p_1 = [345,\ 355]\ ;\ p_2 = [49,\ 51].$$

The nonlinear function $f(\bullet)$ represents a soft saturation:

$$f(u) = 12\ \tan^{-1}\left(\frac{u}{9}\right).$$

The control input is expected to remain less than 15 volts, so this nonlinearity can be modeled as an uncertain gain.

Design a control system for the above plant using μ-synthesis. Assume that $|v(t)| \le 0.1$, $|w(t)| \le 1$, and minimize the error subject to the constraint that $|u(t)| \le 15$. What is the minimum achievable bound on the error for sinusoidal inputs? Note that all bounds should be achieved robustly. Compute and plot the structured singular value of the weighted, closed-loop system to evaluate robust performance. Comment on the results.

Repeat the design in step 1 assuming that the disturbance torque is lowpass with a bandwidth of 0.1 rad/sec. Further, assume that the measurement noise is bounded as given and highpass, with a cutoff frequency of 10 rad/sec. What is the optimal bound on the gain between the disturbance inputs and the error? Comment on the difference between the performance of this design and the previous design.

11 Controller Order Reduction

State-space control system design methodologies (especially μ-synthesis via $\mathfrak{D}$-$\mathfrak{K}$ iteration) tend to yield controllers of excessively high order. The basic LQG and $\mathcal{H}_\infty$ design methodologies generate controllers with orders equal to the plant order. Frequency-shaped loop transfer recovery, integral control, and other modifications of the controller's frequency-domain characteristics increase the controller order. The appending of $\mathcal{D}$-scales to the plant, as done during $\mathfrak{D}$-$\mathfrak{K}$ iteration, increases (often greatly) the order of the subsequent controller design.

High-order controllers increase the hardware quantity and complexity for hard-wired controllers, which translates directly to increased cost. For digital control, high-order controllers increase the computational burden and therefore the speed of the required processor. A less obvious effect of high-order controllers is that the required sampling rate is usually considerably faster, since high-order controllers typically contain faster poles than low-order controllers. These high-speed poles are the result of two factors: (1) The design methodologies presented in previous chapters frequently add high-speed poles to the controller; (2) faster poles may occur in a high-order controller simply because more poles are present. A faster sampling rate tends to compound the computational burden of a high-order controller, and may also increase the cost of A/D and D/A conversion. In addition, high-order controllers require increased software complexity, which increases coding and debugging time. For these reasons, it is desirable to use low-order controllers whenever possible.

Controller order can be reduced in three ways: (1) A reduced-order approximation of the plant can be generated before designing the controller; (2) the order of the controller can be constrained during the design; and (3) a reduced-order approximation of the controller can be generated after controller design.

Reduced-order approximation of the plant dynamics is standard engineering practice. All mathematical models are reduced-order approximations of the true system generated by ignoring "minor" effects during modeling. While it is desirable to use relatively low-order plant models whenever possible, plant model order reduction tends to be application-specific and is therefore not overtly discussed in this section. Note that the techniques presented in this section for reduced-order controller synthesis can also be applied to reduce the plant model order, but this reduction is usually best accomplished when based on a sound understanding of the plant's physics.

Constraints on the controller order are frequently employed during classical controller design. For example, using a proportional control constrains the controller to have an order of zero, and using a lead compensator constrains the order to be 1. These classical control ideas allow the design of non-optimal, low-order controllers. Unfortunately, the design of a fourth-order controller for a MIMO system may be difficult using classical techniques. The development of optimal methods of designing controllers with order constraints is an interesting topic of current research but is not included in this text.

The generation of a reduced-order controller is addressed in the remainder of this section. A wide variety of techniques exist for generating reduced-order controllers, two of which are presented below. Reduced-order approximation of SISO controllers can be accomplished by removing poles and zeros. This technique can be extended to MIMO systems, but is presented primarily to increase intuitive understanding of controller approximation. A systematic method of reduced-order controller approximation based on balanced realizations is also presented. This method can be directly applied to MIMO controllers.

In summary, controllers should be designed based on plant models of reasonably small order. After controller design, the controller order can be reduced, if desired. The resulting closed-loop system should be analyzed to insure stability and to evaluate the performance degradation that results when using the reduced-order controller. An intelligent trade-off can then be made between controller order and performance.

11.1 Perturbation Analysis

The performance of the reduced-order controller can be evaluated directly or by using perturbation theory. When using perturbation theory, the approximation error is assumed to be stable. This is a reasonable assumption, since the controller can be written as a parallel combination of stable $\mathbf{G}_s(s)$ and unstable $\mathbf{G}_u(s)$ parts:

$$\mathbf{G}_c(s) = \mathbf{G}_s(s) + \mathbf{G}_u(s).$$

The reduced-order controller is then obtained by reduced-order approximation of the stable portion of the controller:

$$\tilde{\mathbf{G}}_c(s) = \tilde{\mathbf{G}}_s(s) + \mathbf{G}_u(s)$$

where the ~ indicates reduced-order approximation.

The unstable portion of the controller is retained during approximation, since deleting unstable terms generates large errors between the controller and the reduced-order approximation. In addition, unstable terms are typically included in a controller for important reasons. For example, an integral control term, which is used to yield good constant reference input tracking, should be included in the reduced-order controller to maintain tracking performance.

The reduced-order controller can be written as the full-order controller plus a perturbation:

$$\tilde{\mathbf{G}}_c(s) = \mathbf{G}_c(s) + \mathbf{\Delta}(s).$$

The perturbation

$$\mathbf{\Delta}(s) = \tilde{\mathbf{G}}_c(s) - \mathbf{G}_c(s) = \tilde{\mathbf{G}}_s(s) - \mathbf{G}_s(s)$$

is assumed to be stable. The robustness analysis techniques presented in Chapter 5 can be used to analyze the stability and robustness of a system subject to stable perturbations. These techniques are not necessary for analyzing reduced-order control system performance, since the perturbation is known exactly. But application of robustness analysis results does yield criteria for evaluating the reduced-order controller without resorting to closed-loop system analysis. These criteria provide insight into the goals of reduced-order approximation.

11.2 Frequency Weighting

The reduced-order controller is required to yield an internally stable closed-loop system, and to yield good transient performance, tracking performance, disturbance rejection, and robustness. The full-order controller is assumed to satisfy these stability and performance requirements. Therefore, the reduced-order controller should satisfy these requirements provided it is an accurate approximation of the full-order controller. The question then arises, How should the approximation error be evaluated? This question is not straightforward, since various performance requirements lead to different criteria for evaluating the approximation error.

Stability of the closed-loop system can be related to a bound on a frequency-weighted ∞-norm of the approximation error. Other performance requirements can also be related to frequency-weighted ∞-norms of the approximation error, but the frequency weights for the various performance criteria are disparate. The frequency weights for closed-loop stability and tracking performance are generated below for a generic unity feedback system. These weights are generated to illustrate the development of frequency-weighted approximation error criteria in reduced-order controller design. In addition, these conditions are used to motivate controller reduction based on pole-zero truncation.

A unity gain feedback system containing a reduced-order controller is shown in Figure 11.1. Figure 11.1b shows this system in standard form; that is, with the approxi-

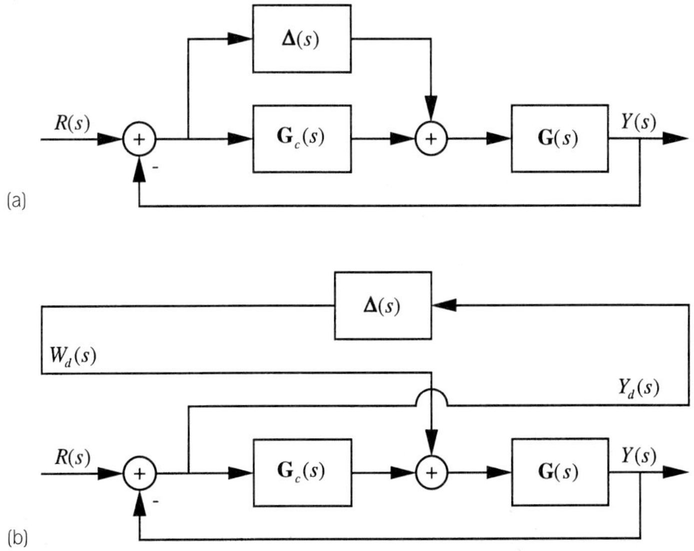

FIGURE 11.1 Perturbation analysis for a reduced-order controller: (a) the closed-loop system; (b) the closed-loop system in standard form

mation error normalized and in a feedback loop. At first glance, the structured singular value can be used to evaluate "robust performance" of this system. This is typically not done, since the reduced-order controller is a known controller, and robustness over the range of all approximation errors is not required. Instead, the stability and performance of the system using the reduced-order controller can be evaluated directly. The robust stability and robust performance conditions for this system are presented below, not as a means of analysis, but to develop insight into the requirements placed on the reduced-order controller.

The stability robustness condition for the system in Figure 11.1 is given by the small-gain theorem. The transfer function between $W_d(s)$ and $Y_d(s)$,

$$\mathbf{G}_{y_d w_d}(s) = -[\mathbf{I} + \mathbf{G}(s)\mathbf{G}_c(s)]^{-1}\mathbf{G}(s),$$

is stable, since the full-order, closed-loop system is assumed to be internally stable. The small-gain theorem states that the closed-loop system in Figure 11.1 is stable if

$$\|\mathbf{G}_{y_d w_d}(s)\boldsymbol{\Delta}(s)\|_\infty = \max_\omega \bar{\sigma}\{-[\mathbf{I} + \mathbf{G}(j\omega)\mathbf{G}_c(j\omega)]^{-1}\mathbf{G}(j\omega)\boldsymbol{\Delta}(j\omega)\} < 1. \tag{11.1}$$

This condition for stability indicates that reduced-order controllers should be selected to minimize this frequency-weighted ∞-norm of the controller approximation error.

Some intuitive ideas concerning the frequencies both where the reduced-order controller must be accurate and where accuracy is not important can be developed for the weighting in (11.1):

$$\mathbf{G}_{y_d w_d}(j\omega) = -[\mathbf{I} + \mathbf{G}(j\omega)\mathbf{G}_c(j\omega)]^{-1}\mathbf{G}(j\omega).$$

The size of the frequency weight can be bounded by the maximum singular value:

$$\bar{\sigma}\{\mathbf{G}_{y_d w_d}(j\omega)\} \le \bar{\sigma}\{[\mathbf{I} + \mathbf{G}(j\omega)\mathbf{G}_c(j\omega)]^{-1}\}\bar{\sigma}\{\mathbf{G}(j\omega)\} = \frac{\bar{\sigma}\{\mathbf{G}(j\omega)\}}{\underline{\sigma}\{\mathbf{I} + \mathbf{G}(j\omega)\mathbf{G}_c(j\omega)\}}, \tag{11.2}$$

where Appendix equation (A3.7) is used in generating this final expression. This gain is potentially large when the minimum singular value of $\{\mathbf{I} + \mathbf{G}(j\omega)\mathbf{G}_c(j\omega)\}$ is small (i.e., when this matrix is nearly singular). For this matrix to be nearly singular, the gain of the loop transfer function must be approximately 1, and the phase shift approximately $-180°$ for some input. Therefore, the reduced-order controller must be accurate at frequencies where the range of gains includes 1. This idea is familiar from classical control, where the frequency response near the gain crossover frequency is particularly important to stability.

The bound (11.2) can be approximated:

$$\bar{\sigma}\{\mathbf{G}_{y_d w_d}(j\omega)\} \le \frac{\bar{\sigma}\{\mathbf{G}(j\omega)\}}{\underline{\sigma}\{\mathbf{I} + \mathbf{G}(j\omega)\mathbf{G}_c(j\omega)\}} \approx \frac{\bar{\sigma}\{\mathbf{G}(j\omega)\}}{\underline{\sigma}\{\mathbf{G}(j\omega)\mathbf{G}_c(j\omega)\}}$$

provided the minimum loop transfer function gain is large. The frequency weighting (11.2) is therefore small if the loop transfer function $\mathbf{G}(j\omega)\mathbf{G}_c(j\omega)$ has a large minimum gain compared to the maximum plant gain. This idea is familiar from classical control, where large loop gains tend to decrease the sensitivity of the closed-loop systems to perturbations. Note that for MIMO systems, the loop gain must be large for all input directions in order to ensure that the sensitivity to perturbations is small.

Lastly, observe that the bound (11.2) is small when the gain of the plant transfer function is small (except possibly at frequencies where $\{\mathbf{I} + \mathbf{G}(j\omega)\mathbf{G}_c(j\omega)\}$ is nearly singular). Therefore, the reduced-order approximation need not be very good outside of the plant bandwidth. The above comments regarding controller approximation are applicable to a fairly broad range of control systems despite the fact that they are developed intuitively for the particular example in Figure 11.1.

Stability is only a minimum control system requirement. Performance, as quantified by the closed-loop transfer function, must also be considered. A reasonable performance requirement is that the change in the closed-loop transfer function due to the reduced-order approximation be small. For the system in Figure 11.1, the closed-loop transfer function from the reference input to the reference output is

$$\mathbf{G}_{yr}(s) = [\mathbf{I} + \mathbf{G}(s)\mathbf{G}_c(s) + \mathbf{G}(s)\boldsymbol{\Delta}(s)]^{-1}\mathbf{G}(s)[\mathbf{G}_c(s) + \boldsymbol{\Delta}(s)].$$

The change in this transfer function due to the reduced-order controller is

$$\boldsymbol{\Delta}_{\mathbf{G}_{yr}} = \mathbf{G}_{y_d w_d} - \mathbf{G}_{y_d w_d 0} = [\mathbf{I} + \mathbf{G}\mathbf{G}_c + \mathbf{G}\boldsymbol{\Delta}]^{-1}\mathbf{G}[\mathbf{G}_c + \boldsymbol{\Delta}] - [\mathbf{I} + \mathbf{G}\mathbf{G}_c]^{-1}\mathbf{G}\mathbf{G}_c,$$

where $\mathbf{G}_{y_d w_d 0}$ is the nominal transfer function, that is, the transfer function with $\boldsymbol{\Delta} = \mathbf{0}$. Note that the Laplace variable s has been deleted in this expression to simplify the notation. The change in the closed-loop transfer function can be expanded by applying the matrix inversion lemma of Appendix equation (A2.8):

$$\begin{aligned}\boldsymbol{\Delta}_{\mathbf{G}_{yr}}\\ &= \{[\mathbf{I} + \mathbf{G}\mathbf{G}_c]^{-1} - [\mathbf{I} + \mathbf{G}\mathbf{G}_c]^{-1}\mathbf{G}(\mathbf{I} + \boldsymbol{\Delta}[\mathbf{I} + \mathbf{G}\mathbf{G}_c]^{-1}\mathbf{G})^{-1}\boldsymbol{\Delta}[\mathbf{I} + \mathbf{G}\mathbf{G}_c]^{-1}\}\mathbf{G}[\mathbf{G}_c + \boldsymbol{\Delta}]\\ &\quad - [\mathbf{I} + \mathbf{G}\mathbf{G}_c]^{-1}\mathbf{G}\mathbf{G}_c.\end{aligned}$$

Rearranging the terms in this expression yields

$$\begin{aligned}\boldsymbol{\Delta}_{\mathbf{G}_{yr}}\\ &= [\mathbf{I} + \mathbf{G}\mathbf{G}_c]^{-1}\{\mathbf{G}\mathbf{G}_c + \mathbf{G}\boldsymbol{\Delta} - \mathbf{G}(\mathbf{I} + \boldsymbol{\Delta}[\mathbf{I} + \mathbf{G}\mathbf{G}_c]^{-1}\mathbf{G})^{-1}\boldsymbol{\Delta}[\mathbf{I} + \mathbf{G}\mathbf{G}_c]^{-1}[\mathbf{G}\mathbf{G}_c + \mathbf{G}\boldsymbol{\Delta}]\}\\ &\quad - [\mathbf{I} + \mathbf{G}\mathbf{G}_c]^{-1}\mathbf{G}\mathbf{G}_c.\end{aligned}$$

Noting that the first term within the curly brackets cancels the term outside the brackets, we have

$$\begin{aligned}\boldsymbol{\Delta}_{\mathbf{G}_{yr}} &= [\mathbf{I} + \mathbf{G}\mathbf{G}_c]^{-1}\,\mathbf{G}\boldsymbol{\Delta}\\ &\quad - \{[\mathbf{I} + \mathbf{G}\mathbf{G}_c]^{-1}\mathbf{G}(\mathbf{I} + \boldsymbol{\Delta}[\mathbf{I} + \mathbf{G}\mathbf{G}_c]^{-1}\mathbf{G})^{-1}\boldsymbol{\Delta}[\mathbf{I} + \mathbf{G}\mathbf{G}_c]^{-1}\}[\mathbf{G}\mathbf{G}_c + \mathbf{G}\boldsymbol{\Delta}].\end{aligned}$$

For small perturbations $(\mathbf{I} + \boldsymbol{\Delta}[\mathbf{I} + \mathbf{G}\mathbf{G}_c]^{-1}\mathbf{G})^{-1} \approx \mathbf{I}$, and the change in the closed-loop transfer function reduces to

$$\boldsymbol{\Delta}_{\mathbf{G}_{yr}} \approx [\mathbf{I} + \mathbf{G}\mathbf{G}_c]^{-1}\mathbf{G}\boldsymbol{\Delta} - \{[\mathbf{I} + \mathbf{G}\mathbf{G}_c]^{-1}\mathbf{G}\boldsymbol{\Delta}[\mathbf{I} + \mathbf{G}\mathbf{G}_c]^{-1}\}[\mathbf{G}\mathbf{G}_c + \mathbf{G}\boldsymbol{\Delta}].$$

Higher-order terms in the perturbation can be ignored when the controller perturbation is small:

$$\boldsymbol{\Delta}_{\mathbf{G}_{yr}} \approx [\mathbf{I} + \mathbf{G}\mathbf{G}_c]^{-1}\mathbf{G}\boldsymbol{\Delta} - [\mathbf{I} + \mathbf{G}\mathbf{G}_c]^{-1}\mathbf{G}\boldsymbol{\Delta}[\mathbf{I} + \mathbf{G}\mathbf{G}_c]^{-1}\mathbf{G}\mathbf{G}_c.$$

Postmultiplying the first term by $[\mathbf{I} + \mathbf{G}\mathbf{G}_c]^{-1}[\mathbf{I} + \mathbf{G}\mathbf{G}_c]$ yields

$$\begin{aligned}\boldsymbol{\Delta}_{\mathbf{G}_{yr}} \approx\ &\{[\mathbf{I} + \mathbf{G}\mathbf{G}_c]^{-1}\mathbf{G}\boldsymbol{\Delta}[\mathbf{I} + \mathbf{G}\mathbf{G}_c]^{-1}\}[\mathbf{I} + \mathbf{G}\mathbf{G}_c] \\ &- \{[\mathbf{I} + \mathbf{G}\mathbf{G}_c]^{-1}\mathbf{G}\boldsymbol{\Delta}[\mathbf{I} + \mathbf{G}\mathbf{G}_c]^{-1}\}\mathbf{G}\mathbf{G}_c.\end{aligned}$$

Combining the two terms in this equation gives the change in the closed-loop transfer function due to the reduced-order controller:

$$\boldsymbol{\Delta}_{\mathbf{G}_{yr}} \approx [\mathbf{I} + \mathbf{G}\mathbf{G}_c]^{-1}\mathbf{G}\boldsymbol{\Delta}[\mathbf{I} + \mathbf{G}\mathbf{G}_c]^{-1}.$$

The reduced-order controller should be selected to make this change small at all frequencies. This is equivalent to requiring that

$$\|\boldsymbol{\Delta}_{\mathbf{G}_{yr}}\|_\infty \approx \|[\mathbf{I} + \mathbf{G}\mathbf{G}_c]^{-1}\mathbf{G}\boldsymbol{\Delta}[\mathbf{I} + \mathbf{G}\mathbf{G}_c]^{-1}\|_\infty \tag{11.3}$$

be small. Equation (11.3) states that the change in the closed-loop transfer function is given by a frequency-weighted ∞-norm of the controller approximation error. Note that this frequency weighting includes gains that both premultiply and postmultiply the approximation error.

The frequency weighting in (11.3) can be bounded using the maximum singular value:

$$\begin{aligned}\|\boldsymbol{\Delta}_{\mathbf{G}_{yr}}\|_\infty &\approx \max_\omega \bar{\sigma}\{[\mathbf{I} + \mathbf{G}(j\omega)\mathbf{G}_c(j\omega)]^{-1}\mathbf{G}(j\omega)\boldsymbol{\Delta}(j\omega)[\mathbf{I} + \mathbf{G}(j\omega)\mathbf{G}_c(j\omega)]^{-1}\} \\ &\le \max_\omega \frac{\bar{\sigma}\{\mathbf{G}(j\omega)\}\bar{\sigma}\{\boldsymbol{\Delta}(j\omega)\}}{\underline{\sigma}\{[\mathbf{I} + \mathbf{G}(j\omega)\mathbf{G}_c(j\omega)]\}^2},\end{aligned}$$

where this final expression is derived using Appendix equation (A3.7). The frequency weighting $\mathbf{W}(j\omega)$ in the norm (11.3) can then be bounded:

$$\bar{\sigma}\{\mathbf{W}(j\omega)\} \le \frac{\bar{\sigma}\{\mathbf{G}(j\omega)\}}{\underline{\sigma}\{[\mathbf{I} + \mathbf{G}(j\omega)\mathbf{G}_c(j\omega)]\}^2}. \tag{11.4}$$

This bound on the frequency weighting indicates that the reduced-order controller must be most accurate at frequencies where the range of gains includes 1. Further, (11.4) indicates that the reduced-order controller need not be very accurate at frequencies outside the plant bandwidth, and large controller gains tend to mitigate the effects of controller approximation error. These intuitive criteria for the reduced-order approximation are identical to those developed for the stability robustness example. This fact tends to support the applicability of these intuitive ideas to a wide range of feedback systems.

11.3 Removing Poles and Zeros from SISO Controllers

A reduced-order SISO controller must match the frequency response of the full-order controller within the plant bandwidth, and especially near the gain and phase crossover frequencies. These facts indicate that terms in the controller transfer function that either have little effect on the frequency response, or only effect the frequency response outside the plant bandwidth, can be deleted. In particular, controller poles and zeros can be deleted if they are far outside of the plant bandwidth. Since most plants tend to be lowpass filters, this means that high-frequency controller poles and zeros can be deleted. The removal of high-frequency poles and zeros has the additional advantage of allowing lower sampling rates during digital implementation. Controller poles and zeros can also be deleted if they are nearly collocated (provided they are stable), since these pole-zero pairs do not significantly affect the frequency response.

Deleting poles and zeros changes the controller gain at all frequencies in addition to modifying the shape of the frequency response. Often it is desirable to hold the gain constant at a specified frequency, usually dc. An additional gain can be added to the reduced-order controller to force the resulting gain to match the full-order gain at the specified frequency.

A proper controller transfer function is desirable to avoid the need to differentiate the measurement. In addition, the difference between the number of poles and the number of zeros determines the gain roll-off at high frequencies. Therefore, it is desirable to maintain this difference when generating a reduced-order controller. This difference is maintained if poles and zeros are deleted in pairs. Deleting pole-zero pairs is usually a good idea, but this is not strictly necessary.

EXAMPLE 11.1 The frequency response of the dc motor

$$G(s) = \frac{50}{s(s+1)(s+8)}$$

is given in Figure 11.2. The unity gain bandwidth of this plant is 2.3 rad/sec. A μ-synthesis controller is designed for this motor to track a constant reference input and to yield good robustness to input-multiplicative perturbations. The resulting controller

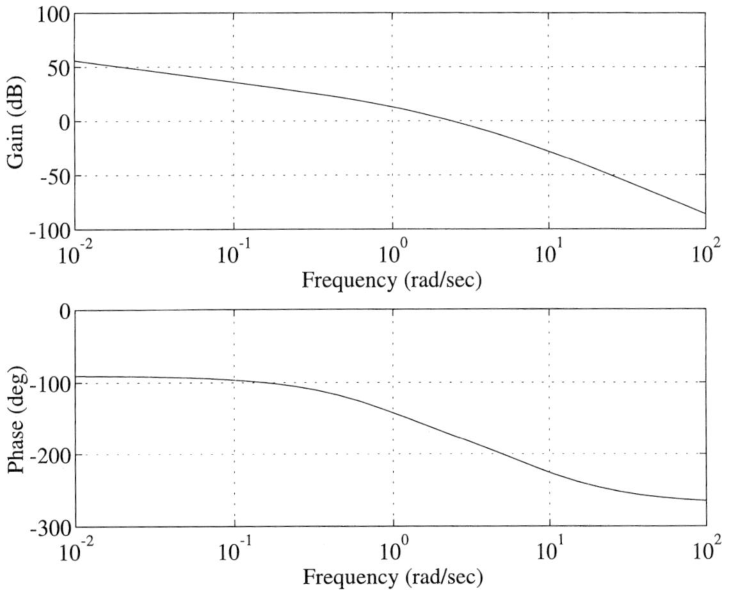

FIGURE 11.2 Frequency response for the dc motor in Example 11.1

is an eleventh-order, SISO unity feedback controller, that is, it operates on the difference between the reference input and the motor position. The dc controller gain is 11.55, and the controller poles and zeros are listed in Table 11.1.

TABLE 11.1 μ-Synthesis Controller Poles and Zeros

Controller Poles	Controller Zeros
−1980	−1980
−1980	−1980
−1080	−125
−124	−124
−122	−5.2 + 3.1j
−44.9 + 41.4j	−5.2 − 3.1j
−44.9 − 41.4j	−2.3 + 5.6j
−2.3 + 5.6j	−2.3 − 5.6j
−2.3 − 5.6j	−2.3 + 5.6j
−2.0 + 5.9j	−2.3 − 5.6j
−2.0 − 5.9j	∞

Note that there are several collocated poles and zeros, and also several high-frequency poles and zeros. This is fairly typical of μ-synthesis controllers, which often incorporate poles and zeros that only weakly effect performance.

A reduced-order controller can be obtained by deleting collocated (or nearly collocated) poles and zeros. For this example, there are five collocated (or nearly collocated) pole-zero pairs at the locations -1980, -1980, -124, $-2.3 + 5.6j$, and $-2.3 - 5.6j$. After deleting these pole-zero pairs, additional nondominant poles and zeros can be deleted.

Consider the poles and zeros that remain after deleting collocated pole-zero pairs. The four poles nearest the imaginary axis are all complex. Two reduced-order controllers are generated that contain one and two complex conjugate pole pairs, respectively. The zeros nearest the imaginary axis are also complex. The two reduced-order controllers then include one and two complex conjugate zero pairs, respectively. In order to maintain the high-frequency roll-off of the controller, an additional real pole is added to both controllers. The dc gain of the reduced-order controllers is set equal to full-order dc gain.

A third-order controller approximation is generated (as discussed above) that has the poles $\{-2.0 + 5.9j, -2.0 - 5.9j, -122\}$ and the zeros $\{-2.3 + 5.6j, -2.3 - 5.6j\}$. This controller is used to generate a closed-loop state model. This closed-loop system is unstable implying that the third-order pole-zero approximation is unacceptable.

A fifth-order controller approximation is generated (as discussed above) that has the poles $\{-2.0 + 5.9j, -2.0 - 5.9j, -44.9 + 41.4j, -44.9 - 41.4j, -122\}$ and the zeros $\{-2.3 + 5.6j, -2.3 - 5.6j, -5.2 + 3.1j, -5.2 - 3.1j\}$. The closed-loop system that results when using this reduced-order controller is stable. The frequency response of this controller is given in Figure 11.3a, along with the frequency response of the full-order controller. The reduced-order frequency response is labeled "p-z5" in this

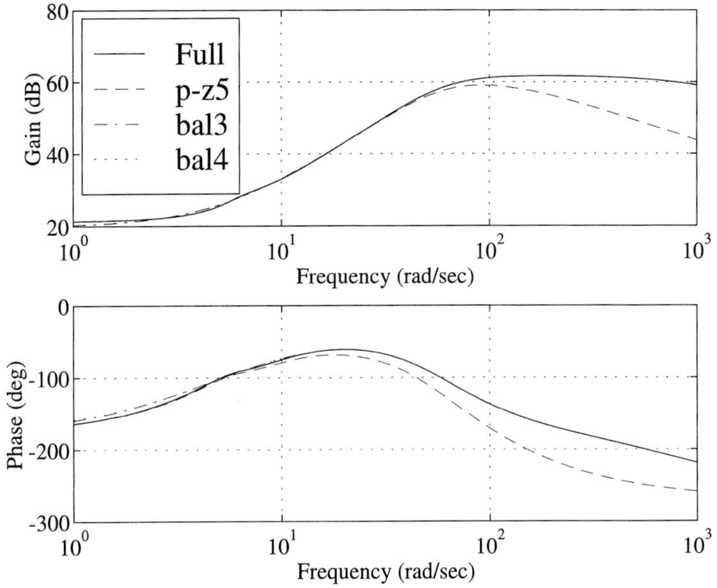

(a)

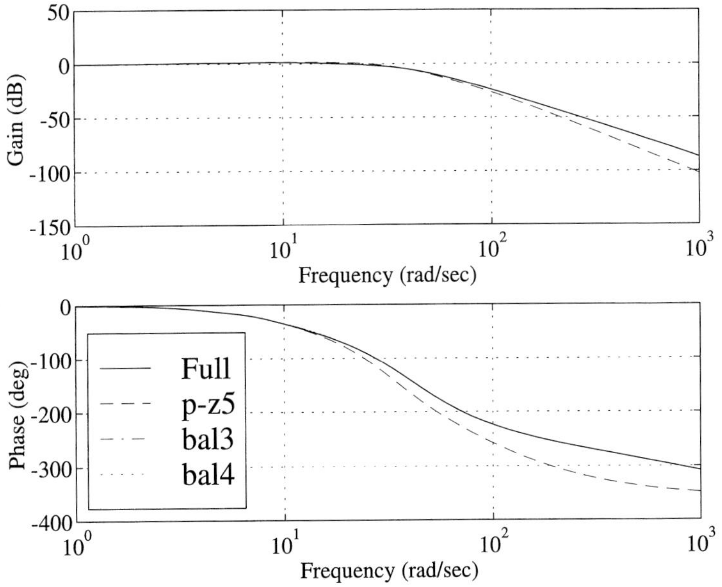

(b)

FIGURE 11.3 Frequency responses for Example 11.1: (a) controller frequency response; (b) closed-loop frequency response figure. The reduced-order controller yields an accurate approximation of the full-order controller over the plant bandwidth. The frequency responses of the closed-loop system with both the reduced-order controller and the full-order controller are given in Figure 11.3b. These two frequency responses are similar, indicating satisfactory performance of the reduced-order controller. In general, stability, noise rejection, and robustness of the closed-loop system should also be evaluated before deciding if this reduced-order controller is acceptable. ◆

The removal of insignificant poles and zeros from SISO controllers provides a useful means of reducing controller order. State-space design methods often produce controllers containing high-frequency poles and zeros, and also often yield near pole-zero cancellations. Therefore, this method of reducing controller order is particularly useful when the full-order controller is designed using state methods. Unfortunately, the removal of poles and zeros is an *ad hoc* approach that can be cumbersome for high-order systems, does not directly address any optimality conditions, and is not easily extended to MIMO controllers. The method of balanced truncation, while not optimal, does provide a much more systematic approach to reduced-order controller design and can be used for both SISO and MIMO controllers.

11.4 Balanced Truncation

Balanced truncation is based on the use of a special state-space realization (the balanced realization), where each state is equally strongly coupled to both the input and the output. The use of this realization allows an ordering of the states based on their input-output coupling. Reduced-order controllers can then be generated by truncating those states that only weakly participate in the input-output behavior of the full-order controller. The definition of the balance realization requires that the coupling of the states to the input and the output be quantified. Such a quantization can be provided by the controllability and observability grammians.

A generic state model of a time-invariant, stable system is

$$\dot{x}(t) = \mathbf{A}x(t) + \mathbf{B}u(t); \tag{11.5a}$$

$$y(t) = \mathbf{C}x(t) + \mathbf{D}u(t). \tag{11.5b}$$

The controllability grammian of this system is defined as

$$\mathbf{L}_c = \int_0^\infty e^{\mathbf{A}t}\mathbf{B}\mathbf{B}^T e^{\mathbf{A}^T t} dt,$$

and can be computed by solving the following Lyapunov equation:

$$\mathbf{A}\mathbf{L}_c + \mathbf{L}_c\mathbf{A}^T + \mathbf{B}\mathbf{B}^T = \mathbf{0}.$$

The controllability grammian can be interpreted as the steady-state, state covariance matrix of (11.5) when this system is subject to a white noise input with a spectral density equal to the identity matrix. States with larger variances can be thought of as being more strongly coupled to the input, while states with smaller variances are more weakly coupled to the input.

The observability grammian is defined as

$$\mathbf{L}_o = \int_0^\infty e^{\mathbf{A}^T t}\mathbf{C}\mathbf{C}^T e^{\mathbf{A}t}dt,$$

and can be computed by solving the following Lyapunov equation:

$$\mathbf{A}^T\mathbf{L}_o + \mathbf{L}_o\mathbf{A} + \mathbf{C}^T\mathbf{C} = \mathbf{0}.$$

The observability grammian can be interpreted as the steady-state, state covariance matrix of the adjoint of the state equation:

$$\dot{\tilde{x}}(t) = \mathbf{A}^T\tilde{x}(t) + \mathbf{C}^T\tilde{y}(t);$$

$$\tilde{u}(t) = \mathbf{B}^T\tilde{x}(t) + \mathbf{D}^T\tilde{y}(t),$$

subject to a white noise input with a spectral density equal to the identity matrix. States with larger variances can be thought of as being more strongly coupled to the output, while states with smaller variances are more weakly coupled to the output.

It is reasonable to assume that states that are only weakly coupled to both the input and the output can be deleted from the model without greatly impacting the input-output behavior of the system. Unfortunately, deleting weakly coupled states is not straightforward. States may be strongly coupled to the input, but only weakly coupled to the output, or the opposite. Therefore, the balanced realization is defined.

The balanced realization is a special state realization where the controllability and observability grammians are both equal and diagonal. The states of a balanced realization are then equally strongly coupled to both the input and the output, and can be deleted when this coupling is weak.

11.4.1 The Balanced Realization

The controllability and observability grammians are not invariant with respect to similarity transformations; that is, these grammians depend on the definition of the states. Defining a new state vector,

$$\tilde{x}(t) = \mathbf{T}^{-1}x(t),$$

results in a new state model related to the original state model by the following similarity transformation:

$$\mathbf{A} \Rightarrow \mathbf{T}^{-1}\mathbf{A}\mathbf{T}\ ;\ \mathbf{B} \Rightarrow \mathbf{T}^{-1}\mathbf{B}\ ;\ \mathbf{C} \Rightarrow \mathbf{C}\mathbf{T}\ ;\ \mathbf{D} \Rightarrow \mathbf{D}. \tag{11.6}$$

The effect of this state transformation on the controllability grammian is

$$\mathbf{L}_c \Rightarrow \mathbf{T}^{-1}\mathbf{L}_c\mathbf{T}^{-T}.$$

This fact can be shown by substituting this quantity into the Lyapunov equation associated with the transformed system:

$$(\mathbf{T}^{-1}\mathbf{A}\mathbf{T})(\mathbf{T}^{-1}\mathbf{L}_c\mathbf{T}^{-T}) + (\mathbf{T}^{-1}\mathbf{L}_c\mathbf{T}^{-T})(\mathbf{T}^{-1}\mathbf{A}\mathbf{T})^T + (\mathbf{T}^{-1}\mathbf{B})(\mathbf{T}^{-1}\mathbf{B})^T = \mathbf{0}.$$

This equation reduces to

$$\mathbf{T}^{-1}(\mathbf{A}\mathbf{L}_c + \mathbf{L}_c\mathbf{A}^T + \mathbf{B}\mathbf{B}^T)\mathbf{T}^{-T} = \mathbf{0}.$$

Premultiplying and postmultiplying by the transformation matrix and its transpose yields the Lyapunov equation of the original system, which is known to equal zero. Analogously, similarity transformations can be shown to modify the observability grammian:

$$\mathbf{L}_o \Rightarrow \mathbf{T}^T\mathbf{L}_o\mathbf{T}.$$

A state vector can be defined or, equivalently, a transformation matrix can be selected, to yield grammians with desirable properties.

A balanced realization has the controllability and observability grammians equal and diagonal:

$$\mathbf{L}_c = \mathbf{L}_o = \boldsymbol{\Sigma} = \begin{bmatrix} \sigma_1 & & & \mathbf{0} \\ & \sigma_2 & & \\ & & \ddots & \\ \mathbf{0} & & & \sigma_n \end{bmatrix}.$$

The diagonal elements in this expression are real, positive, and ordered from largest to smallest as a matter of convention. The notation σ is used for these elements, since they are the singular values of both grammians.

The balanced realization is generated by performing a similarity transformation (11.6) on the original system, where the transformation matrix is selected as follows:

1. Factor the original controllability grammian:

$$\mathbf{L}_c = \mathbf{R}^T\mathbf{R}, \tag{11.7}$$

where the matrix $\mathbf{R}$ is positive definite.[1] The system (11.5) is assumed to be minimal, which implies controllability and the fact that the controllability grammian is positive definite. If (11.5) is not minimal, a lower-order model can (and should) be generated that has input-output behavior identical to the original model. Further model reduction is then performed with reference to this minimal model.

2. Generate the singular value decomposition of the matrix $\mathbf{R}\mathbf{L}_o\mathbf{R}^T$:

$$\mathbf{R}\mathbf{L}_o\mathbf{R}^T = \mathbf{U}\boldsymbol{\Sigma}^2\mathbf{U}^T, \tag{11.8}$$

where $\boldsymbol{\Sigma}^2$ is the singular value matrix, and $\mathbf{U}$ is an orthogonal matrix.[2] Note that the left and right singular vectors of $\mathbf{R}\mathbf{L}_o\mathbf{R}^T$ are equal since $\mathbf{R}\mathbf{L}_o\mathbf{R}^T$ is real and symmetric.

3. Generate the transformation matrix:

$$\mathbf{T} = \mathbf{R}^T\mathbf{U}\boldsymbol{\Sigma}^{-1/2}. \tag{11.9}$$

This transformation can be shown to yield a balanced realization by evaluating the transformed controllability and observability grammians.

---

[1] The Cholesky factorization is of this form, can be simply computed, and exists for all positive definite matrices.

[2] An orthogonal matrix is a real matrix with the property that $\mathbf{U}\mathbf{U}^T = \mathbf{I}$. Note that an orthogonal matrix is also unitary.

The transformed controllability grammian is

$$\begin{aligned}\mathbf{T}^{-1}\mathbf{L}_c\mathbf{T}^{-T} &= (\mathbf{R}^T\mathbf{U}\boldsymbol{\Sigma}^{-1/2})^{-1}\mathbf{L}_c(\mathbf{R}^T\mathbf{U}\boldsymbol{\Sigma}^{-1/2})^{-T} \\ &= \boldsymbol{\Sigma}^{-1/2}\mathbf{U}^T\mathbf{R}^{-T}\mathbf{L}_c(\mathbf{R}^{-1}\mathbf{U}\boldsymbol{\Sigma}^{1/2})^{-T}.\end{aligned}$$

Substituting for the original controllability grammian using the factorization (11.7) yields

$$\mathbf{T}^{-1}\mathbf{L}_c\mathbf{T}^{-T} = \boldsymbol{\Sigma}^{1/2}\mathbf{U}^T\mathbf{R}^{-T}\mathbf{R}^T\mathbf{R}\mathbf{R}^{-1}\mathbf{U}\boldsymbol{\Sigma}^{1/2} = \boldsymbol{\Sigma}, \tag{11.10}$$

where the fact that $\mathbf{U}$ is an orthogonal matrix is used in generating this result.

The transformed observability grammian is

$$\mathbf{T}^T\mathbf{L}_o\mathbf{T} = (\mathbf{R}^T\mathbf{U}\boldsymbol{\Sigma}^{-1/2})^T\mathbf{L}_o(\mathbf{R}^T\mathbf{U}\boldsymbol{\Sigma}^{-1/2}) = \boldsymbol{\Sigma}^{-1/2}\mathbf{U}^T\mathbf{R}\mathbf{L}_o\mathbf{R}^T\mathbf{U}\boldsymbol{\Sigma}^{-1/2}.$$

Substituting for $\mathbf{R}\mathbf{L}_o\mathbf{R}^T$ using the singular value decomposition (11.8) yields

$$\mathbf{T}^T\mathbf{L}_o\mathbf{T} = \boldsymbol{\Sigma}^{-1/2}\mathbf{U}^T\mathbf{U}\boldsymbol{\Sigma}^2\mathbf{U}^T\mathbf{U}\boldsymbol{\Sigma}^{-1/2} = \boldsymbol{\Sigma}. \tag{11.11}$$

The fact that $\mathbf{U}$ is an orthogonal matrix is again used in generating this result. The combination of (11.10) and (11.11) shows that performing the similarity transformation (11.6) yields a balanced realization. Further, the controllability and observability grammians of this realization are both equal to the real, diagonal matrix $\boldsymbol{\Sigma}$.

A balanced realization exists whenever the system is stable and minimal. The balance realization is not strictly unique, but the subspace of the state space associated with each singular value in (11.8) is invariant for all balanced realizations. Multiple balanced realizations can only be obtained by sign changes in the state basis, and by rotations of the bases for subspaces associated with repeated singular values. Therefore, the reduced-order controller generated by truncating the states of a balance realization associated with a given set of singular values is unique.

11.4.2 Balanced Truncation

Balanced truncation eliminates those states in the balanced realization associated with the smallest elements in the grammians. For example, let the balanced realization be given as

$$\begin{bmatrix}\dot{\tilde{x}}_1(t) \\ \hline \dot{\tilde{x}}_2(t)\end{bmatrix} = \left[\begin{array}{c|c}\mathbf{A}_{11} & \mathbf{A}_{12} \\ \hline \mathbf{A}_{21} & \mathbf{A}_{22}\end{array}\right]\begin{bmatrix}\tilde{x}_1(t) \\ \hline \tilde{x}_2(t)\end{bmatrix} + \begin{bmatrix}\mathbf{B}_1 \\ \hline \mathbf{B}_2\end{bmatrix}u(t);$$

$$y(t) = [\mathbf{C}_1 \mid \mathbf{C}_2]\begin{bmatrix}\tilde{x}_1(t) \\ \hline \tilde{x}_2(t)\end{bmatrix} + \mathbf{D}u(t),$$

where $\tilde{x}_1(t)$ is the portion of the state vector to be retained, and $\tilde{x}_2(t)$ is the portion of the state vector to be truncated. The reduced-order system is then

$$\begin{aligned}\dot{\tilde{x}}_1(t) &= \mathbf{A}_{11}\tilde{x}_1(t) + \mathbf{B}_1u(t); \\ y(t) &= \mathbf{C}_1\tilde{x}_1(t) + \mathbf{D}u(t).\end{aligned}$$

An analogous procedure can be used to generate reduced-order models when the state model is not balanced. This more general procedure is known as state truncation.

The generation of a reduced-order controller using balanced truncation is straightforward and can be applied to both SISO and MIMO systems. The order of the reduced-order controller must still be determined. This order should be selected so that the reduced-order controller provides a sufficiently accurate approximation of the full-order controller. In addition, the closed-loop system (with the reduced-order control) should be stable and have acceptable performance.

Balanced truncation always yields a stable reduced-order controller provided the full-order controller is stable and minimal. Note that this is not true of state truncation, in general. Stability of the reduced-order controller can be shown by considering the Lyapunov equation for the controllability grammian of the full-order system:

$$\begin{bmatrix} \mathbf{A}_{11} & \mathbf{A}_{12} \\ \mathbf{A}_{21} & \mathbf{A}_{22} \end{bmatrix} \begin{bmatrix} \boldsymbol{\Sigma}_1 & \mathbf{0} \\ \mathbf{0} & \boldsymbol{\Sigma}_2 \end{bmatrix} + \begin{bmatrix} \boldsymbol{\Sigma}_1 & \mathbf{0} \\ \mathbf{0} & \boldsymbol{\Sigma}_2 \end{bmatrix} \begin{bmatrix} \mathbf{A}_{11}^T & \mathbf{A}_{21}^T \\ \mathbf{A}_{12}^T & \mathbf{A}_{22}^T \end{bmatrix} + \begin{bmatrix} \mathbf{B}_1 \\ \mathbf{B}_2 \end{bmatrix} [\mathbf{B}_1^T \; \mathbf{B}_2^T] = \mathbf{0}.$$

The upper left block of this matrix equation is

$$\mathbf{A}_{11}\boldsymbol{\Sigma}_1 + \boldsymbol{\Sigma}_1\mathbf{A}_{11}^T + \mathbf{B}_1\mathbf{B}_1^T = \mathbf{0},$$

which is another Lyapunov equation. The existence of a positive definite solution to this equation implies, see Appendix equation (A7.6):

$$\lim_{t\to\infty} e^{\mathbf{A}_{11}t}\mathbf{B}_1 = \mathbf{0}. \tag{11.12}$$

Further, the reduced-order system is controllable, since the reduced-order controllability grammian is nonsingular. Therefore, (11.12) implies that the reduced-order system is stable.

The quality of a reduced-order controller can be quantified by the ∞-norm or by a frequency-weighted ∞-norm of the approximation error. The approximation error is

$$\begin{aligned} \boldsymbol{\Delta}(s) = \tilde{\mathbf{G}}(s) - \mathbf{G}(s) &= \mathbf{C}_1(s\mathbf{I} - \mathbf{A}_{11})^{-1}\mathbf{B}_1 + \mathbf{D} \\ &\quad - [\mathbf{C}_1 \; \mathbf{C}_1] \begin{bmatrix} s\mathbf{I} - \mathbf{A}_{11} & \mathbf{A}_{12} \\ \mathbf{A}_{21} & s\mathbf{I} - \mathbf{A}_{22} \end{bmatrix}^{-1} \begin{bmatrix} \mathbf{B}_1 \\ \mathbf{B}_2 \end{bmatrix} - \mathbf{D}. \end{aligned}$$

The inverse of the block matrix in this expression can be expanded using Appendix equation (A2.9):

$$\begin{aligned} &\begin{bmatrix} s\mathbf{I} - \mathbf{A}_{11} & \mathbf{A}_{12} \\ \mathbf{A}_{21} & s\mathbf{I} - \mathbf{A}_{22} \end{bmatrix}^{-1} \\ &= \begin{bmatrix} \boldsymbol{\Omega} + \boldsymbol{\Omega}\mathbf{A}_{12}(s\mathbf{I} - \mathbf{A}_{22} - \mathbf{A}_{21}\boldsymbol{\Omega}\mathbf{A}_{12})^{-1}\mathbf{A}_{21}\boldsymbol{\Omega} & -\boldsymbol{\Omega}\mathbf{A}_{12}(s\mathbf{I} - \mathbf{A}_{22} - \mathbf{A}_{21}\boldsymbol{\Omega}\mathbf{A}_{12})^{-1} \\ -(s\mathbf{I} - \mathbf{A}_{22} - \mathbf{A}_{21}\boldsymbol{\Omega}\mathbf{A}_{12})^{-1}\mathbf{A}_{21}\boldsymbol{\Omega} & (s\mathbf{I} - \mathbf{A}_{22} - \mathbf{A}_{21}\boldsymbol{\Omega}\mathbf{A}_{12})^{-1} \end{bmatrix}, \end{aligned}$$

where

$$\boldsymbol{\Omega} = (s\mathbf{I} - \mathbf{A}_{11})^{-1}.$$

Substituting this expression into the equation for the approximation error yields

$$\begin{aligned}\Delta = &-[\mathbf{C}_1(s\mathbf{I} - \mathbf{A}_{11})^{-1}\mathbf{A}_{12} + \mathbf{C}_2][s\mathbf{I} - \mathbf{A}_{22} - \mathbf{A}_{21}(s\mathbf{I} - \mathbf{A}_{11})^{-1}\mathbf{A}_{12}]^{-1} \\ &\times [\mathbf{A}_{21}(s\mathbf{I} - \mathbf{A}_{11})^{-1}\mathbf{B}_1 + \mathbf{B}_2].\end{aligned}$$

The infinity norm, or frequency-weighted infinity norm, of this approximation error can be easily evaluated numerically.

A bound on the ∞-norm of the approximation error resulting from balanced truncation is

$$\|\Delta(s)\|_\infty \leq 2(\sigma_{n_{\tilde{x}}+1} + \sigma_{n_{\tilde{x}}+2} + \cdots + \sigma_{n_x}),$$

where $n_{\tilde{x}}$ is the order of the reduced-order controller, and n_x is the order of the original controller. A derivation of this result can be found in [1], page 159. This bound provides guidance in selecting a reasonable controller order. But final evaluation of the controller approximation should be based on the exact error norm or frequency-weighted error norm. In addition, a thorough analysis of closed-loop system stability, performance, and robustness should be performed with the reduced-order control before committing to a design.

EXAMPLE 11.1 CONTINUED

The balance realization is generated for the μ-synthesis motor controller in Example 11.1. The formula given for the balanced realization is only valid for minimal realizations. Therefore, collocated pole-zero pairs must be deleted from the controller prior to generating the balanced realization. The balanced realization of this minimal controller model has the following grammians:

$$\mathbf{L}_c = \mathbf{L}_o = \begin{bmatrix} 756 & 0 & 0 & 0 & 0 & 0 \\ 0 & 601 & 0 & 0 & 0 & 0 \\ 0 & 0 & 160 & 0 & 0 & 0 \\ 0 & 0 & 0 & 1.23 & 0 & 0 \\ 0 & 0 & 0 & 0 & 1.15 & 0 \\ 0 & 0 & 0 & 0 & 0 & 0.88 \end{bmatrix}.$$

The drastic reduction of the singular values after σ_3 indicates that a third-order approximation is probably desirable. To test this idea, first- through fourth-order controller approximations are generated using balanced truncation. The first- and second-order controller approximations result in unstable closed-loop systems and are unacceptable. The third- and fourth-order controller approximations result in stable closed-loop systems. The frequency responses of these reduced-order controllers are labeled "bal3" and "bal4" in Figure 11.3a. This figure also includes the frequency responses of the full-order controller and the fifth-order controller approximation generated using pole-zero truncation. Observe that the frequency response of both balanced, reduced-order controllers is a very good approximation of the full-order controller. Also note that these frequency responses are most accurate at high frequency, which is typical of balanced approximation.

Figure 11.3b gives the frequency responses of the closed-loop systems obtained using the reduced-order balanced controllers. The closed-loop frequency responses for both the third-order and the fourth-order balanced approximations are almost indistinguishable from the original system. The fact that these two balanced approximations yield such similar performance indicates that a third-order approximation is sufficient for this controller. This controller order is consistent with that indicated by the grammian singular values as discussed above.

The balanced approximations are also more accurate than the approximation obtained using pole-zero truncation. This improved approximation accuracy occurs even though the controller order using balanced truncation is lower than the fifth-order approximation obtained using pole-zero truncation.

In summary, balanced truncation provides a useful and systematic means of generating reduced-order controllers. Further, the size of the elements in the grammians can be used to determine an approximate controller order. ◆

A number of modifications to the balanced truncation method are presented in the literature. Modifications to the balanced truncation method that allow the incorporation of frequency weighting were first presented in [2]. A more accessible description of frequency-weighted balance truncation can be found in [1], page 163, and [3], page 294. Balanced singular perturbation uses balanced truncation on a frequency inverted model of the controller [4], [5], and [6], page 332. The resulting reduced-order controller is then inverted to yield an approximation to the original controller. This procedure has the advantage of yielding zero dc error, which is desirable for maintaining steady-state performance.

11.5 Summary

One drawback of the $\mathfrak{D}$-$\mathcal{K}$ iteration algorithm is that it typically yields high-order controllers. This drawback can be partially obviated by reducing the order of the controller prior to implementation. Controller order can be reduced by removing nearly collocated, stable pole-zero pairs, and also by removing high-frequency poles and zeros.

A more systematic approach to controller order reduction is based on the balanced realization. In the balanced realization, the individual states are equally coupled to both the input and the output. Truncating states with little coupling to the input and output therefore has little effect on the controller. Many additional techniques exist for controller order reduction. These include optimal model reduction techniques, which minimize the weighted ∞-norm of the approximation error; see [1], page 505, and [6], page 341. When using any approach, we must verify that the reduced-order controller yields a stable closed-loop system, acceptable performance, and acceptable robustness.

Reduced-order controllers can also be generated in discrete-time. Stable, nearly collocated Z-transform poles and zeros can typically be deleted from the controller. In addition, poles and zeros located near the origin of the Z-plane and poles and zeros that only affect frequencies much higher than the plant bandwidth can often be deleted. These poles and zeros are analogous to the high-frequency poles and zeros encountered in continuous time. Note that deleting high-frequency poles and zeros is not as widely applicable in discrete time as it is in continuous time due to the limitations imposed by

TABLE 11.2 Balanced Realization Equations for Continuous-Time and Discrete-Time Systems

Formula	Continuous-Time	Discrete-Time
Controllability grammian	$\mathbf{L}_c = \int_0^\infty e^{\mathbf{A}t}\mathbf{B}\mathbf{B}^T e^{\mathbf{A}^T t}dt$	$\mathbf{L}_{cd} = \sum_{k=0}^{\infty} \mathbf{\Phi}^k \mathbf{B}\mathbf{B}^T(\mathbf{\Phi}^T)^k$
	$\mathbf{A}\mathbf{L}_c + \mathbf{L}_c\mathbf{A}^T + \mathbf{B}\mathbf{B}^T = \mathbf{0}$	$\mathbf{L}_{cd} = \mathbf{\Phi}\mathbf{L}_{cd}\mathbf{\Phi}^T + \mathbf{\Gamma}\mathbf{\Gamma}^T$
Observability grammian	$\mathbf{L}_o = \int_0^\infty e^{\mathbf{A}^T t}\mathbf{C}\mathbf{C}^T e^{\mathbf{A}t}dt$	$\mathbf{L}_{od} = \sum_{k=0}^{\infty} \mathbf{\Phi}^k \mathbf{C}\mathbf{C}^T(\mathbf{\Phi}^T)^k$
	$\mathbf{A}^T\mathbf{L}_o + \mathbf{L}_o\mathbf{A} + \mathbf{C}^T\mathbf{C} = \mathbf{0}$	$\mathbf{L}_{od} = \mathbf{\Phi}\mathbf{L}_{od}\mathbf{\Phi}^T + \mathbf{C}^T\mathbf{C}$
Transformation to a balanced realization	$\tilde{x}(t) = \mathbf{T}^{-1}x(t)$ where $\mathbf{L}_c = \mathbf{R}^T\mathbf{R}$; $\mathbf{R}\mathbf{L}_o\mathbf{R}^T = \mathbf{U}\mathbf{\Sigma}^2\mathbf{U}^T$; $\mathbf{T} = \mathbf{R}^T\mathbf{U}\mathbf{\Sigma}^{-1/2}$	$\tilde{x}(k) = \mathbf{T}^{-1}x(k)$ where $\mathbf{L}_{cd}\mathbf{R}^T\mathbf{R}$; $\mathbf{R}\mathbf{L}_{od}\mathbf{R}^T = \mathbf{U}\mathbf{\Sigma}^2\mathbf{U}^T$; $\mathbf{T} = \mathbf{R}^T\mathbf{U}\mathbf{\Sigma}^{-1/2}$

the Nyquist frequency. Balanced truncation can be directly extended to discrete-time systems. The fundamental equations for the balance realization are summarized for both continuous-time and discrete-time systems in Table 11.2.

REFERENCES

[1] K. Zhou, with J. C. Doyle, and K. Glover, *Robust and Optimal Control,* Prentice-Hall, Upper Saddle River, NJ, 1996.

[2] D. F. Enns, "Model reduction with balance realizations: An error bound and frequency weighted generalization," *Proceedings of the IEEE Conference on Decision and Control,* pp. 127–32, 1984.

[3] B. D. O. Anderson and J. B. Moore, *Optimal Control: Linear Quadratic Methods,* Prentice-Hall, Englewood Cliffs, NJ, 1990.

[4] K. V. Fernando and H. Nicholson, "Singular perturbational model reduction of balanced systems," *IEEE Transactions on Automatic Control,* 27 (1982): 466–68.

[5] Y. Liu and B. D. O. Anderson, "Singular perturbation approximation of balanced systems," *International Journal of Control,* 50 (1989): 1379–1405.

[6] M. Green and D. J. N. Limebeer, *Linear Robust Control,* Prentice-Hall, Englewood Cliffs, NJ, 1995.

Some additional references on model reduction follow:

[7] B. C. Moore, "Principal component analysis in linear systems: Controllability, observability, and model reduction," *IEEE Transactions on Automatic Control,* 26, no. 1 (1981): 17–32.

[8] D. S. Bernstein and D. C. Hyland, "The optimal projection equations for fixed-order dynamic compensation," *IEEE Transactions on Automatic Control,* 29, no. 2 (1984): 1034–37.

[9] K. Glover, "All optimal Hankel-norm approximation of linear multivariable systems and their $\mathcal{L}_\infty$-error bounds," *International Journal of Control,* 39 (1984): 1115–93.

[10] A. Yousuff and R. E. Skelton, "Controller reduction by component cost analysis," *IEEE Transactions on Automatic Control,* 29, no. 6 (1981): 520–30.

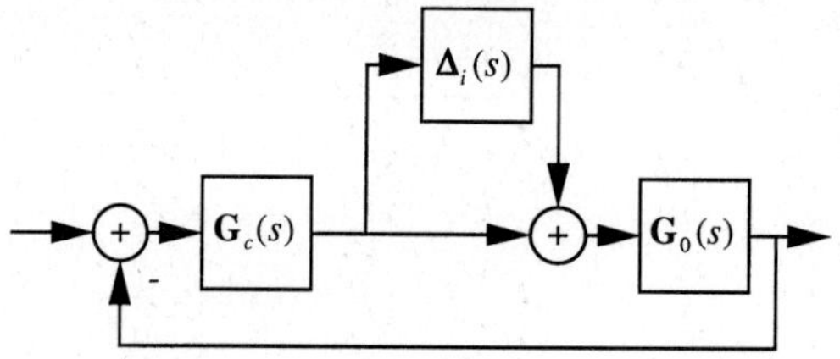

FIGURE P11.1 Feedback control with input-multiplicative uncertainty

EXERCISES

11.1 Develop a frequency-weighted norm for accessing the impact of controller order reduction on the sensitivity of the closed-loop transfer function to input-multiplicative perturbations. The feedback system is shown in Figure P11.1. Under what conditions can the controller error be large without greatly impacting this sensitivity?

11.2 You are given the following system:

$$\mathbf{G}(s) = \frac{500(s+3)}{(s+20)(s+2.8)(s+5)}.$$

Remove poles and zeros to generate first- and second-order approximations of this transfer function. For each approximation (and the original system), generate and plot the impulse response, the step response, and the frequency response (magnitude and phase). Compare the results and comment. Please plot each response (e.g., the step response) of the three systems on a single graph for easy comparison.

11.3 You are given the following system:

$$\mathbf{G}(s) = \frac{100(s^2+1s+3)}{(s+10)(s^2+2s+1)}.$$

Use balanced truncation to generate first- and second-order approximations to this transfer function. For each approximation (and the original system), generate and plot the impulse response, the step response, and the frequency response (magnitude and phase). Compare the results and comment. Please plot each response (e.g., the step response) of the three systems on a single graph for easy comparison.

11.4 You are given the following system:

$$\dot{x}(t) = \begin{bmatrix} 0 & 1 & 0 \\ 0 & 0 & 1 \\ -1 & -2 & -8 \end{bmatrix} x(t) + \begin{bmatrix} 0 & 1 \\ 0 & 0 \\ 1 & 1 \end{bmatrix} u(t);\; y(t) = \begin{bmatrix} 1 & 2 & 3 \\ 4 & 5 & 6 \end{bmatrix} x(t) + \begin{bmatrix} 1 & 2 \\ 3 & 4 \end{bmatrix} u(t).$$

Generate a balanced realization for this system. What are the diagonal elements of the grammians for this balanced realization? Truncate the balanced realization to generate first- and second-order approximations of this system. For each approximation (and the original system), generate and plot the step response from each input, the maximum principle gains, and the minimum principle gains. Compare the results and comment. Please plot each response (e.g., the step response for input one) of the three systems on a single graph for easy comparison.

COMPUTER EXERCISES

11.1 Reduced-Order Control of an Unmanned Surveillance Vehicle

An unmanned surveillance system is required to maintain a constant altitude while imaging a given location. A state equation for the height of this vehicle is given:

$$\begin{bmatrix} \dot{h}(t) \\ \ddot{h}(t) \\ \dddot{h}(t) \end{bmatrix} = \begin{bmatrix} 0 & 1 & 0 \\ 0 & 0 & 1 \\ 0 & 0 & -2 \end{bmatrix} \begin{bmatrix} h(t) \\ \dot{h}(t) \\ \ddot{h}(t) \end{bmatrix} + \begin{bmatrix} 0 \\ 0 \\ 2 \end{bmatrix} u(t) + \begin{bmatrix} 0 \\ 0 \\ 2 \end{bmatrix} w(t),$$

where the spectral density of the actuator noise is 1. The height of this vehicle is measured using a noisy radar altimeter:

$$m(t) = [1 \quad 0 \quad 0] \begin{bmatrix} h(t) \\ \dot{h}(t) \\ \ddot{h}(t) \end{bmatrix} + v(t),$$

where the spectral density of the measurement noise is 1. Generate a control system to minimize the following cost:

$$J = E[h^2(t) + 2u^2(t)].$$

For the closed-loop system, compute the variance of the height, the variance of the control, and the cost. Simulate the system and plot the height and the control.

Generate reduced-order controllers with orders of one and two via the balance truncation method. Compare the frequency response of these controllers with that of the original controller design, and compute the ∞-norm of the approximation errors. Determine if the closed-loop system is stable when using each of the reduced-order controllers. For the stable closed-loop systems, (1) compute the variance of the height, the variance of the control, and the cost; (2) simulate the closed-loop system and plot the height and the control. Compare these results to those obtained for the full-order controller.

11.2 Reduced-Order Control of a DC Motor

The block diagram of a dc motor is shown in Figure P11.2, where $w(t)$ is a disturbance torque, $v(t)$ is measurement error, and $e(t)$ is the motor shaft angle error. Note that the coordinate system is chosen so that the desired motor shaft angle is zero (i.e., a regulator is desired). The gains are $k_1 = 3500$, $k_2 = 1$, $k_3 = 50$, and the poles have uncertain locations:

$$p_1 = [345,\ 355]\ ;\ p_2 = [49,\ 51].$$

There is an input-multiplicative uncertainty for the system, with

$$|\Delta_i|_\infty \leq 0.25.$$

The specifications for this control system follow: The magnitude of the steady-state motor shaft angle should be less than 0.05, and the magnitude of the steady-state control input should be less than 15. Assume that the disturbance torque is bounded $|w(t)| \leq 1$ and lowpass, with a bandwidth of 0.1 rad/sec. Further, the measurement error is bounded $|v(t)| \leq 0.1$ and highpass with a cutoff frequency of 10 rad/sec.

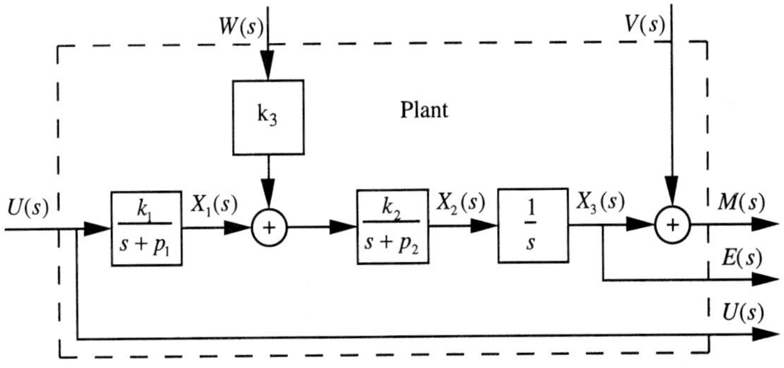

FIGURE P11.2 Block diagram of a dc motor

A control system has been designed for this plant:

$$\dot{x}_c(t) = \mathbf{A}_c x_c(t) + \mathbf{B}_c m(t) \ ; \ u(t) = \mathbf{C}_c x_c(t),$$

where

$$\mathbf{A}_c = \begin{bmatrix} -942 & 680 & -230 & 910 & -320 \\ -326 & 231 & -78 & 328 & -126 \\ -835 & 610 & -208 & 787 & -261 \\ -1058 & 771 & -262 & 999 & -333 \\ -1092 & 776 & -262 & 1094 & -419 \end{bmatrix};$$

$$\mathbf{B}_c = [43 \quad 16 \quad 39 \quad 48 \quad 60]^T;$$

$$\mathbf{C}_c = [-229 \quad 168 \quad -60 \quad 230 \quad -85].$$

Compute and plot the structured singular value of the weighted closed-loop system to evaluate robust performance.

Generate reduced-order controllers with orders of one, two, and three via the balance truncation method. Compare the frequency response of these controllers with that of the original controller design, and compute the ∞-norm of the approximation errors. Compute and plot the structured singular value of the weighted closed-loop system to evaluate robust performance for the nominal and reduced-order controllers. Comment on the results.

APPENDIX
Mathematical Notes

A number of definitions, theorems, and formulas from various mathematical fields have been employed in deriving expressions in this book. This appendix summarizes these results. The proofs of most theorems are included for the interested reader.

A1 Calculus of Vectors and Matrices

The differentiation and integration of time functions involving vectors and matrices arises in solving state equations, optimization, and so on. This section summarizes the basic definitions of differentiation and integration on vectors and matrices. A number of formulas for the derivative of vector-matrix products are also included.

A1.1 Calculus of Vector-Matrix Functions of a Scalar

The derivative of a matrix (or, as a special case, a vector) function of a scalar is the matrix of the derivatives of each element in the matrix:

$$\frac{d\mathbf{M}(t)}{dt} = \begin{bmatrix} \frac{dM_{11}(t)}{dt} & \cdots & \frac{dM_{1n}(t)}{dt} \\ \vdots & \ddots & \vdots \\ \frac{dM_{m1}(t)}{dt} & \cdots & \frac{dM_{mn}(t)}{dt} \end{bmatrix}. \tag{A1.1}$$

The integral of a matrix (or, as a special case, a vector) function of a scalar is the matrix of the integrals of each element in the matrix:

$$\int_a^b \mathbf{M}(t)dt = \begin{bmatrix} \int_a^b M_{11}(t)dt & \cdots & \int_a^b M_{1n}(t)dt \\ \vdots & \ddots & \vdots \\ \int_a^b M_{m1}(t)dt & \cdots & \int_a^b M_{mn}(t)dt \end{bmatrix}. \tag{A1.2}$$

The Laplace transform of a matrix (or, as a special case, a vector) function of time is the matrix of the Laplace transforms of each element in the matrix:

$$\int_0^\infty \mathbf{m}(t)e^{-st}dt = \begin{bmatrix} \int_0^\infty m_{11}(t)e^{-st}dt & \cdots & \int_0^\infty m_{1n}(t)e^{-st}dt \\ \vdots & \ddots & \vdots \\ \int_0^\infty m_{m1}(t)e^{-st}dt & \cdots & \int_0^\infty m_{mn}(t)e^{-st}dt \end{bmatrix}. \quad \text{(A1.3)}$$

THEOREM: The scalar derivative of the product of two matrix time functions is

$$\frac{d\mathbf{A}(t)\mathbf{B}(t)}{dt} = \frac{d\mathbf{A}(t)}{dt}\mathbf{B}(t) + \mathbf{A}(t)\frac{d\mathbf{B}(t)}{dt}. \quad \text{(A1.4)}$$

PROOF: This theorem is analogous to the derivative of a product of two scalar functions of a scalar, except caution must be used in preserving the order of the product. The proof of this theorem is analogous to the case of scalar functions and is omitted. ◆

THEOREM: The scalar derivative of the inverse of a matrix time function is

$$\frac{d\mathbf{A}^{-1}(t)}{dt} = -\mathbf{A}^{-1}(t)\frac{d\mathbf{A}(t)}{dt}\mathbf{A}^{-1}(t). \quad \text{(A1.5)}$$

This theorem is similar to the derivative of the inverse of a scalar time function, except the square of the inverse is divided into two terms that left- and right-multiply the matrix derivative. Additionally, caution must be used in preserving the order of the products.

PROOF: This theorem can be proved by noting that

$$\mathbf{0} = \frac{d\mathbf{I}}{dt} = \frac{d\mathbf{A}^{-1}(t)\mathbf{A}(t)}{dt} = \frac{d\mathbf{A}^{-1}(t)}{dt}\mathbf{A}(t) + \mathbf{A}^{-1}(t)\frac{d\mathbf{A}(t)}{dt}.$$

Solving for the derivative of $\mathbf{A}^{-1}(t)$ in this expression yields (A1.5). ◆

A1.2 Derivatives of Vector-Matrix Products

A vector function of a vector is given:

$$v(u) = \begin{bmatrix} v_1(u) \\ \vdots \\ v_n(u) \end{bmatrix}, \quad \text{(A1.6)}$$

where $v_i(u)$ is a function of the n_u dimensional vector u. The derivative of a vector function of a vector (the Jacobian) is defined as follows:

$$\frac{\partial v(u)}{\partial u} = \begin{bmatrix} \frac{\partial v_1(u)}{\partial u_1} & \cdots & \frac{\partial v_1(u)}{\partial u_m} \\ \vdots & \ddots & \vdots \\ \frac{\partial v_n(u)}{\partial u_1} & \cdots & \frac{\partial v_n(u)}{\partial u_m} \end{bmatrix}. \quad \text{(A1.7)}$$

Note that the Jacobian is sometimes defined as the transpose of the matrix in (A1.7).

THEOREM:
$$\frac{\partial \mathbf{R}u}{\partial u} = \mathbf{R}. \tag{A1.8}$$

PROOF: Consider the second-order example:

$$v(u) = \mathbf{R}u = \begin{bmatrix} r_{11} & r_{12} \\ r_{21} & r_{22} \end{bmatrix} \begin{bmatrix} u_1 \\ u_2 \end{bmatrix} = \begin{bmatrix} r_{11}u_1 + r_{12}u_2 \\ r_{21}u_1 + r_{22}u_2 \end{bmatrix}.$$

The derivative of the above vector-matrix product is

$$\frac{\partial \mathbf{R}u}{\partial u} = \frac{\partial v(u)}{\partial u} = \begin{bmatrix} \frac{\partial v_1(u)}{\partial u_1} & \frac{\partial v_1(u)}{\partial u_2} \\ \frac{\partial v_2(u)}{\partial u_1} & \frac{\partial v_2(u)}{\partial u_2} \end{bmatrix} = \begin{bmatrix} r_{11} & r_{12} \\ r_{21} & r_{22} \end{bmatrix} = \mathbf{R}.$$

This result can be easily generalized to vectors and matrices of arbitrary dimension. ◆

The derivative of a scalar function of a vector (the gradient) is a special case of (A1.7) and is defined as follows:

$$\frac{\partial J(u)}{\partial u} = \begin{bmatrix} \frac{\partial J(u)}{\partial u_1} \cdots \frac{\partial J(u)}{\partial u_{n_u}} \end{bmatrix}. \tag{A1.9}$$

The gradient is often defined as the transpose of (A1.9) in the literature.

THEOREM: For $\mathbf{R}$ symmetric,

$$\frac{\partial u^T \mathbf{R} u}{\partial u} = 2u^T \mathbf{R}. \tag{A1.10}$$

PROOF: Consider the second-order example:

$$J(u) = u^T \mathbf{R} u = [u_1 \quad u_2] \begin{bmatrix} r_{11} & r_{12} \\ r_{12} & r_{22} \end{bmatrix} \begin{bmatrix} u_1 \\ u_2 \end{bmatrix} = r_{11}u_1^2 + 2r_{12}u_1u_2 + r_{22}u_2^2.$$

Taking the gradient of $J(u)$, we have

$$\begin{aligned} \frac{\partial u^T \mathbf{R} u}{\partial u} = \frac{\partial J(u)}{\partial u} &= \begin{bmatrix} \frac{\partial J(u)}{\partial u_1} & \frac{\partial J(u)}{\partial u_2} \end{bmatrix} = [2r_{11}u_1 + 2r_{12}u_2 \quad 2r_{12}u_1 + 2r_{22}u_2] \\ &= 2u^T \mathbf{R}. \end{aligned}$$

This result can be easily generalized to vectors and matrices of arbitrary dimension. ◆

A2 Useful Relations from Linear Algebra

The following are useful definitions and theorems from linear algebra. The proofs of the theorems are also given.

A2.1 Positive Definite and Positive Semidefinite Matrices

A matrix $\mathbf{P}$ is positive definite if $\mathbf{P}$ is real, symmetric, and

$$x^T \mathbf{P} x > 0 \text{ for all } x \neq 0, \tag{A2.1}$$

or, equivalently, if all the eigenvalues of **P** have positive real parts.[1] A matrix **S** is positive semidefinite if **S** is real, symmetric, and

$$x^T\mathbf{S}x \geq 0 \text{ for all } x \neq 0, \tag{A2.2}$$

or, equivalently, if all the eigenvalues of **S** have nonnegative real parts.

A2.2 Relations Involving the Trace

THEOREM: The trace is invariant under cyclic perturbations:

$$\operatorname{tr}(\mathbf{AB}) = \operatorname{tr}(\mathbf{BA}), \tag{A2.3a}$$

where **AB** is square, and therefore the trace is defined.

PROOF: Let $\mathbf{A} \in \mathscr{C}^{n\times l}$ and $\mathbf{B} \in \mathscr{C}^{l\times n}$; then $\mathbf{AB} \in \mathscr{C}^{n\times n}$ and $\mathbf{BA} \in \mathscr{C}^{l\times l}$. Expanding the matrix multiplication and noting that the trace is the sum of the diagonal elements of the matrix, we have

$$\operatorname{tr}(\mathbf{AB}) = \sum_{k=1}^{n}\sum_{i=1}^{l} a_{ki}b_{ik} = \sum_{i=1}^{l}\sum_{k=1}^{n} b_{ik}a_{ki} = \operatorname{tr}(\mathbf{BA}).$$ ◆

COROLLARY: Successive applications of the above theorem yield

$$\operatorname{tr}(\mathbf{ABC}) = \operatorname{tr}(\mathbf{CAB}) = \operatorname{tr}(\mathbf{BCA}) \tag{A2.3b}$$

provided **ABC** is square.

THEOREM:

$$\operatorname{tr}(\mathbf{A}^T\mathbf{BA}) = \sum_{k=1}^{l} a_k^T\mathbf{B}a_k, \tag{A2.4}$$

where $\mathbf{A} \in \mathscr{R}^{n\times l}$, $\mathbf{B} \in \mathscr{R}^{n\times n}$, and $\{a_k\}$ are the columns of **A**.

PROOF:

$$\operatorname{tr}(\mathbf{A}^T\mathbf{BA}) = \operatorname{tr}\left(\begin{bmatrix} a_1^T \\ a_2^T \\ \vdots \\ a_m^T \end{bmatrix} \mathbf{B}[a_1 \quad a_2 \cdots a_m]\right)$$

$$= \operatorname{tr}\left(\begin{bmatrix} a_1^T\mathbf{B}a_1 & a_1^T\mathbf{B}a_2 & \cdots & a_1^T\mathbf{B}a_m \\ a_2^T\mathbf{B}a_1 & a_2^T\mathbf{B}a_2 & \cdots & a_2^T\mathbf{B}a_m \\ \vdots & \vdots & \ddots & \vdots \\ a_m^T\mathbf{B}a_1 & a_m^T\mathbf{B}a_2 & \cdots & a_m^T\mathbf{B}a_m \end{bmatrix}\right) = \sum_{k=1}^{n} a_k^T\mathbf{B}a_k.$$ ◆

A2.3 Determinants of Block Matrices

THEOREM:

$$\det\left(\left[\begin{array}{c|c} \mathbf{A} & \mathbf{B} \\ \hline \mathbf{0} & \mathbf{C} \end{array}\right]\right) = \det(\mathbf{A})\det(\mathbf{C}), \tag{A2.5}$$

where **A** and **C** are square.

---

[1]For complex matrices, the conjugate transpose is used and **P** is required to be Hermitian. The definition of positive semidefinite can be similarly modified. All positive and semipositive definite matrices encountered in this book are strictly real.

PROOF:
$$\det\left(\begin{bmatrix} \mathbf{A} & \mathbf{B} \\ \mathbf{0} & \mathbf{C} \end{bmatrix}\right) = \det\left(\begin{bmatrix} \mathbf{A} & \mathbf{0} \\ \mathbf{0} & \mathbf{I} \end{bmatrix}\begin{bmatrix} \mathbf{I} & \mathbf{0} \\ \mathbf{0} & \mathbf{C} \end{bmatrix}\begin{bmatrix} \mathbf{I} & \mathbf{A}^{-1}\mathbf{B} \\ \mathbf{0} & \mathbf{I} \end{bmatrix}\right)$$
$$= \det\left(\begin{bmatrix} \mathbf{A} & \mathbf{0} \\ \mathbf{0} & \mathbf{I} \end{bmatrix}\right)\det\left(\begin{bmatrix} \mathbf{I} & \mathbf{0} \\ \mathbf{0} & \mathbf{C} \end{bmatrix}\right)\det\left(\begin{bmatrix} \mathbf{I} & \mathbf{A}^{-1}\mathbf{B} \\ \mathbf{0} & \mathbf{I} \end{bmatrix}\right)$$
$$= \det(\mathbf{A})\det(\mathbf{C}).$$

Note that the existence of $\mathbf{A}^{-1}$ is required above. If $\mathbf{A}^{-1}$ does not exist, the $\det(\mathbf{A}) = 0$, and the theorem

$$\det\left(\begin{bmatrix} \mathbf{A} & \mathbf{B} \\ \mathbf{0} & \mathbf{C} \end{bmatrix}\right) = \det(\mathbf{A})\det(\mathbf{C}) = 0$$

is correct, since the columns of **A** are linearly dependent. ◆

THEOREM:
$$\det\left(\begin{bmatrix} \mathbf{A} & \mathbf{B} \\ \mathbf{C} & \mathbf{D} \end{bmatrix}\right) = \det(\mathbf{A})\det(\mathbf{D} - \mathbf{C}\mathbf{A}^{-1}\mathbf{B}) \qquad \text{(A2.6)}$$

if **A** is invertible.

PROOF: The determinant is invariant under row operations:

$$\det\left(\begin{bmatrix} \mathbf{A} & \mathbf{B} \\ \mathbf{C} & \mathbf{D} \end{bmatrix}\right) = \det\left(\begin{bmatrix} \mathbf{A} & \mathbf{B} \\ \mathbf{C} - \mathbf{C}\mathbf{A}^{-1}\mathbf{A} & \mathbf{D} - \mathbf{C}\mathbf{A}^{-1}\mathbf{B} \end{bmatrix}\right)$$
$$= \det\left(\begin{bmatrix} \mathbf{A} & \mathbf{B} \\ \mathbf{0} & \mathbf{D} - \mathbf{C}\mathbf{A}^{-1}\mathbf{B} \end{bmatrix}\right)$$
$$= \det(\mathbf{A})\det(\mathbf{D} - \mathbf{C}\mathbf{A}^{-1}\mathbf{B}).$$
◆

THEOREM:
$$\det\left(\begin{bmatrix} \mathbf{A} & \mathbf{B} \\ \mathbf{C} & \mathbf{D} \end{bmatrix}\right) = \det(\mathbf{D})\det(\mathbf{A} - \mathbf{B}\mathbf{D}^{-1}\mathbf{C}). \qquad \text{(A2.7)}$$

if **D** is invertible.

PROOF: The proof is analogous to the proof of the previous theorem. ◆

A2.4 The Matrix Inversion Lemma

THEOREM: Given that **A**, **C**, and $(\mathbf{A} + \mathbf{BCD})$ are nonsingular,

$$(\mathbf{A} + \mathbf{BCD})^{-1} = \mathbf{A}^{-1} - \mathbf{A}^{-1}\mathbf{B}(\mathbf{D}\mathbf{A}^{-1}\mathbf{B} + \mathbf{C}^{-1})^{-1}\mathbf{D}\mathbf{A}^{-1}. \qquad \text{(A2.8)}$$

PROOF: Taking the product of the matrix and its inverse, we have

$$(\mathbf{A} + \mathbf{BCD})\{\mathbf{A}^{-1} - \mathbf{A}^{-1}\mathbf{B}(\mathbf{D}\mathbf{A}^{-1}\mathbf{B} + \mathbf{C}^{-1})^{-1}\mathbf{D}\mathbf{A}^{-1}\}$$
$$= \mathbf{I} + \mathbf{BCD}\mathbf{A}^{-1} - \mathbf{B}(\mathbf{D}\mathbf{A}^{-1}\mathbf{B} + \mathbf{C}^{-1})^{-1}\mathbf{D}\mathbf{A}^{-1} - \mathbf{BCD}\mathbf{A}^{-1}\mathbf{B}(\mathbf{D}\mathbf{A}^{-1}\mathbf{B} + \mathbf{C}^{-1})^{-1}\mathbf{D}\mathbf{A}^{-1}$$
$$= \mathbf{I} + \mathbf{BCD}\mathbf{A}^{-1} - \mathbf{B}(\mathbf{I} + \mathbf{CD}\mathbf{A}^{-1}\mathbf{B})(\mathbf{D}\mathbf{A}^{-1}\mathbf{B} + \mathbf{C}^{-1})^{-1}\mathbf{D}\mathbf{A}^{-1}$$
$$= \mathbf{I} + \mathbf{BCD}\mathbf{A}^{-1} - \mathbf{BC}(\mathbf{C}^{-1} + \mathbf{D}\mathbf{A}^{-1}\mathbf{B})(\mathbf{D}\mathbf{A}^{-1}\mathbf{B} + \mathbf{C}^{-1})^{-1}\mathbf{D}\mathbf{A}^{-1} = \mathbf{I}.$$
◆

A2.5 Block Matrix Inversion

THEOREM: The inverse of a block matrix (with four blocks) is

$$\begin{bmatrix} \mathbf{A} & \mathbf{B} \\ \mathbf{C} & \mathbf{D} \end{bmatrix}^{-1} = \begin{bmatrix} \mathbf{A}^{-1} + \mathbf{A}^{-1}\mathbf{B}(\mathbf{D} - \mathbf{C}\mathbf{A}^{-1}\mathbf{B})^{-1}\mathbf{C}\mathbf{A}^{-1} & -\mathbf{A}^{-1}\mathbf{B}(\mathbf{D} - \mathbf{C}\mathbf{A}^{-1}\mathbf{B})^{-1} \\ -(\mathbf{D} - \mathbf{C}\mathbf{A}^{-1}\mathbf{B})^{-1}\mathbf{C}\mathbf{A}^{-1} & (\mathbf{D} - \mathbf{C}\mathbf{A}^{-1}\mathbf{B})^{-1} \end{bmatrix} \tag{A2.9}$$

if **A** is invertible.

PROOF: Taking the product of the matrix and the inverse, and doing a little algebra, we have

$$\begin{bmatrix} \mathbf{A}^{-1} + \mathbf{A}^{-1}\mathbf{B}(\mathbf{D} - \mathbf{C}\mathbf{A}^{-1}\mathbf{B})^{-1}\mathbf{C}\mathbf{A}^{-1} & -\mathbf{A}^{-1}\mathbf{B}(\mathbf{D} - \mathbf{C}\mathbf{A}^{-1}\mathbf{B})^{-1} \\ -(\mathbf{D} - \mathbf{C}\mathbf{A}^{-1}\mathbf{B})^{-1}\mathbf{C}\mathbf{A}^{-1} & (\mathbf{D} - \mathbf{C}\mathbf{A}^{-1}\mathbf{B})^{-1} \end{bmatrix}\begin{bmatrix} \mathbf{A} & \mathbf{B} \\ \mathbf{C} & \mathbf{D} \end{bmatrix} = \begin{bmatrix} \mathbf{I} & \mathbf{0} \\ \mathbf{0} & \mathbf{I} \end{bmatrix}.$$

◆

A3 The Singular Value Decomposition

The singular value decomposition (SVD) is a matrix factorization that has found a number of applications to engineering problems. The SVD of a matrix $\mathbf{M} \in \mathscr{C}^{r \times m}$ is

$$\mathbf{M} = \mathbf{U}\mathbf{S}\mathbf{V}^{\dagger} = \sum_{i=1}^{p} \sigma_i U_i V_i^{\dagger}, \tag{A3.1}$$

where $\mathbf{U} \in \mathscr{C}^{n_y \times n_y}$, and $\mathbf{V} \in \mathscr{C}^{n_u \times n_u}$ are unitary matrices ($\mathbf{U}^{\dagger}\mathbf{U} = \mathbf{I}$, and $\mathbf{V}^{\dagger}\mathbf{V} = \mathbf{I}$); $\mathbf{S} \in \mathscr{R}^{n_y \times n_u}$ is diagonal (but not necessarily square); and p equals the minimum of n_y and n_u. The singular values $\{\sigma_1, \sigma_2, \cdots, \sigma_{n_u}\}$ of **M** are defined as the positive square roots of the diagonal elements of $\mathbf{S}^T\mathbf{S}$, and are ordered from largest to smallest.

The singular value decomposition can be used to provide bounds on the gain of a matrix. These bounds are demonstrated after first presenting a theorem concerning unitary matrices.

THEOREM: If **U** is a unitary matrix ($\mathbf{U}^{\dagger}\mathbf{U} = \mathbf{I}$), then the transformation **U** preserves length

$$\|\mathbf{U}x\| = \|x\|. \tag{A3.2}$$

PROOF: $\|\mathbf{U}x\| = \sqrt{(\mathbf{U}x)^{\dagger}(\mathbf{U}x)} = \sqrt{x^{\dagger}\mathbf{U}^{\dagger}\mathbf{U}x} = \sqrt{x^{\dagger}x} = \|x\|.$ ◆

THEOREM: The maximum gain of a matrix is given by the maximum singular value σ_1:

$$\max_{\|x\|=1} \|\mathbf{M}x\| = \sigma_1. \tag{A3.3}$$

PROOF: The norm of $\mathbf{M}x$ can be written

$$\|\mathbf{M}x\| = \sqrt{x^{\dagger}\mathbf{M}^{\dagger}\mathbf{M}x} = \sqrt{x^{\dagger}\mathbf{V}\mathbf{S}^T\mathbf{U}^{\dagger}\mathbf{U}\mathbf{S}\mathbf{V}^{\dagger}x} = \sqrt{x^{\dagger}\mathbf{V}\mathbf{S}^T\mathbf{S}\mathbf{V}^{\dagger}x}, \tag{A3.4}$$

where the ordinary transpose is used for $\mathbf{S}$, since this matrix is real. The maximum of the norm (A3.4) is

$$\max_{\|x\|=1} \|\mathbf{M}x\| = \max_{\|x\|=1} \sqrt{x^\dagger \mathbf{V}\mathbf{S}^T\mathbf{S}\mathbf{V}^\dagger x} = \max_{\|\tilde{x}\|=1} \sqrt{\tilde{x}^\dagger \mathbf{S}^T\mathbf{S}\tilde{x}}.$$

Note that maximization over $\tilde{x} = \mathbf{V}x$ is equivalent to maximizing over x since $\mathbf{V}$ is invertible and preserves the norm (1 in this case). Expanding the norm yields

$$\max_{\|x\|=1} \|\mathbf{M}x\| = \max_{\|\tilde{x}\|=1} \sqrt{\tilde{x}^\dagger \mathbf{S}^T\mathbf{S}\tilde{x}} = \max_{\|\tilde{x}\|=1} \sqrt{\sigma_1^2|\tilde{x}_1|^2 + \sigma_2^2|\tilde{x}_2|^2 + \cdots + \sigma_{n_u}^2|\tilde{x}_{n_u}|^2}. \tag{A3.5}$$

This expression is maximized, given the constraint $\|\tilde{x}\| = 1$, when $\tilde{x}$ is concentrated at the largest singular value; that is $|\tilde{x}| = [1 \quad 0 \quad \cdots \quad 0]^T$. The maximum gain is then

$$\max_{\|x\|=1} \|\mathbf{M}x\| = \sqrt{\sigma_1^2|1|^2 + \sigma_2^2|0|^2 + \cdots + \sigma_{n_u}^2|0|^2} = \sigma_1 = \bar{\sigma}.$$

◆

THEOREM: The minimum gain of a matrix is given the smallest singular value:

$$\min_{\|x\|=1} \|\mathbf{M}x\| = \sigma_{n_u} = \underline{\sigma} = \begin{cases} \sigma_p & n_y \geq n_u \\ 0 & n_y < n_u \end{cases}. \tag{A3.6}$$

PROOF: Substituting the minimum operator for the maximum operator in (A3.5), the minimum gain is seen to occur when $\tilde{x}$ is concentrated at the smallest singular value; that is, $|\tilde{x}| = [0 \quad \cdots \quad 0 \quad 1]^T$. The minimum gain in this case is

$$\begin{aligned} \min_{\|x\|=1} \|\mathbf{M}x\| &= \sqrt{\sigma_1^2|0|^2 + \cdots + \sigma_{n_u-1}^2|0|^2 + \sigma_{n_u}^2|1|^2} \\ &= \sigma_{n_u} = \underline{\sigma} = \begin{cases} \sigma_p & n_y \geq n_u \\ 0 & n_y < n_u \end{cases}. \end{aligned}$$

◆

THEOREM: The maximum singular value of a square matrix's inverse equals the inverse of the matrix's minimum singular value:

$$\bar{\sigma}(\mathbf{M}^{-1}) = \frac{1}{\underline{\sigma}(\mathbf{M})}. \tag{A3.7}$$

PROOF: The SVD of $\mathbf{M}^{-1}$ is given (provided all inverses exist):

$$\mathbf{M}^{-1} = (\mathbf{U}\mathbf{S}^{-1}\mathbf{V}^\dagger)^{-1} = \mathbf{V}\mathbf{S}^{-1}\mathbf{U}^\dagger = \sum_{i=1}^{p} \sigma_i^{-1} V_i U_i^\dagger,$$

where σ_i, U_i, and V_i are the singular values, left singular vectors, and right singular vectors of $\mathbf{M}$, respectively. Note that the inverse of a unitary matrix exists and equals the conjugate transpose of the matrix. The matrix $\mathbf{S}$ is a diagonal matrix with no zeros on the diagonal, since $\mathbf{M}$ is full rank. Therefore, the inverse of $\mathbf{S}$ exists and is a diagonal matrix with the inverses of the singular values on the diagonal. The maximum singular value of $\mathbf{M}^{-1}$ is then $1/\underline{\sigma}$.

◆

A4 Spectral Theory of Matrices

The eigenvalues of a function of a matrix are simply related to the eigenvalues of the original matrix.

THEOREM: Let **B** be formed as a polynomial of a square matrix:

$$\mathbf{B} = \sum_{k=0}^{m} \alpha_k \mathbf{A}^k, \tag{A4.1a}$$

where $\mathbf{A} \in \mathscr{C}^{n\times n}$ and $\mathbf{A}^0 = \mathbf{I}$. Let λ be an eigenvalue of **A**. Then

$$\sum_{k=0}^{m} \alpha_k \lambda^k \tag{A4.1b}$$

is an eigenvalue of **B**.

PROOF: Let λ and ϕ be an eigenvalue/eigenvector pair of **A**. Then λ^k and ϕ are an eigenvalue/eigenvector pair of $\mathbf{A}^k$:

$$\mathbf{A}^k\phi = \mathbf{A}^{k-1}\lambda\phi = \mathbf{A}^{k-2}\lambda^2\phi = \cdots = \lambda^k\phi.$$

Using this result,

$$\mathbf{B}\phi = \sum_{k=0}^{m} \alpha_k \mathbf{A}^k\phi = \left(\sum_{k=0}^{m} \alpha_k \lambda^k\right)\phi.$$

◆

THEOREM: Let **A** be a square matrix with an eigenvalue λ; then

$$e^{\lambda t} \text{ is an eigenvalue of } e^{\mathbf{A}t}. \tag{A4.2}$$

PROOF: Let λ and ϕ be an eigenvalue/eigenvector pair of **A**; then

$$\begin{aligned} e^{\mathbf{A}t}\phi &= \mathbf{I}\phi + \mathbf{A}\phi t + \mathbf{A}^2\phi\frac{t^2}{2!} + \cdots = 1\phi + \lambda\phi t + \lambda^2\phi\frac{t^2}{2!} + \cdots \\ &= \left(1 + \lambda t + \lambda^2\frac{t^2}{2!} + \cdots\right)\phi = e^{\lambda t}\phi. \end{aligned}$$

◆

THEOREM: The square matrices **M** and $(\mathbf{I} + \mathbf{M})^{-1}$ commute, given that the inverse exists:

$$\mathbf{M}(\mathbf{I} + \mathbf{M})^{-1} = (\mathbf{I} + \mathbf{M})^{-1}\mathbf{M}. \tag{A4.3}$$

PROOF: Two matrices commute if they have the same eigenvectors. Therefore, it is sufficient to show that the two matrices **M** and $(\mathbf{I} + \mathbf{M})^{-1}$ have the same eigenvectors. To show this, let λ and ϕ be any eigenvalue/eigenvector pair of **M**. By the previous theorem, $(1 + \lambda)$ and ϕ are an eigenvalue/eigenvector pair of $(\mathbf{I} + \mathbf{M})$. Further,

$$\begin{aligned} \phi &= (\mathbf{I} + \mathbf{M})^{-1}(\mathbf{I} + \mathbf{M})\phi = (\mathbf{I} + \mathbf{M})^{-1}(1 + \lambda)\phi \\ \Rightarrow \frac{1}{(1 + \lambda)}\phi &= (\mathbf{I} + \mathbf{M})^{-1}\phi. \end{aligned}$$

◆

A5 $\mathscr{L}_2$ Stability

An alternative definition of input-output stability can be given in terms of the system gain, where the gain is defined in terms of the signal 2-norm.

DEFINITION: A system is $\mathscr{L}_2$ stable if for all inputs with finite signal 2-norm, the system gain is bounded:

$$\frac{\|y(t)\|_2}{\|u(t)\|_2} < M.$$

THEOREM: A causal, linear time-invariant system is $\mathscr{L}_2$ stable if and only if all of its poles have negative real parts.[2]

PROOF: The proof begins by showing that the system is $\mathscr{L}_2$ stable if all of the poles have negative real parts. The 2-norm of the output can be written in terms of the impulse response by using (2.8):

$$\begin{aligned}\|y(t)\|_2^2 &= \int_{-\infty}^{\infty}\left\{\int_{-\infty}^{\infty}\mathbf{g}(\tau_1)u(t-\tau_1)d\tau_1\right\}^T\left\{\int_{-\infty}^{\infty}\mathbf{g}(\tau_2)u(t-\tau_2)d\tau_2\right\}dt\\ &= \mathrm{tr}\left\{\int_{-\infty}^{\infty}\int_{-\infty}^{\infty}\int_{-\infty}^{\infty}\mathbf{g}^T(\tau_1)\mathbf{g}(\tau_2)u(t-\tau_2)u^T(t-\tau_1)d\tau_2 d\tau_1 dt\right\}.\end{aligned}$$

The fact that the trace is invariant under cyclic permutations (A2.3) has been used in generating the expression containing the triple integral. An integral is bounded from above by the integral of the absolute value:

$$\|y(t)\|_2^2 \le \mathrm{tr}\left\{\int_{-\infty}^{\infty}\int_{-\infty}^{\infty}\int_{-\infty}^{\infty}|\mathbf{g}^T(\tau_1)||\mathbf{g}(\tau_2)||u(t-\tau_2)||u^T(t-\tau_1)|d\tau_2 d\tau_1 dt\right\}.$$

Interchanging the order of integration yields

$$\|y(t)\|_2^2 \le \mathrm{tr}\left\{\int_{-\infty}^{\infty}|\mathbf{g}^T(\tau_1)|\int_{-\infty}^{\infty}|\mathbf{g}(\tau_2)|\left\{\int_{-\infty}^{\infty}|u(t-\tau_2)||u^T(t-\tau_1)|dt\right\}d\tau_2 d\tau_1\right\}.$$

The final integral is a correlation that is maximized when $\tau_1 = \tau_2$, implying that

$$\|y(t)\|_2^2 \le \mathrm{tr}\left\{\int_{-\infty}^{\infty}|\mathbf{g}^T(\tau_1)|d\tau_1\int_{-\infty}^{\infty}|\mathbf{g}(\tau_2)|d\tau_2\left\{\int_{-\infty}^{\infty}|u(t)||u^T(t)|dt\right\}\right\}.$$

[2] This theorem is only strictly correct for causal, linear, time-invariant systems described by a state model. An additional requirement that the transfer function be proper (the order of the numerator be less than or equal to the order of the denominator) must be added to this test for general, linear, time-invariant systems described by transfer functions. Note that state models always result in transfer function models that are proper.

Given that the system is BIBO stable (i.e., all poles have negative real parts), every element in the impulse response is absolutely integrable:

$$\begin{aligned}\|y(t)\|_2^2 &\le \operatorname{tr}\left\{M^T M \int_{-\infty}^{\infty} |u(t)\|u^T(t)|dt\right\} = \int_{-\infty}^{\infty} |u^T(t)|M^T M|u(t)|dt \\ &\le \int_{-\infty}^{\infty} |u^T(t)|\bar{\sigma}(M^T M)|u(t)|dt = \bar{\sigma}(M^T M)\int_{-\infty}^{\infty} u^T(t)u(t)dt \\ &= \bar{\sigma}(M^T M)\|u(t)\|_2^2,\end{aligned} \tag{A5.1}$$

where

$$\mathbf{M} = \int_{-\infty}^{\infty} |\mathbf{g}(\tau)|d\tau.$$

Taking the square root of (A5.1), we have

$$\|y(t)\|_2 \le \sqrt{\bar{\sigma}(\mathbf{M}^T\mathbf{M})}\|u(t)\|_2,$$

which shows the system is $\mathcal{L}_2$ stable when all poles have negative real parts.

It remains to show that $\mathcal{L}_2$ stability implies that all poles have negative real parts. Let

$$u(t) = u_0\delta(t).$$

Then

$$\|y(t)\|_2^2 = \int_0^{\infty} u_0^T \mathbf{g}^T(t)\mathbf{g}(t)u_0 dt.$$

If the system has an unstable pole, then the impulse response contains an unstable mode (it is assumed that there are no pole-zero cancellations), and some element of $\mathbf{g}(t)$ does not approach zero as time goes to infinity. Therefore, the above integral is infinite for some u_0. ◆

A6 Change of Basis (Time-Varying Transformations)

A change of variables can be performed using a time-varying transformation matrix. In this case, an additional term involving the derivative of the transformation matrix appears in the resulting state model.

Given the state model

$$\dot{x}(t) = \mathbf{A}x(t) + \mathbf{B}u(t); \tag{A6.1a}$$

$$y(t) = \mathbf{C}x(t) + \mathbf{D}u(t), \tag{A6.1b}$$

a new state vector can be defined:

$$\tilde{x}(t) = \mathbf{T}^{-1}(t)x(t),$$

where $\mathbf{T}(t)$ is a time-varying, invertible transformation matrix. The fact that $\mathbf{T}(t)$ is invertible means that $x(t)$ can be computed from $\tilde{x}(t)$:

$$x(t) = \mathbf{T}(t)\tilde{x}(t).$$

Further, the derivatives of the two state vectors are related:

$$\dot{x}(t) = \mathbf{T}\dot{\tilde{x}}(t) + \dot{\mathbf{T}}\tilde{x}(t).$$

Making these substitutions in the state model (A6.1) yields

$$\mathbf{T}(t)\dot{\tilde{x}}(t) + \dot{\mathbf{T}}(t)\tilde{x}(t) = \mathbf{AT}(t)\tilde{x}(t) + \mathbf{B}u(t);$$

$$y(t) = \mathbf{CT}(t)\tilde{x}(t) + \mathbf{D}u(t).$$

Multiplying this state equation by $\mathbf{T}^{-1}(t)$ and rearranging yields a state model in terms of the new state $\tilde{x}(t)$:

$$\dot{\tilde{x}}(t) = [\mathbf{T}^{-1}(t)\mathbf{AT}(t) - \mathbf{T}^{-1}(t)\dot{\mathbf{T}}(t)]\tilde{x}(t) + \mathbf{T}^{-1}(t)\mathbf{B}u(t); \quad \text{(A6.2a)}$$

$$y(t) = \mathbf{CT}(t)\tilde{x}(t) + \mathbf{D}u(t). \quad \text{(A6.2b)}$$

THEOREM: Let $\mathbf{\Phi}(t)$ be the state transition matrix of the system (A6.1). Then the state transition matrix of the system (A6.2) is

$$\tilde{\mathbf{\Phi}}(t) = \mathbf{T}^{-1}(t)\mathbf{\Phi}(t). \quad \text{(A6.3)}$$

PROOF: Differentiating (A6.3) yields

$$\begin{aligned}\dot{\tilde{\mathbf{\Phi}}}(t) &= \mathbf{T}^{-1}(t)\dot{\mathbf{\Phi}}(t) - \mathbf{T}^{-1}(t)\dot{\mathbf{T}}(t)\mathbf{T}^{-1}(t)\mathbf{\Phi}(t)\\ &= \mathbf{T}^{-1}(t)\mathbf{A\Phi}(t) - \mathbf{T}^{-1}(t)\dot{\mathbf{T}}(t)\mathbf{T}^{-1}(t)\mathbf{\Phi}(t)\\ &= [\mathbf{T}^{-1}(t)\mathbf{AT}(t) - \mathbf{T}^{-1}(t)\dot{\mathbf{T}}(t)]\mathbf{T}^{-1}(t)\mathbf{\Phi}(t) = [\mathbf{T}^{-1}(t)\mathbf{AT}(t) - \mathbf{T}^{-1}(t)\dot{\mathbf{T}}(t)]\tilde{\mathbf{\Phi}}(t).\end{aligned}$$

◆

A7 Controllability and Observability Grammians

Given the stable system

$$\dot{x}(t) = \mathbf{A}x(t) + \mathbf{B}u(t) \; ; \; y(t) = \mathbf{C}x(t), \quad \text{(A7.1)}$$

the controllability grammian is

$$\mathbf{L}_c = \int_0^\infty e^{\mathbf{A}t}\mathbf{BB}^T e^{\mathbf{A}^T t} dt. \quad \text{(A7.2)}$$

THEOREM: The controllability grammian can be computed by solving the following Lyapunov equation:

$$\mathbf{AL}_c + \mathbf{L}_c\mathbf{A}^T = -\mathbf{BB}^T. \quad \text{(A7.3)}$$

PROOF: The following proof follows that presented by Dailey [1]. Differentiating the integrand yields

$$\frac{de^{\mathbf{A}t}\mathbf{BB}^T e^{\mathbf{A}^T t}}{dt} = \mathbf{A}e^{\mathbf{A}t}\mathbf{BB}^T e^{\mathbf{A}^T t} + e^{\mathbf{A}t}\mathbf{BB}^T e^{\mathbf{A}^T t}\mathbf{A}^T.$$

Integrating both sides of this equation yields

$$\begin{aligned}\int_0^\infty \frac{de^{\mathbf{A}\tau}\mathbf{BB}^T e^{\mathbf{A}^T \tau}}{d\tau} d\tau &= \mathbf{A}\int_0^\infty e^{\mathbf{A}t}\mathbf{BB}^T e^{\mathbf{A}^T \tau} d\tau + \int_0^\infty e^{\mathbf{A}\tau}\mathbf{BB}^T e^{\mathbf{A}^T \tau} d\tau \mathbf{A}^T\\ &= \mathbf{AL}_c + \mathbf{L}_c\mathbf{A}^T;\\ e^{\mathbf{A}\tau}\mathbf{BB}^T e^{\mathbf{A}^T \tau}\Big|_{\tau=0}^{\infty} &= \mathbf{AL}_c + \mathbf{L}_c\mathbf{A}^T.\end{aligned}$$

The expression on the left equals zero at the upper limit, since (A7.1) is stable. Evaluating at the lower limit then yields

$$-\mathbf{B}\mathbf{B}^T = \mathbf{A}\mathbf{L}_c + \mathbf{L}_c\mathbf{A}^T.$$

◆

The observability grammian of the system (A7.1) is defined as

$$\mathbf{L}_o = \int_0^\infty e^{\mathbf{A}^T t}\mathbf{C}^T\mathbf{C}e^{\mathbf{A}t}dt. \tag{A7.4}$$

THEOREM: This observability grammian can be computed by solving the following Lyapunov equation:

$$\mathbf{A}^T\mathbf{L}_o + \mathbf{L}_o\mathbf{A} = -\mathbf{C}^T\mathbf{C}. \tag{A7.5}$$

PROOF: The proof is similar to that presented for the controllability grammian and is therefore omitted. ◆

THEOREM: If there exists a solution to (A7.3) such that $\mathbf{L}_c$ is positive semidefinite, then

$$\lim_{t\to\infty} e^{\mathbf{A}t}\mathbf{B} = \mathbf{0}. \tag{A7.6}$$

PROOF: The positive semidefinite solution of (A7.3) is given by (A7.2), which can be written

$$\mathbf{L}_c = \int_0^\infty \|e^{\mathbf{A}t}\mathbf{B}\|_E^2 dt.$$

The existence of this integral requires that the integrand approach zero as time goes to infinity, which implies (A7.6). ◆

THEOREM: If there exists a solution to (A7.5) such that $\mathbf{L}_o$ is positive semidefinite, then

$$\lim_{t\to\infty} \mathbf{C}e^{\mathbf{A}t} = \mathbf{0}. \tag{A7.7}$$

PROOF: The proof is similar to that presented for (A7.6) and is therefore omitted. ◆

THEOREM: We are given a system with the state matrix $\mathbf{A}$ and the output matrix $\mathbf{C}$. If the system is observable and

$$\lim_{t\to\infty} \mathbf{C}e^{\mathbf{A}t} = \mathbf{0}, \tag{A7.8}$$

then the system is stable.

PROOF: Assume that $\mathbf{A}$ has an eigenvalue λ with a non-negative real part, and an associated eigenvector ϕ:

$$\mathbf{A}\phi = \lambda\phi.$$

The spectral theory of matrices, Appendix equation (A4.2), states that

$$e^{\mathbf{A}t}\phi = e^{\lambda t}\phi.$$

Premultiplying by $\mathbf{C}$ and taking the limit as time approaches infinity yields

$$0 = \lim_{t\to\infty} (\mathbf{C}e^{\mathbf{A}t})\phi = \lim_{t\to\infty} e^{\lambda t}(\mathbf{C}\phi).$$

This implies that $\mathbf{C}\phi = 0$, since

$$\lim_{t\to\infty} e^{\lambda t} \neq 0.$$

If $\mathbf{C}\phi = \mathbf{0}$, the homogeneous (no input) system response with the initial condition $x(0) = \phi$ is

$$y(t) = \mathbf{C}e^{\mathbf{A}t}\phi = e^{\lambda t}(\mathbf{C}\phi) = 0.$$

Further, the homogeneous system response with the initial condition $x(0) = 2\phi$ is identical to the above response:

$$y(t) = \mathbf{C}e^{\mathbf{A}t}2\phi = 2e^{\lambda t}(\mathbf{C}\phi) = 0.$$

Therefore, the state cannot be found from a record of the output, and the system unobservable. This contradiction indicates that $\mathbf{A}$ has no eigenvalues with non-negative real parts; that is, the system is stable. ◆

A8 Useful Relations Involving (I + GK) and (I + KG)

The following are useful theorems involving the matrices $(\mathbf{I} + \mathbf{GK})$ and $(\mathbf{I} + \mathbf{KG})$. The proofs of these theorems are also given.

A8.1 Equivalence of the Determinants

THEOREM:
$$\det(\mathbf{I} + \mathbf{GK}) = \det(\mathbf{I} + \mathbf{KG}). \tag{A8.1}$$

PROOF: Let $\mathbf{G} \in \mathscr{C}^{n\times l}$, $\mathbf{K} \in \mathscr{C}^{l\times n}$, and without loss of generality $n \geq l$ (otherwise interchange $\mathbf{G}$ and $\mathbf{K}$). Now, consider the eigenequation of $\mathbf{KG}$:

$$\mathbf{KG}\phi_i = \lambda_i\phi_i;$$

$$\Rightarrow \qquad \mathbf{GK}(\mathbf{G}\phi_i) = \mathbf{G}\lambda_i\phi_i = \lambda_i(\mathbf{G}\phi_i).$$

Defining $\psi_i = \mathbf{G}\phi_i$,

$$\Rightarrow \qquad \mathbf{GK}\psi_i = \lambda_i\psi_i.$$

Therefore, any eigenvalue of $\mathbf{KG}$ is also an eigenvalue of $\mathbf{GK}$ (note that the eigenvectors are different). $\mathbf{GK}$ also has $(n - l)$ additional zero eigenvalues. Now, the eigenvalues of $(\mathbf{I} + \mathbf{KG})$ equal the eigenvalues of $(\mathbf{KG})$ plus 1:

$$(\mathbf{I} + \mathbf{KG})\phi_i = \phi_i + \lambda_i\phi_i = (1 + \lambda_i)\phi_i.$$

The $\det(\mathbf{I} + \mathbf{GK})$ is the product of the eigenvalues:

$$\det(\mathbf{I} + \mathbf{GK}) = \prod_{i=1}^{m}(1 + \lambda_i) \prod_{i=m+1}^{n} (1) = \prod_{i=1}^{m}(1 + \lambda_i) = \det(\mathbf{I} + \mathbf{KG}).$$ ◆

A8.2 The Push-Through Theorem

THEOREM: $$(\mathbf{I} + \mathbf{GK})^{-1}\mathbf{G} = \mathbf{G}(\mathbf{I} + \mathbf{KG})^{-1}. \qquad \text{(A8.2)}$$

PROOF: $$\mathbf{G} + \mathbf{GKG} = \mathbf{G} + \mathbf{GKG};$$

$$\Rightarrow \quad (\mathbf{I} + \mathbf{GK})\mathbf{G} = \mathbf{G}(\mathbf{I} + \mathbf{KG});$$

$$\Rightarrow \quad (\mathbf{I} + \mathbf{GK})^{-1}(\mathbf{I} + \mathbf{GK})\mathbf{G}(\mathbf{I} + \mathbf{KG})^{-1} = (\mathbf{I} + \mathbf{GK})^{-1}\mathbf{G}(\mathbf{I} + \mathbf{KG})(\mathbf{I} + \mathbf{KG})^{-1};$$

$$\Rightarrow \quad \mathbf{G}(\mathbf{I} + \mathbf{KG})^{-1} = (\mathbf{I} + \mathbf{GK})^{-1}\mathbf{G}.$$ ◆

A8.3 Miscellaneous

THEOREM: $$\mathbf{I} - \mathbf{GK}(\mathbf{I} + \mathbf{GK})^{-1} = (\mathbf{I} + \mathbf{GK})^{-1}. \qquad \text{(A8.3)}$$

PROOF:
$$\begin{aligned}\mathbf{I} - \mathbf{GK}(\mathbf{I} + \mathbf{GK})^{-1} &= (\mathbf{I} + \mathbf{GK})(\mathbf{I} + \mathbf{GK})^{-1} - \mathbf{GK}(\mathbf{I} + \mathbf{GK})^{-1}\\ &= (\mathbf{I} + \mathbf{GK})^{-1} + \mathbf{GK}(\mathbf{I} + \mathbf{GK})^{-1} - \mathbf{GK}(\mathbf{I} + \mathbf{GK})^{-1}\\ &= (\mathbf{I} + \mathbf{GK})^{-1}.\end{aligned}$$ ◆

A9 Properties of the System ∞-Norm

The system ∞-norm is defined in terms of the transfer function matrix of the system:

$$\|\mathbf{G}\|_\infty = \sup_\omega \bar{\sigma}(\mathbf{G}(j\omega)). \qquad \text{(A9.1)}$$

The ∞-norm represents the maximum gain of the system over all possible sinusoidal inputs.

THEOREM: The 2-norm of the output can be bounded by the product of the system ∞-norm and the 2-norm of the input:

$$\|y\|_2 = \|\mathbf{G}u\|_2 \le \|\mathbf{G}\|_\infty \|u\|_2. \qquad \text{(A9.2)}$$

PROOF: The signal 2-norm of the output is given as

$$\|y\|_2 = \sqrt{\frac{1}{2\pi}\int_{-\infty}^{\infty} y^\dagger(j\omega)y(j\omega)d\omega} = \sqrt{\frac{1}{2\pi}\int_{-\infty}^{\infty} \|y(j\omega)\|_E^2 d\omega},$$

where $\|\bullet\|_E$ is the Euclidean norm. The Euclidean norm of the output is bounded:

$$\|y(j\omega)\|_E^2 \le \bar{\sigma}(j\omega)\|u(j\omega)\|_E^2.$$

The signal 2-norm of the output can then be bounded:

$$\|y\|_2 \le \sqrt{\frac{1}{2\pi}\int_{-\infty}^{\infty} \bar{\sigma}(j\omega)\|u(j\omega)\|_E^2 d\omega}.$$

The maximum singular value at a specific frequency is bounded by the supremum over all frequencies:

$$\|y\|_2 \leq \sqrt{\frac{1}{2\pi}\int_{-\infty}^{\infty} \bar{\sigma}(j\omega)\|u(j\omega)\|_E^2 d\omega} \leq \sqrt{\frac{1}{2\pi}\int_{-\infty}^{\infty} \left\{\sup_{\omega} \bar{\sigma}(j\omega)\right\}\|u(j\omega)\|_E^2 d\omega}$$

$$= \sqrt{\frac{1}{2\pi}\int_{-\infty}^{\infty} \|\mathbf{G}\|_\infty^2\|u(j\omega)\|_E^2 d\omega} = \|\mathbf{G}\|_\infty\sqrt{\frac{1}{2\pi}\int_{-\infty}^{\infty} \|u(j\omega)\|_E^2 d\omega} = \|\mathbf{G}\|_\infty\|u\|_2.$$

◆

The bound of the preceding proof is nearly achieved for some input. In fact, the ∞-norm is given by the supremum:

THEOREM:

$$\|\mathbf{G}\|_\infty = \sup_{u\neq 0} \frac{\|\mathbf{G}u\|_2}{\|u\|_2}. \quad \text{(A9.3)}$$

PROOF: Using (A9.2),

$$\|\mathbf{G}\|_\infty \geq \frac{\|\mathbf{G}u\|_2}{\|u\|_2} \quad \forall u \neq 0,$$

which implies

$$\|\mathbf{G}\|_\infty \geq \sup_{u\neq 0} \frac{\|\mathbf{G}u\|_2}{\|u\|_2}.$$

The input direction and frequency that yield the maximums in the ∞-norm are defined as follows:

$$\|\mathbf{G}\|_\infty = \sup_{\omega} \bar{\sigma}(\mathbf{G}(j\omega)) = \bar{\sigma}(\mathbf{G}(j\omega_0)) = \max_{u(j\omega_0)} \frac{\mathbf{G}(j\omega_0)u(j\omega_0)}{u(j\omega_0)} = \frac{\mathbf{G}(j\omega_0)u_0}{u_0}$$

where u_0 is a unit vector. Note that u_0 is the right singular vector associated with the largest singular value. For the input

$$u(j\omega) = u_0\delta(\omega - \omega_0),$$

$$\frac{\|\mathbf{G}u\|_2}{\|u\|_2} = \frac{\|\mathbf{G}(j\omega_0)u_0\delta(\omega - \omega_0)\|_2}{\|u_0\delta(\omega - \omega_0)\|_2} = \frac{\mathbf{G}(j\omega_0)u_0\|\delta(\omega - \omega_0)\|_2}{u_0\|\delta(\omega - \omega_0)\|_2}$$

$$= \frac{\mathbf{G}(j\omega_0)u_0}{u_0} = \|\mathbf{G}\|_\infty.$$

◆

The above proof is not rigorous, since the supremum has been treated as a maximum. The supremum says only that the limit can be approached arbitrarily closely, not that the value can be achieved. The theorem remains unchanged when using the actual supremum, but the proof must include a limiting argument that has been omitted.

In addition to the ordinary properties of norms, the infinity norm has the following property:

THEOREM:
$$\|\mathbf{GH}\|_\infty \le \|\mathbf{G}\|_\infty \|\mathbf{H}\|_\infty. \tag{A9.4}$$

PROOF: By equation (A9.3), for any $u \neq 0$,

$$\|\mathbf{GH}u\|_2 \le \|\mathbf{G}\|_\infty \|\mathbf{H}u\|_2 \le \|\mathbf{G}\|_\infty \|\mathbf{H}\|_\infty \|u\|_2.$$

Rearranging and taking the supremum of both sides of this equation concludes the proof:

$$\|\mathbf{GH}\|_\infty = \sup_{u\neq 0} \frac{\|\mathbf{GH}u\|_2}{\|u\|_2} \le \|\mathbf{G}\|_\infty \|\mathbf{H}\|_\infty.$$

◆

The infinity norm can be generalized to operate over finite time intervals:

$$\|\mathbf{G}\|_{\infty,[t_0,t_f]} = \sup_{u\neq 0} \frac{\|\mathbf{G}u\|_{2,[t_0,t_f]}}{\|u\|_{2,[t_0,t_f]}}. \tag{A9.5}$$

THEOREM: The finite-time ∞-norm of the time-varying system described by the state model

$$\dot{x}(t) = \mathbf{A}(t)x(t) + \mathbf{B}(t)u(t) \ ; \ y(t) = \mathbf{C}(t)x(t) + \mathbf{D}(t)u(t)$$

is finite:

$$\|\mathbf{G}\|_{\infty,[t_0,t_f]} < \infty, \tag{A9.6}$$

provided all matrices are continuous.

PROOF: The solution of this state model (assuming zero initial conditions) is

$$\mathbf{G}u = y(t) = \int_{t_0}^{t_f} \mathbf{g}(t, \tau)u(\tau)d\tau + \mathbf{D}(t)u(t),$$

where

$$\mathbf{g}(t, \tau) = \mathbf{C}(t)\mathbf{\Phi}(t, \tau)\mathbf{B}(t)$$

is the impulse response of the system without the input-to-output coupling term $\mathbf{D}(t)$. The signal 2-norm of the output is

$$\|\mathbf{G}u\|_{2,[t_0,t_f]} = \left\| \int_{t_0}^{t_f} \mathbf{g}(t, \tau)u(\tau)d\tau + \mathbf{D}(t)u(t) \right\|_{2,[t_0,t_f]}.$$

Using the triangle inequality, and noting that the integral is simply a sum, we have

$$\begin{aligned}\|\mathbf{G}u\|_{2,[t_0,t_f]} &\le \int_{t_0}^{t_f} \|\mathbf{g}(t, \tau)u(\tau)\|_{2,[t_0,t_f]} d\tau + \|\mathbf{D}(t)u(t)\|_{2,[t_0,t_f]} \\ &= \int_{t_0}^{t_f} \sqrt{\int_{t_0}^{t_f} \|\mathbf{g}(t, \tau)u(\tau)\|_E^2 dt d\tau} + \sqrt{\int_{t_0}^{t_f} \|\mathbf{D}(t)u(t)\|_E^2 dt},\end{aligned}$$

where the norms in the final expression are Euclidean vector norms. The maximum singular value provides a bound on the gain of a matrix, which implies

$$\|\mathbf{G}u\|_{2,[t_0,t_f]} \leq \int_{t_0}^{t_f} \sqrt{\int_{t_0}^{t_f} \bar{\sigma}^2[\mathbf{g}(t,\,\tau)]\|u(\tau)\|_E^2 dt d\tau} + \sqrt{\int_{t_0}^{t_f} \bar{\sigma}^2[\mathbf{D}(t)]\|u(t)\|_E^2 dt}.$$

Noting that $\mathbf{D}(t)$ and $\mathbf{g}(t,\,\tau)$ are continuous and therefore bounded over a finite time interval, a maximum value for the maximum singular values in this expression exists:

$$\|\mathbf{G}u\|_{2,[t_0,t_f]} \leq \int_{t_0}^{t_f} \sqrt{\int_{t_0}^{t_f} \max_{t,\tau\in[t_0,t_f]} \{\bar{\sigma}^2[\mathbf{g}(t,\,\tau)]\}\|u(\tau)\|_E^2 dt d\tau}$$
$$+ \sqrt{\int_{t_0}^{t_f} \max_{t\in[t_0,t_f]} \{\bar{\sigma}^2[\mathbf{D}(\mathrm{t})]\}\|u(t)\|_E^2 dt}.$$

Since these maximums are both independent of t and τ, they can be removed from the integrals:

$$\|\mathbf{G}u\|_{2,[t_0,t_f]} \leq \sqrt{(t_f - t_0) \max_{t,\tau\in[t_0,t_f]} \{\bar{\sigma}^2[\mathbf{g}(t,\,\tau)]\}} \int_{t_0}^{t_f} \|u(\tau)\|_2 dt$$
$$+ \sqrt{\max_{t\in[t_0,t_f]} \{\bar{\sigma}^2[\mathbf{D}(t)]\}} \sqrt{\int_{t_0}^{t_f} \|u(t)\|_2^2 dt}.$$

Now, the integral of the norm can be bounded using the Cauchy-Schwarz inequality (see for example [2], page 337):

$$\int_{t_0}^{t_f} \|u(\tau)\|_2 d\tau = \int_{t_0}^{t_f} \|u(\tau)\|_2 \cdot 1 d\tau \leq \sqrt{t_f - t_0} \sqrt{\int_{t_0}^{t_f} \|u(\tau)\|_2^2 d\tau}.$$

Substituting this result into the above inequality yields

$$\|\mathbf{G}u\|_{2,[t_0,t_f]} \leq \left((t_f - t_0) \sqrt{\max_{t,\tau\in[t_0,t_f]} \{\bar{\sigma}^2[\mathbf{g}(t,\,\tau)]\}} + \sqrt{\max_{t\in[t_0,t_f]} \{\bar{\sigma}^2[\mathbf{D}(t)]\}} \right) \|u(t)\|_{2,[t_0,t_f]}^2.$$

Dividing by the signal 2-norm of the input yields

$$\frac{\|\mathbf{G}u\|_{2,[t_0,t_f]}}{\|u(t)\|_{2,[t_0,t_f]}} \leq (t_f - t_0) \sqrt{\max_{t,\tau\in[t_0,t_f]} \{\bar{\sigma}^2[\mathbf{g}(t,\,\tau)]\}} + \sqrt{\max_{t\in[t_0,t_f]} \{\bar{\sigma}^2[\mathbf{D}(t)]\}},$$

which shows that the finite time system ∞-norm is bounded. ◆

A10 A Bound on the System ∞-Norm

A bound on the system ∞-norm can be given based on the eigenvalues of a Hamiltonian matrix. A related bound can be given in terms of the solution of an algebraic Riccati equation. These theorems are both versions of the bounded real lemma.

THEOREM: The infinity norm of the generic stable system,

$$\dot{x}(t) = \mathbf{A}x(t) + \mathbf{B}u(t); \tag{A10.1a}$$

$$y(t) = \mathbf{C}x(t) + \mathbf{D}u(t), \tag{A10.1b}$$

is bounded:

$$\|\mathbf{G}(s)\|_\infty = \sup_\omega \bar{\sigma}[\mathbf{G}(j\omega)] = \sup_\omega \bar{\sigma}[\mathbf{C}(j\omega\mathbf{I} - \mathbf{A})^{-1}\mathbf{B} + \mathbf{D}] < \gamma \tag{A10.2}$$

if and only if

$$\bar{\sigma}(\mathbf{D}) < \gamma, \tag{A10.3a}$$

and the Hamiltonian matrix

$$\mathcal{H} = \left[\begin{array}{c|c} \mathbf{A} + \mathbf{B}(\gamma^2\mathbf{I} - \mathbf{D}^T\mathbf{D})^{-1}\mathbf{D}^T\mathbf{C} & \mathbf{B}(\gamma^2\mathbf{I} - \mathbf{D}^T\mathbf{D})^{-1}\mathbf{B}^T \\ \hline -\mathbf{C}^T(\mathbf{I} + \mathbf{D}(\gamma^2\mathbf{I} - \mathbf{D}^T\mathbf{D})^{-1}\mathbf{D}^T)\mathbf{C} & -\mathbf{A}^T - \mathbf{C}^T\mathbf{D}(\gamma^2\mathbf{I} - \mathbf{D}^T\mathbf{D})^{-1}\mathbf{B}^T \end{array}\right] \tag{A10.3b}$$

has no eigenvalues on the imaginary axis.

PROOF: Condition (A10.3a) is obtained by noting that **D** is the high-frequency limit of the transfer function:

$$\lim_{\omega\to\infty} \mathbf{G}(j\omega) = \lim_{\omega\to\infty} \{\mathbf{C}(j\omega\mathbf{I} - \mathbf{A})^{-1}\mathbf{B} + \mathbf{D}\} = \mathbf{D}.$$

Taking the maximum singular value yields

$$\lim_{\omega\to\infty} \bar{\sigma}[\mathbf{G}(j\omega)] = \bar{\sigma}(\mathbf{D}), \tag{A10.4}$$

and the infinity norm is bounded from below:

$$\|\mathbf{G}\|_\infty \geq \bar{\sigma}(\mathbf{D}).$$

Therefore, the infinity norm may be less than γ only if the maximum singular value of **D** is less than γ.

Condition (A10.3b) is then derived assuming that (A10.3a) is satisfied. The maximum singular value of the transfer function is bounded:

$$\bar{\sigma}[\mathbf{G}(j\omega)] < \gamma$$

if and only if

$$\text{eig}[\mathbf{G}^\dagger(j\omega)\mathbf{G}(j\omega)] < \gamma^2,$$

where eig[•] indicates the eigenvalues, and the inequality implies that all of the eigenvalues are less than γ^2. Note that the eigenvalues of this product are strictly real, and the singular values of $\mathbf{G}(j\omega)$ are the positive square roots of these eigenvalues. Further, the eigenvalues of this product are less than γ^2 if and only if (A4.1):

$$\text{eig}[\gamma^2\mathbf{I} - \mathbf{G}^\dagger(j\omega)\mathbf{G}(j\omega)] > 0. \tag{A10.5}$$

This bound is satisfied in the limit as ω approaches infinity (A10.4) provided (A10.3a) is assumed. Further, the eigenvalues in (A10.5) are continuous functions of ω. Therefore, the bound (A10.5) is satisfied for all frequencies if and only if the eigenvalues in (A10.5) do not cross the origin; that is

$$\det[\gamma^2\mathbf{I} - \mathbf{G}^\dagger(j\omega)\mathbf{G}(j\omega)] \neq 0$$

for all ω. This statement is equivalent to the condition that $[\gamma^2\mathbf{I} - \mathbf{G}^\dagger(j\omega)\mathbf{G}(j\omega)]$ has no zeros on the imaginary axis.

The system zeros can be found from the state model. The state model of $\mathbf{G}^\dagger(j\omega)$ is

$$\dot{\tilde{x}}(t) = -\mathbf{A}^T\tilde{x}(t) - \mathbf{C}^T u(t);$$

$$y(t) = \mathbf{B}^T\tilde{x}(t) + \mathbf{D}^T u(t),$$

which can be verified by computing the transfer function of this system and comparing it to $\mathbf{G}(j\omega)$. The state model of $[\gamma^2\mathbf{I} - \mathbf{G}^\dagger(j\omega)\mathbf{G}(j\omega)]$ is then

$$\begin{bmatrix} \dot{x}(t) \\ \hline \dot{\tilde{x}}(t) \end{bmatrix} = \left[\begin{array}{c|c} \mathbf{A} & \mathbf{0} \\ \hline -\mathbf{C}^T\mathbf{C} & -\mathbf{A}^T \end{array}\right] \begin{bmatrix} x(t) \\ \hline \tilde{x}(t) \end{bmatrix} + \begin{bmatrix} -\mathbf{B} \\ \hline \mathbf{C}^T\mathbf{D} \end{bmatrix} u(t);$$

$$y(t) = [\mathbf{D}^T\mathbf{C} \mid \mathbf{B}^T] \begin{bmatrix} x(t) \\ \hline z(t) \end{bmatrix} + (\gamma^2\mathbf{I} - \mathbf{D}^T\mathbf{D})u(t).$$

The zeros of this system are the solutions of

$$\det \left[\begin{array}{c|c} s\mathbf{I} - \left[\begin{array}{c|c} \mathbf{A} & \mathbf{0} \\ \hline -\mathbf{C}^T\mathbf{C} & -\mathbf{A}^T \end{array}\right] & \begin{bmatrix} -\mathbf{B} \\ \hline \mathbf{C}^T\mathbf{D} \end{bmatrix} \\ \hline [\mathbf{D}^T\mathbf{C} \mid \mathbf{B}^T] & \gamma^2\mathbf{I} - \mathbf{D}^T\mathbf{D} \end{array}\right] = 0.$$

The determinant of this block matrix can be expanded using (A2.7), which yields

$$\det(\gamma^2\mathbf{I} - \mathbf{D}^T\mathbf{D})\det\left(s\mathbf{I} - \left[\begin{array}{c|c} \mathbf{A} & \mathbf{0} \\ \hline -\mathbf{C}^T\mathbf{C} & -\mathbf{A}^T \end{array}\right] - \begin{bmatrix} -\mathbf{B} \\ \hline \mathbf{C}^T\mathbf{D} \end{bmatrix}(\gamma^2\mathbf{I} - \mathbf{D}^T\mathbf{D})^{-1}[\mathbf{D}^T\mathbf{C} \mid \mathbf{B}^T]\right)$$
$$= 0.$$

The first determinant is not equal to zero due to the bound on $\mathbf{D}$, so the second determinant must equal zero:

$$\det\left(s\mathbf{I} - \left[\begin{array}{c|c} \mathbf{A} & \mathbf{0} \\ \hline -\mathbf{C}^T\mathbf{C} & -\mathbf{A}^T \end{array}\right] - \begin{bmatrix} -\mathbf{B} \\ \hline \mathbf{C}^T\mathbf{D} \end{bmatrix}(\gamma^2\mathbf{I} - \mathbf{D}^T\mathbf{D})^{-1}[\mathbf{D}^T\mathbf{C} \mid \mathbf{B}^T]\right) = 0.$$

Note that this is an eigenequation, and the system $[\gamma^2\mathbf{I} - \mathbf{G}^\dagger(j\omega)\mathbf{G}(j\omega)]$ has no zeros on the imaginary axis if and only if the matrix

$$\mathcal{X} = \left[\begin{array}{c|c} \mathbf{A} & \mathbf{0} \\ \hline -\mathbf{C}^T\mathbf{C} & -\mathbf{A}^T \end{array}\right] - \begin{bmatrix} -\mathbf{B} \\ \hline \mathbf{C}^T\mathbf{D} \end{bmatrix}(\gamma^2\mathbf{I} - \mathbf{D}^T\mathbf{D})^{-1}[\mathbf{D}^T\mathbf{C} \mid \mathbf{B}^T]$$

has no eigenvalues on the imaginary axis. ◆

The eigensolution of a Hamiltonian matrix is related to the solution of an algebraic Riccati equation in the previous chapters. This is also the case for the Hamiltonian matrix appearing in the bound on the infinity norm. The second condition for this bound can be given in terms of this Riccati equation.

THEOREM: The infinity norm of the generic stable system (A10.1) is bounded:

$$\|\mathbf{G}(s)\|_\infty < \gamma \tag{A10.6}$$

if and only if

$$\bar{\sigma}(\mathbf{D}) < \gamma, \tag{A10.7a}$$

and there exists a symmetric matrix $\mathbf{P}$ that satisfies the following algebraic Riccati equation:

$$\mathbf{P}(\mathbf{A}+\mathbf{B}\mathbf{R}^{-1}\mathbf{D}^T\mathbf{C}) + (\mathbf{A}+\mathbf{B}\mathbf{R}^{-1}\mathbf{D}^T\mathbf{C})^T\mathbf{P} + \mathbf{P}\mathbf{B}\mathbf{R}^{-1}\mathbf{B}^T\mathbf{P} + \mathbf{C}^T(\mathbf{I}+\mathbf{D}\mathbf{R}^{-1}\mathbf{D}^T)\mathbf{C} = \mathbf{0}, \tag{A10.7b}$$

such that

$$\mathbf{A} + \mathbf{B}\mathbf{R}^{-1}\mathbf{D}^T\mathbf{C} + \mathbf{B}\mathbf{R}^{-1}\mathbf{B}^T\mathbf{P} \tag{A10.7c}$$

is stable (i.e., has all eigenvalues with negative real parts). Note that $\mathbf{R} = \gamma^2\mathbf{I} + \mathbf{D}^T\mathbf{D}$.

PROOF OF IF: Suppose a matrix $\mathbf{P}$ exists that satisfies (A10.7b). Performing a similarity transformation on the Hamiltonian matrix (A10.3b) yields

$$\left[\begin{array}{c|c} \mathbf{I} & \mathbf{0} \\ \hline -\mathbf{P} & \mathbf{I} \end{array}\right] \mathcal{H} \left[\begin{array}{c|c} \mathbf{I} & \mathbf{0} \\ \hline \mathbf{P} & \mathbf{I} \end{array}\right]$$
$$= \left[\begin{array}{c|c} \mathbf{A} + \mathbf{B}\mathbf{R}^{-1}\mathbf{D}^T\mathbf{C} + \mathbf{B}\mathbf{R}^{-1}\mathbf{B}^T\mathbf{P} & \mathbf{B}\mathbf{R}^{-1}\mathbf{B}^T \\ \hline \mathbf{X} & -(\mathbf{A} + \mathbf{B}\mathbf{R}^{-1}\mathbf{D}^T\mathbf{C} + \mathbf{B}\mathbf{R}^{-1}\mathbf{B}^T\mathbf{P})^T \end{array}\right],$$

where

$$\mathbf{X} = \mathbf{P}\mathbf{A} + \mathbf{A}^T\mathbf{P} + \mathbf{C}^T\mathbf{C} + \mathbf{P}\mathbf{B}\mathbf{R}^{-1}\mathbf{D}^T\mathbf{C} + \mathbf{C}^T\mathbf{D}\mathbf{R}^{-1}\mathbf{B}^T\mathbf{P}$$
$$+ \mathbf{P}\mathbf{B}\mathbf{R}^{-1}\mathbf{B}^T\mathbf{P} + \mathbf{C}^T\mathbf{D}\mathbf{R}^{-1}\mathbf{D}^T\mathbf{C}.$$

Note that $\mathbf{X}$ equals the left side of (A10.7b) and is therefore equal to zero, making the above matrix block triangular. The eigenvalues of this block triangular matrix equal the sum of the eigenvalues of the blocks on the diagonal, none of which lie on the imaginary axis. Further, since the eigenvalues of a matrix are invariant under similarity transformation, none of the eigenvalues of $\mathcal{H}$ lie on the imaginary axis, which implies $\|\mathbf{G}(s)\|_\infty < \gamma$. ◆

A11 The Adjoint System

The method of adjoint differential equations has been used extensively in the study of time-varying terminal control problems. In addition, Lagrange multipliers in quadratic optimal control problems and the duality between control and estimation can be precisely defined in terms of the adjoint system.

Given the causal time-varying system

$$\dot{x}(t) = \mathbf{A}(t)x(t) + \mathbf{B}(t)u(t); \tag{A11.1a}$$

$$y(t) = \mathbf{C}(t)x(t) + \mathbf{D}(t)u(t), \tag{A11.1b}$$

the adjoint system is defined as[3]

$$\dot{\tilde{x}}(\tau) = \mathbf{A}^T(t_f - \tau)\tilde{x}(\tau) + \mathbf{C}^T(t_f - \tau)\tilde{y}(\tau); \tag{A11.2a}$$

$$\tilde{u}(\tau) = \mathbf{B}^T(t_f - \tau)\tilde{x}(\tau) + \mathbf{D}^T(t_f - \tau)\tilde{y}(\tau). \tag{A11.2b}$$

THEOREM: The state-transition matrices of the adjoint and original systems are related:

$$\tilde{\mathbf{\Phi}}(t, \tau) = \mathbf{\Phi}^T(t_f - \tau, t_f - t). \tag{A11.3}$$

PROOF: Note that for any state-transition matrix, we have

$$\mathbf{\Phi}^{-1}(t, \tau) = \mathbf{\Phi}(\tau, t);$$

$$\mathbf{\Phi}(t, \tau)\mathbf{\Phi}(\tau, t) = \mathbf{I}.$$

Differentiating both sides of this last equation yields

$$\frac{d}{dt}\{\mathbf{\Phi}(t, \tau)\mathbf{\Phi}(\tau, t)\} = \dot{\mathbf{\Phi}}(t, \tau)\mathbf{\Phi}(\tau, t) + \mathbf{\Phi}(t, \tau)\dot{\mathbf{\Phi}}(\tau, t) = \mathbf{0}.$$

Solving for $\dot{\mathbf{\Phi}}(\tau, t)$ yields

$$\dot{\mathbf{\Phi}}(\tau, t) = -\mathbf{\Phi}^{-1}(t, \tau)\dot{\mathbf{\Phi}}(t, \tau)\mathbf{\Phi}(\tau, t).$$

Noting that $\dot{\mathbf{\Phi}}(t, \tau) = \mathbf{A}(t)\mathbf{\Phi}(t, \tau)$, we have

$$\dot{\mathbf{\Phi}}(\tau, t) = -\mathbf{\Phi}^{-1}(t, \tau)\mathbf{A}(t)\mathbf{\Phi}(t, \tau)\mathbf{\Phi}(\tau, t) = -\mathbf{\Phi}(\tau, t)\mathbf{A}(t).$$

Taking the transpose of this expression yields

$$\dot{\mathbf{\Phi}}^T(\tau, t) = -\mathbf{A}^T(t)\mathbf{\Phi}^T(\tau, t).$$

Performing the change of variables $\tau = t_f - \tau$ and $t = t_f - t$ gives

$$-\dot{\mathbf{\Phi}}^T(t_f - \tau, t_f - t) = -\mathbf{A}^T(t_f - t)\mathbf{\Phi}^T(t_f - \tau, t_f - t),$$

which shows that $\mathbf{\Phi}^T(t_f - \tau, t_f - t)$ is the state-transition matrix of the adjoint system. ◆

THEOREM: The impulse response matrices of the adjoint and original systems are related:

$$\tilde{\mathbf{g}}(t, \tau) = \mathbf{g}^T(t_f - \tau, t_f - t). \tag{A11.4}$$

PROOF: This theorem is demonstrated by noting that the impulse response of the adjoint system (A11.2) is

$$\tilde{\mathbf{g}}(t, \tau) = \mathbf{B}^T(t_f - t)\tilde{\mathbf{\Phi}}(t, \tau)\mathbf{C}^T(t_f - \tau) + \mathbf{D}^T(t_f - t)\delta(t - \tau).$$

[3] Adjoints are typically defined by the relationship $\langle \mathcal{G}x, y\rangle = \langle x, \tilde{\mathcal{G}}y\rangle$, where $\mathcal{G}$ is a linear operator and $\langle \bullet, \bullet \rangle$ is an inner product. The given definition of the adjoint system is not a true adjoint, but is closely related to the true adjoint. Some authors refer to the system (A11.2) as the modified adjoint system.

Using (A11.3),

$$\begin{aligned}\tilde{\mathbf{g}}(t,\tau) &= \mathbf{B}^T(t_f - t)\mathbf{\Phi}^T(t_f - \tau, t_f - t)\mathbf{C}^T(t_f - \tau) + \mathbf{D}^T(t_f - t)\delta(t - \tau) \\ &= \mathbf{g}^T(t_f - \tau, t_f - t).\end{aligned}$$

◆

THEOREM: The system 2-norms of the original and the adjoint systems are related:

$$\|\mathbf{g}\|_{2,[t_0,t_f]} = \|\tilde{\mathbf{g}}\|_{2,[0,t_f-t_0]}. \tag{A11.5}$$

PROOF: The system 2-norm of the adjoint system is

$$\begin{aligned}\|\tilde{\mathbf{g}}\|_{2,[0,t_f-t_0]} &= \frac{1}{t_f - t_0}\int_0^{t_f-t_0}\int_0^{t_f-t_0} trace\{\mathbf{g}^T(t_f - \tau, t_f - t)\mathbf{g}(t_f - \tau, t_f - t)\}d\tau dt \\ &= \frac{1}{t_f - t_0}\int_{t_0}^{t_f}\int_{t_0}^{t_f} trace\{\mathbf{g}^T(\tau, t)\mathbf{g}(\tau, t)\}d\tau dt = \|\mathbf{g}\|_{2,[t_f,t_0]},\end{aligned}$$

where the second integral expression is arrived at by using the fact that the trace is invariant under cyclic permutations (A2.3), changing the order of integration, and performing the change of variables $t \to t_f - \tau$; $\tau \to t_f - t$. ◆

THEOREM: The system ∞-norms of the original and the adjoint systems are related:

$$\|\mathbf{g}\|_{\infty,[t_0,t_f]} = \|\tilde{\mathbf{g}}\|_{\infty,[0,t_f-t_0]}. \tag{A11.6}$$

PROOF: The ∞-norm of the adjoint system is

$$\|\tilde{\mathbf{g}}\|^2_{\infty,[0,t_f-t_0]} = \sup_{\|\tilde{y}\|=1}\int_0^{t_f-t_0}\int_0^{t_f-t_0}\int_0^{t_f-t_0} \tilde{y}^T(\tau_1)\tilde{\mathbf{g}}^T(t_f, \tau_1)\tilde{\mathbf{g}}(t_f, \tau_2)\tilde{y}(\tau_2)d\tau_1 d\tau_2 dt.$$

Substituting for the impulse response of the adjoint system using (A11.4) and performing the change of variables $t \to t_f - t$, $\tau_1 \to t_f - \tau_1$, and $\tau_2 \to t_f - \tau_2$ yields

$$\begin{aligned}\|\tilde{\mathbf{g}}\|^2_{\infty,[0,t_f-t_0]} &= \sup_{\|\tilde{y}\|=1}\int_{t_0}^{t_f}\int_{t_0}^{t_f}\int_{t_0}^{t_f} \tilde{y}^T(\tau_1)\mathbf{g}(\tau_1, t_f)\mathbf{g}^T(\tau_2, t_f)\tilde{y}(\tau_2)d\tau_1 d\tau_2 dt \\ &= \|\mathbf{g}\|^2_{\infty,[t_0,t_f]}.\end{aligned}$$

◆

The theorems (A11.5) and (A11.6) are also valid in the infinite-time case.

A12 The Kalman Filter Innovations

The innovations (or residuals) of the Kalman filter are the errors between the measurement and the predicted measurement:

$$\check{m}(t) = m(t) - \mathbf{C}_m\hat{x}(t).$$

THEOREM: The innovations form a white noise random process with spectral density equal to the measurement noise spectral density $\mathbf{S}_v$.

PROOF: The innovations can be written in terms of the state estimation error:

$$\check{m}(t) = \mathbf{C}_m x(t) + v(t) - \mathbf{C}_m \hat{x}(t) = \mathbf{C}_m e(t) + v(t).$$

The correlation function of the innovations is then

$$\begin{aligned}\mathbf{R}_{\check{m}}(t_1, t_2) &= \mathbf{C}_m E[e(t_1)e^T(t_2)]\mathbf{C}_m^T + E[v(t_1)e^T(t_2)]\mathbf{C}_m^T \\ &\quad + \mathbf{C}_m E[e(t_1)v^T(t_2)] + E[v(t_1)v^T(t_2)] \\ &= \mathbf{C}_m E[e(t_1)e^T(t_2)]\mathbf{C}_m^T + E[v(t_1)e^T(t_2)]\mathbf{C}_m^T \\ &\quad + \mathbf{C}_m E[e(t_1)v^T(t_2)] + \mathbf{S}_v \delta(t_2 - t_1).\end{aligned} \tag{A12.1}$$

The expected values in this expression can be evaluated by noting that the state estimation error is given by the error model (7.29):

$$\dot{e}(t) = [\mathbf{A} - \mathbf{G}(t)\mathbf{C}_m]e(t) + \mathbf{B}_w w(t) - \mathbf{G}(t)v(t).$$

The solution of this state equation is

$$e(t) = \mathbf{\Phi}(t, t_0)e(t_0) + \int_{t_0}^{t} \mathbf{\Phi}(t, \tau)\{\mathbf{B}_w w(\tau) - \mathbf{G}(\tau)v(\tau)\}d\tau, \tag{A12.2}$$

assuming an initial condition at time t_0.

The expected value $E[e(t_1)e^T(t_2)]$ is evaluated by substituting for $e^T(t_2)$ using (A12.2), assuming an initial condition at $t_1 + \varepsilon$, and assuming that $t_2 \geq t_1 + \varepsilon$:

$$\begin{aligned}E[e(t_1)e^T(t_2)] &= E[e(t_1)e^T(t_1 + \varepsilon)]\mathbf{\Phi}^T(t_2, t_1 + \varepsilon) \\ &\quad + \int_{t_1+\varepsilon}^{t_2} \{E[e(t_1)w^T(\tau)]\mathbf{B}_w^T - E[e(t_1)v^T(\tau)]\mathbf{G}^T(\tau)\}\mathbf{\Phi}^T(t_2, \tau)d\tau \\ &= E[e(t_1)e^T(t_1 + \varepsilon)]\mathbf{\Phi}^T(t_2, t_1 + \varepsilon).\end{aligned}$$

The expected values within the integral are zero, since the estimation error is uncorrelated with both future plant and future measurement noise. Taking the limit of this expression as ε approaches zero yields

$$\begin{aligned}E[e(t_1)e^T(t_2)] &= \lim_{\varepsilon \to 0} E[e(t_1)e^T(t_1 + \varepsilon)]\mathbf{\Phi}^T(t_2, t_1 + \varepsilon) \\ &= \mathbf{\Sigma}_e(t)\mathbf{\Phi}^T(t_2, t_1),\end{aligned} \tag{A12.3}$$

where $\mathbf{\Sigma}_e(t) = E[e(t)e^T(t)]$ is the state estimation error covariance matrix.

The expectation $E[v(t_1)e^T(t_2)]$ is evaluated by substituting (A12.2) into this expression with the initial condition at $t = 0$:

$$\begin{aligned}E[v(t_1)e^T(t_2)] &= E[v(t_1)e^T(0)]\mathbf{\Phi}^T(t_2, 0) \\ &\quad + \int_0^{t_2} \{E[v(t_1)w^T(\tau)]\mathbf{B}_w^T - E[v(t_1)v^T(\tau)]\mathbf{G}^T(\tau)\}\mathbf{\Phi}^T(t_2, \tau)d\tau.\end{aligned}$$

The first expectation in this expression is zero, since the initial estimation error is uncorrelated with the measurement noise. In addition, the first expectation within the integral is zero, since the measurement noise and plant noise are uncorrelated. Using these results, we have

$$E[v(t_1)e^T(t_2)] = -\int_0^{t_2} \mathbf{S}_v \delta(t_2 - t_1)\mathbf{G}^T(\tau)\mathbf{\Phi}^T(t_2, \tau)d\tau$$
$$= \begin{cases} -\mathbf{S}_v\mathbf{G}^T(t_1)\mathbf{\Phi}^T(t_2, t_1) & t_2 > t_1 \\ -\frac{1}{2}\mathbf{S}_v\mathbf{G}^T(t_1)\mathbf{\Phi}^T(t_2, t_1) & t_2 = t_1, \\ 0 & t_2 < t_1 \end{cases} \tag{A12.4}$$

where the impulse function is included within the integral for $t_2 > t_1$, and the impulse function is outside the integral for $t_2 > t_1$. For $t_2 = t_1$, the impulse function is on the boundary of the integral. Note that the impulse function must be symmetric, since it appears in the correlation function of the noise, and the integral yields the factor of 1/2.[4]

The expectation $E[e(t_1)v^T(t_2)]$ can be evaluated in a similar manner:

$$E[e(t_1)v^T(t_2)] = \begin{cases} 0 & t_2 > t_1 \\ -\frac{1}{2}\mathbf{S}_v\mathbf{G}^T(t_1)\mathbf{\Phi}^T(t_2, t_1) & t_2 = t_1. \\ -\mathbf{S}_v\mathbf{G}^T(t_1)\mathbf{\Phi}^T(t_2, t_1) & t_2 < t_1 \end{cases} \tag{A12.5}$$

Substituting (A12.3), (A12.4), and (A12.5) into (A12.1) yields

$$\mathbf{R}_{\tilde{m}}(t_1, t_2) = \mathbf{C}_m\mathbf{\Sigma}_e(t_1)\mathbf{\Phi}^T(t_2, t_1)\mathbf{C}_m^T + \mathbf{S}_v\delta(t_2 - t_1) - \mathbf{S}_v\mathbf{G}^T(t_1)\mathbf{\Phi}^T(t_2, t_1)\mathbf{C}_m^T. \tag{A12.6}$$

Using the relationship between the Kalman gain and the estimation error covariance matrix (7.26),

$$\mathbf{\Sigma}_e(t)\mathbf{C}_m^T = \mathbf{G}(t)\mathbf{S}_v,$$

the first and last terms in (A12.6) cancel, and the correlation function of the innovations becomes

$$\mathbf{R}_{\tilde{m}}(t_1, t_2) = \mathbf{S}_v\delta(t_2 - t_1).$$ ◆

The innovations form a white noise random process provided that the plant model, the spectral density of the plant noise, and the spectral density of the measurement noise are all known perfectly. Errors in these parameters can be detected by monitoring the innovations and observing how closely they correspond to white noise.

A13 The Phase-Gain Relationship

The phase of a scalar, stable, minimum-phase system (with positive dc gain) is uniquely determined by the gain. This fact can be used to reconstruct the phase of frequency response data from gain-only measurements.

[4] This result can be generated formally using distribution theory.

THEOREM: The phase of a scalar, stable, minimum phase transfer function $G(j\omega)$ with $|G(0)| > 0$ can be uniquely computed from the gain:

$$\angle G(j\omega_l) = \frac{2\omega_l}{\pi} \int_0^\infty \frac{\ln\{G(j\omega)\} - \ln\{G(j\omega_l)\}}{\omega^2 - \omega_l^2} d\omega. \qquad \text{(A13.1)}$$

PROOF: The proof of this theorem can be found in [3], page 109. ◆

REFERENCES

[1] R. L. Dailey, "Lecture notes for the workshop on $\mathcal{H}_\infty$ and μ methods for robust control," American Control Conference, San Diego, CA, 1990.

[2] J. E. Marsden, *Elementary Classical Analysis,* W. H. Freeman, New York, 1974.

[3] J. C. Doyle, B. A. Francis, and A. R. Tannenbaum, *Feedback Control Theory,* Macmillian, New York, 1992.

INDEX